VIRUSES

VIRUSES

From Understanding to Investigation

SUSAN PAYNE

Department of Veterinary Medicine and Biomedical Sciences, Texas A&M University, College Station, Texas, United States

ACADEMIC PRESS

An imprint of Elsevier

Academic Press is an imprint of Elsevier
125 London Wall, London EC2Y 5AS, United Kingdom
525 B Street, Suite 1800, San Diego, CA 92101-4495, United States
50 Hampshire Street, 5th Floor, Cambridge, MA 02139, United States
The Boulevard, Langford Lane, Kidlington, Oxford OX5 1GB, United Kingdom

British Library Cataloguing-in-Publication Data
A catalogue record for this book is available from the British Library

Library of Congress Cataloging-in-Publication Data
A catalog record for this book is available from the Library of Congress

ISBN: 978-0-12-803109-4

For Information on all Academic Press publications
visit our website at https://www.elsevier.com/books-and-journals

Publisher: Mica Haley
Acquisitions Editor: Linda Versteeg-Buschman
Editorial Project Manager: Fenton Coulthurst
Production Project Manager: Chris Wortley
Designer: Mark Rogers

Typeset by MPS Limited, Chennai, India

Illustrations: Illustrations for this book were provided by Marcy Edelstein, to whom the publisher would like to extend their thanks.

Contents

About the Author

Dr. Susan Payne is an associate professor in the Department of Veterinary Pathobiology at the Texas A&M University, United States. During her career, she has mentored graduate and undergraduate students at three universities and has taught virology to undergraduate, graduate, medical, and veterinary students. Those courses are the basis for this textbook. She has also had an active research career and has written over 40 peer reviewed research and review articles. She serves as an ad hoc reviewer for several virology journals. She currently lives in Caldwell, Texas with her husband, mom, five cats, one dog, nine goats, one donkey, eight chickens (if the dog has not eaten one recently), and eight guinea fowls. She is most easily available in email at SPayne@cvm.tamu.edu.

Preface

This book, *Viruses: From Understanding to Investigation*, was inspired by a long career of teaching and research. My students have included undergraduate, graduate, medical, and veterinary students.

As regards the book title, my intent is to lead students of virology from a basic understanding to an interest in the investigations that have provided the information contained herein. The focus of this textbook is on animal and human viruses, only because these have been the focus of my research and teaching for many years. The viruses of plants, fungi, bacteria, and single-celled organisms are certainly no less interesting.

There is a huge amount of information about viruses available online, in journals, books, websites, and blogs. So why the need for another virology textbook? My intent was to organize and present a thoughtful, understandable, and up-to-date summary of the volumes of information available for consumption elsewhere. While every textbook, including this one, contains many facts, I have tried to emphasize general concepts.

With 38 chapters, this book contains more than enough material for a semester long course in introductory virology. The book is geared toward students with some background in cell biology, microbiology, immunology, and/or biochemistry, and I hope that it will be useful for both undergraduate and beginning graduate students. I also hope that no instructor will try to cover all of the material contained herein during a single semester. The book is organized into two parts, the first nine chapters cover topics including an introduction to viruses (containing information on replication cycle, diversity, taxonomy, and outcomes of virus infection), structure, interactions with the host cell, methods for studying viruses, immunity to viruses, and introductions to viral epidemiology, evolution, and pathogenesis. There are also chapters that serve as introductions to RNA and DNA viruses. I imagine that this will be more than enough information for many instructors and students.

The remaining chapters present viruses by family, with information about structure, genome organization, replication strategies, and disease. I have tried to be up-to-date and include virus families that are relatively new (hence these chapters are short). While each chapter includes basic information about a particular virus family, I am fond of narratives that tie the molecular basis of virus replication to pathogenesis, and have provided examples from a variety of animals, including human animals. The inclusion of "animal diseases" specifically serves as a reminder that companion and food animals play integral roles in human health and well-being. (As do plant and bacterial viruses, but those are subjects for other authors to address.)

I encourage instructors to review the material on virus families and choose a handful of these chapters to use in their courses. Positive-strand RNA viruses are presented first followed by negative and dsRNA viruses. The DNA viruses are presented from the smallest to the largest. Last, but certainly not least, are chapters covering the reverse transcribing retroviruses and hepadnaviruses. I have included some taxonomic information in each chapter, sometimes more, sometimes less. I imagine this to be a reference resource and starting point for students who wish to know more. (And I ask my colleagues not to make these boxes a giant exercise in memorization.)

Throughout the book, I have included brief discussions of both a historical nature (for example, oncogenic retroviruses and an account of the discovery of hepatitis B virus) and current issues such as the recent initiative of the World Health Organization and the World Organization for Animal Health (OIE) to collaborate to reduce human deaths by rabies virus in underdeveloped countries. In the mix are also topics relevant to basic research such as use of vesicular stomatitis virus G protein for pseudotyping and lymphocytic choriomeningitis virus (LCMV) as a model for pathogenesis.

For instructors and colleagues, a final word. You will find the depth of coverage somewhat mixed throughout, and I may have neglected your favorite virus or disease. I am also quite sure that I have presented ideas with which you disagree. Share these with your students, start a conversation, and call me out if necessary. My research manuscripts have always been improved by thoughtful criticism, and if this book is to have a life beyond the first edition, I expect that the same will be true in this case.

Acknowledgments

I cannot overstate the contributions of my illustrator, and wonderful sister, Marcy Edelstein. As I expected, she went "above and beyond" in assisting with this project. In addition to creating illustrations, she learned virology and gently pushed this project to completion with constant help and advice. Many thanks are also owed to my husband, Ross Payne, and to my mom for their patience and understanding during this project. I also wish to thank Texas A&M University, United States, for providing an incredible work and learning environment and my virology colleagues at the College of Veterinary Medicine and Biomedical Sciences and the Health Science Center for their support and inspiration. Finally, many thanks to my mentors over the years and to the dedicated and imaginative researchers who work to unravel the complex and beautiful world of viruses.

CHAPTER

1

Introduction to Animal Viruses

OUTLINE

After studying this chapter, you should be able to:

- Provide a meaningful definition of a virus.
- Explain difference between cell division and virus replication.
- Explain the correct usage of "virion" versus "virus."
- Describe the basic steps in a virus replication-cycle.
- Draw, label, and describe each part of a "one-step" growth curve.
- List possible outcomes of a virus infection (1) at the level of the individual cell and (2) at the level of the host animal.
- Define the term "host range" as regards viruses.

WHAT IS A VIRUS?

Most of us are familiar with the term virus and know viruses as disease causing agents, transmitted from one person or animal to another. We are familiar with "cold" and "flu" viruses; we fear a worldwide pandemic of Ebola. We may even be aware that viruses are used to deliver genes to cells for the purposes of gene therapy or genetic engineering. But what are viruses?

- Viruses are infectious agents that are *not* cellular in nature.
- Viruses must enter a living host cell in order to replicate, thus *all* viruses are obligate intracellular parasites. Synthesis of the proteins and nucleic acids (DNA and RNA) for assembly into new virus particles (virions) requires an energy source (ATP), building materials (amino acids and nucleotides), and protein synthesis machinery (ribosomes) supplied by the host cell. The cell also provides scaffolds (microtubules, filaments, membranes) on which virus particles replicate their genomes and assemble. Thus the cell is a factory providing working machinery and raw materials. The infected cell may or may not continue normal cellular processes (host cell mRNA and protein synthesis) during a viral infection.
- Viruses have nucleic acid genomes that are surrounded by and protected by protein coats called capsids. Capsids protect genomes from environmental hazards and are needed for efficient delivery of viral genomes into new host cells. Some viruses have lipid membranes, called envelopes that surround the capsid (Fig. 1.1).
- Viruses are structurally much simpler than cells. Some viruses can be crystalized. Viruses do not increase in number by cell division; instead they assemble from *newly synthesized* protein and nucleic

Viruses. DOI: http://dx.doi.org/10.1016/B978-0-12-803109-4.00001-5

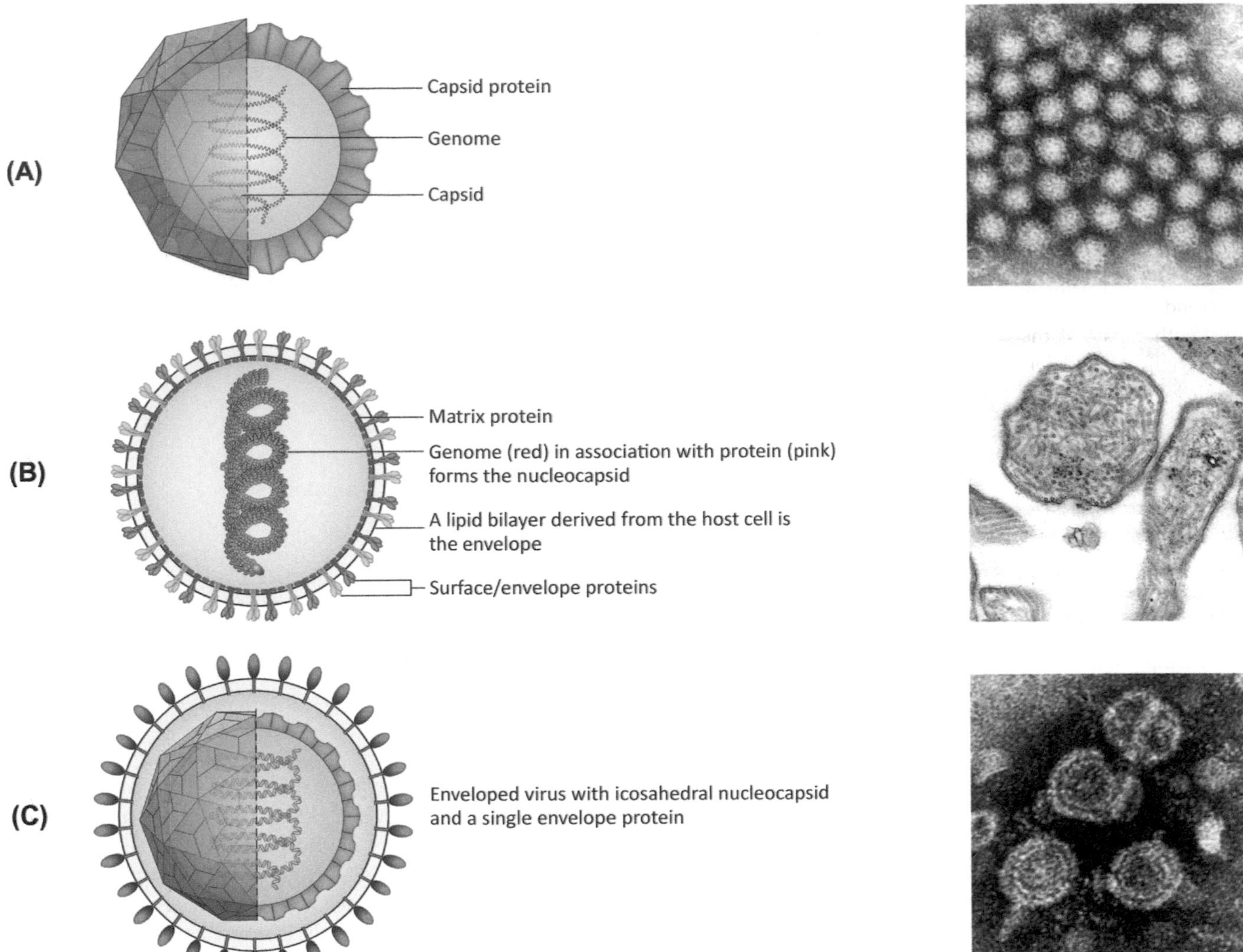

FIGURE 1.1 Basic features of virions. Panel A. On left, simple diagram of an unenveloped virus with icosahedral symmetry; on right, electron micrograph of calicivirus (Chapter 12: Family *Caliciviridae*). Panel B. On left, simple diagram of enveloped virus with a helical nucleocapsid; on right, electron micrograph of measles virus, a paramyxovirus (Chapter 20: Families *Paramyxoviridae* and *Pneumoviridae*). Panel C. On left a simple diagram of an enveloped virus with an icosahedral nucleocapsid; on right, electron micrograph of hepadnavirus (Chapter 38: Family *Hepadnaviridae*).

acid parts (building blocks). As viruses are not cells, they have none the organelles associated with cells. A sample of purified virions has no metabolic activity.
- Viruses are packages designed to deliver nucleic acids to cells; they are excellent examples of "selfish genes."

The preceding description might suggest uninteresting, inanimate particles, but examining virus replication strategies and interactions with host cells provides a diverse and dynamic view into cellular and molecular processes. Viruses are not a homogenous group. They are an extremely diverse group of infectious agents. It is highly unlikely that they arose from a single common ancestor (Box 1.1).

DIVERSITY IN THE WORLD OF VIRUSES

- All viruses have nucleic acid genomes, but some utilize DNA as genetic material, while others have RNA genomes. Viral genomes are not always double-stranded molecules; there are single-stranded viral RNA and DNA genomes. There are viral genomes that consist of a single molecule of nucleic acid, but some genomes are segmented. For example, reoviruses (Chapter 26: Family *Reoviridae*) package 11–12 different pieces of double-stranded RNA and each genome segment encodes a different gene.
- Some viruses have lipid envelopes in addition to a genome and protein coat. Viral envelopes are not

BOX 1.1

WHAT IS IN A DEFINITION?

By the late 19th century, the term "virus" was used to describe infectious agents that could pass through filters designed to remove bacteria from liquids. Thus viruses were "smaller than bacteria" and size became an important part of the definition of a virus. Today we know of a few viruses that are larger than many bacteria, so the trend has been to drop "smallness" from the definition. Another part of the definition of a virus is that they are all obligate intracellular parasites. This is certainly true of all viruses, but there are also bacteria and protozoan parasites that are obligate intracellular parasites. When the biochemical nature of viruses was discovered, it became clear that viruses lack many of the complex structures common to cells. This resulted in definitions of viruses based on comparisons to cells. While these comparisons emphasize the many ways that viruses are different from cells, they do not help us understand these unique infectious agents. So, what are viruses? Very simply, they are genes packaged within a protein coat. Their replication process begins when the virion delivers its genome to a cell. The viral genome encodes proteins required for the synthesis of new viral genomes. New viral proteins plus new viral genomes assemble to form new particles or virions, and so the cycle continues.

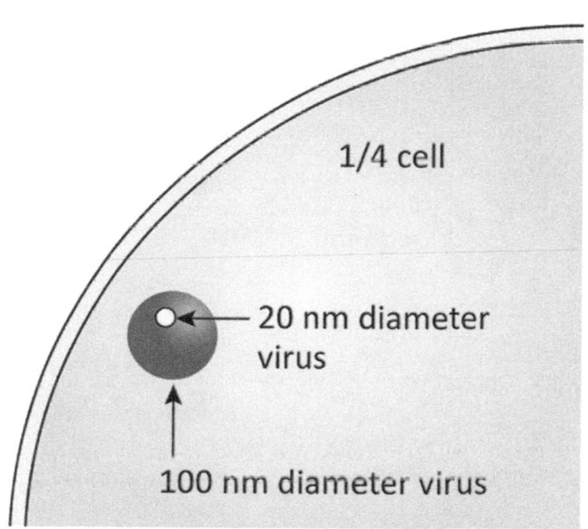

FIGURE 1.2 Relative sizes of an animal cells and virions.

homogenous. Different types of host membranes may be utilized, and their specific lipid and protein components can differ.

- Viruses range in size from 10 to 1000 nm is size (Fig. 1.2).
- Viral genomes range in size from 3000 nucleotides (nt) to over 1,000,000 base pairs.
- Outcomes of viral infections are diverse. Infection does not always result in cell or host death. Some host genes are derived from viruses and have played key roles in evolution. (Some plant viruses are beneficial in extreme environments.)
- Some viruses complete their replication cycles in minutes while others take days. Some viruses are transiently associated with an infected host (days or weeks) while others (for example, herpesviruses) are life-long residents.
- Where did viruses come from? Three general scenarios for virus evolution have been proposed:
  - Retrograde evolution: Intracellular parasites lost the ability for independent metabolism keeping only those genes necessary for replication. Poxviruses are very large complex viruses that *may* have evolved in this manner.
  - Origins from cellular DNA and RNA components: Some DNA genomes resemble plasmids or episomes. Did these DNAs acquire protein coats and the ability to be transferred from cell to cell efficiently?
  - Descendants of primitive precellular life forms: Viruses originated and evolved along with primitive, self-replicating molecules. This is the likeliest origin of the RNA viruses described in this text.

For the most part, names of specific viruses have been omitted in this section, to emphasize the general subject of viral diversity. Throughout this text, the details will be forthcoming. But I hope that now, when reading about any virus, you will want to learn its place in the complex world of viruses. (Big? small? friend? foe? transient visitor? life-long partner?)

ARE VIRUSES ALIVE?

Viruses parasitize every known form of life on this planet and they have both short-term and long-term impacts on their hosts. But are viruses alive? This question is the subject of ongoing debate, but the

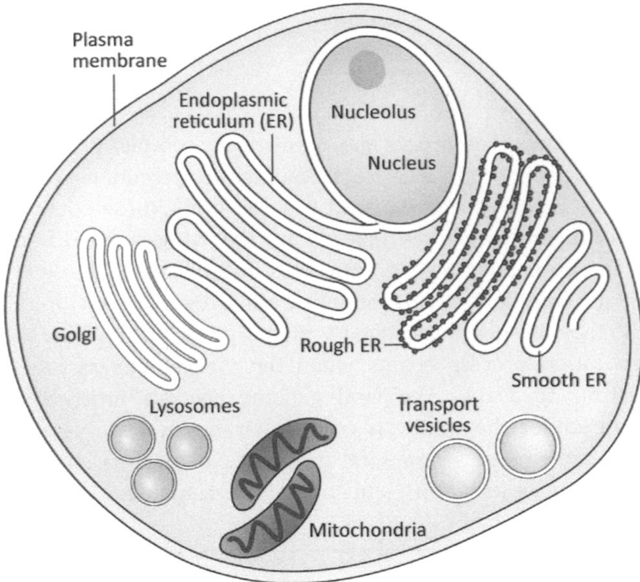

FIGURE 1.3 Simple schematic of a eukaryotic cell identifying some major organelles.

answer does not change the *nature* of the virus. As we discuss and describe viruses it is easy to assume that they are alive. They replicate to increase in number and the terms "virus replication-cycle" and "virus life-cycle" are often used interchangeably. Viruses also evolve (change their genomes), sometimes very rapidly. In this manner they adapt to new hosts and environments.

In contrast, the virion (the physical package that we view with an electron microscope) has no metabolic activity. Some viruses can be assembled simply by mixing purified genomes and proteins in a test tube. The genomes may have been synthesized by machine and the viral proteins may have been produced in bacteria. If those component parts combine under suitable conditions, a fully infectious virion can be produced. To avoid the question of living versus nonliving, the term "infectious agent" is both appropriate and descriptive. We can then speak of *infectious* virions that are *capable of entering a cell and initiating a replication-cycle*, or inactivated virions that cannot

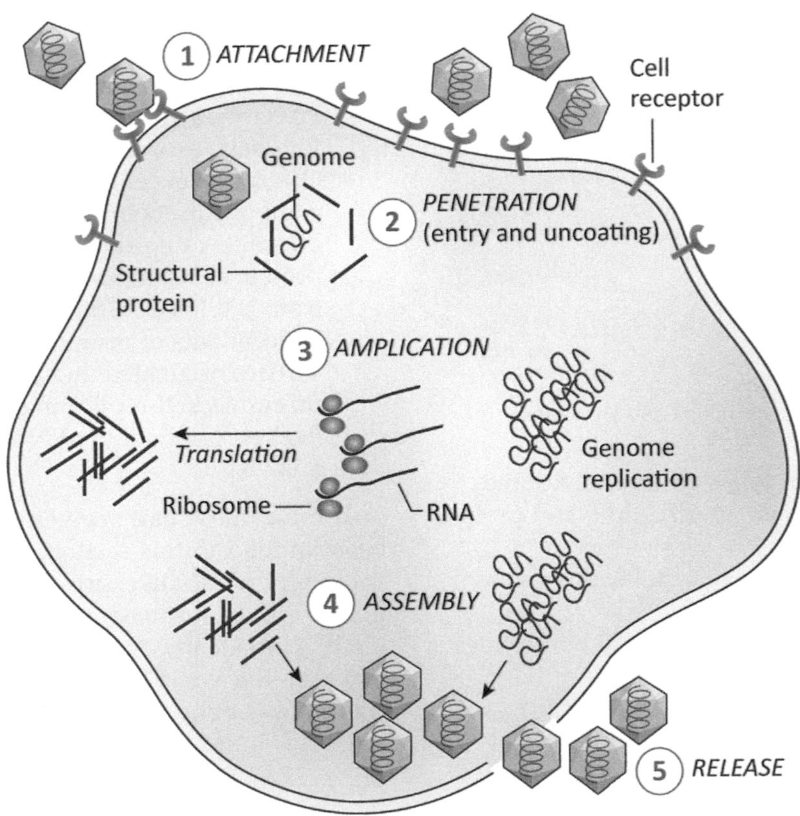

FIGURE 1.4 The basic virus life-cycle is shown in a generic cells. (For simplicity no cell organelles are shown but the processes of virus replication are intimately associated with cell organelles and structures.) The basic virus life-cycle begins with: (1) Attachment of the virion to receptors on a cell. (2) The genome is delivered into cytoplasm (penetration). (3) Viral proteins and nucleic acids are synthesized (amplification). (4) Genomes and proteins assemble to form new virions. (5) Virions are released from the cell.

complete a replication-cycle. As we will see in later chapters, the difference between an infectious and a noninfectious virion may be as small as the cleavage of a single peptide bond.

BASIC STEPS IN THE VIRUS REPLICATION-CYCLE

The first step in a virus life cycle is attachment (or binding) to the host cell (Figs. 1.3 and 1.4). Attachment results from very specific interactions between viral proteins and molecules on the surface of the host cell. The interactions are usually hydrophobic and ionic, rather than covalent bonds. Thus attachment is influenced by environmental conditions such as pH and salt concentration. Attachment becomes stronger as many copies of a viral surface protein interact with multiple copies of the host cell receptor molecules.

The next step in the virus life-cycle is penetration of the viral genome into the host cell cytoplasm or nucleoplasm. After penetration, there may be a further rearrangement of viral proteins to release the viral genome, a process called uncoating. Penetration and uncoating are two distinct steps for some viruses while for others the viral genome is uncoated during the process of penetration. The processes of penetration and uncoating are irreversible, the infecting virion cannot reassemble.

The next phase in the virus life-cycle is synthesis of the new viral proteins and genomes. This is a complex process that requires transcription (synthesis of mRNA), translation (protein synthesis), and genome replication to generate the parts that will assemble into new virions. Synthesis of viral proteins and genomes occurs in close association with, and depends upon, many host cell proteins and structures. The great diversity among viruses will be evident as we examine processes that regulate transcription, translation, genome replication and the specific virus–host cell interactions that shape these processes.

The next step in the virus replication-cycle is assembly of new virions. New particles assemble from the genome and protein components that accumulate in the infected cell. Viruses are assembled at different sites in host cells; sometime large areas of the cell become virus factories, concentrated regions of viral proteins and genomes from which host cell organelles are excluded.

The final step(s) in the virus replication-cycle are release from the host cell and maturation of the released virions. Virion release may occur upon cell rupture or lysis. Enveloped viruses must acquire their envelopes from cellular membranes in a process called budding. Some enveloped viruses bud through the plasma membrane, but budding can occur at other, intracellular membranes. The budding process can, but does not always, kill the host cell. Other viruses obtain their lipid enveloped by budding into cellular vesicles. These vesicles then fuse with the plasma membrane to release the virions; this is process called exocytosis.

Maturation is the term used to describe changes in virus structure that occur after a virus is released from the host cell. Maturation may be required before a virus is able to infect a new cell; maturation may involve cleavage or rearrangement of viral proteins. Viruses assemble in the cell (under conditions of favorable energy) but when the released virions encounter new cells they must be able to disassemble (uncoating). Maturation events that occur after virus release set the stage for a productive encounter with the next cell. Maturation processes are well understood for several important animal viruses and examples will be presented in future chapters.

It is important to stress that each step in the virus replication-cycle requires specific interactions between viral proteins and host cell proteins. Some viruses can infect many different cell types and organisms because they interact with proteins found on, and in, many cell types. These viruses are said to have a broad host range. Other viruses have a very narrow host range due to their need to interact with specific cellular proteins that are expressed only in a few cell types. Factors that impact virus replication include the presence or absence of receptors, the metabolic state of the cell, the presence or absence of any number of intracellular proteins required to complete the virus replication-cycle.

Another way to view the replication-cycle of a virus is the one-step growth curve (Fig. 1.5). This graph illustrates the concept that penetration of a virus into the host cell is not reversible. During the so-called eclipse phase infectious virions cannot be detected, even if cells are broken open (lysed), there are no infectious particles to be found!

GROWING VIRUSES

Viruses are obligate intracellular parasites; they replicate only within living cells. Thus in the laboratory, susceptible cells or organisms are required to study virus replication. For the virologist, ideal host cells are easily grown and maintained in the laboratory. Animal virologists often use cell and (less often) organ cultures. To culture animal cells, tissues or organs are harvested and disrupted (using mechanical and enzymatic methods) to obtain individual cells. Often cells are derived from tumors that grow robustly in culture. Cells circulating in the blood, such as lymphocytes, can be obtained directly from animal blood samples. If cells are provided with the appropriate environment (growth media, temperature, pH, and CO_2), they will remain metabolically active and may undergo cell divisions. Cell cultures will be described in more detail in a later chapter (Box 1.2).

Often the best-studied viruses are those that have been adapted for robust growth in a culture system. However, cell or organ cultures may be very different from the natural environment of the human or animal host. The biggest difference is that the cultured cells lack the many antiviral defenses encountered in an organism. Thus it is not uncommon for a virus highly adapted to cell cultures to perform poorly when used to infect an animal. In fact, propagation in culture is a common method for producing attenuated (weakened) live viral vaccines. Attenuated viruses replicate in a host, but do not cause disease. When considering experiments with viruses, it is very important to understand both the host system and the origins of the virus being studied.

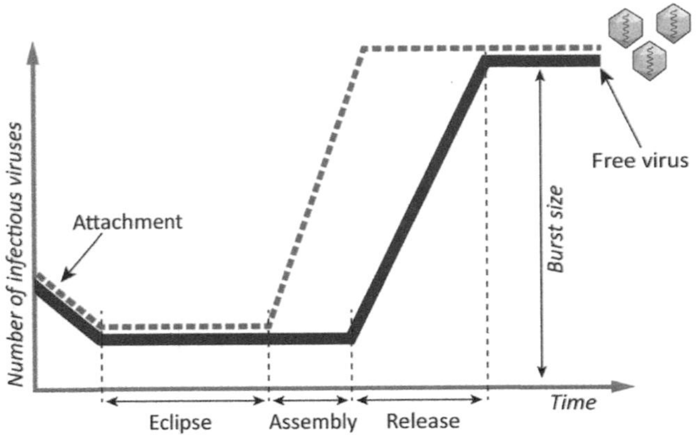

FIGURE 1.5 One-step virus growth curve. The red curve represents infectious virions released from the infected cells. The blue curve represents infectious virions released if the cells are lysed. Key to understanding the one-step growth curve is to note that after attachment, the number of virions detected in media and within cells *decreases*. These virions have penetrated cells and their genomes have uncoated, thus they are no longer "infectious." New virions are detected only after amplification and assembly.

CATEGORIZING VIRUSES (TAXONOMY)

The most widely accepted method to group viruses is by the type of nucleic acid (RNA or DNA) that serves as the viral genome. Within this scheme, there are three overarching groups of viruses:

- DNA viruses: Package DNA genomes synthesized by a DNA-dependent DNA polymerase.
- RNA viruses: Package RNA genomes synthesized by an RNA-dependent RNA polymerase (RdRp).
- The third group of viruses uses the enzyme reverse transcriptase (RT) during the replication-cycle. RT is an RNA-dependent DNA polymerase as it synthesizes a DNA copy of an RNA molecule. Reverse transcribing viruses (examples are the retroviruses and hepadnaviruses) use both RNA and DNA versions of their genomes (at different times) during their replication cycles.

The DNA and RNA viruses are further differentiated by the physical makeup of their genomes (single stranded, double stranded, unsegmented, segmented, linear, circular). The importance of genome type, and how it influences virus replication will be covered in upcoming chapters.

In addition to genome type, other physical traits are used to subdivide viruses into smaller groups. Some viruses have lipid envelopes (enveloped viruses) while others do not (naked viruses). Capsids also come in different shapes and sizes. The goal of viral taxonomy is to categorize viruses using groups of traits. Borrowing nomenclature from the Linnaean classification system, viruses are grouped into orders, families, genera, and species (Fig. 1.6). Orders contain two or more related families, and families can be subdivided into multiple genera. A genus is further subdivided into species (or strains). The family is often called the fundamental unit of viral taxonomy. Viruses in the

BOX 1.2

PERMISSIVE OR NOT?

Some viruses replicate very poorly when first introduced into cultured cells. There may be no visible signs of virus infection, but upon prolonged incubation or "blind passage" (often over a period of weeks or months) the virus will adapt to the new environment. This "cell culture adapted" virus now grows well in cultured cells. Therefore the initial virus infection was permissive, although very poorly so. After becoming adapted to cell culture conditions, the virus may be attenuated (replicate poorly or become incapable of causing disease) in the animal host.

Viral species

May contain **strains** that differ by up to 10% at the nucleotide sequence level

Viral genus

Group of similar species constitutes a genus. Members share nucleotide homology and are likely serologically related

Orders

Related families may be grouped into orders

FIGURE 1.6 Viral taxonomy is based on groups of characteristics such as genome type, genome organization, capsid structure, presence/absence of an envelope. The virus family is often considered the focal point of virus taxonomy. Viruses in a family share genome type, overall genome organization, size, and shape. Related families can be group into orders. Families are also subdivided into smaller groups of more closely related viruses (genera) within the family. A genus can contain a number of different species or strains. These may differ by up to 10% at the nucleotide sequence level. Closely related strains may sometimes be quite phenotypically distinct.

same family are considerably more closely related than viruses from different families. Placement of viruses into families is accomplished by examining shared characteristics such as genome type, presence or absence of an envelope, shape of the capsid, arrangement of genes on the viral genome, etc. All viruses within a family share a core set of properties. Thus, if one knows the major characteristics of any single member of the family *Picornaviridae* (for example, poliovirus), one know the genome type, general genome organization, approximate size, and shape of all picornaviruses. One needs only to learn the characteristics of a handful of virus families, rather than thousands of individual viruses.

Viral taxonomy is determined by groups of expert virologists from around the world who volunteer to serve on the International Committee on the Taxonomy of Viruses (ICTV). Visit the ICTV website at http://ictvonline.org/virusTaxonomy.asp to find the most recent virus classification schemes. The site also provides a helpful history of virus names.

Before it was possible to generate genome sequences quickly and cheaply, classifying viruses was often done using phenotypic traits such as host range, or tissue tropism. Now it is standard practice to use genome sequences to categorize or classify viruses. Genome sequences provide detailed and objective criteria to subdivide viruses into related groups. Genome sequences from many different viruses can be compared to generate phylogenies that provide a visual "map" of relationships among viruses (Fig. 1.7). In some cases, many thousands of viral genome sequences are compared in order to generate detailed phylogenies. Such is the case with human immunodeficiency viruses (HIV).

The recent explosion viral in genome sequence data has necessitated extensive taxonomic changes in some virus families. For example, until recently the site of infection (respiratory versus enteric) was used as a

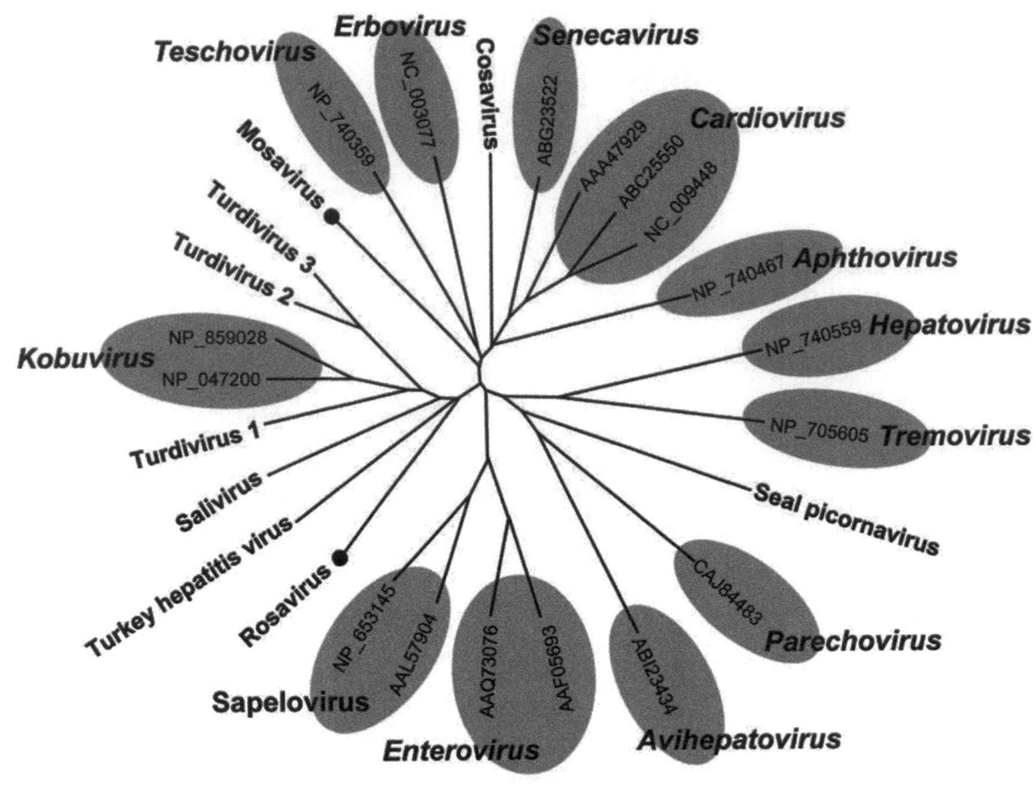

FIGURE 1.7 Phylogeny of the family *Picornaviridae* showing recognized genera. In some cases a genus contains only one virus isolate or strain. *Phan, T. G., Kapusinszky, B., Wang, C., Rose, R., Lipton, H., E. Delaware. 2011. The Fecal Flora of Wild Rodents. PLOS Pathogen 7:e1002218.*

criterion to define genera within the family *Picornaviridae*. However a phylogeny based on *genome sequences* does not split the picornaviruses cleanly along these lines. Thus the family *Picornaviridae* still contains the genus *Enterovirus*, but there is no longer a genus *Rhinovirus*, although you will see frequent reference to it in older literature.

Alternatives to ICTV taxonomy are sometimes used to group viruses that share common phenotypic characteristics. Hepatitis viruses are so named because they share the phenotype of replicating in the liver. However the hepatitis B virus (HBV) and the hepatitis C virus (HCV) are not related, either structurally or genetically, and vaccines and antiviral treatments developed for HCV are not effective for treating, or preventing, HBV infection. Another common phenotypic grouping is use of the term arbovirus (meaning arthropod-borne virus) to describe viruses that are transmitted by insects. Members of many different virus families can properly be called arboviruses; the term does not imply genetic relatedness among the diverse members of this "group".

You might ask if it is useful to generate or understand, phylogenies of viruses. The answer is a resounding yes. For example, the origins of a disease outbreak can be determined using detailed genetic information. Information from genome sequencing can be used to analyze past outbreaks and track the transmission of viruses from one person or animal to another in order to determine the best methods to curb virus transmission during an epidemic.

OUTCOMES OF VIRAL INFECTION

Virus infection impacts individual cells, and these cellular changes may or may not noticeably influence the health and fitness of the organism. There are four general outcomes when a virus encounters a cell:

- Productive or permissive infection. Viral proteins and nucleic acids are synthesized and virions are assembled and released.
- Nonpermissive infection. The cell is completely resistant to infection.
- Abortive or nonproductive infection. The virus enters the cell, but replication becomes irreversibly blocked at some step before particles are produced.
- Latent infection. Describes a situation where a viral genome is present in the cell, but no or only a few

BOX 1.3

LATENT VERSUS CHRONIC INFECTIONS: WHERE IS THE BOUNDARY?

A latent infection is one in which viral genomes are present in cells but virions are not produced. The term chronic infection describes one where virions can be routinely detected. Thus the sensitivity of the assays used for virus detection becomes an important factor in the distinction. As virus detection methods become more sensitive, the distinction between latency and chronic infection has become blurred. Consider genital herpes, caused by human herpesviruses 1 and 2 (HHV1 and 2). These viruses are abundant in visible lesions but also can be transmitted when there are no visible lesions. So is the infection latent or is it chronic? How often are the HHVs found on the skin in the absence of lesions? How often must a latent virus reactivate before it is considered chronic? From a public health standpoint calling genital herpes, a chronic infection might better convey the fact that herpesvirus can be transmitted in the absence of lesions.

viral proteins are produced. Latency implies that the virus can productively replicate given the right conditions (Box 1.3).

Both productive and nonproductive infections can impact the cell. The effects of infection can range from no apparent change, to cell death, to transformation (immortalization). Productive infection often results in cell death (lytic or cytopathic infection), but this is not always the case. Some viruses can replicate without damaging the cell, resulting in an inapparent infection. Viruses that cause inapparent infections are often produced in small amounts for the life of the cell. Sometimes an inapparent infection results from latency. A much less frequent outcome of infection is transformation or immortalization that allows the cell to divide without restriction. Immortalized cells may be productively infected (virus is released) or the condition may result from a nonproductive infection.

In the preceding paragraphs we learned that cells can be inapparently infected by a virus. Inapparent infection also occurs at the level of the animal host. Some viruses replicate in hosts without causing disease. After all, the "job" of a virus is replicate and infect another host; disease is not a required side effect. Until very recently it was hard to find viruses that caused inapparent infections. But many inapparent infections are now being identified through large-scale sequencing of host nucleic acids. (Methods for virus detection and discovery will be discussed in a later chapter).

Disease is the result of damage to tissues or organs. Many viral infections cause disease, and diseases can be described as acute, chronic, or latent (Fig. 1.8). Acute disease has a rapid onset, lasts from days to months, and the virus is either controlled or cleared, or causes death of the host. There are many examples of acute viral infections, the common cold being one. From a public health standpoint, it is important to know that virus replication and spread may begin well before symptoms develop and virus may be shed for days or weeks after symptoms have resolved. The peak of clinical signs and symptoms may or may not correspond to peak virus titers, or the time of maximum transmissibility.

Chronic viral infections have a slower progression and the time to resolution is years to a lifetime. These viral infections may, but do not always, lead to death of the host. Chronic infections are also called persistent infections. Virus is produced and shed continuously (albeit sometimes at very low levels). Examples of viruses that may cause chronic or persistent infections of humans are hepatitis C virus (HCV), hepatitis B virus (HBV), and human immunodeficiency virus (HIV). It should be noted that a chronic viral infection can be without symptoms (inapparent) for years.

Latent infection describes the maintenance of a viral genome without the production of detectable virus. Herpesviruses are a good example of viruses that cause latent infections. The chickenpox/shingles virus, formerly known as varicella-zoster virus, but recently renamed human herpesvirus 3 (HHV3) is an instructive example. Prior to 1995 chickenpox was a common childhood infection in the United States. Chickenpox infection is usually mild, characterized by blister-like pustules that resolve in about a week. However, HHV3 remains in the body long after the pustules have disappeared. HHV3 genomes are silently maintained in neurons, for decades. Shingles, a very painful and debilitating disease of adults, occurs when HHV3 exits latency and travels down neurons to the skin to produce blister-like lesions. These lesions contain infectious virus, thus a person with shingles can transmit chickenpox to a nonimmune person. HHV3 reactivates (breaks out of latency) when the host's immune

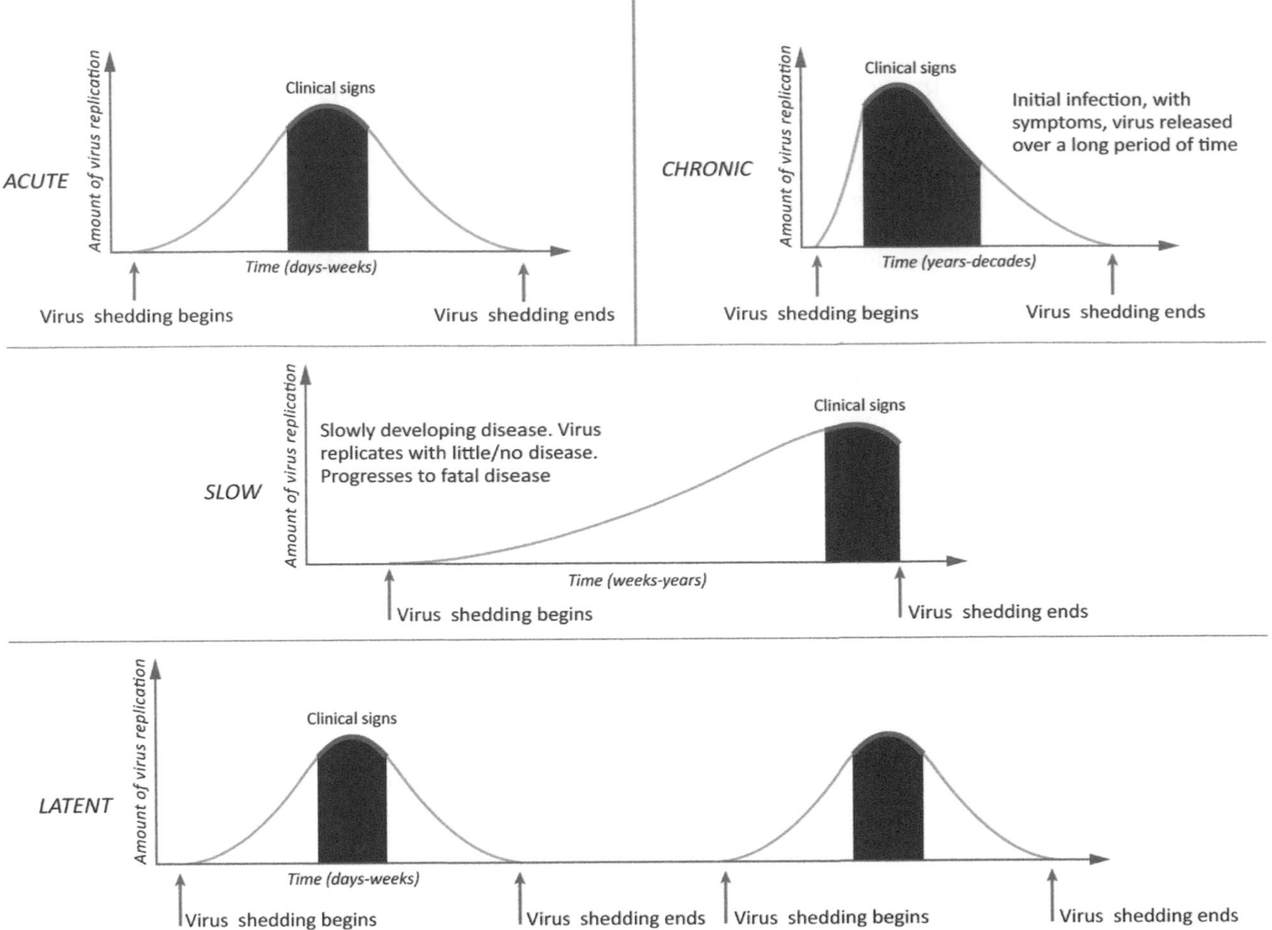

FIGURE 1.8 Outcomes of viral infection at the level of the animal host are quite variable. The red areas under the curves depict periods of clinical disease. In the examples depicted here, virus shedding begins before the onset of symptoms and ends after symptoms have resolved. Note that periods of virus shedding vary. Shedding may begin at the time of onset of clinical symptoms and may end prior to the resolution of disease. During latent infection, there may be intermittent virus shedding without clinical symptoms.

system is impaired (by advancing age or stress, for example). Shingles vaccines boost immune responses to HHV3, reducing the likelihood of virus reactivation.

INTRODUCTION TO VIRAL PATHOGENESIS

Viral pathogenesis is defined as the mechanism by which viruses causes disease. A simple view of viral pathogenesis is that viruses replicate and kill cells, thus causing disease. For example, death of liver cells (hepatocytes) causes hepatitis, death of enterocytes may cause diarrhea, death of respiratory epithelial cells may cause severe respiratory tract disease. However loss of cell function, without death, can also produce disease. During HIV infection, immunodeficiency is not simply caused by cell death; the virus also alters the function of some cells needed to maintain a healthy immune system.

Signs and symptoms of disease can also result from tissue damage caused by host immune responses. Inflammation, killing of virus-infected cells by the immune system, or deposition of immune complexes are examples. Of course, like any biological event, disease is often a complex combination of direct damage by virus in concert with host immune responses. Understanding viral pathogenesis, the mechanism by which disease develops, is an important consideration in developing effective treatments.

INTRODUCTION TO VIRUS TRANSMISSION

How are viruses transmitted from one animal to another? Common routes of infection include:

- fecal-oral,
- respiratory droplets,

- contact with contaminated fomites,
- exchange of infected bodily fluids, tissues, or organs,
- airborne,
- insect vectors.

Fecal-oral transmission occurs via ingestion of contaminated food or water. Virus enters the body through epithelial cells or lymphoid in the gastrointestinal tract. Examples include rotaviruses and the Norwalk-like viruses (noroviruses). Noroviruses have caused notable outbreaks on cruise ships, sickening hundreds of guests and crew in a matter of days. Human hepatitis A virus is also transmitted by the fecal-oral route via contaminated produce or uncooked shellfish. Fomites (objects contaminated with infectious organisms) can also play a role in fecal-oral transmission.

Respiratory transmission occurs when viruses in the respiratory tract are expelled as droplets. The transmission may be directly from one individual to another (please do not cough in my face) or may occur through fomites, hence the advice to wash your hands often! Viruses expelled from the respiratory tract may also be transmitted by contact with mucosal surfaces such as the eye. Health care providers and infectious disease researchers must remember to keep gloved hands away from their eyes. Examples of viruses that can be spread by the respiratory route are influenza viruses, rhinoviruses (one of the common cold viruses), and the severe acute respiratory syndrome (SARS) and Middle East respiratory syndrome (MERS) coronaviruses.

Transmission of viruses via exchange of bodily fluids can result from blood transfusions, use of dirty needles, trauma (bleeding), organ or tissue transplantation, sexual contact, or artificial insemination. The human viruses HIV, HBV, and HCV are all transmitted via contaminated blood. But these viruses can also be transmitted through contact with other bodily fluids such as semen or saliva. HIV can be transmitted via breast milk. Rabies virus is transmitted by saliva.

A few viruses, such as foot-and-mouth disease virus of livestock, can be transmitted over long distances through the air, a process called airborne transmission. Measles virus is also known for airborne transmission. Simply sitting in a room with a measles-infected individual can lead to infection! It should be noted that airborne transmission is distinct from aerosol transmission. In airborne transmission, particle sizes are very small and remain suspended in the air for long periods. The importance of understanding the distinction between these two types of transmission is exemplified by the 2014 Ebola virus epidemic. Ebola virus is transmitted through contact with body fluids of an infected individual. Transmission occurs when the patient is clearly symptomatic and virus titers are highest. Ebola can be transmitted via respiratory droplets but there is no evidence that the virus is transmitted in the absence of direct contact with respiratory droplets, or other secretions, thus Ebola is not considered to be an "airborne" virus.

Many viruses (West Nile virus, the equine encephalitis viruses, dengue virus, chikungunya virus, and zika virus, for example) are transmitted from one host to another primarily via an *insect* intermediary. Blood-feeding insects such as mosquitos, ticks, and midges are common vectors. Viruses transmitted by insect vectors are collectively called arboviruses.

It should be emphasized that a virus can be transmitted by more than one route. The SARS coronavirus, considered primarily a respiratory virus, is also transmitted by the fecal-oral route. Blood transfusions, dirty needles, and organ transplants may facilitate transmission of viruses usually spread by other routes. Mucosal surfaces, such as the eye, can be entry points for transmission of virus in present in blood or other bodily fluids. Some mosquito-vectored viruses (West Nile, chikungunya, yellow fever, and equine encephalitis viruses) require special precautions to avoid transmission in a research setting, where these viruses can be transmitted via aerosols.

Finally, a discussion of virus transmission should also include brief mention of virus transmissibility. Transmissibility is the ease of virus spread from one host to another. Measles virus is highly transmissible by the airborne route, and outbreaks can quickly become widespread in nonimmune population. Transmissibility is not related to the ability of a virus to cause disease (virulence). A virus may be relatively difficult to transmit, but highly virulent if transmission does occur. It is easy to overestimate the transmissibility of a highly virulent virus.

In this chapter we have learned that:

- Viruses are infectious agents (but are not cells).
- Viruses are obligate intracellular parasites that require host cells for their replication.
- Virions are the packages that contain the viral genome.
- Virions assemble from viral proteins and genomes synthesized within the infected cell.
- In the laboratory viruses are cultured or grown in cell or organ culture.
- Viruses can change or adapt to new growth conditions.
- Viruses have different genome types, capsid types, routes of infection, and diverse interactions with host cells.
- Virus infection may but does not always lead to cell death or host disease.
- Virus infections may be relatively short lived (acute infections) or may be life-long (chronic or persistent).

CHAPTER

2

Virus Structure

OUTLINE

After studying this chapter, you should be able to:

- Define "capsid" and explain its functions.
- Define "nucleocapsid," "envelope," and "envelope protein."
- Describe the difference between "structural" and "nonstructural" proteins.
- Be able to distinguish between icosahedral and helical capsids.
- Indicate the twofold, threefold, and fivefold axes of symmetry on an icosahedral capsid.
- Describe the activities of envelope glycoproteins.
- Describe the location and a common function of matrix (MA) proteins of enveloped viruses.

ANATOMY OF A VIRUS

The simplest viruses consist of genome packaged in a protein shell or capsid (see Fig. 1.1). Capsids are assembled from many copies of a single, or a few types of capsid proteins. Some viruses surround their capsids with a lipid bilayer called the envelope. Envelopes may be derived from the cell plasma membrane, nuclear membrane, or other intracellular membranes. All enveloped viruses encode proteins that are associated with the lipid bilayer. They are usually glycosylated (thus are envelope *glyco*proteins) and often contain transmembrane anchoring domains. They often project out from the surface of the envelope forming distinct spikes. Many enveloped viruses have a matrix protein positioned inside, and associated with the envelope (via direct membrane interactions or through interactions with the cytosolic tails of envelope glycoproteins). The matrix protein often forms a link between the membrane and the nucleocapsid. The term nucleocapsid refers to a complex of viral nucleic acid and protein. The term is most often used to refer to the assemblage of protein and nucleic acid within an *enveloped* virus. If viral envelopes are gently lysed, the nucleocapsids are released.

The proteins that assemble to form the virion (the extracellular particle) are called structural proteins. Additional proteins may be encoded by a virus, but are not present in the virion. These so-called nonstructural proteins have a variety of functions in the virus replication cycle. For example, nonstructural proteins may modulate cell and host antiviral responses; others are enzymes such as proteases or polymerases. It is important to note: nonstructural does not mean

Viruses. DOI: http://dx.doi.org/10.1016/B978-0-12-803109-4.00002-7

nonfunctional, or unimportant. Most nonstructural proteins are in fact essential for virus replication.

CAPSID STRUCTURE AND FUNCTION

In the simplest terms, viral capsids are protein packages that protect the genome. However capsids should not be considered static boxes, as they are dynamic structures that have other important functions. In addition to simply providing "packaging," the capsids of unenveloped viruses mediate attachment to, and penetration into, the host cell. Capsids must also be able to assemble, specifically package viral genomes and direct budding or release from cells.

Capsids come in two basic shapes: helical (rod shaped) or icosahedral (spherical). The simplest capsids are assemblies of many copies of a single protein (often called the capsid protein). As capsids assemble, they are stabilized by the repeated interactions (largely electrostatic) of the capsid protein building blocks. It should come as no surprise that there are few covalent bonds between these building blocks, because the genome must be released from the capsid at a later time!

The repeated occurrence of similar protein–protein interfaces leads to construction of a symmetrical capsid. The simple helical capsid of tobacco mosaic virus (TMV), a plant virus, is assembled from many copies of a single capsid protein. Each capsid protein forms identical same side-to-side and top to bottom interactions with its neighbors, as indicated in Fig. 2.1. In addition to interacting with neighboring proteins, each TMV capsid protein interacts with three nucleotides of the viral (RNA) genome. The capsid proteins are tightly packed around the RNA and form a rigid rod whose length is determined by the length of the genome. Not all helical capsids are rigid rods, many enveloped animal viruses have very flexible, helical nucleocapsids surrounded by an envelope.

Viruses with spherical capsids have icosahedral symmetry. An icosahedron is a closed cube with twofold, threefold, and fivefold axes of symmetry (Fig. 2.1). The simplest icosahedron can be assembled from 20 equilateral triangles. More complex, and larger, icosahedra can be built by assembling more than 20 triangular subunits.

Capsids that vary from the definitions of a helix or an icosahedron are sometimes called "complex." For example, there are viruses of bacteria (bacteriophage) with icosahedral "heads," rod like "tails" and long "fibers" extending from the tails.

CAPSIDS ARE BUILT FROM MANY COPIES OF ONE OR A FEW TYPES OF PROTEIN

Biological constraints require that capsids be assembled from multiple copies of one, or a few, small (usually in the range of 20–60 kDa) proteins. A key consideration of capsid assembly is the relative size of an amino acid and the triplet codon for that amino acid. The average size of an amino acid is 110 daltons (Da) while the average size of a triplet codon is ~330 Da × 3, or nearly 1000 Da. The codon is considerably larger than the amino acid, so it requires many copies of any single amino acid to package it.

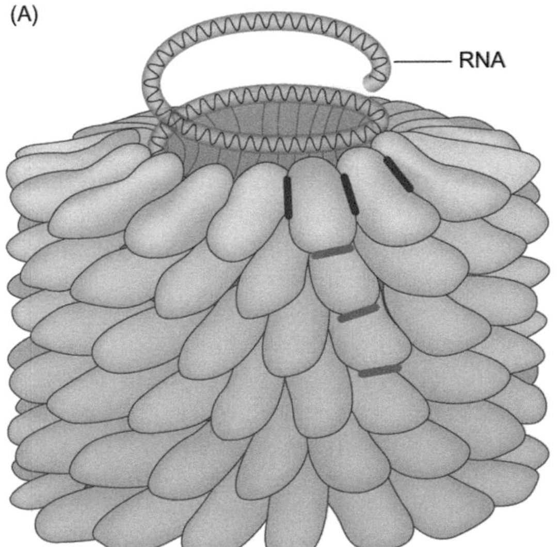

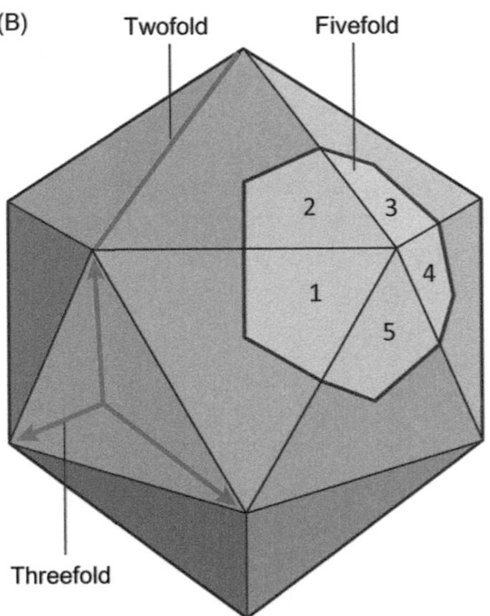

FIGURE 2.1 Capsids come in two basic shapes helices (rods shown on left) and spherical particles with icosahedral symmetry (shown on right). (A) The red and blue bars on the helical virus indicate sites of repeating contacts between subunits. Red bars indicate side-to-side contacts. Blue bars indicated top to bottom contacts. (B) The two, three and five fold axes symmetry are indicated for this simple icosahedral capsid.

A single polypeptide is always physically smaller than the gene that encodes it. Luckily, translation produces many copies of a protein from each mRNA.

Another biological constraint to the size of a capsid protein is its ability to fold. Small polypeptides can fold tightly while larger ones are unable to fold without leaving gaps in the structure; gaps could render the genome susceptible to environmental damage. Larger proteins also need chaperones to help them fold correctly. Indeed, there are some viral capsids that assemble with the help of chaperones.

Fidelity of protein synthesis is another issue that constrains the size of a capsid protein. The observed error rate of protein synthesis is $\sim 10^{-4}$, or 1 mistake per 1000 amino acids polymerized. Thus on average, a 1000 amino acid long protein would contain one mistake. In contrast, synthesis of 10 copies of a 100 amino acid protein would produce, on average, nine perfect proteins and one with a mistake.

SIMPLE ICOSAHEDRAL CAPSIDS

In the previous section the icosahedron was described as an assembly of equilateral triangles. However, capsids are not assembled from "triangular" or even symmetrical, proteins. Among animal viruses, the most common type of capsid protein structure is called an eight-stranded jelly roll β-barrel motif. This structure consists of four pairs of antiparallel β sheets (Fig. 2.2).

The most common icosahdral capsids can be envisioned as an arrangement of three capsid proteins into a triangle. Thus, as shown in Fig. 2.2, the simplest icosahedral capsid would contain 60 capsid proteins (20 triangular faces, each having three capsid proteins). This structure is called a $T=1$ capsid, and T is the triangulation number (Box 2.1). Without reviewing the mathematics involved, a $T=1$ icosahedron is an assembly of 20 triangular faces, a $T=\underline{3}$ icosahedron has $\underline{3}\times 20$ or $\underline{60}$ triangular faces, a $T=\underline{4}$ icosahedron has $\underline{4}\times 20$ or $\underline{80}$ triangular faces and so on. The T number can be used to determine the theoretical number of capsid proteins required to assemble a shell, if each triangular face is constructed from three capsid proteins. A $T=1$ virus (for example, a parvovirus) has a shell assembled from 60 capsid proteins. Caliciviruses have $T=3$ capsids, assembled from 180 (3×60) molecules of a capsid protein. Togaviruses have $T=4$ capsids that are assembled from 240 (3×80) molecules of capsid protein (Table 2.1). Thus larger capsids are assembled by using more building blocks (Fig. 2.3).

How do capsid proteins assemble to form a simple icosahedral shell? The process is stepwise and sometimes the substructures can be identified. $T=1$ viruses assemble from 60 capsid proteins. A general model is that three capsid proteins assemble to form a capsomere. Five capsomers form a pentamer and 12 pentamers assemble to form the closed shell. $T=3$ and $T=4$ capsids are built with pentamers and hexamers (hexamers are assemblies of six capsid proteins). $T=3$ capsids assemble from 12 pentamers and 20 hexamers while $T=4$ viruses assemble from 12 pentamers and 30 hexamers.

The surfaces of icosahedral capsids can be quite variable. Some capsids are relatively smooth, while others have prominent projections and/or deep canyons or pits (Fig. 2.4). Electron micrographs of unenveloped icosahedral viruses are shown in Fig. 2.5. In the case of unenveloped viruses, attachment proteins may

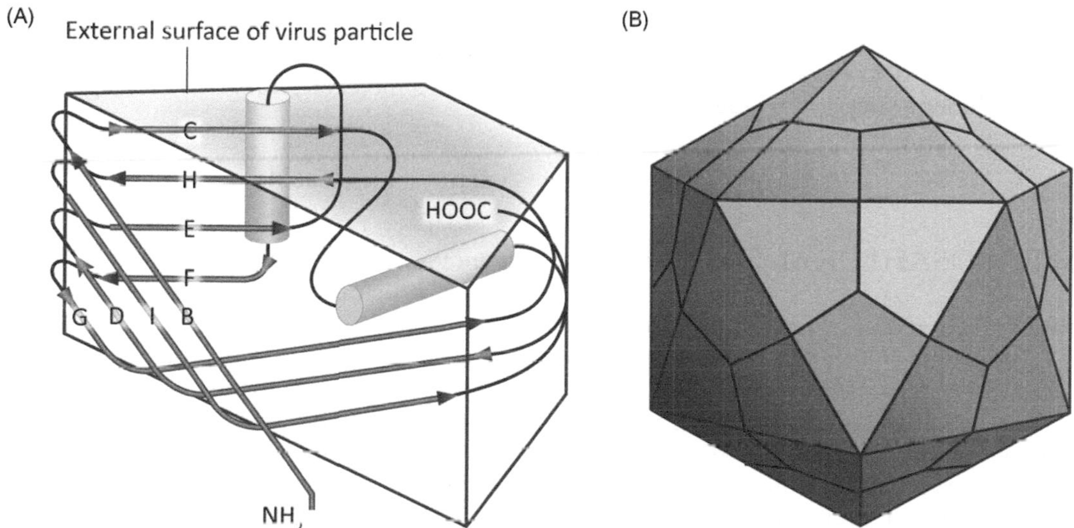

FIGURE 2.2 (A) The icosahedral capsids of most animal viruses are assembled from proteins that have a common shape: an eight-stranded jelly roll β-barrel motif. (B) Each triangular 'face' of the capsid is assembled from three copies of an eight-stranded jelly roll β-barrel motif.

BOX 2.1

TRIANGULATION NUMBERS AND STRUCTURAL SUBUNITS

The T number was first described by Caspar and Klug in 1962 to explain the structural basis for icosahedral capsids of different sizes. In their models, the T number predicted the number of capsid proteins in a structure, and their predictions proved true for many small viruses. However, as more icosahedral capsids were analyzed it became apparent that T numbers more often predicted numbers of "structural subunits" rather than numbers of polypeptides in a capsid. The most common structural subunits of icosahedral animal virus capsids are β-barrels (eight-stranded jelly roll β-barrel motifs). This common structure is shown in Fig. 2.3. Capsid proteins of different viruses do not have conserved amino acid sequences; however, they do assume this conserved structure. The distinction between a structural subunit and a polypeptide in building capsids can be demonstrated by comparing the capsids of caliciviruses, picornaviruses, and cowpea mosaic virus. All three virus groups have $T=3$ capsid architecture. The calicivirus capsid is assembled from 180 copies of a single capsid (C) protein while picornaviruses are assembled from 60 copies each, of three capsid proteins (VP1, VP2, and VP3). The three picornavirus capsid proteins have different primary amino acid sequences but fold to assume similar structures. Thus the capsids of picornaviruses are assembled from 180 equivalent structural subunits. The $T=3$ capsids of picornaviruses are sometimes called pseudo $T=3$ or $P=3$ capsids as they do not strictly adhere to the original predictions put forth by Caspar and Klug. In the case of cowpea mosaic virus the capsid is assembled from two different polypeptides, 60 copies each of a large (L) and a small (S) capsid protein. However, a close look at capsid architecture shows that L folds into two independent β-barrels connected by a hinge region. Thus the cowpea mosaic virus capsid is assembled from 180 "structural" domains, where the structural unit is the β-barrel.

TABLE 2.1 Construction of Simple Icosahedral Capsids

T number	Triangular faces/ icosahedron	Number of proteins (structural subunits)/capsid	Number of pentamers + number of hexamers
1	20	60	12 + 0
3	60	180	12 + 20
4	80	240	12 + 30

project out from the capsid to engage a receptor, while in other cases attachment sites are inside a canyon or pit.

LARGER ICOSAHEDRAL CAPSIDS

Large icosahedral capsids can also be envisioned as assemblages of triangular faces. For example, one could construct a $T=7$ icosahedron using 140 equilateral triangles (420 structural subunits) or a $T=13$ icosahedron from 260 equilateral triangles (780 structural subunits). However cryo-electron microscopic visualization of viruses with larger icosahedral capsids reveals that some of these structures (described below) are not simply larger versions of a $T=1$ or $T=3$ virus.

Polyomavirus and Papillomavirus Capsids

Based on the predictions of Caspar and Klug, these $T=7$ capsids (~50 nm diameter for polyomavirus and ~60 nm diameter for papillomavirus) would contain 420 structural subunits assembled into 12 pentamers and 80 hexamers, but this is not exactly the case. Instead the capsids of polyomaviruses and papillomaviruses are assembled from 360 structural subunits organized as 72 pentamers (Fig. 2.3). These capsids structures are referred to at $T=7$d (rather than $T=7$). Papillomavirus capsids are assembled from two structural proteins, L1 and L2. Each pentamer assembles from five molecules of L1 and a single molecule of L2.

Adenovirus Capsids

The large (~95 nm diameter) icosahedral capsids of adenoviruses contain 12 different polypeptides. Their capsid structure is described as pseudo $T=25$. The major capsid protein (MCP) is called the hexon protein and 720 copies of hexon protein assemble to form 240 trimers. Each trimer forms a hexamer subunit. But how can a hexamer be assembled from a trimer of capsid proteins? It turns out that hexon protein is a large capsid protein that contains two structural domains. Adenovirus capsids also contain 12 pentamers or vertices (Fig. 2.3, Panel E) assembled from three additional structural proteins, the penton base protein

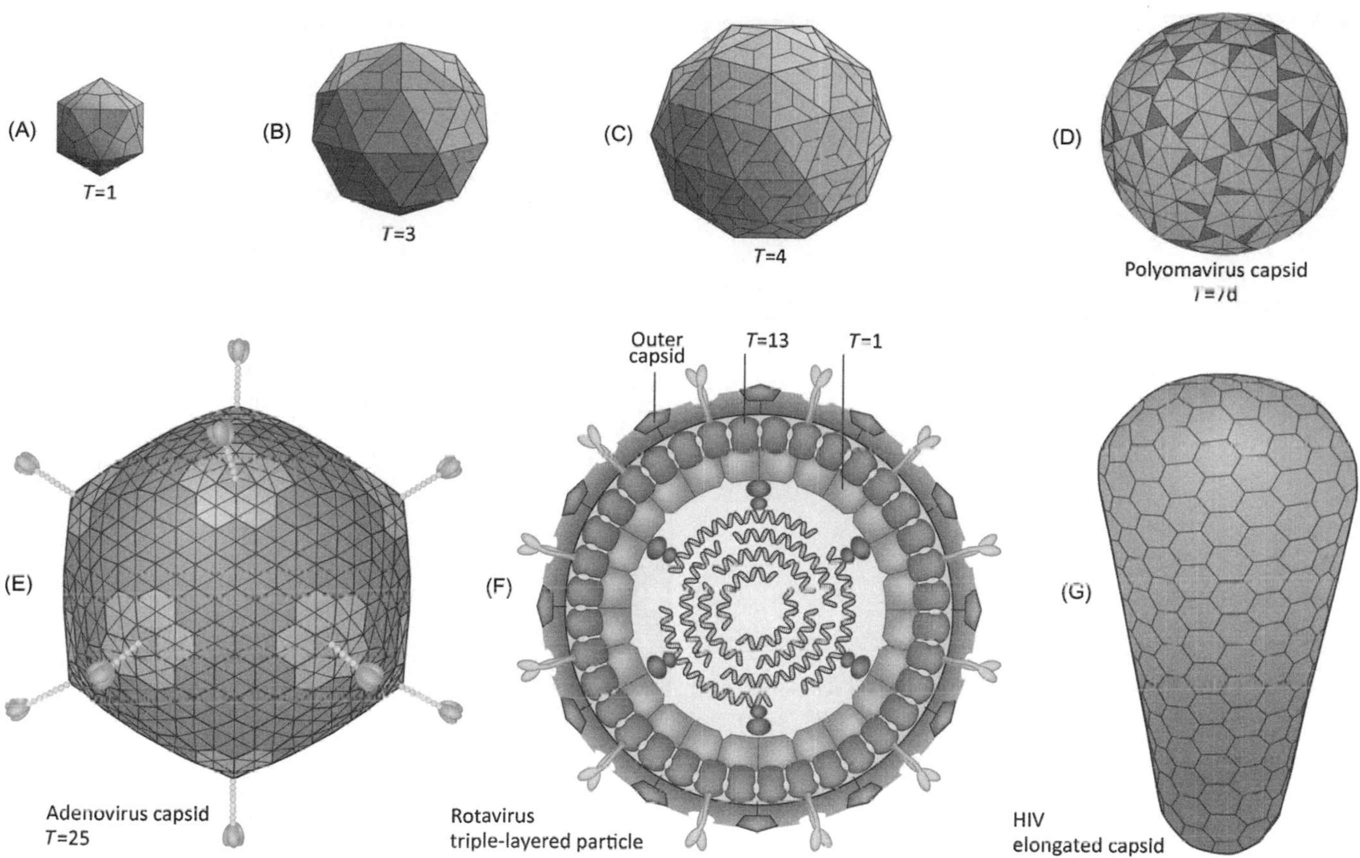

FIGURE 2.3 Examples of large and small icosahedral capsids. (A) T = 1 icosahedral capsid; (B) T = 3 icosahedral capsid; (C) T = 4 icosahedral capsid; (D) T = 7d polyomavirus capsid; (E) T = 25 adenovirus capsid; (F) Triple layer particle of rotavirus contains a T = 1 core surrounded by T = 13 shells. (G) HIV capsid. Elongated icosahedral capsids are formed by adding rows of hexamers to the middle of the structure.

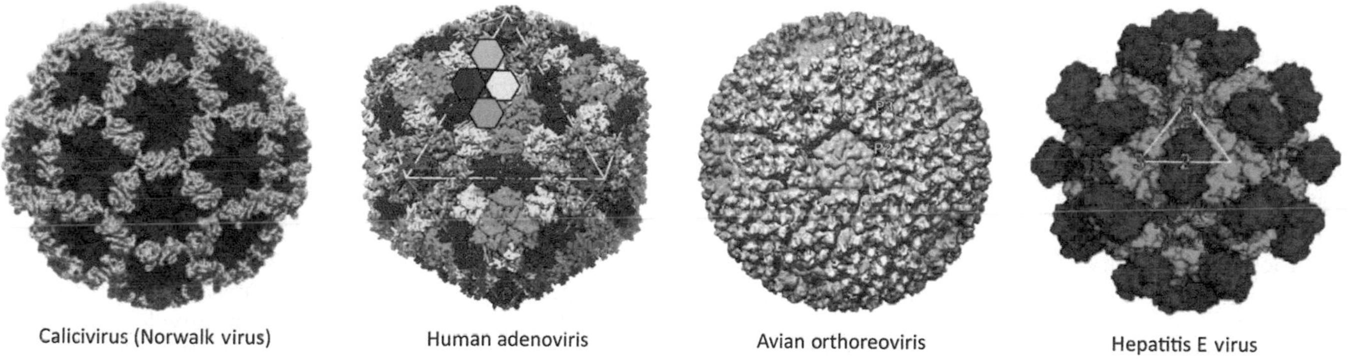

FIGURE 2.4 Molecular models of icosahedral viruses.

(polypeptide III), the fiber protein (polypeptide IV), and polypeptide IIIa. The long spikes that extend from each vertex are homotrimers of the fiber protein. They are anchored in place by the penton protein. Five copies of polypeptide IIIa are arranged in a ring beneath each vertex. Thus pentons and hexons are assembled from different types of capsid proteins.

Reoviruses

Reoviruses have multilayered icosahedral capsids. Viruses in the genus *Rotavirus* have triple-layered particles (Fig. 2.3, Panel F). The smallest, innermost layer is a $T = 1$ capsid formed from 60 copies of viral protein (VP) 2. The middle layer is a $T = 13$ capsid assembled

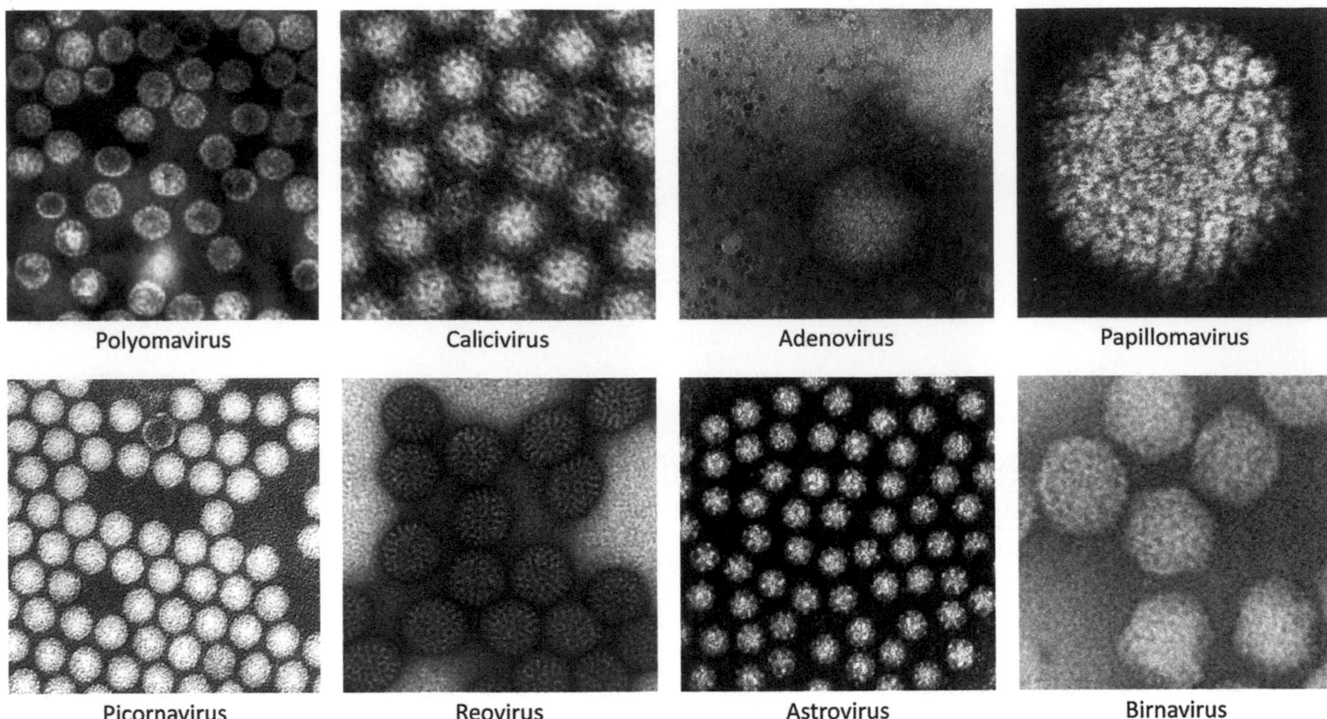

FIGURE 2.5 Examples of unenveloped viruses.

from the ~45 kDa VP6. The outmost layer of the rotavirus capsid is assembled from two additional structural proteins, VP4 and VP7. Members of the genus *Reovirus* have two capsids layers, an inner $T=1$ core surrounded by an outer $T=13$ shell.

Herpesvirus Capsids

Herpesviruses are large, enveloped viruses with icosahedral capsids. The major capsid protein or MCP is quite large (~1300 amino acids) and folds to assume a structure that is distinct from the eight-stranded jelly roll β-barrel motif of other animal viruses. MCP assumes a shape very similar to the capsid proteins of double-stranded (ds) DNA bacteriophage such as P22 or HK97. Herpesviral capsids are assembled from a total of 162 capsomers (150 hexons and 12 pentons). MCP forms all of the hexons and 11 of 12 pentons. The 12th penton is formed from a different herpesvirus protein called the portal protein to form the portal "complex." The portal complex is cylindrical and contains a channel. Bacterio phage portal complexes actively transport DNA into and out of bacteriophage capsids. Herpesvirus portals likely serve a similar function. The structure of herpesvirus capsids suggests an evolutionary relationship between herpesviruses and dsDNA phage.

Poxvirus Structure

Poxviruses are among the largest of the animal viruses. Virions are brick shaped or ovoid. They are enveloped and range in size from 140 to 260 nm in height and 220 to 450 nm in length. The envelope surrounds a dumb bell-shaped core (capsid) and two globular protein structures called lateral bodies. Morphogenesis of poxvirions is described in Chapter 35, Family *Poxviridae*.

VIRAL ENVELOPES

Many important human and animal pathogens are enveloped viruses. Their helical or icosahedral capsids are enclosed within a lipid bilayer. Enveloped viruses come in a variety of shapes, ranging from the long, filamentous ebolaviruses (Chapter 21: Family *Filoviridae*) to the icosahedral togaviruses (Chapter 16: Family *Togaviridae*). Most enveloped viruses acquire their lipid bilayer by budding through a cellular membrane (for example, plasma membrane, endoplasmic reticulum (ER), Golgi, or nuclear membranes). However some large viruses, such as poxviruses apparently build their membranes from crescent-shaped lipid fragments associated with protein scaffolds (Chapter 35: Family *Poxviridae*).

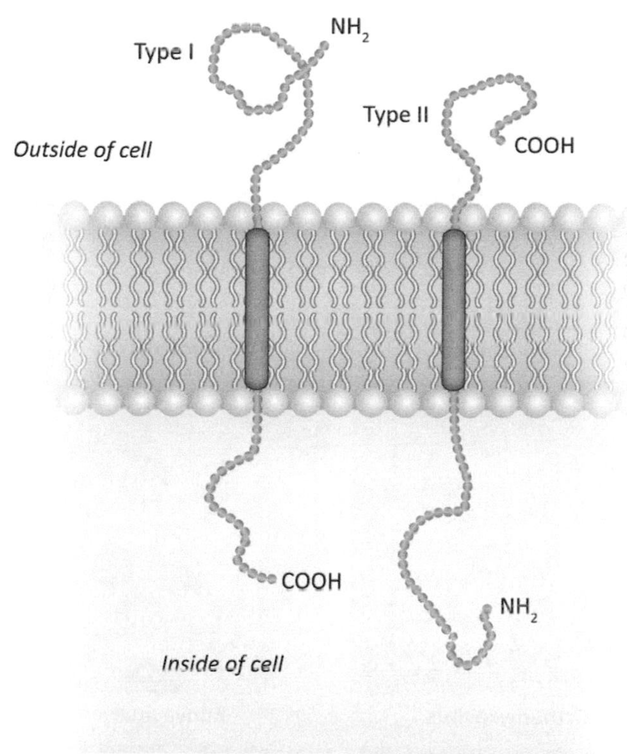

FIGURE 2.6 Arrangement of types I and II membrane proteins.

Enveloped viruses (with the exception of poxviruses) have proteins that are anchored into or across the lipid bilayer. Thus they are membrane proteins and can be categorized as type I or type II depending on their orientation in the membrane (Fig. 2.6). Types I and II membrane proteins span the lipid bilayer a single time, via an α-helical TM domain. Type I proteins are inserted into membranes with their amino-terminal domain to the outside and their carboxyl-terminal domain to the inside of the cell or virion. Type II membrane proteins have the opposite polarity with respect to the membrane Envelope proteins often take the form of long spikes. Spikes are usually homo- or heterodimers or trimers and many have distinct globular heads. Examples of viral type I membrane proteins include: influenza virus HA, rabies virus G protein, paramyxovirus fusion proteins, and hepatitis C virus E1 and E2. Paramyxovirus attachment proteins are type II membrane proteins.

Virus envelope proteins must carry out at least two functions: receptor binding and fusion. These activities may be carried out by a single protein (influenza virus HA, rabies virus G) or by two distinct proteins in the case of the paramyxoviruses. In the case of influenza virus, the HA glycoprotein is cleaved by cellular proteases to generate a fusion-active version of the protein but the two halves of HA (HA1 and HA2) remain associated via disulfide bonds. Rabies virus G protein also has both attachment and fusion activities; however, rabies virus G remains uncleaved. Human immunodeficiency virus (HIV), a retrovirus, also encodes a single glycoprotein precursor, but it is cleaved during synthesis to produce two distinct structural proteins. The amino-terminal domain of the envelope precursor is the surface unit (SU) protein, which functions in attachment. The carboxyl-terminal half of the precursor is the transmembrane (TM) protein. As its name implies, it has a membrane-spanning domain, anchoring it into the envelope. TM is the fusion protein. SU and TM are not covalently linked. SU is on the outside of the virion, held thereby noncovalent interactions with ectodomain of TM (Chapter 37: Replication and Pathogenesis of Human Immunodeficiency Virus). Viruses in the family *Paramyxoviridae* (see Chapter 20: Families Paramyxoviridae and Pneumoviridae) produce two envelope glycoproteins using distinct open reading frames. Both are membrane-anchored proteins. The attachment protein is a type II integral membrane protein while the fusion protein is a type I integral membrane protein.

Glycosylation

The ectodomains of envelope proteins are usually glycosylated (they have polysaccharides linked to the protein backbone). Carbohydrates can be linked to the peptide backbone via the nitrogen atom on an asparagine side chain (N)-linked or to the oxygen atoms of serine or threonine side chains (O-linked). Glycosylation occurs during transit of envelope proteins through the ER and Golgi. The amount of carbohydrate decorating envelope proteins varies. Some viral envelope proteins have one molecule of polysaccharide per polypeptide chain. In the case of the HIV SU, the molecular mass of carbohydrate equals that of the polypeptide backbone. HIV SU protein has an apparent molecular weight of 120 kDa (as determined by SDS-PAGE) but the calculated mass of the amino acid backbone is <50 kDa.

OTHER MEMBRANE PROTEINS

A few enveloped viruses also have membrane-associated proteins with other roles in the replication cycle. Influenza viruses have a receptor-destroying protein called neuraminidase (NA) that is required for efficient release of virions from the surface of the infected cell. Drugs that block the enzymatic activity of NA do not block virion formation, but virions remain tightly associated with the infected cell, thereby

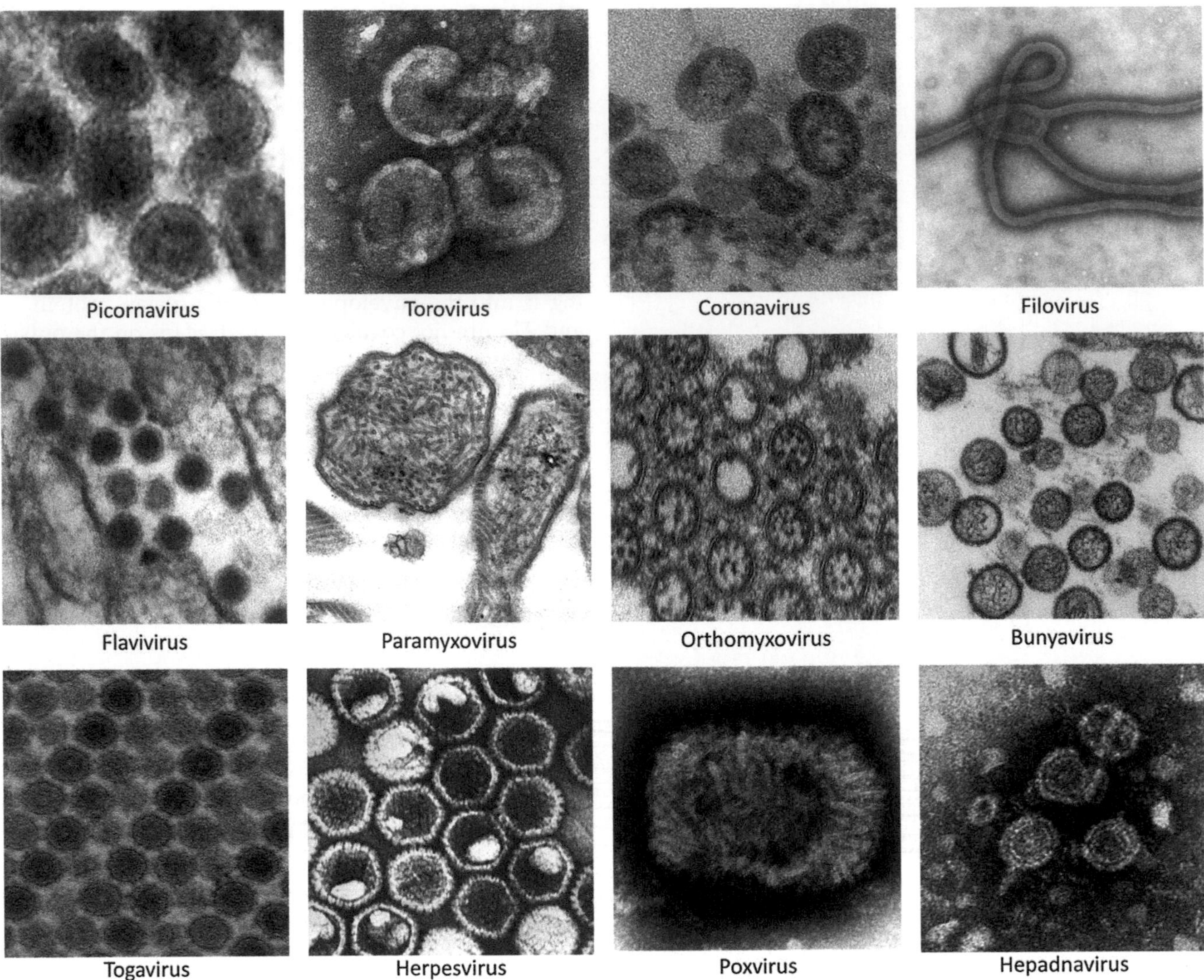

FIGURE 2.7 Examples of enveloped viruses.

limiting their ability to infect other cells (Chapter 23: Family *Orthomyxovirdae*).

Another role for viral membrane-associated proteins is ion transport. Influenza viruses encode a small integral membrane protein called M2. The M2 ion channel is a low pH-activated homotetramer that facilitates transport of H^+ ions across the viral envelope to acidify the core during virus entry (Chapter 23: Family *Orthomyxovirdae*). Fig. 2.7 shows micrographs of some enveloped viruses.

MATRIX PROTEINS

Matrix proteins are not conserved across virus families. The term matrix protein is used to describe a protein that forms layer on the inside of the viral envelope. Matrix proteins play important roles in virus assembly, as they form links or bridge between nucleocapsids/cores and the envelope. Retroviral matrix(MA) proteins are fatty-acylated, allowing them to interact tightly with cellular membranes to form budding sites. This interaction anchors the remainder of uncleaved retroviral GAG polyprotein to the membrane (Chapter 36: Family *Retroviridae*). The influenza virus matrix protein (M or M1) first makes contact with nucleocapsids in the nucleus of the infected cell and may participate in nucleocapsid assembly; M is required for transport of nuclecapsids from the nucleus. Additional M is found at the plasma membrane where it associates with the cytoplasmic domains of influenza virus envelope glycoproteins. Thus M organizes both glycoproteins and the nucleocapsids. The matrix (M) proteins of the paramyxoviruses are highly basic proteins and paramyxovirus M is the most abundant protein in the virion. M associates with nucleocapsids and the plasma membrane (probably via interactions with cytosolic domains envelope glycoproteins) and is likely the driving force in budding.

NUCLEOCAPSID STRUCTURE

When discussing enveloped viruses, the term nucleocapsid is commonly used instead of capsid. The nucleocapsid refers to the assembly of protein and nucleic acid (the genome) that remains after the viral envelope is removed. In the case of the negative-strand RNA viruses, the main structural protein of the nucleocapsid is an RNA-binding nucleoprotein (N) but it also contains several molecules of the viral RNA-dependent RNA polymerase (Chapter 10: Introduction to RNA Viruses). Hepadnaviruses and retroviruses have icosahedral capsids surrounded by an envelope. These structures are often called "cores."

In this chapter we learned that:

- Capsids are protein coats that package viral genomes. Simple capsids are either rods (helical) or spheres. Spherical viruses have icosahedral symmetry. An icosahedron is a shell defined by its symmetry. An icosahedral shell has twofold, threefold, and fivefold axes of symmetry.
- Capsids are not static packages. They have important functions such as genome packaging, attachment, and entry. Structural flexibility of capsids allows them to carry out multiple functions in a highly regulated manner.
- Virions are either naked (unenveloped) or have a lipid envelope. Naked virions use capsid proteins to mediate attachment and entry.
- Enveloped viruses obtain their lipid membranes during budding. Viral envelopes may be derived from plasma membranes, nuclear membranes, or other internal membranes such as Golgi, ER, or transport vesicles.
- Enveloped viruses encode one or more envelope proteins with transmembrane-domain anchors. Envelope proteins often form spikes extending from the surface of the virion and most are glycosylated. Envelope proteins mediate attachment and fusion.
- The assembly of genome (nucleic acid) and capsid proteins found inside of the viral envelope is often called the nucleocapsid (particularly in the case of negative-strand RNA viruses). In the case of retroviruses, hepandaviruses, and more complex viruses, the structure is often called the core. Gently lysing the envelope will release nucleocapsids or cores. Viral polymerases may also be present in nucleocapsids/cores.
- Matrix proteins are found beneath the lipid membrane of enveloped viruses. They often function to drive virion assembly and budding.
- The group of proteins associated with the extracellular virus particle or virion are collectively called *structural proteins*. All viruses encode one or more structural proteins.
- The term *nonstructural protein* describes any virus proteins produced in the infected cell, but not packaged in the virion. Nonstructural proteins have many critical roles in the virus replication cycle.

CHAPTER

3

Virus Interactions With the Cell

OUTLINE

After studying this chapter, you should be able to:

- List conditions that impact virus attachment.
- Explain how to set up a synchronized infection in the laboratory.
- Define "penetration" and "uncoating" as regards the virus replication cycle.
- Describe the cellular machinery that viruses use to move through the cell.
- Describe the general structure of a eukaryotic gene, including definitions of promoter, intron, and exon.
- Explain the differences between DNA-dependent DNA polymerases, RNA-dependent RNA polymerases (RdRp), and reverse transcriptase (RT).
- Explain why DNA virus replication is often linked to cell cycle.

This chapter examines the major steps in virus replication within the context of cellular structures and processes. The emphasis is on virus interactions with eukaryotic, primarily animal, cells. Recall that the major steps in virus replication are: (1) attachment, (2) penetration and uncoating, (3) synthesis of viral genomes and proteins, (4) assembly of new virions, (5) virion release and maturation.

VIRUS INTERACTIONS WITH THE CELL

The Extracellular Space

Most animal cells are embedded in an extracellular matrix that helps form the architecture of tissues and organs. Much of the extracellular matrix is proteinaceous in nature; however, in the case of epithelial cells, the plasma membrane (PM) is surrounded by a coat of polysaccharides called the glycocalyx. Thus there are molecules outside of the cell that can interact with infectious agents such as viruses, before they reach the plasma membrane (PM). For example, the glycocalyx can serve as a barrier, but some viruses use the glycocalyx to their advantage by binding to its polysaccharides.

Viruses. DOI: http://dx.doi.org/10.1016/B978-0-12-803109-4.00003-9

The Plasma Membrane

The PM is both a barrier to, and required for, virus attachment and entry. The PM is a selectively permeable lipid bilayer that contains many proteins; it is a complex and dynamic structure. Small molecules such as carbon dioxide and oxygen cross the PM by diffusion. Sugars and amino acids cross the PM using protein channels or transporters. Larger molecules must be endocytosed in order to enter the cell. PM-associated proteins include receptors, signaling molecules, enzymes, and adherence proteins. As shown in Fig. 3.1, some proteins associated with the PM have cytoplasmic, membrane, and extracellular domains while others are embedded entirely with the bilayer. The proteins in the PM are mobile; they move laterally through the membrane, can reorganize and form complexes as the result of signaling. PM proteins are often glycosylated (carbohydrate chains are attached to the protein backbone). The lipids of the outer leaflet of the PM can also be glycosylated. Finally, the lipids within the PM not homogeneous; cholesterol can be found in discrete regions or microdomains called "rafts." Specific proteins are associated with these lipid rafts.

Attachment Occurs at the Plasma Membrane

The first step in the virus replication cycle is attachment to PM-associated molecules. The process requires the interactions of receptor molecules on the host cell with attachment proteins on the infecting virus. Receptors are often proteins, but viruses also bind to the sugar residues found on glycoproteins or glycolipids on the cell surface. Influenza virus is an example of a virus that attaches to carbohydrate receptors. Naked (unenveloped) virions use capsid proteins for attachment while enveloped viruses use an envelope-associated protein.

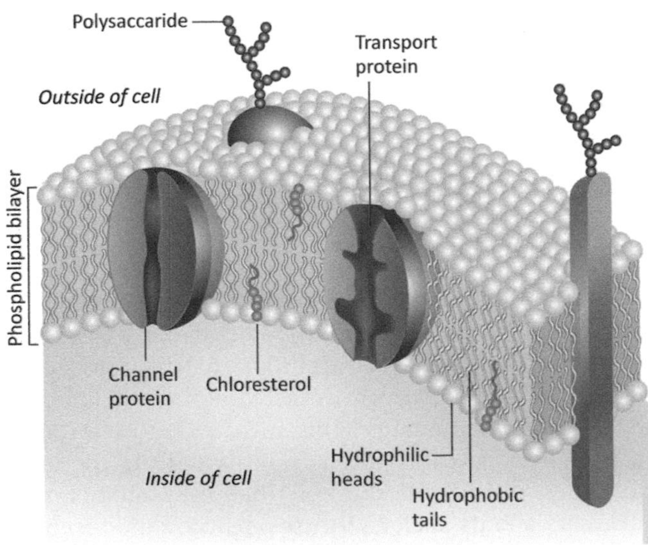

FIGURE 3.1 The PM is a selectively permeable lipid bilayer. Proteins associated with the PM include receptors, signaling molecules, enzymes, and adherence proteins.

Attachment is achieved by interactions of small subdomains of molecules. Interaction faces usually comprise just a few amino acids or sugar residues, and the interactions are usually electrostatic in nature. Thus the initial contacts between virus and receptor are weak and reversible. However, as multiple viral attachment proteins interact with multiple receptor molecules, binding becomes strong and irreversible. Thus it follows that cells with a higher density or number of receptors are more readily infected (Fig. 3.2). Because attachment is electrostatic, it can be affected by pH, ion concentration, and types of ions in the extracellular space. When we propagate viruses in the laboratory, the type of media used, and its pH, are important. Attachment does not require energy and can take place in the cold (4°C). A virus infection can be synchronized by allowing attachment to occur in the cold. Once sufficient time has passed the culture is quickly warmed up and the attached virions penetrate at the same time.

The presence or absence of receptors is a major *host range* determinant, as absence of receptors excludes viruses from a cell. Different cells within an organism display different surface molecules, thus one cell or tissue type may be permissive for virus attachment while others are not. Epithelial cells are an example of polarized cells. They display different molecules on their apical (facing the lumen) and basolateral (facing the inside) surfaces (Fig. 3.3). Receptor molecules may be expressed on only one surface of the polarized cell, such that passage through those cells is a one-way process.

The receptors for many human and animal pathogens have been identified, but it is important to note that viruses adapted to cell cultures may use different receptors than those used during a natural infection. Even within an organism, a virus may utilize a variety of receptors. Attachment can also require interaction of

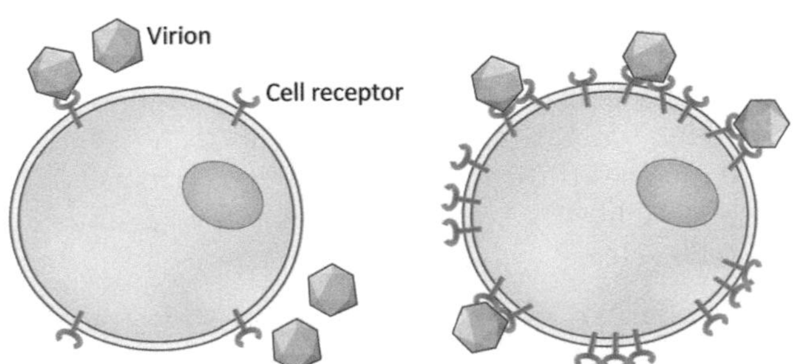

FIGURE 3.2 Receptor density and attachment. Cells with a higher density or number of receptors are more readily infected.

the virus with more than one type of receptor molecule. An example is human immunodeficiency virus (HIV). The surface unit (SU) glycoprotein of HIV (also called gp120) attaches to the CD4 protein present on helper T-lymphocytes, macrophages, and dendritic cells. After initial interactions between SU and CD4, SU then binds to a second receptor, one of several chemokine receptors on the cell surface. The chemokine receptors are the coreceptors for HIV (Chapter 37: Replication and Pathogenesis of Human Immunodeficiency Virus).

Virus Penetration and Uncoating at the Plasma Membrane

Once a virus attaches to a cell, the next critical step is delivery of the viral genome into the cytoplasm or nucleoplasm. Different viruses exploit different strategies to accomplish this. The RNA genome of a picornavirus crosses the PM through a channel or pore formed by capsid proteins (Fig. 3.4). In the case of the reoviruses, the whole capsid crosses a channel through the PM.

Enveloped viruses must transport their genomes across two sets of membranes, both the viral envelope and a cell membrane. The process sometimes occurs at the PM (Fig. 3.5). Unfavorable energetic barriers must be overcome to bring the two membranes into close enough proximity to allow formation of a pore (Box 3.1). If fusion occurs at the PM, viral envelope proteins remain on the cell surface.

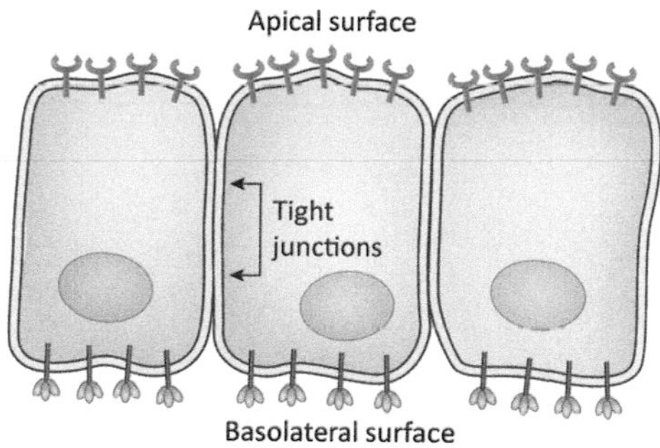

FIGURE 3.3 Polarized cells have discrete apical and basolateral surfaces. Apical surfaces face the outside (for example, the airspace in the lung or the lumen of the intestine) while basolateral surfaces face the inside of the body. Polarized cells display distinct proteins on their apical versus basolateral sides. Materials, such as viruses, can be transported directionally through polarized cells.

Virus Penetration and Uncoating From Membrane Bound Vesicles

Many viruses exploit cell processes designed to bring cargo into cells via membrane bound vesicles. These processes include phagocytosis, macropinocytosis, and receptor-mediated endocytosis. There are different mechanisms by which endocytosis can occur, such as clatherin-dependent or caveolin-1-dependent pathways (Fig. 3.6). The environment inside of endosomes is distinct from the cell cytosol. Of importance to our discussion of viruses, endosomes become acidified as they mature and low pH often triggers penetration. As mentioned above, some picornaviruses form membrane channels at the PM upon attachment. However other picornaviruses are endocytosed, and low pH triggers the capsid rearrangements that form membrane channels. Enveloped viruses often use low pH to trigger rearrangement of their surface proteins, with the effect of releasing previously hidden hydrophobic (fusion) domains.

Membrane Fusion

Membrane fusion results in lipid mixing or formation of a pore through both the viral envelope and a host cell membrane (Fig. 3.7). Key to the process, the viral envelope must be in very close proximity to the

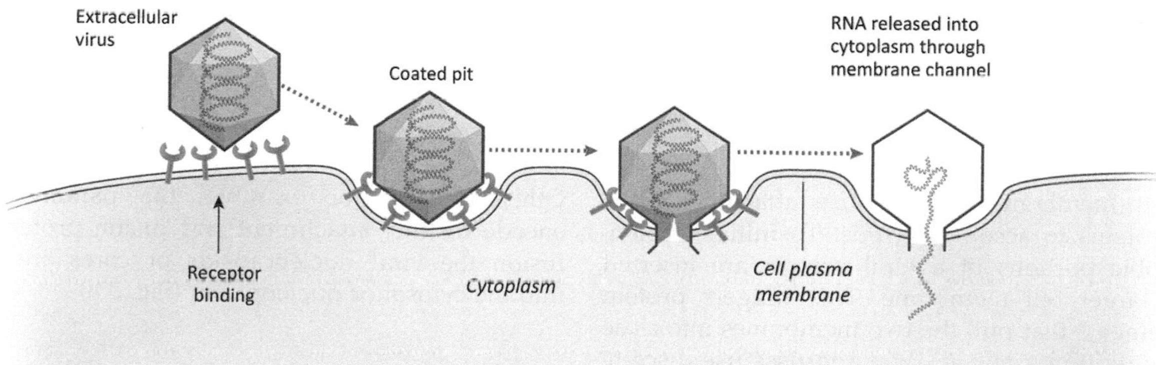

FIGURE 3.4 Illustration depicting the penetration of a viral genome across the PM. In this example the RNA genome of a picornavirus crosses the PM through a channel or pore formed by capsid proteins.

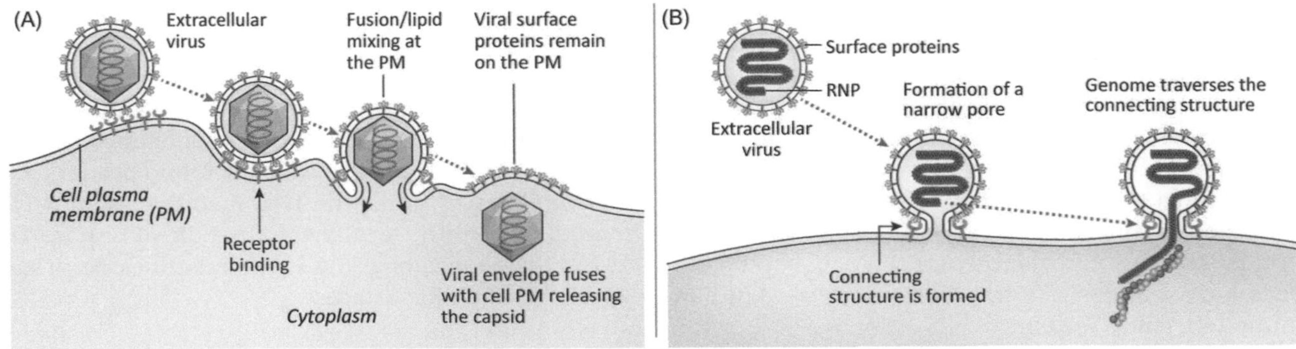

FIGURE 3.5 Illustration depicting the penetration of a viral genome after fusion of an enveloped virus with the PM. Panel A shows a process in which the viral envelope and PM become contiguous. Panel B shows a process in which a small pore is formed through the PM.

BOX 3.1

FUSION PROTEINS

Fusion of a viral membrane to a cellular membrane requires that the high kinetic barrier to membrane fusion be breached. Fusion proteins serve as the catalysts in this process. A hallmark of fusion proteins is that they undergo structural changes as a result of attachment and/or changes in pH (during endocytosis). Often these structural changes expose a hydrophobic segment called the fusion loop or fusion peptide whose function is to engage the target cell membrane. The fusion protein becomes a bridge between the two membranes, drawing them together (see Fig. 36.3). Viral fusion proteins are suicide enzymes that function only once.

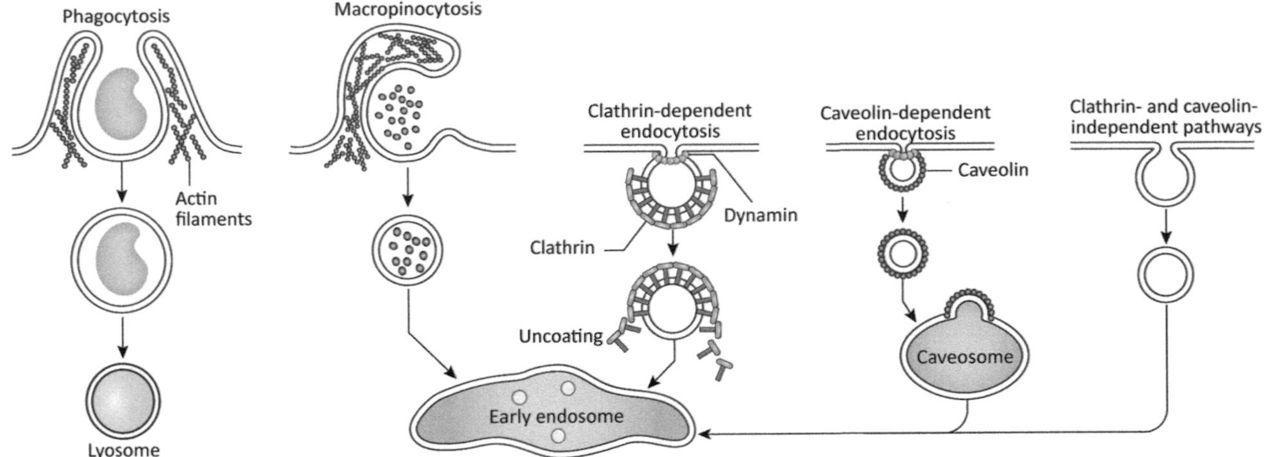

FIGURE 3.6 Cellular uptake of macromolecules by phagocytosis, macropinocytosis, and various endocytic pathways.

target cell membrane. Viruses use attachment and fusion proteins to accomplish this. To initiate fusion, hydrophobic portions of a viral protein are inserted into the target cell membrane. This triggers protein rearrangements that pull the two membranes into close proximity (within a few Å). Some viruses use discrete domains of a single protein to accomplish attachment and fusion. Influenza viruses have attachment and hydrophobic fusion domains in different regions of a single molecule, the hemagglutinin (HA) protein. Other viruses (for example, the paramyxoviruses) encode distinct attachment and fusion proteins. After fusion the viral nucleocapsids or cores are released into the cytosol or nucleoplasm (Fig. 3.8).

Uncoating the Genome

Uncoating is the removal of viral proteins from the genome to allow for translation, transcription, and replication. For some viruses the processes of penetration

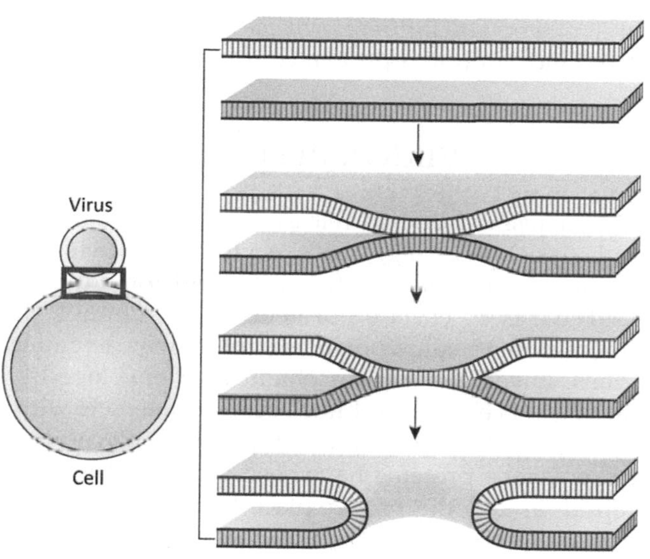

FIGURE 3.7 General illustration of the membrane fusion process. Membrane fusion or pore formation require membranes to be in very close proximity.

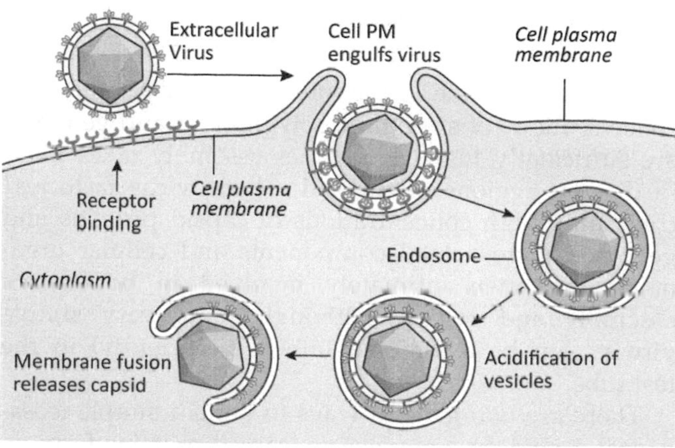

FIGURE 3.8 Many viruses, both enveloped and unenveloped, are brought into cells by endocytosis. The low pH environment in the endosome triggers molecular rearrangements of capsid or envelope proteins. In this example an enveloped virus is fusing with an endosomal membrane to release the capsid into the cytosol.

BOX 3.2

CELL CYTOSKELETON

The cytoskeleton is a dynamic and interconnected network of filaments. There are three major types of filaments. Actin filaments control cell shape, locomotion, and cytokinesis (separation of daughter cells after cell division). Intermediate filaments provide mechanical strength. Microtubules function in intracellular transport and chromosome segregation. Microtubules also control the position of organelles. The cytoskeleton is not static. It is constantly reorganizing via the assembly and disassembly of filaments. Regulation of the cytoskeleton is quite complex, requiring many molecular players. Molecular motors are key to moving materials along filaments. Members of the myosin superfamily of proteins are actin-based motors. For example, myosins facilitate movement of cargo along tracks of actin filaments and also move actin filaments relative to the PM.

and uncoating cannot be separated, but in other cases, uncoating requires additional steps. Some viral genomes are uncoated within the cytosol while in other cases uncoating occurs in the nucleus. Reoviruses are an interesting exception, their genomes are never uncoated but instead are transcribed within the capsid; mRNAs are released to the cytosol through pores in the capsid (Chapter 26: Family *Reoviridae*).

CYTOSKELETON

Movement Through the Cell, Interactions With the Cytoskeleton

The cytosol is a dense network of filaments, organelles, and molecular assemblies (see Box 3.2). It is a highly viscous environment that restricts diffusion of molecules larger than 500 kDa or particles larger than 20 nm. (Recall that most viruses are larger than 20 nm.) Cells have evolved highly regulated processes to move and sort cargo in order to maintain the highly organized and complex environment of the cell. These processes can potentially inhibit virus movement and replication but are often exploited by viruses to enhance their replication. For example, viruses use actin filaments and microtubules to move through the cell as well as from cell to cell. To aid in their movement, viruses highjack the cellular molecular motors associated with actin filaments and microtubules. In fact specific interactions with the cell cytoskeleton are important in every step of the virus replication cycle, from attachment to replication to assembly to release.

VIRUS ASSEMBLY

Viruses assemble in the infected cell when the local concentrations of structural polypeptides and genomes are sufficiently high. Sometimes assembly takes place in discrete regions of the cell called "virus factories" containing high concentrations of capsid proteins and genomes. Cytoskeletal components and cellular organelles are often intimately involved in both virus assembly and release, although some very simple viruses can be assembled (albeit inefficiently) in the test tube.

There are two general ways to build a simple icosahedral capsid. In one scenario, capsid proteins form an empty shell *into which* the genome is inserted. Picornaviruses use this strategy and empty capsids can be seen in virus preparations. Empty particles are less dense than complete virions and can be separated from them by density centrifugation. In the second scenario, an encapsidation signal on the genome interacts with one or more capsid proteins, followed by recruitment of additional capsid proteins.

Most animal viruses with helical nucleocapsids are RNA viruses. They usually encode a basic protein (often called the nucleocapsid (N) protein) that interacts with viral RNA. Initial interactions are specific such that cellular RNAs are not packaged indiscriminately. Genome packaging signals often include sequences at both the 5′ and 3′ ends of the viral genomic RNA. This is a mechanism whereby genomic RNA can be "distinguished" from subgenomic mRNAs. In some cases the N protein surrounds the viral RNA, but in others (i.e., influenza viruses) it appears that the RNA winds around a protein core.

VIRION RELEASE

Mechanisms for virus release from cells include cell death (lysis), budding, and exocytosis. The cytoskeleton can present a barrier to release and some unenveloped viruses encode proteins that disrupt the cytoskeleton to allow dispersal of newly assembled virions. Enveloped viruses obtain their envelope by a budding process. A viral nucleocapsid interacts with a region of host cell membrane into which glycosylated viral envelope proteins have been inserted. The nucleocapsid "finds" the proper place to bud from the cell by forming specific interactions with the cytoplasmic tail(s) of the envelope proteins. Cell membranes that can serve as sites of budding include the PM, endosomal, and nuclear membranes. Viruses released by budding from the PM (for example, the HIV) are released individually (Fig. 3.9). Viruses with envelopes derived from endosomal or nuclear membranes may bud into vesicles that traffic to the PM and fuse, releasing their cargo of virions by a process called exocytosis (Fig. 3.10).

It is easy to assume that release always results in free virions escaping into the extracellular environment. And while this certainly does happen, virions can also be transmitted directly from one cell to another. Vaccinia virus (family *Poxviridae*) uses actin tails to move between cells. We can visualize this process using

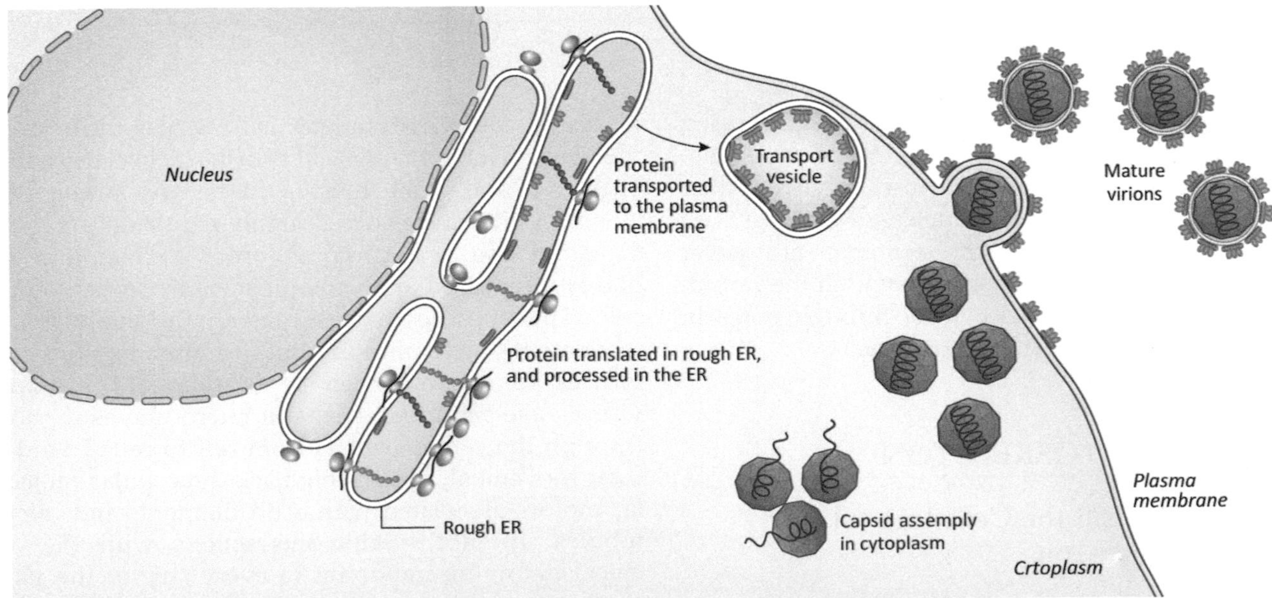

FIGURE 3.9 Illustration showing the process of assembly and budding of a virus particle from the PM.

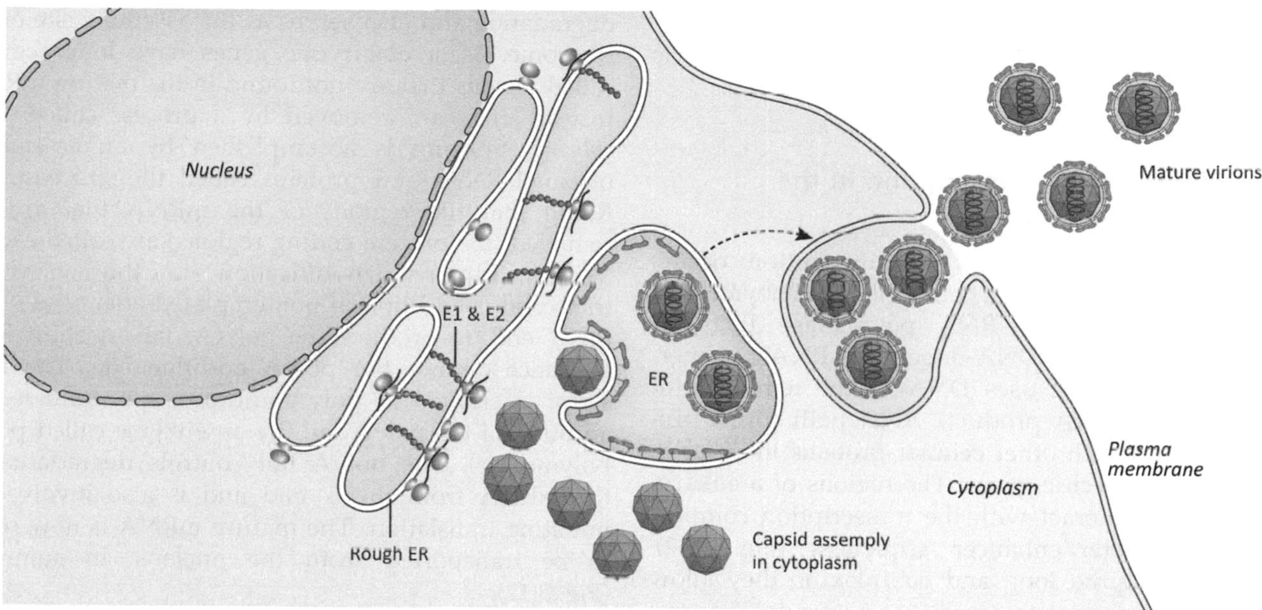

FIGURE 3.10 Illustration showing release of enveloped virions by exocytosis.

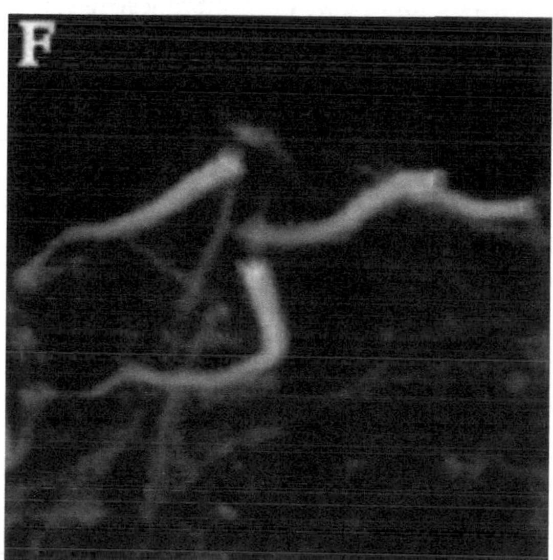

FIGURE 3.11 Actin projections (green) extending from a vaccinia virus infected cell. There is virus particle (red) at the tip of each projection. *From Figure 6 (panel F) of Cudmore, S., Reckmann, I., Griffiths, G., Way, M. 1996. Vaccinia virus: a model system for actin-membrane interactions. J. Cell Sci. 109, 1739–1747.*

cells with fluorescent-tagged actin and virions tagged with a different fluorescent tag (Fig. 3.11).

VIRION MATURATION

Maturation cleavages of virion proteins may occur after particle release, and these may be required to produce infectious virus (Fig. 3.9). Why do some viruses mature after release from a cell? The virion has two very different roles in the replication cycle: To assemble under conditions of favorable energy and to disassemble during the processes of attachment and penetration into a new cell. Maturation cleavages prepare the newly released virion to attach to a new cell and disassemble. For example, picornaviruses include in their capsids, an uncleaved precursor protein. The precursor is cleaved within the assembled capsid. If this cleavage is blocked, the capsid is unable to deliver the genome to a new cell. Retroviruses require a set of cleavages that convert long precursor polyproteins into individual structural proteins. The cleavages are made by a retroviral protease (Chapter 36: Family *Retroviridae*) and inhibitors of HIV protease activity are powerful antiviral drugs. Protease inhibitors are small molecules that interact with the HIV protease to prevent it from cleaving the polyproteins in the immature particle. For many enveloped viruses, maturation involves cleavage of glycoproteins that fuse the viral envelope to a host cell membrane.

AMPLIFICATION OF VIRAL PROTEINS AND NUCLEIC ACIDS IN THE CONTEXT OF THE INFECTED CELL

After penetration and uncoating have been achieved, the next events in the viral replication cycle are the synthesis (amplification) of viral proteins and genomes. In the absence of successful amplification, the proteins and genomes needed to assemble new virions are not made. To better appreciate the processes used by viruses to synthesize mRNAs, genomes, and proteins let us review some of the most basic aspects of these processes in the host cell.

A Short Review of Transcription in the Eukaryotic Host Cell

Synthesis of cellular mRNAs (transcription) occurs in the nucleus of the eukaryotic cell. The enzyme that synthesizes mRNAs is RNA polymerase II (RNA polII). RNA polII is a DNA-dependent RNA polymerase (an enzyme that uses DNA as the template for synthesis of an RNA product). RNA polII forms protein complexes with other cellular proteins in order to be directed to specific genes. The regions of a eukaryotic gene that interact with the transcription complex are the promoter/enhancer sequences (Fig. 3.12). These can be quite long and complex as they allow graded cell responses to a variety of stimuli. The promoter/enhancer region of a gene defines the conditions under which a gene product will be synthesized. The promoter-enhancer sequences themselves are not transcribed.

Shortly after initiation of a transcript, the RNA is modified by addition of a 5′-methyl guanosine "cap." The cap protects the 5′ end of the mRNA from degradation and also serves as the assembly site of the ribosome. Most eukaryotic genes have long regions, called introns that are not found in the mature mRNA. Instead, they are removed by a process called *RNA splicing*. Splicing is accomplished by an assemblies of small RNAs and proteins called the spliceosomes. Recall that the regions of the mRNA that are not removed (the protein coding regions) are called exons.

An additional modification to the eukaryotic transcript is addition of nontemplated adenosines near the 3′ end, to produce the poly(A) tail. A short RNA sequence defines the polyA addition site. The transcript is cleaved at polyA addition site, followed by addition of the polyA tail (by an enzyme called polyA polymerase). The polyA tail controls degradation of the mRNA from the 3′ end and is also involved in initiating translation. The mature mRNA is now ready to be transported from the nucleus. In summary (Fig. 3.12):

- Many eukaryotic genes have nontranscribed promoter/enhancer sequences.
- The 5′ end of the mRNA is capped.
- Many eukaryotic genes have introns that are removed from RNA by splicing.
- The 3′ ends of mRNAs are cleaved and polyadenylated.

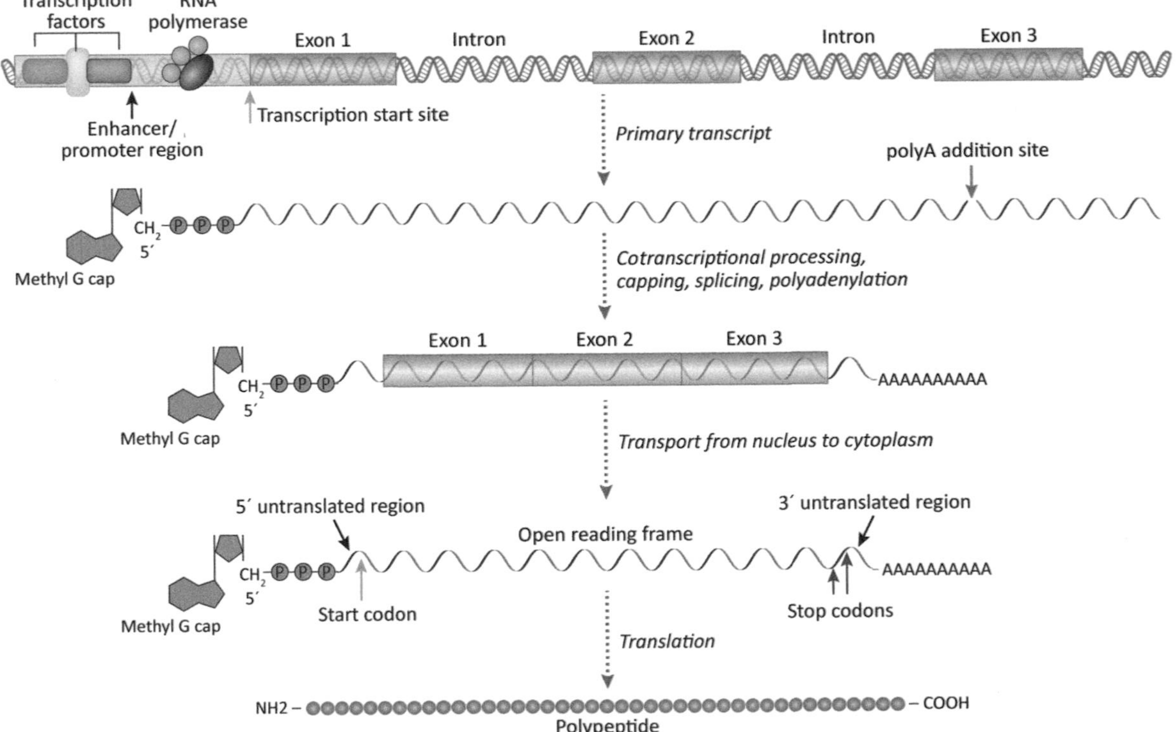

FIGURE 3.12 Organization of a eukaryotic gene showing a promoter region followed by introns and exons. The primary transcript is produced, followed by splicing to remove introns. After export to the cytoplasm, the capped and polyadenylated transcript is translated to produce the protein product of the gene.

- Capped, polyadenylated, spliced mRNAs are transported out of the nucleus.
- The open reading frame or coding region of the mRNA is usually flanked by 5′ and 3′ nontranslated regions.
- For the most part, one mRNA encodes one protein.

Transcription of Viral mRNAs

Do viruses follow the rules of gene organization, transcription and RNA processing used by the eukaryotic cell? Some do, but many do not. As we will see throughout this text, viruses use a variety of unique strategies to control synthesis of their mRNAs, as the abundance of an mRNA can directly impact the amount of protein product produced. Why have some viruses adopted unique strategies?

- The coding capacity of many viruses is small because their genomes are small (compared to the host cell genome), thus they cannot have long complex promoter/enhancer regions and/or long nontranscribed introns.
- Most RNA viruses transcribe their mRNAs in the cell cytoplasm, therefore have no access to spliceosomes or cellular RNA polII.
- Many viruses inhibit host cell transcription and/or translation, therefore must use alternatives to "normal" cellular processes to produce viral proteins.
- An infecting virus usually brings one copy of its genome into a very crowded cell. Viral mRNA synthesis and protein synthesis must be very efficient in order to compete for building materials.

Thus the organization and expression strategies of viral genes may differ from host genes:

- Many viral genes have no introns.
- Some viral mRNAs do not have 5′ caps.
- Some viral mRNAs do not have poly(A) tails.
- Some viral mRNAs have overlapping open reading frames such that more than one type of protein can be produced from a single transcript.
- Viral transcripts with introns can be *alternatively* spliced to generate multiple, different mRNAs. (Obviously these must be viruses that are replicating in cell nucleus where the splicing machinery is present.)
- Some viral mRNAs are not exact copies of the genome. In a process called *RNA editing* or *pseudotemplated transcription*, the polymerases of a few RNA viruses add nucleotides not present in the genome sequence. The term pseudotemplated suggests that there are some specific signals in the genome that instruct the addition of these extra residues.

A Short Review of Translation in the Eukaryotic Host Cell

Let us review a few basics of translation in the cell, focusing on initial interactions of the translation apparatus with the mRNA. Both the 5′ cap and the 3′ poly(A) tail are involved in translation initiation. The poly(A) tail is bound by polyA-binding protein (PAPB). A complex of initiation proteins (eIF4F complex) binds to the 5′ cap (Fig. 3.13) and interacts with PAPB. This is followed by association with a preinitiation complex that includes the 40S ribosomal subunit, $tRNA^{met}$, and other initiation factors. The preinitiation complex moves or scans along the untranslated region of the mRNA until it encounters an AUG codon with the appropriate surrounding sequences (Kozac sequence). Now the 60S subunit binds at the AUG codon to generate the 80S initiation complex with $tRNA^{met}$ in the A site. A second tRNA enters the P site and a peptide bond is formed. The ribosome moves along the mRNA (translocates) one codon (3 nt) at a time. At the end of an open reading frame, a ribosome will encounter stop codons that trigger termination. In eukaryotes transcription termination is facilitated by two release factors (eRF1 and eRF3). eRF1 recognizes stop codons (UAA, UGA, or UAG) in the A site.

Translation of Viral Proteins

We noted above that most eukaryotic mRNAs code for a single protein. But one hallmark of viruses is their ability to efficiently exploit a small genome. So it is not uncommon for individual viral mRNAs to encode different versions of a protein, or two or more completely different proteins. Common strategies are illustrated in Fig. 3.14 and include:

- Use of alternative start codons, a process often called leaky scanning.
- Suppression of stop codons to produce a longer protein product.
- Frame-shifting moves a ribosome to another reading frame to produce a longer protein product.
- Termination-reinitiation or stop-start.

These mechanisms all function at the level of the ribosome and serve to control relative amounts of protein products. For example, use of alternate start codons will result in production of a greater amount of the protein with the best Kozak sequence and lesser amounts of proteins that initiate at alternative start codons. In the case of ribosomal frame-shifting, folding

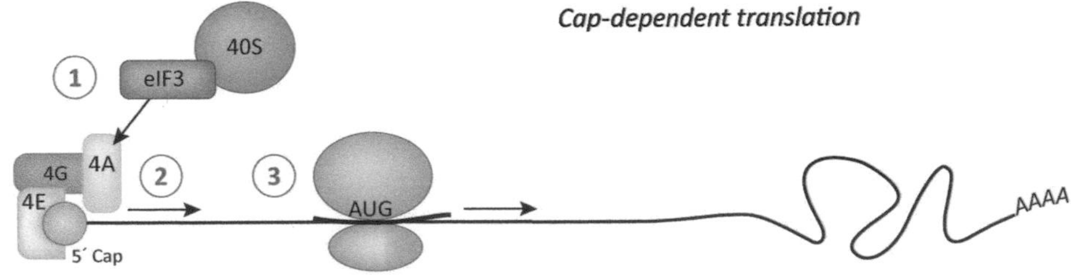

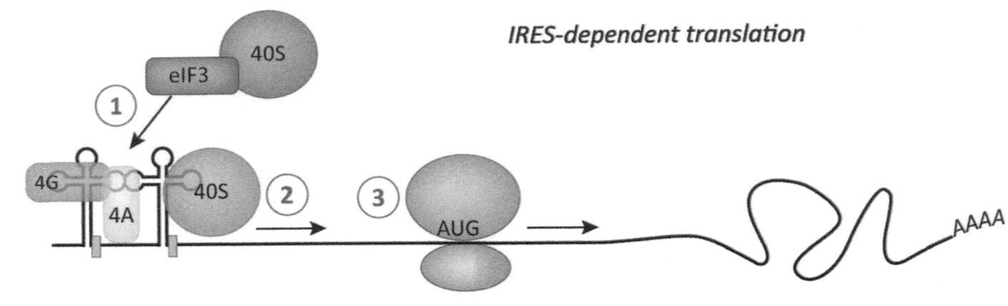

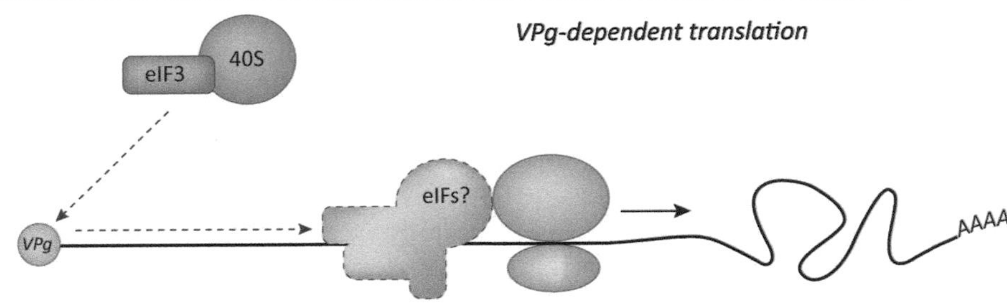

FIGURE 3.13 Methods of translation initiation. Most cellular transcripts have a 5′ cap and a 3′ poly(A) sequence that are key players in ribosome assembly. Many types of viruses adhere to this host cell strategy. However, several positive strand RNA viruses use cap-independent translation. Picornaviruses and some flaviviruses use an RNA structure called the internal ribosome entry site (IRES) to direct ribosome assembly. Another strategy is ribosome assembly directed by a viral protein (VpG) covalently linked to an RNA transcript (see Ch. 12: Family Caliciviridae).

of the mRNA into structures called pseudoknots modifies the translation process (Fig. 3.15) (See Box 3.3).

Many RNA viruses produce polyproteins that are cleaved by viral proteases to generate smaller, functional proteins (Fig. 3.16). The process is exemplified by picornaviruses. The picornaviral genome contains a single, long open reading frame that is translated to produce a large precursor polyprotein. Protease domains within the polyprotein are enzymatically active immediately after being translated and work to quickly cleave the large precursor into mature products. The proteolytic cleavages occur in an ordered sequence. Intermediate cleavage products may have activities distinct from those of the final cleavage products.

As viruses must use the host cell translational apparatus, their viral mRNAs must complete with cellular mRNAs for ribosomes and amino acids. One way to compete is simply to inhibit synthesis of host cell transcription and/or translation. (This also limits the ability of the cell to respond to infection.) Another strategy is to efficiently compete for the translational machinery. Some viruses use a combination of both methods. But how can a virus inhibit translation of host proteins without effecting viral protein synthesis? An example is provided in Box 3.4.

Synthesis of Viral Genomes

As noted in Chapter 1, Introduction to Animal Viruses, viruses can be placed into one of three major groups based on genome type/genome replication strategy. The groups are (1) DNA viruses, (2) RNA viruses, and (3) viruses that use RT.

There are many families of DNA viruses. The genomes of DNA viruses range from 3000 nt to well over 1 million base pairs. Some DNA viruses have circular genomes, others linear. But all DNA viruses use a DNA-dependent DNA polymerase to synthesize additional genome copies. Some DNA viruses use a cellular DNA polymerase but others encode their own DNA polymerases for genome synthesis. There is also obviously a need for a sufficient pool of dNTPs. Eukaryotic DNA synthesis is a highly regulated process. Some

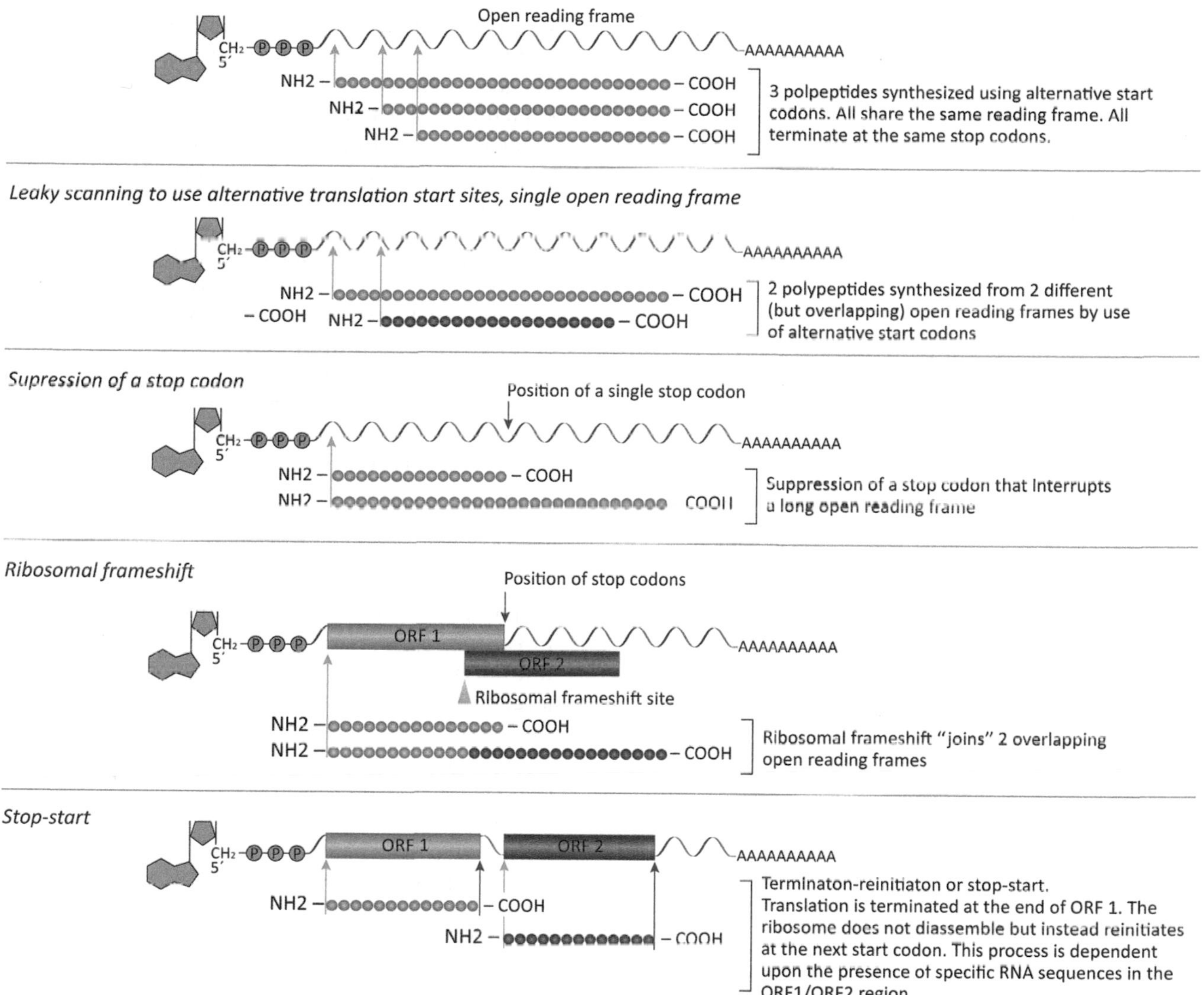

FIGURE 3.14 Viruses use a number of strategies to produce more than one protein from a transcript.

types of animal cells divide regularly (for example, epithelial cells) but others are quiescent, seldom dividing except in response to damage (for example, hepatocytes (liver cells)). Nondividing cells have limited DNA replication machinery and very limited dNTP pools. Strict regulation of cell division causes a problem for some DNA viruses. Some can only replicate in mitotically active cells. But others can stimulate quiescent cells to divide (see Chapter 28: Introduction to DNA Viruses). Some large DNA viruses encode enzymes required for dNTP synthesis, thereby increasing the cellular pools of these building blocks in nondividing cells. Animal DNA viruses (with the exception of the poxviruses) replicate their genomes in the nucleus.

RNA viruses were originally defined as viruses with an RNA genome packaged within the virion. But more critical to the definition is that the RNA genome found in the virion is used as the template for the synthesis of additional RNA genomes. Animal RNA viruses encode an RdRp for genome synthesis. The RdRp is also called the "replicase." The RdRp is used both for genome synthesis and transcription. The RNA viruses never use a host-encoded RNA polymerase to replicate their genomes. As all living cells continuously produce RNAs for "housekeeping" purposes, the synthesis of viral RNA genomes can occur in dividing or nondividing cells.

Members of the families *Retroviridae* and *Hepadnaviridae* use a polymerase called reverse transcriptase (RT) to synthesize their genomes. Retroviruses package an RNA genome while hepadnaviruses package a DNA genome. Both virus families use the enzyme RT to make a DNA copy of an RNA molecule. The RNA genome of a reverse transcribing virus is an mRNA, transcribed from the viral DNA genome by host cell RNA polII. (In the case of retroviruses the DNA form of the viral genome is integrated into the cell DNA)

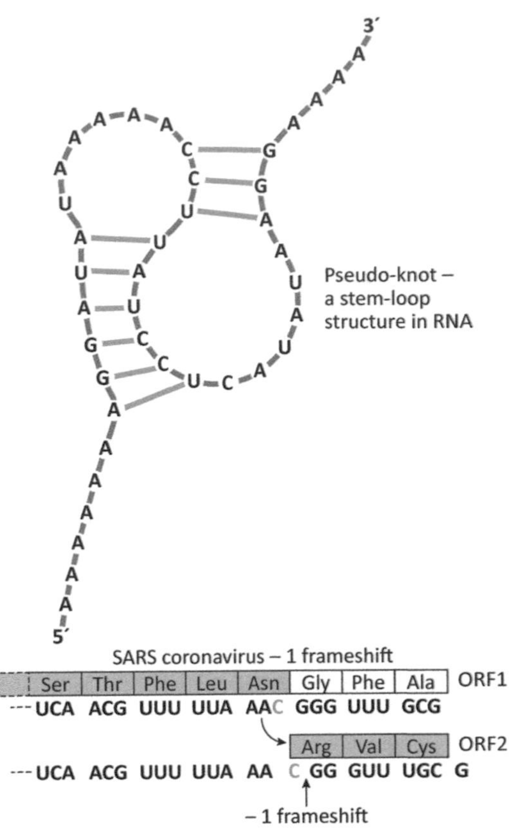

FIGURE 3.15 An RNA pseudo-knot directs ribosomal frame shifting.

In this chapter we have learned that:

- Virus attachment requires specific host receptors and the process is influenced by receptor density, pH, and ions.
- Attachment does not require energy, thus can occur in the cold. Adding virions to cells in the cold is a way to synchronize a viral infection. Penetration is an energy-requiring step that occurs only after the culture has been warmed to physiologic temperature (37°C for most animal cells).
- In order to penetrate into the host cell, a virion or genome must cross a lipid membrane. This can occur at the PM or after endocytosis.
- Uncoating is a process that can occur during or after penetration. Uncoating provides the viral genome access to cytoplasm or nucleoplasm.
- Enveloped viruses use fusion proteins to overcome the kinetic barriers to membrane fusion. Fusion proteins are suicide enzymes that catalyze membrane fusion.
- The cell has a complex and dynamic cytoskeletal system that is often highjacked by viruses.
- Eukaryotic genes are large. They are preceded by untranscribed promoter regions and most have large introns. RNA polII transcribes cellular mRNAs. Cellular mRNAs are highly processed in the nucleus, undergoing capping, splicing (removal of introns), and polyadenylation.
- Viruses maximize use of small genomes by minimizing the size of promoters and encoding multiple proteins from a single mRNA. Some viruses make extensive use of alternative splicing. Some viruses modulate the process of translation by inducing ribosomal frame-shifting and suppression of stop codons.

BOX 3.3

RIBOSOMAL FRAME-SHIFTING AND STOP CODON SUPPRESSION

Ribosomal frame-shifting is common among retroviruses and coronaviruses. It is a process whereby the ribosome moves from one reading frame (protein 1) to a second reading frame (protein 2). Ribosomes frame-shift when they stall during protein synthesis. They stall at specific sites on viral mRNAs called "pseudoknots." The pseudoknot forms as a result of base pairing within the mRNA. In most cases the pseudoknot causes frame-shifting about 20% of the time. A surprisingly efficient process!

Suppression of a stop codon is a process whereby a ribosome fails to terminate protein synthesis at a stop codon. Most eukaryotic genes terminate with multiple stop codons, but if there is a single stop codon, an amino acid can be inserted into the growing polypeptide and translation continues. This is estimated to occur as often as 20% of the time. Thus a protein coding sequence downstream of a pseudoknot or a single stop codon will be synthesized at a lower quantity (~20%) than the protein encoded when the signal is ignored (~80% of the time).

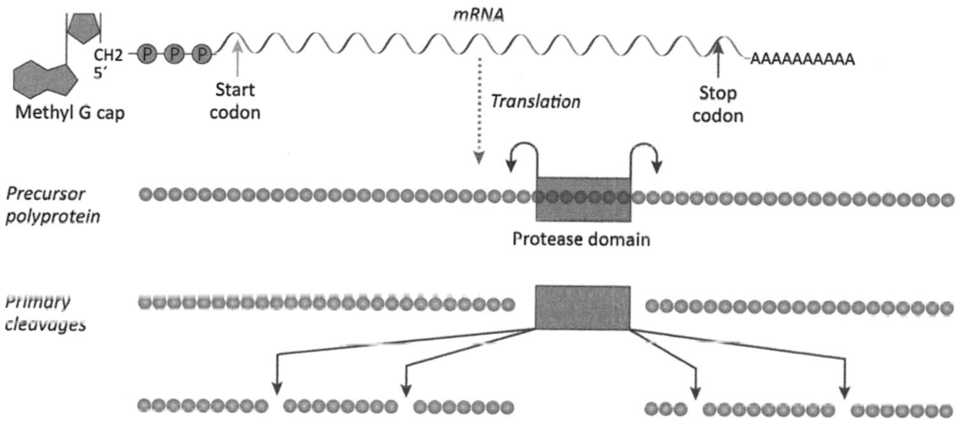

FIGURE 3.16 Proteolytic cleavage of a viral polyprotein.

BOX 3.4

POLIOVIRUS (PV) AND CAP-INDEPENDENT TRANSLATION

PV inhibits cap-dependent translation in the infected host cell. It does so by cleaving the eukaryotic initiation factor 4G (eIF-4G). A function of eIF-4G is to interact with both the 5′-cap on the mRNA and the 40S ribosomal subunit, thereby bringing mRNA and ribosome together. A protease encoded by PV cuts eIF-4G into two pieces. One piece binds to eIF-3 on the 40S subunit and the other piece binds to the cap-binding protein eIF-4E. As the two business ends of eIF-4G are now separate polypeptides, the 40S subunit cannot bind the cap. PV RNA is not capped so the cleavage does not interfere with viral protein synthesis. PV uses a cap-independent mechanism for ribosome assembly onto its mRNA.

- DNA viruses use cellular or virally encoded DNA-dependent DNA polymerases for genome replication. Most DNA viruses require that the cellular DNA synthesis machinery be active to provide adequate dNTP pools.
- RNA viruses use virally encoded RdRp to transcribe mRNAs and to replicate genomes. RNA viruses can replicate in dividing or nondividing cells.

CHAPTER

4

Methods to Study Viruses

OUTLINE

This chapter describes methods for growing, purifying, counting, and characterizing viruses. It also presents general principles of diagnostic virology. After studying this chapter, you should be able to:

- Describe general requirements for culturing cells and tissues.
- Describe differences between cultures of primary and transformed cells.
- Describe how centrifugation is used to purify viruses.
- Understand the types of information provided by negative staining electron microscopy (EM), thin sectioning EM, cryo-EM, and confocal microscopy.
- Understand what is being measured by each of the following techniques: plaque assays, PCR, ELISA, hemagglutination, and hemagglutination inhibition assays.

GROWING VIRUSES

Viruses replicate only within living cells, thus many early studies of viruses were done in bacteria or plants. Tobacco mosaic virus (TMV) was an early "model virus" as it replicates in a variety of plants, at levels sufficient for biochemical analysis and imaging. Growing TMV is as simple as applying virus to abraded leaves of a susceptible plant. The earliest studies of animal viruses were limited to using whole animals. When possible animal pathogens were adapted to small animals such as mice, rats, and rabbits. These small animal models provided a means to study viral pathogenesis and vaccine efficacy. Fertile chicken and duck eggs were, and continue to be, widely used for propagating viruses. In the 1940s and 1950s development of robust cell culture techniques revolutionized the study of animal viruses. Today, most animal viruses are grown in cultured cells.

Generating Cell Cultures

The following steps describe an overall strategy for generating primary cell cultures. It is of utmost importance that all work is done under sterile conditions (Fig. 4.1):

1. The desired tissue is removed from the animal and is chopped or minced.

Viruses. DOI: http://dx.doi.org/10.1016/B978-0-12-803109-4.00004-0

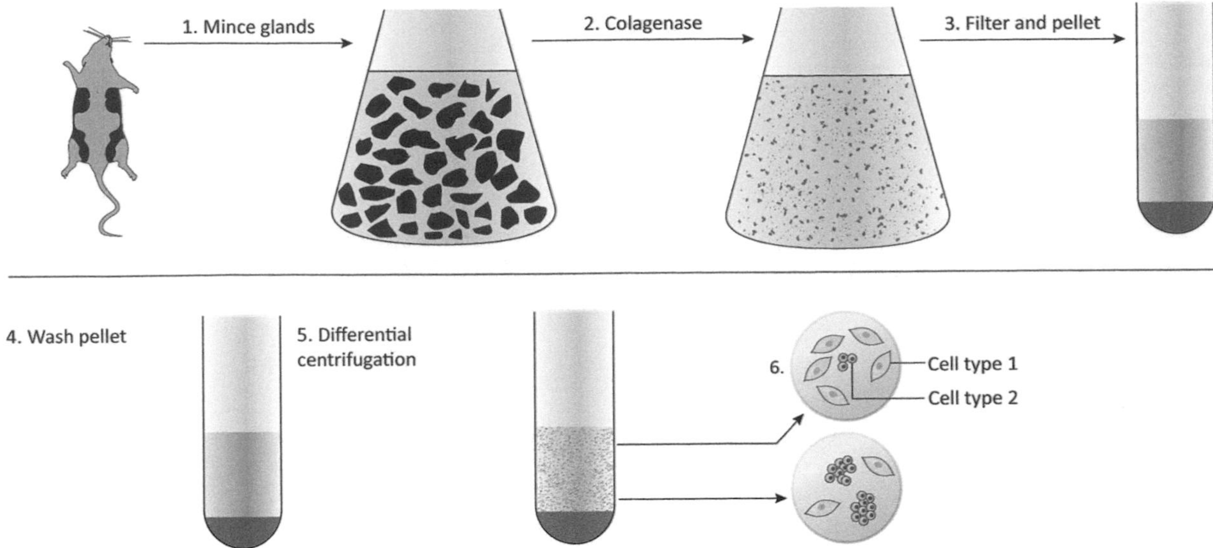

FIGURE 4.1 Generating cell cultures begins with removing tissues (normal or tumor) from an animal. Tissues are minced and treated with enzymes to degrade the extracellular matrix. Centrifugation is used to pellet the cells. Cells are resuspended in media and placed in culture vessels.

2. Tissue fragments are treated with enzymes such as collagenase to degrade the extracellular matrix and release single cells and small aggregates of cells.
3. Cells are pelleted by centrifugation and are resuspended in buffered saline or cell culture media.
4. Additional centrifugation steps may be performed to separate single cells from cell aggregates.
5. Cells and growth media are added to culture dishes and are maintained in a humidified incubator (37°C, 5% CO_2).
6. Cells attach to the bottom of the dish where they grow and divide to form a monolayer.
7. The cells can be removed with trypsin, washed, and divided among new culture plates or dishes. This is called a passage, and is done to increase cell number.

Primary cells can be propagated for only a limited number of passages before the cells undergo a crisis and the culture dies. Embryonic cells can be passaged many more times than cells taken from adults. Some types of cells (for example, fibroblasts) divide more readily than do cells that are normally nondividing in the adult animal (for example, neurons). Tumors provide another source of cells for virus culture. Tumor-derived cells can often be passaged indefinitely. These immortalized cells are excellent tools for the virologist. They are relatively easy to culture, many types are commercially available and they can be genetically modified. Multiple genes can be introduced, mutated, or deleted to generate an unlimited supply of "designer" cells.

PURIFYING VIRUSES

Viruses grown in cultured cells can be purified, quantitated, imaged, and biochemically analyzed. The higher the initial virus concentration the easier it is to purify virus away from cell debris and media components. If a virus is cytopathic it is present amongst the cell debris and media from the dish or plate. In contrast, if the virus is highly cell-associated, the cells must be gently lysed to release the virus. Mixtures of virus and cell debris are subjected to low-speed ($\sim 5000 \times g$) centrifugation to pellet cell debris, but not the much smaller virions. At the end of the low-speed centrifugation, the liquid supernatant, containing the virus, is saved and the pellet (containing the cell debris) is discarded. Separating larger from smaller molecules is called differential centrifugation (Fig. 4.2).

The virus-containing supernatant can then be recentrifuging at a much higher speed ($\sim 30{,}000-100{,}000 \times g$) to pellet the virus. After the centrifugation is completed, the supernatant is discarded and the virus pellet is saved. If desired, the virus can be further purified by centrifugation through a density gradient. Sucrose and glycerol gradients are commonly used; a sucrose gradient might range from 40% sucrose at the bottom of the tube to 5% sucrose at the top of the tube. The gradient is prepared and the virus sample layered carefully onto the top. During centrifugation, the components of the sample will separate into layers depending on their buoyant density. By using a gradient, one can achieve a finer separation of the different macromolecules within a sample. If sufficient virus is present in the sample, a

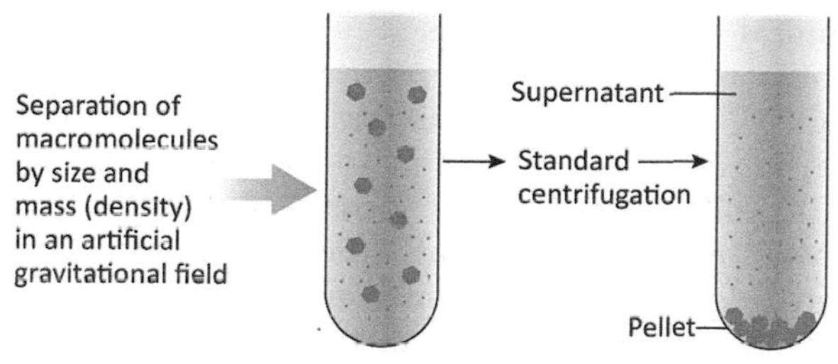

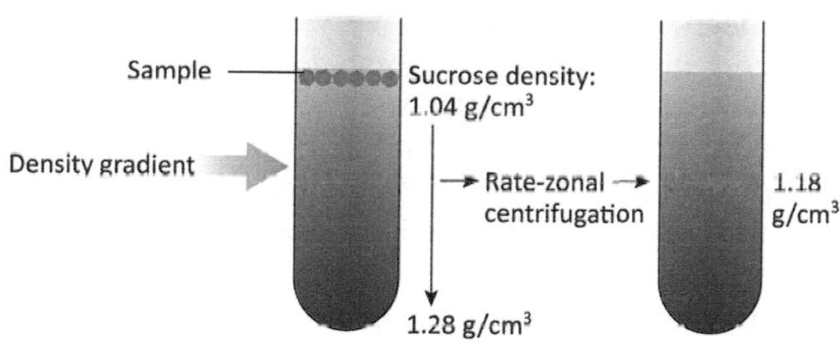

FIGURE 4.2 Centrifugation is a common technique used to purify viruses away from cell debris and other small molecules. Low-speed centrifugation ($\sim 5000 \times g$) is used to remove cells and large cell debris, leaving virions in the supernatant. Virus can be further purified and concentrated by centrifugation at higher speeds. A gradient of sucrose or glycerol can be used to achieve further purification as materials in the sample will separate into layers depending on their buoyant density.

visible band forms, and can be carefully removed from the tube.

VISUALIZING VIRUSES

Optical Microscopy

Light Microscopes

Light microscopes use visible light (400–700 nm wavelengths) to image objects. They are seldom used to visualize viruses, as they lack sufficient magnifying or resolving power to do so. However, the largest known viruses can be seen with a light microscope. Examples of these very large viruses include Mimivirus, Pithovirus, Megavirus, and Pandoravirus, all of which infect amoebas.

Even though many viruses are far too small to be seen with the light microscope one can often observe virally induced changes to infected cells. A classic example is the Negri body, which was once used as a diagnostic test for rabies virus infection. The Negri body is not "a virus" but is a structure (a type of inclusion body) that is seen in rabies virus-infected neurons. Some inclusion bodies are in fact densely packed virus factories that form at discrete locations in the infected cell.

Fluorescence Microscopy

Fluorescence is the emission of light by a substance that has absorbed light (or other electromagnetic radiation). The basic function of a fluorescence microscope is to irradiate a specimen with high-energy excitation light and detect the much weaker emitted fluorescence. The microscope is designed so that only the emission light (fluorescence) reaches the eye or detector, allowing fluorescent structures to be seen with high contrast against a dark background. This is achieved by use of filters of specific wavelength. The advantage of fluorescence microscopy is that a single fluorescein molecule can emit as many as 300,000 photons before it is destroyed, thus making it theoretically possible to detect a single molecule in a field of view.

Fluorochromes (also termed fluorophores) are stains used in fluorescence microscopy. As in light microscopy, the most useful of these stains attach themselves to specific targets in the cell. For example, DAPI (4′,6-diamidino-2-phenylindole) is a fluorescent stain that binds strongly to A-T rich regions in DNA. DAPI (Fig. 4.3) can pass through an intact cell membrane thus can be used to stain both live and fixed cells. It is used extensively in fluorescence microscopy to label the cell nucleus because it binds to ds DNA. Different fluorochromes are excited by specific wavelengths of irradiating light and emit light of defined wavelength and intensity. Thus multiple fluorochromes can be used simultaneously to identify different target molecules within a cell.

Alternatively, fluorochromes can be chemically attached to molecules, such as antibodies, to specifically label their ligands. Alexa dyes (Fig. 4.4) are a series of fluorescent molecules that are widely used as fluorochromes. Among the Alexa dyes, there are choices of molecules with a variety of excitation and emission spectra. If different Alexa dyes are attached to different molecules, cell images can be collected using filters specific for each color. The relationship of one macromolecule to another can be determined by overlaying the images. For example, if a green image and a red image are superimposed, yellow pixels are seen where the tagged macromolecules colocalize in the cell.

A viral diagnostic assay called an immunofluorescence assay (IFA) uses tagged antibodies to detect viral proteins in infected cells. IFAs often include DAPI to stain cell nuclei. The result is that all cells have blue DAPI-stained nuclei and infected cells glow green. If cells have fused together due to virus infection, one may see a large cell (syncytium) containing many blue nuclei.

Another way to add a fluorescent tag to a protein is to use genetic engineering (cloning) to add the coding sequence for a fluorescent protein to the coding sequence of a protein of interest. Green fluorescent protein (GFP), isolated from a jellyfish, is a 238 amino acid protein (Fig. 4.5) that folds to create a fluorescent center that emits green light. Genes encoding cellular or viral proteins can be modified by addition of the GFP gene to create a "tagged" protein. If the engineered gene is introduced into a cell, the protein product is visible when cells are exposed to UV light. This is a very powerful imaging technique; however, one drawback is that the addition of a relatively large protein tag can alter the trafficking, localization, or enzymatic activity of the protein of interest.

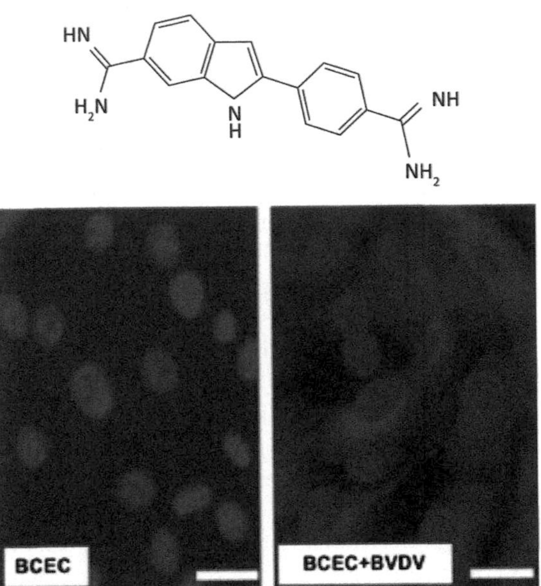

FIGURE 4.3 Chemical structure of DAPI (top) a fluorescent stain used to stain nuclei. Bottom left, bovine caruncular epithelial cell line (BCEC) stained with DAPI and showing blue nuclei. Bottom right, BCEC infected with bovine viral diarrhea virus (BVDV) showing cell nuclei stained with DAPI and fluorescence cy3-conjugated antibody to detect BVDV (red). *Bridger et al. 2007.*

Confocal microscopy offers several advantages over conventional fluorescent microscopy. Confocal microscopy uses "point-like" illumination and detection to reduce out-of-focus glare. The method provides the ability to collect serial *optical* sections through a thick specimen by collecting and measuring the light intensity originating from numerous thin focal areas in a cell. Confocal microscopy can image either fixed or living cells that have been labeled with one or more fluorescent probes. The excitation energies required to reach deep into the cells are very high and specimen damage can result. In contrast *deconvolution methods* employ conventional widefield fluorescence microscopes for image acquisition. Excitation intensities can be kept low, resulting in less damage to the specimen. Deconvolution is often used to image monolayers of living cells. Deconvolution methods have high computational demands as the images are computationally processed.

Electron Microscopy

The development of EM in the 1930s allowed individual viruses to be seen for the first time. EM uses an electron beam in place of light, and the beam is focused by electromagnets, rather than by glass lenses. EM has great resolving power and can magnify objects by up to ~10,000,000 times. Disadvantages of EM are that fixed and processed samples are dead, and biological samples are heavily damaged by the electron beam even as they are being imaged (sample damage results from the interaction of electrons with organic matter). However, the simple structures and crystalline nature of viruses were first revealed by EM and it remains a common tool of the virologist. In a process called negative staining, heavy metals are used to coat the surface of virus particles to make a "cast" or "relief map" (Fig. 4.6). The heavy metal coating is much less

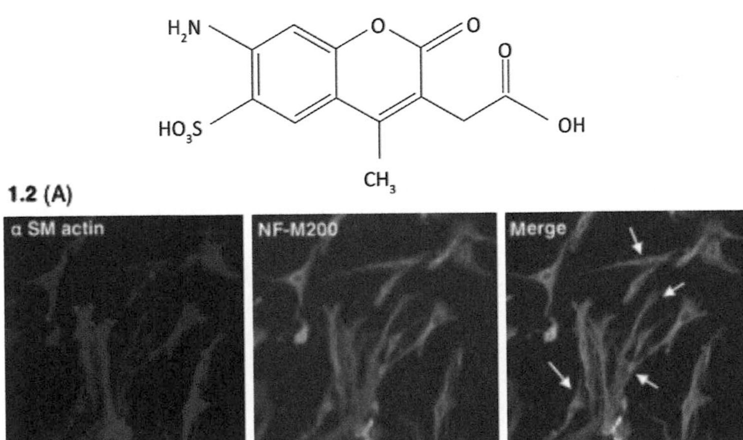

FIGURE 4.4 Chemical structure of Alexa dye (top). Smooth muscle actin (αSM actin, in red Alexa 594) (bottom left) and neurofilament M200 (NF-M200, in green Alexa 488) (bottom middle); merged shows αSM actin-negative cells (white arrows) (bottom right) (magnification 20×). *From Bouchez et al. 2008.*

sensitive to radiation damage than the biological sample, thus the coated surface structures are well preserved. However, information about internal structures is poor.

Thin sectioning is a method that reveals structures inside viruses and cells. Samples are fixed, embedded in resin, and sliced with a diamond knife and a microtome. The sections are carefully placed onto small carbon grids and can be processed in a variety of ways. Samples can be treated with electron dense stains such as uranyl acetate or lead citrate to provide contrast, or samples might be incubated with gold-labeled antibodies. Use of gold-labeled antibodies is a powerful technique that allows the localization of specific molecules within a virus or cell. Micrographs showing the process of virus budding from a cell are produced by thin sectioning (Fig. 4.7 need permissions).

Electron cryomicroscopy (cryo-EM) uses ultralow temperatures (liquid nitrogen or liquid helium) to preserve biological samples. Cryofixation of samples produces a glass-like (vitrified) material rather than crystalline ice. Cryo-EM allows the observation of specimens that have not been chemically stained or fixed, thus showing them in their native aqueous environment. Low temperatures provide for better images with less specimen damage, but cryo-EM micrographs have less contrast than stained images. Cryo-EM images are more complex than negative stains because the entire particle (including the interior) is seen in the image. However, thicker frozen specimens can be cut into thin sections using a cryoultramicrotome. Images generated by cryo-EM are often enhanced by collecting and computationally averaging multiple images. This allows use of lower and less damaging doses of electrons. Tomography is a process whereby a series of images are taken with the specimen tilted at different angles relative to the direction of the electron beam. The images are computationally combined to create 3D images (tomograms) of the sample. Extremely high-resolution (<4 Å) images have been achieved for icosahedral viruses because they are constructed from many repeating subunits that can be computationally averaged. Cryo-EM, while a powerful technique, requires expensive and specialized equipment and highly trained personnel (Fig. 4.8).

COUNTING VIRUSES

Methods of counting or quantitating viruses fall into two discrete categories. Infectivity assays, as the name implies, measure virions that can successfully infect a cell to produce infectious progeny. Inactivated

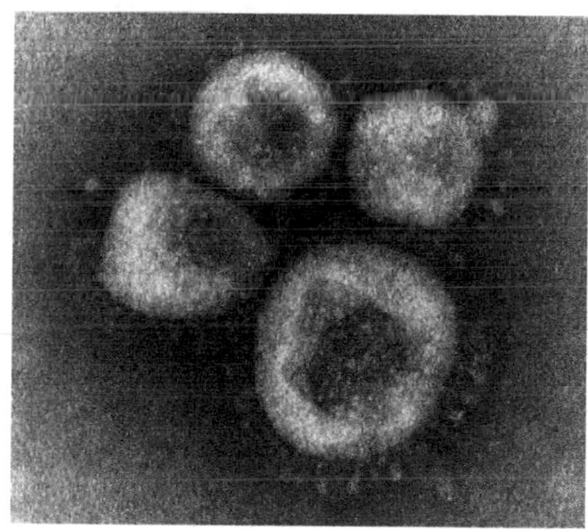

FIGURE 4.6 Negative stain of bovine coronavirus. *Courtesy: HR Payne.*

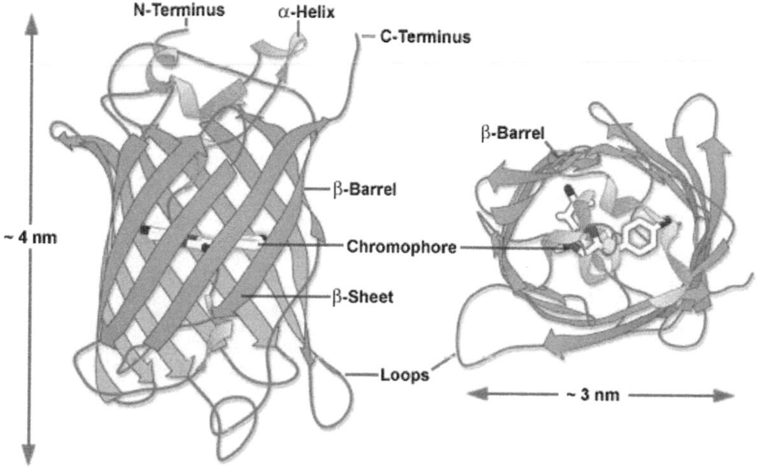

FIGURE 4.5 Ribbon diagram showing the structure of the GFP protein. *From Zeiss Education in Microscopy and Digital Imaging. Contributing Authors Richard N. Day and Michael W. Davidson.*

(noninfectious) virions are not counted. The second category of techniques measures specific virion components, often a specific viral protein, or the viral genome. These techniques are chemical/physical measures of virus quantification and they include serologic assays, polymerase chain reaction (PCR), and hemagglutination assays (HA). Negative staining EM can also be used as a chemical/physical assay to detect and count virus particles. Chemical/physical assays do not distinguish between infectious and inactivated (noninfectious) virions. For example, if one were to take two identical samples and expose one to UV radiation (to damage viral genomes), the amount of viral protein measured would be the same for both samples, even though the irradiated sample contains no infectious virus.

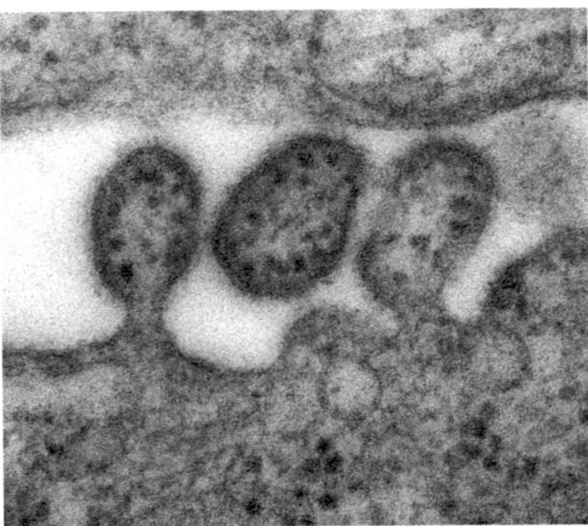

FIGURE 4.7 Highly magnified transmission electron microscopic (TEM) image depicting some ultrastructural details of three Lassa virus virions. The virion in the middle has completely budded from its host cell, while the two other particles are in the process of budding. Content provider: CDC/C.S Goldsmith, D. Auperin. Public Health Image Library, Image 8699.

Infectivity Assays

Infectivity assays require cell cultures, embryonated eggs, or animals (or plants or bacteria). An infectivity assay measures particles capable of replicating in a particular cell type or animal.

Plaque assays are a common type of infectivity assay, used to count discrete "infectious centers." Samples containing virus are serially diluted and aliquots of each dilution are added to a dish of cultured cells (or a plant leaf in the case of a plant virus) (Fig. 4.9). Each dilution is usually tested in triplicate. The process is:

- An individual virion infects a cell.
- The virus replicates, producing progeny virions.
- Second generation virions infect and kill surrounding cells to generate a hole or plaque in the cell monolayer that is visible to the naked eye (or at low magnification). Often the live cells are stained, providing a dark background for the clear plaques (Fig. 4.10).

The purpose of making and testing serial dilutions is to achieve a "countable" number of plaques in the cell monolayer. If too many infectious particles are present, individual cells are likely to be infected with more than one virion or individual plaques will merge to form large areas lacking cells. Plots of the number of plaques versus the dilution factor are linear if a single virion is sufficient to initiate a replication cycle ("single hit" kinetics). Results of plaque assays are reported as plaque forming units (PFU) per volume of measure (usually a mL). The term PFU acknowledges that we cannot know, with certainty, that the visualized plaques were formed by infection with a single virus of interest. Use of the term PFU should also remind us that variables such

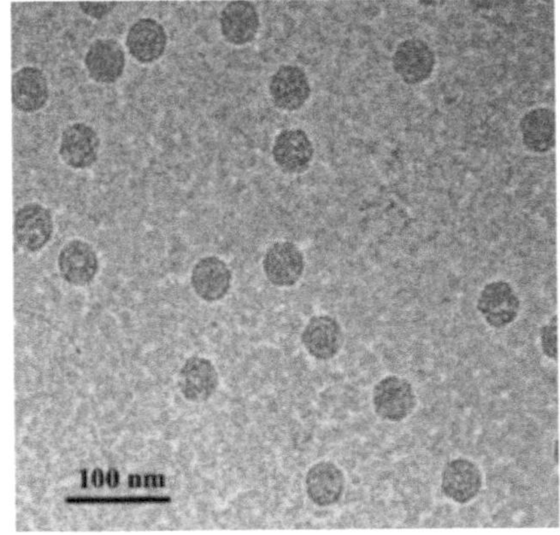

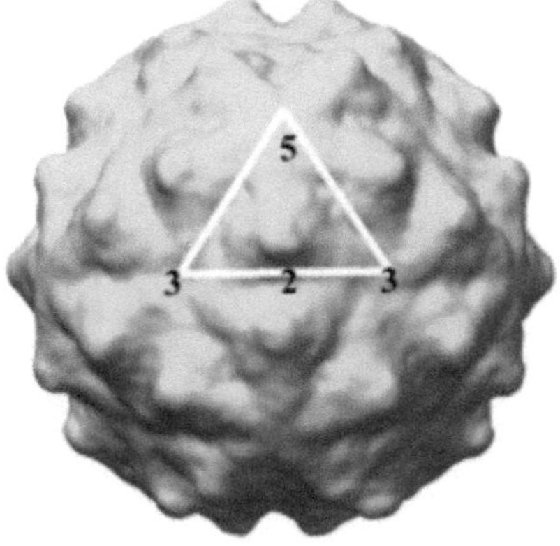

FIGURE 4.8 Cryo-EM structure of Flock house virus, virus like particle (VLP). Left panel: Electron cryomicrograph of frozen hydrated VLPs. Right panel: 3D reconstruction using image processing and viewed down the 2-fold axis of symmetry. The icosahedral asymmetric unit is highlighted by the white triangle, showing the 5-, 2-, and 3-fold axes of symmetry. *From: Bajaj et al. 2016.*

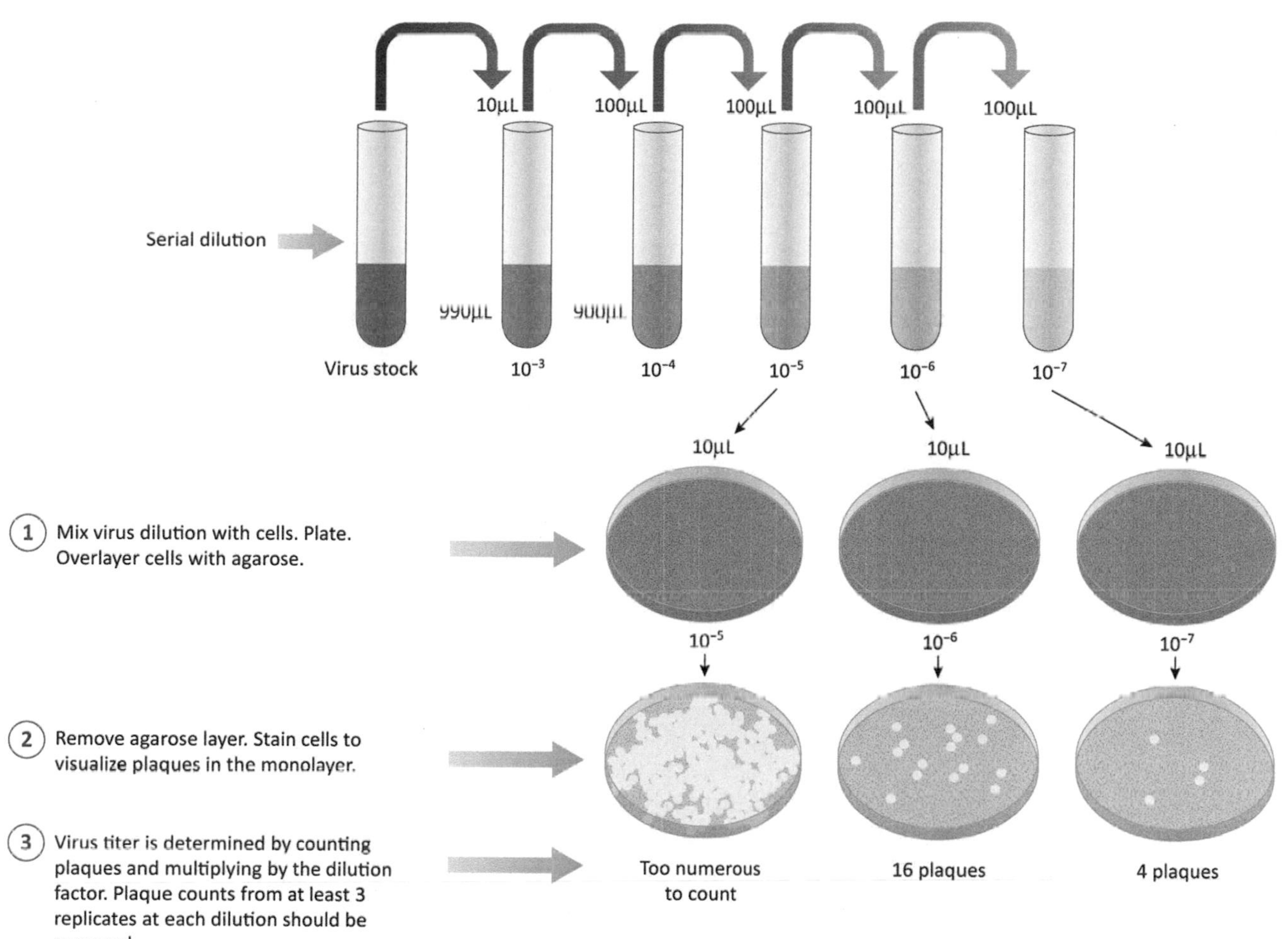

FIGURE 4.9 Plaque assays are used to count infectious particles. Samples are diluted and aliquots of each dilution are added to cultured cells. The cells are covered with an agaroseoverlay. Virus produced from an infected cell can infect nearby cells. If infected cells are killed, a region free of cells (the plaque) develops.

as type of host cell, type of culture media, ion concentration, or pH might affect the apparent concentration of virus. For example, a sample with a titer of 10^3 PFU/mL on mouse cells might have a titer of 10^6 PFU/mL on human cells.

What if the virus of interest does not kill infected cells? Can plaque assays be performed? The answer is yes, if there is a method for counting groups of infected cells. Groups of infected cells are called foci and the assays are focus-forming assays. Results are reported as focus-forming units (FFU). It is common to take advantage of antiviral antibodies to stain groups of infected cells. The antibodies can be detected if they are tagged with a either a fluorescent molecule or with an enzyme that can cleave an uncolored substrate to produce a colored product. There are several variations on the focus-forming assays but all have in common the identification of small, discrete groups of infected cells in a monolayer culture (Figs. 4.11 and 4.12).

Endpoint dilution assays are another type of infectivity assay. The general protocol is as follows:

1. Prepare serial dilutions of the virus stock.
2. Use aliquots of each dilution to infect 3–6 test units. A test unit might be a culture dish, a single well on a multiwell plate, a leaf on a plant, or an animal.
3. Incubate test units to allow virus replication.
4. After a predetermined period of time, check each test unit and determine if virus has replicated. If virus can be detected, the unit is scored positive. If no virus is detected, the unit is scored negative.
5. Count the total number of infected and uninfected units at various dilutions.

When the endpoint dilution assay is done in cell cultures, the titer is reported in Tissue Culture

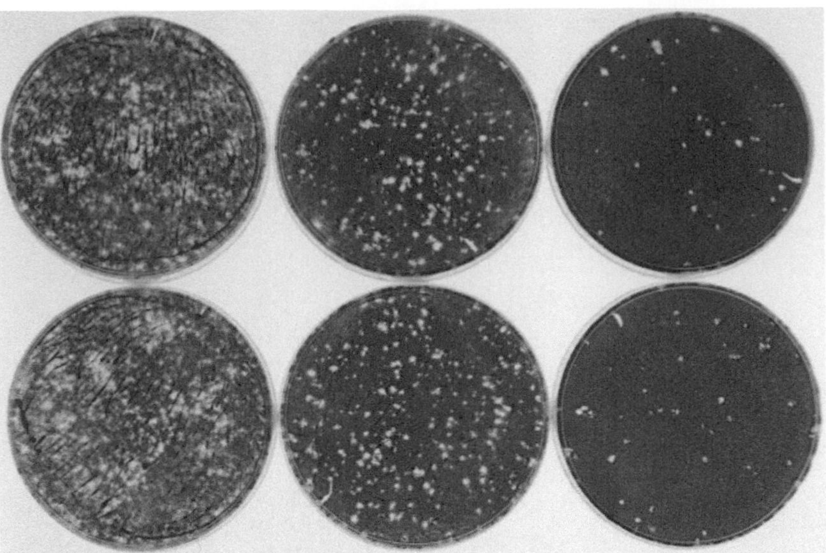

FIGURE 4.10 Stained cell monolayer with plaques. Living cells are stained purple leaving clear areas (plaques) where cells are absent. *From: Cromeans et al. 2008.*

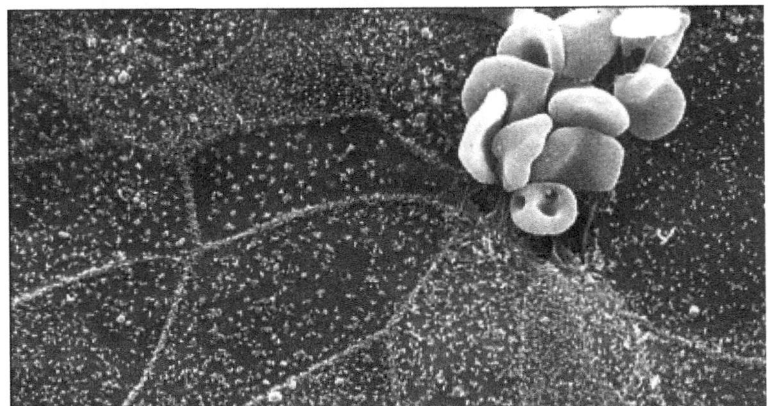

FIGURE 4.11 Red blood cells (RBCs) binding to coronavirus infected cells. Bovine coronavirus infected cells are not killed but do express the viral envelope protein on their surface. The viral glycoprotein binds to RBCs. The RBCs serve to "mark" infected area in this electron micrograph. *Courtesy: HR Payne.*

Infectious Doses 50% ($TCID_{50}$) where one $TCID_{50}$ is the amount of sample that will infect 50% of the test units. For example, if 1 mL of virus sample is added to each of six units, and three of those units are positive at the end of the assay, the $TCID_{50}$ is 1. However, the results of endpoint dilution assays are seldom "perfect" and calculations based on summing all of the data should be used to derive a titer. End-point dilution assays can be performed using animals and results are reported as Infectious Dose 50% (ID_{50}) or Lethal Dose 50% (LD_{50}), if death is the endpoint. For a given virus sample, the $TCID_{50}$ titer is often higher than the ID_{50} titer which is often greater than LD_{50} titer. Note that virus titers are not absolute values, but rather depend on the type of cells or animals used for the assay.

Chemical/Physical Methods of Virus Quantitation

Chemical/physical methods of virus quantitation measure the amount (or relative amount) of a viral protein, genome, or enzyme, in a sample. Types of chemical/physical methods include:

- Direct visualization of virions by EM.
- Hemagglutination (HA) assay.
- Serological assays (based on antigen–antibody interactions, see Box 4.1). Examples of serologic assays include enzyme-linked immunosorbent assays (ELISA), fluorescent-tagged antibody assays, and precipitation assays.
- Genome quantification by PCR.

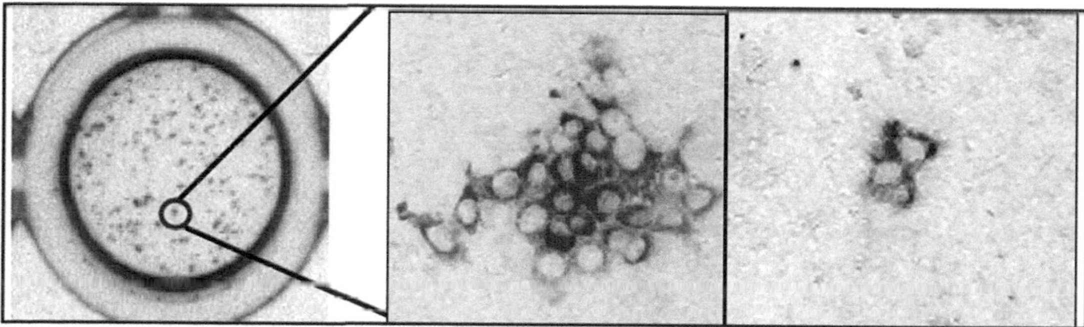

FIGURE 4.12 Focus-forming assay for hepatitis C virus (HCV). Infected cells were stained with an HCV-specific primary antibody followed by a horse radish peroxidase (HRP)-conjugated secondary antibody. *From: Barretto, N. and Uprichard, S. L. (2014).*

BOX 4.1

SEROLOGIC ASSAYS

Serologic assays use the power of antibody–antigen interactions (the ability of an antibody to bind its cognate antigen with high specificity).

A primary antibody is one that binds a specific antigen. If HIV is the antigen, the primary antibodies are those that bind to HIV proteins. For a direct ELISA, the primary antibodies are chemically linked to a fluorescent tag or an enzyme tag.

A secondary antibody is one that binds to a primary antibody. If the primary antibody is human-anti-HIV (from an infected patient, for example), the secondary antibody recognizes the human antibody. The secondary antibody might be rabbit, mouse, or goat antihuman IgG. The benefit of using a secondary antibody is that a single aliquot of secondary antibody can be used to detect many different human antibodies, regardless of what antigen those antibodies bind.

The readouts from chemical/physical assays provide no information about the amount of *infectious* virus in a sample, but they are often convenient, quick, and quite reproducible. They can often be correlated back to infectivity assays as a quick way to estimate the infectivity of a sample. However, if a virus sample is prepared or stored incorrectly, the protein or genome concentration might remain unchanged despite a significant decrease in the infectivity titer.

Hemagglutination Assay

Some viruses bind to red blood cells (RBCs). Hemagglutinating viruses bind to sialic acid residues on the RBCs. A single virion can bind to several different RBCs, and an RBC can be bound by multiple virions to form a large network, or web, of cell and virus that is easily visualized. The HA is fast and inexpensive and does not require either sophisticated instrumentation or extensive training. It is done by preparing serial dilutions of a virus sample. An aliquot of each dilution is added to RBCs in a microtiter plate well or test tube. One well contains RBCs and saline (negative control) and another contains a known positive reference sample of virus. The samples are gently mixed and allowed to sit at room temperature. In the negative wells the RBCs will slide down to form a tight button at the bottom of the tube. In positive wells the RBCs and virions will bind to each other to form a mesh of cells on the bottom of the tube. The reciprocal of the highest dilution of virus that give a positive HA is the HA titer (Fig. 4.13).

Serologic Techniques

HEMAGGLUTINATION INHIBITION ASSAY

HIA is a serologic assay that be used either to detect antibody to a virus or to identify a suspect virus. The HIA is performed by first mixing virus samples with dilutions of serum. Antibody is allowed time to bind the virus and then RBCs are added to the mix. Viruses that have bound to antibody will be unable to bind to RBCs. Thus in the HIA, the absence of hemagglutination is a positive result. If a hemagglutinating virus is

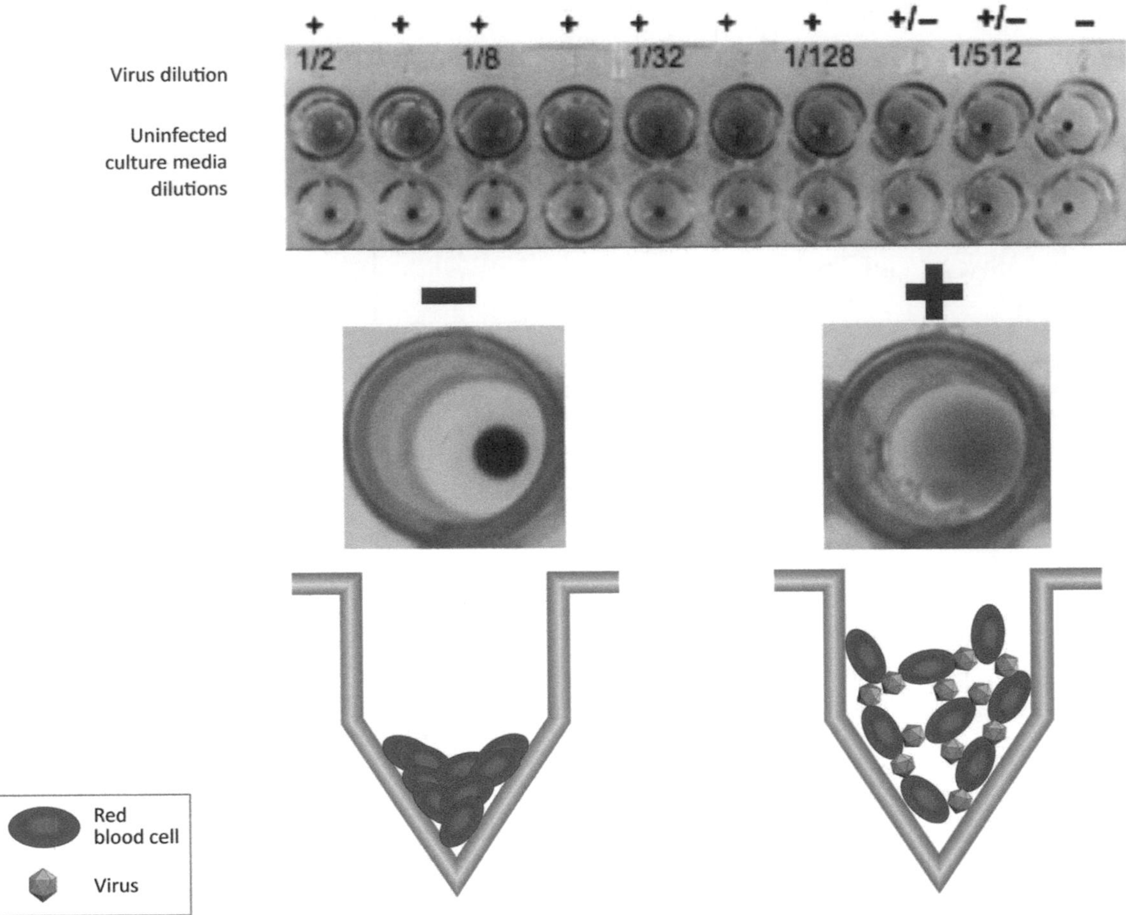

FIGURE 4.13 Some viruses bind to RBCs, causing the cells to form a lattice. Note that positive hemagglutination is the presence of a lacey layer of RBCs. If virus is not present, the RBCs slide to the bottom of the tube to form a tight "button."

the known reagent, the HI assay can be used to detect antibody. If the hemagglutinating virus is unknown, it can be identified by using a panel of known antibodies.

VIRUS NEUTRALIZATION ASSAY

A virus neutralization assay is used in conjunction with an infectivity assay, such as the plaque assay described above. This assay detects antibody that is capable of inhibiting virus replication (or in other words, antibody that can neutralize virus infection). Virus neutralization is a specialized type of immunoassay because it does not detect all antigen–antibody reactions. It only detects antibody that can block virus replication. This is important because related groups of viruses may share common antigens, but only a fraction of these antigens are targets of neutralizing antibody. A virus serotype is usually based on virus neutralization (although this is not always specified). For example, there are three major poliovirus serotypes (neutralization serotypes). In order to protect against poliovirus infection, a successful vaccine must induce neutralizing antibodies to poliovirus Types 1, 2, and 3.

ENZYME-LINKED IMMUNOSORBENT ASSAYS

Interaction of antibodies with their cognate antigens can be visualized if the antibody is "tagged" (Fig. 4.14). The tag can be an enzyme that cleaves a substrate (substrate turns color upon cleavage), a radioactive isotope, or a fluorescent molecule. Serologic assays are commonly used in research and diagnostic laboratories. ELISAs (See Box 4.2) require that either antigen or antibody be adsorbed onto a plate or tube (often plastic). As proteins readily bind to many types of glass or plastic, the adsorption part of the assay is straightforward. The enzyme used for detection is often either horseradish peroxidase (HRP) or alkaline phosphatase (AP). These enzymes relatively stable, cheap, and easy to purify, and they can be chemically linked to an antibody to aid in detection of an immune complex. There are

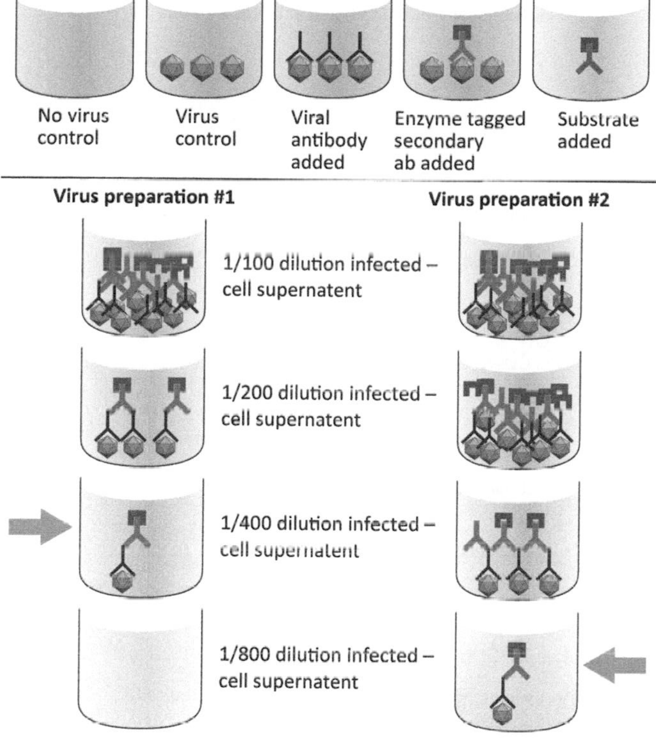

FIGURE 4.14 An ELISA can be used to determine the relative titer of a virus stock. The titer is expressed as the reciprocal of the highest dilution at produces a positive color.

inexpensive substrates that change from colorless to colored when cleaved by these enzymes. ELISAs can be used to detect either antigen (a virus, for example) or antibody (from a potentially infected individual). Antigen detection ELISAs can be used to quantitate the amount of virus in a cell culture supernatant or patient sample. In this case the primary antibody is a purchased and/or standardized reagent. Antibody detection ELISAs demonstrate the presence of antibodies to a specific pathogen (i.e., human immunodeficiency virus, HIV) in patient sera. In this case the antigen is the standardized reagent. The presence of antibody signifies that the patient has either been infected (or vaccinated) with the agent in question.

CELL-BASED FLUORESCENT ANTIBODY ASSAYS

Highly specific antibodies tagged with fluorescent molecules can be used to detect the presence of viral antigens in cells. Assays can be direct (using labeled primary antibody) or indirect (using labeled secondary antibody) (Fig. 4.15). The technique allows one to distinguish infected from uninfected cells, thus can be used to perform focus-forming assays.

WESTERN BLOTS

Western blots are labor intensive and expensive, but provide a method of confirming the identity of a reactive antigen. Western blots are done by electrophoresis of antigen (i.e., purified virus) by sodium dodecyl sulfate (SDS) polyacrylamide gel electrophoresis (PAGE). The proteins in a mixture are separated by size in the gel and are transferred from the gel to a solid (paper-like) substrate often called a membrane. The membrane is incubated with patient sera and the presence of patient antibodies is detected by subsequently incubating the membrane with labeled secondary antibodies. The specificity of the western blot lies in its ability to demonstrate the molecular weights (sizes) of the proteins recognized by patient sera. False positive reactions can be identified when the immunoreactive

BOX 4.2

VARIATIONS OF THE ELISA CAN BE USED FOR ANTIGEN OR ANTIBODY DETECTION

Direct ELISA. Primary antibodies (i.e., anti-influenza virus antibodies) are chemically linked to enzymes such as horseradish peroxidase (HRP) or alkaline phosphatase (AP). These reagents facilitate detection of viral antigens in a sample. If a fluorescent molecule is attached to the primary antibody, the assay is a fluorescence-linked immunosorbent assay.

Indirect ELISA. The primary antibody is untagged, but a tagged secondary antibody is used to detect the presence of primary antibody. For example, if the primary antibody is developed in an immunized rabbit, the secondary antibody is anti-rabbit IgG (perhaps obtained from mice immunized with rabbit IgG). Indirect ELISAs are common because it is cheaper to tag one batch of antirabbit IgG than to tag the specific antibodies directed against hundreds of different viral (or other) proteins.

Sandwich ELISA. The antigen of interest is "sandwiched" between two antibodies. An antibody or a serum sample is used to coat the dish or tube. A sample containing an antigen is added and allowed to bind to the antibody. A second antibody is added to the wells. It will bind to the antigen (if present). In this example the antigen must be able to interact with two antibodies. Requiring that two different antibodies react with the same antigen can improve the specificity of the assay.

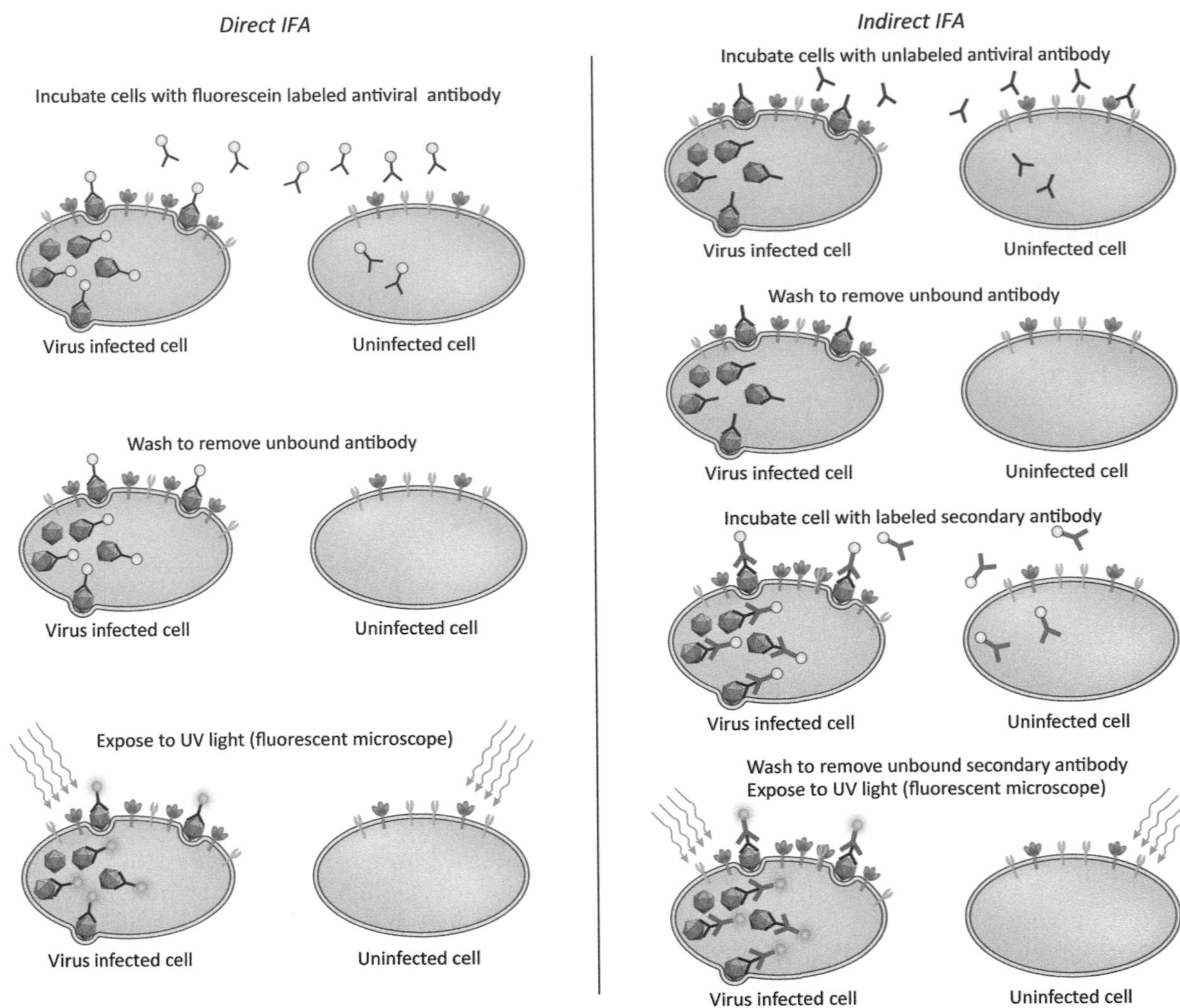

FIGURE 4.15 Cell-based IFAs use tagged antibodies to mark infected cells. Direct immunofluorescence used tagged primary (antiviral) antibodies. Indirect immunofluorescence uses a secondary antibody that is tagged.

protein band does not correspond in size to known viral proteins. For diagnostic purposes, a positive western blot may require that more than one viral protein be bound by patient antibodies.

IMMUNOHISTOCHEMISTRY

The basis of immunohistochemistry is that a tissue section is incubated with enzyme-tagged antibodies. A colorless substrate is added to the sample. If enzyme-tagged antibodies are present, the substrate is cleaved to produce a colored precipitate. This is a powerful technique as it allows one to examine individual virus-infected cells in a tissue section. Patient samples (biopsies) are often preserved in formaldehyde or are stored frozen at ultracold temperatures. If these samples are archived, they can be tested for the presence of viral antigen even after years or decades (Fig. 4.16).

Detecting Viral Nucleic Acids

Polymerase Chain Reaction

PCR can be used to identify and/or quantitate viral genomes in a sample. PCR is a very sensitive method and uses oligonucleotide primers designed to detect suspect viruses. Advantages of PCR include: (1) The PCR product (DNA) can be rapidly sequenced providing genetic information about the virus. (2) Primer sets can be designed to recognize sequences common among groups of related viruses or can be used detect a specific member of a virus group. (3) Multiple primer sets can be used to look for more than one suspect virus in a sample.

PCR assays are very sensitive, but sensitivity can be a disadvantage as well as an advantage. When performing PCR for diagnostic purposes, it is essential that every precaution be taken to avoid contaminating patient samples. This often requires separate

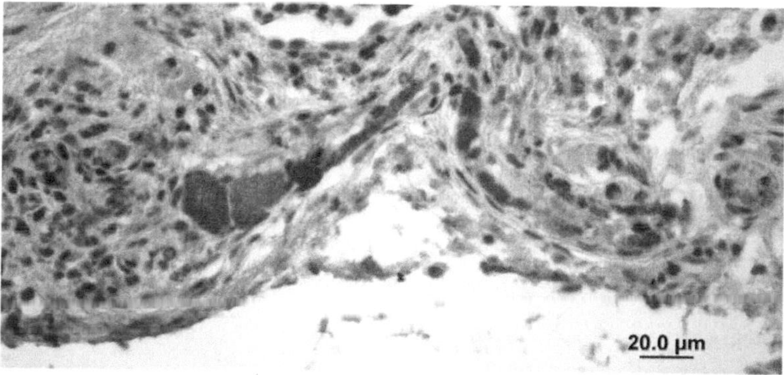

FIGURE 4.16 Immunohistochemistry is a technique used to detect viral antigens in tissue sections. In this example a brain section has been treated with antibodies to a bornavirus. The antibodies are tagged with an enzyme, producing a brown precipitate that indicates bornavirus infected cells. *Courtesy: I. Tizard.*

equipment and work areas. For example, a "clean" area to process the patient sample, another area to set up the assays and a third area where the PCR products are synthesized and analyzed. It is also important to test all purchased reagents for the presence of contaminating nucleic acids. This requires multiple negative controls. For example, one might apply sterile buffer or water to a nucleic acid purification column to check the column for contamination with viruses that might have been introduced during the manufacturing process.

High Throughput Sequencing

While PCR amplification requires some prior knowledge of a viral sequence, it is now routine to sequence all nucleic acid (DNA and RNA) in a sample, using high throughput, unbiased sequencing techniques. Sequencing technologies can provide a genome's worth of data (billions of nucleotides) in a day. Powerful algorithms are used to analyze the data and compare it to information stored in public databases. Sequences of no interest (human DNA in a patient sample) can be ignored, allowing the investigator to quickly focus on any viral sequences that may be present. As in the case of PCR, the sensitivity of the assay is both a positive and a negative, and the most careful researchers will include multiple negative controls in their assays. Use of unbiased sequencing has resulted in an explosion of new viruses from humans, animals, and environmental samples. The current challenge is to develop an understanding of which viruses might be threats and which are part of our normal viral flora.

BASIC PRINCIPLES OF DIAGNOSTIC VIROLOGY

The purpose of diagnostic virology is to identify the agent most likely responsible for causing disease in a human or animal patient. Virus identification can be used to:

- Determine treatment strategies (although there are only a handful of antiviral medications are available).
- Predict disease course and expected outcome.
- Predict the potential for virus spread.
- Allow identification of, and vaccination of, susceptible individuals.
- Trace the movement of a virus through a community, or world-wide.

For the medical practitioner, methods for identifying the virus in an infected patient ideally should be sensitive, specific, and rapid, as once a patient has recovered (or died) diagnosis has less practical value (although a diagnosis could benefit family and community members). On the other hand, epidemiologic studies may include hundreds or thousands of samples requiring use of low cost, high throughput modalities.

Common targets for diagnostic tests are viral proteins (antigens), viral genomes, and/or antiviral antibodies. Some are designed to detect viral proteins, enzymes, or genomes directly from a patient sample (blood, throat swab) and these are useful in a clinical setting. One type of rapid diagnostic assay design (lateral flow immunoassay) uses the process of diffusion to move a sample across a test chamber. The liquid sample contacts various dried reagents as it flows through the chamber. Among the strengths of the lateral flow immunoassay is the ability provide rapid diagnostic capabilities in a medical/veterinary setting. Tests can be designed to detect either antigen or antibody from a patient sample. An example of an *antigen capture assay* is shown in Fig. 4.17. The test strip contains three key assay reagents (dried on the strip). Closest to the sample addition chamber is an antiviral antibody. In our example the antibody is conjugated to gold nanoparticles. When the liquid from the patient

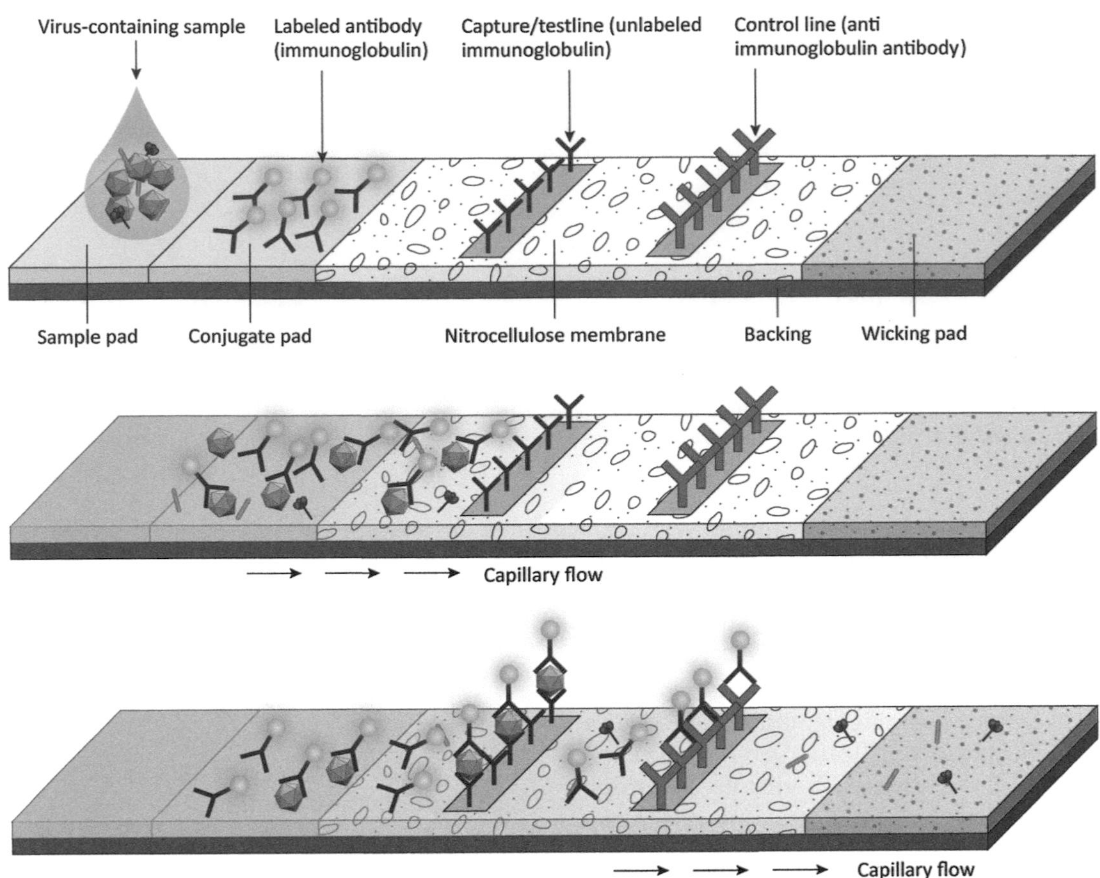

FIGURE 4.17 An antigen capture, lateral flow immunoassay. The test strip contains three key assay reagents. In this example a gold-conjugated antibody binds to virus in the test sample. The liquid moves across the slide by capillary action to the Test (T) strip. The T strip contains membrane bound antiviral antibody. It will bind to capture the virus particles and their associated gold tagged antibodies to generate a visible signal at the T strip. If no virus is present in the sample, the gold-labeled antibody travels past the T strip and binds to anti-immunoglobulin antibodies bound at the C strip. A positive sample must show color reactions at both the T strip and the C strip. A negative sample must show a color reaction at the C strip (to validate that the test worked correctly).

sample is applied, it flows, by capillary action, across the slide and will encounter the labeled antibody. The labeled antibody is picked up in the flowing liquid and will bind to any viral antigen present in the sample. The liquid continues to move across the slide (by capillary action) until it hits the Test (T) strip. In our example the T strip contains antibody to the virus in question. If the labeled antibody is bound to virus, it will be stopped (captured) at the test strip. If the labeled antibody is not bound to virus, it will move past the T strip and reach the control (C) strip. Anti-immunoglobulin antibodies are bound at the C strip and will capture any labeled immunoglobulin in the liquid. Thus if the sample is positive for virus, a colored line should appear at the T strip and the C strip (because there is excess labeled antibody). If the sample is negative for virus, the T strip will remain colorless but the C strip will be positive.

Often there is not enough virus present in the infected host to allow for *direct detection* in a patient sample. In that case the sample may be sent to a diagnostic laboratory for inoculation into cultured cells (or fertile eggs) to generate higher concentrations of virus. The infected cultures are closely observed for visible changes, such as cell killing (cytopathic effects), changes in cell morphology, or formation of syncytia (fused cells). Any of these changes in the infected cell cultures (as compared to uninfected controls) provides an indication that an agent is replicating, and that further testing is warranted to identify the agent. Growing viruses is labor intensive, takes days to weeks, and presents biosafety issues. Patient samples are often inoculated into several different types of cells in the hopes that at least one type will be susceptible. Infected cultures can provide abundant material for detection of viral antigens by ELISA, IFA, immunohistochemistry, or western blots. They can also provide material for PCR or direct sequencing, or EM can be used to look for the presence of virus particles in samples. (EM is not usually helpful for direct examination of patient

BOX 4.3

COMMON BIOCHEMICAL AND MOLECULAR TECHNIQUES FOR INVESTIGATIVE VIROLOGY (1)

In addition to the techniques described above, the virologist has a wide array of molecular and biochemical tools that can be used to study virus replication and/or pathogenesis. The scope of this text does not allow a complete discussion of these techniques but short descriptions of some common techniques are provided.

Electrophoresis. Charged molecules can be separated in an electric field. Often the electric field is applied to a semisolid gel (agarose or polyacrylamide) so the components in a sample can be separated by charge and/or size. Addition of SDS to a protein sample coats the proteins with a uniform negative charge so they are separated based on size alone. Nucleic acids have a uniform net negative charge, thus are easily separated by size using agarose or polyacrylamide gels. Gels are treated with specific stains to detect the presence of proteins or nucleic acids. Materials in the semisolid gel can be transferred to a solid support, or membrane, for additional manipulations, such as incubation with antibodies (western blot).

Chromatography is the collective term for a group of techniques used to separate molecules in a mixture. A liquid mixture is applied to material packed in a column (column chromatography), or layered onto plate (thin layer chromatography). Separation is based on the relative ability of the components in the mixture to move through or across the support materials. Chromatography can be applied to any mixture of molecules but the research virologist most often uses column chromatography to separate viral proteins.

Size exclusion chromatography employs beads with pores of defined sizes. Larger molecules are excluded from the beads, thus move quickly through the column. Smaller molecules enter the pores in the beads and this slows down their passage through the column. The result is that larger molecules are eluted from the column before smaller molecules. Thus molecules (i.e. different viral proteins) are separated by size.

Ion exchange chromatography uses charged materials in a column to separate macromolecules based on their charge. If the material in the column is negatively charged, molecules with a positive charge will be retained in the column. By changing buffer conditions (pH and ion concentrations), macromolecules can be separated on the basis of charge.

The basis of *affinity chromatography* is the very specific interaction between two molecules (for example, the interaction of an antibody to its cognate antigen). If an antibody is attached to a solid support, it can be used to capture antigen from a dilute solution. After washing away unbound material, the antigen can be released from the column by changing salt and or pH of the buffer. Proteins A and G are bacterial proteins that bind to immunoglobulins. They are used to affinity purify antibodies and antibody/antigen complexes.

---

samples as the amount of sample on one microscope grid is very small and it can be tedious to survey more than a handful cells at the highest magnifications. (However, some enteric viruses, for example, rotaviruses, are present in high enough concentrations in stool samples and can be visualized by EM.)

Recent or past exposure to viruses (or immunization) can be determined by detecting antiviral antibodies in patient samples. This type of assay requires a panel of viral antigens against which antisera can be tested. It is possible to detect not only antiviral antibody, but also to determine the class of antibodies in the sample. Detection of IgM indicates recent or acute infection, as these are the first antibodies produced in response to infection. Detection of antiviral IgG indicates that the patient is, or has been, infected with a particular virus; however, IgG may also be present due to prior immunization. Measuring IgG levels can be helpful for diagnostic purposes if *two* patient samples are available: one collected early in the disease process (acute serum sample) and one collected after recovery (convalescent serum sample). If the IgG titer in the convalescent serum is higher (>fourfold higher) than in the acute sample, a diagnosis can be made.

Immunoglobulin titers can be measured using ELISA or IFA. Immunoglobulin can also be measured by its ability to precipitate a particulate antigen (precipitation assays), to block hemagglutination (HIA) or to inhibit virus infectivity (virus neutralization assay).

A few biochemical and molecular techniques commonly used by virologists are described in Boxes 4.3 and 4.4.

In this chapter we have learned:

- A general method for obtaining and culturing cells. Primary cultures contain cells obtained from an animal and they have a limited lifespan. Tumors are

BOX 4.4

COMMON BIOCHEMICAL AND MOLECULAR TECHNIQUES FOR INVESTIGATIVE VIROLOGY (2)

Flow Cytometry. Cells are suspended in a stream of fluid and flow past an electronic detection apparatus. The number of cells passing the detector is counted. Thousands of cells per second can be counted. Often cells are labeled with fluorescent dyes (tagged antibodies, for example) and are not only counted, but are separated into different collection tubes based on labeling patterns. The process is often used to get pure populations of cells from a mixture. Immunologists take advantage of the fact that different types of proteins are found on the surfaces of different types of lymphocytes. Thus labeled antibodies can be used to quantitate and separate different lymphocyte subsets from a blood sample. Not only can the presence of a molecule on the cell surface be detected, but the relative amounts can be determined as well. Flow cytometry is also used for viral diagnostics. Cells infected with an unknown agent can be incubated with panels of antibodies.

Reverse Genetics. Virologists study viral genes to determine their functions. Historically this was accomplished by finding and purifying mutant viruses to determine how they differed from their "wild-type" parents. With the development of gene cloning technologies, virologists are able to clone entire viral genomes, manipulate them in the laboratory (introduce specific mutations) and examine the effects on virus replication or disease. Introducing specific mutation is the "reverse" of traditional genetic method of identifying a mutant phenotype and then determining the genetic cause. In order to do reverse genetics, the virologist usually starts with a cloned viral genome that can be introduced into cells to produce infectious virus.

Designer Cells and Designer Animals. In addition to mutating viral genes, virologists manipulate the hosts as well. It is relatively easy to add, delete, or modify the genetic makeup of cultured cells. Animal genomes can be modified as well. Systems to create designer mice are reasonably efficient and hundreds of types of genetically modified mice are commercially available. What is gained by modifying the host? At every step in the virus life cycle, viral proteins interact with cellular proteins and a single viral protein may interact with dozens of cellular proteins. To analyze the effects of just one type of interaction, it may be preferable to modify the host genome. It is also possible to genetically alter a resistant host to render it susceptible to a virus. For example, the receptor for a human virus could be added to mouse cells to generate a tractable animal model. It can be challenging for a student of virology, new to reading scientific literature, to determine the types of virus or host mutations that have been used to obtain data for a specific study. But this is critical to understanding and evaluating experimental results.

a source of immortalized or transformed cells that can divide continuously.
- A basic method for virus purification using centrifugation.
- Methods to visualize viruses including negative staining EM, thin sectioning. EM, cryo-EM, and confocal microscopy.
- Methods to count infectious virus particles (plaque assays and endpoint dilution).
- The basic features of common techniques used to detect viruses and antibodies.

References

Bajaj, S., Dey, D., Bhukar, R., Kumar, M., Banerjee, M., 2016. Non-enveloped virus entry: structural determinants and mechanism of functioning of a viral lytic particle. J. Mol. Biol. 428, 3540–3556. Available from: http://dx.doi.org/10.1016/j.jmb.2016.06.006.

Barretto, N., Uprichard, S.L., 2014. Hepatitis C virus cell-to-cell spread assay. Bio-Protocol. 4 (24), e1365. Available from: http://dx.doi.org/10.21769/BioProtoc.1365.

Bouchez, G., Sensebe, L., Vourc'h, P., Garreau, L., Bodard, S., Rico, A., et al., 2008. Partial recovery of dopaminergic pathway after graft of adult mesenchymal stem cells in a rat model of Parkinson's disease. Neurochem. Int. 52, 1332–1342.

Bridger, P.S., Menge, C., Leiser, R., Tinneberg, H.-R., Pfarrer, C.D., 2007. Bovine caruncular epithelial cell line (BCEC-1) isolated from the placenta forms a functional epithelial barrier in a polarised cell culture model. Placenta. 28, 1110–1117.

Cromeans, T.L., Lu, X., Erdman, D.D., Humphrey, C.D., Hill, V.R., 2008. Development of plaque assays for adenoviruses 40 and 41. J. Virol. Methods 151, 140–145. Available from: http://dx.doi.org/10.1016/j.jviromet.2008.03.007.

C H A P T E R

5

Virus Transmission and Epidemiology

O U T L I N E

After studying this chapter, you should be able to:

- List some factors that impact virus transmission.
- Define or explain the terms virulence, pathogenicity, prevalence, incidence, incubation period, latent period, and infectious period.
- Explain how serosurveys provide information needed to determine the pathogenicity of a virus.
- List some of the factors that impact the spread of a virus.
- List some of the factors that impact the outcome of a viral infection.
- List modes of virus transmission.

Infectious disease epidemiology (which includes the epidemiology of viruses) is the study of the complex relationships among hosts and infectious agents. Epidemiologists are interested in virus spread or transmission, with or without disease. Viral epidemiologists try to predict the potential for development of epidemics, and a very important part of their job is to define the kinds of interventions that could contain a virus outbreak. Veterinarians are often concerned with threats to food animals (how a disease of food animals might be spread, or be introduced into a disease-free area). In order to model virus transmission, epidemiologists must try to account for a variety of factors involving both host and virus. Factors that can impact virus transmission and spread include:

- Prevalence of the agent within the population.
- Mode or method of transmission of the agent.
- Duration of the infection and the window of transmissibility.
- Numbers of susceptible and nonsusceptible individuals in the population.
- Population density.
- Patterns of travel or associations (for example, schoolchildren and their families form interconnected networks).
- Living conditions.
- Climate and/or season.

Predicting the course of an outbreak is particularly challenging if a new or novel virus is involved, as we often do not have adequate information about modes of virus transmission, duration of infection, window of transmissibility, or stability of the virus in the environment. This can be further complicated by additional factors that may impact the outcome of infection. For example, differences in age, gender, nutrition, and genetic susceptibility of the host are important factors in the outcome of infection. Thus some infected individuals may have an asymptomatic infection while others develop severe, life-threatening disease.

Two key terms used by epidemiologists are "incidence" and "prevalence." These terms have very specific meanings and are not interchangeable.

- *Incidence rate* (also called the attack rate for acute infectious diseases) is a count of the number of

Viruses. DOI: http://dx.doi.org/10.1016/B978-0-12-803109-4.00005-2

new infections during a specific time period. A specific population and a time frame are defined, and the number of new cases is counted in order to arrive at the numerator of the ratio. The denominator of the ratio includes both the size of population and time frame, and is often expressed as person-years.

- *Prevalence* refers to the total number of cases present or counted. In the case of a persistent infection such as human immunodeficiency virus (HIV), the numerator includes patients infected for many years or decades. Prevalence is expressed as a ratio such as "cases per million." Note that there is no time parameter in this ratio.

For an acute viral infection, such as measles, the incidence rate, or attack rate may be similar to the prevalence of the virus, given that virus is present and transmissible for a relatively short period of time. In contrast, when considering persistent or chronic infections, such as those caused by the HIV or hepatitis C virus (HCV), incidence and prevalence may actually be quite different. In the case of HIV, education about safer sex practices can decrease incidence (new infections), while the availability of antiviral drugs results in significantly longer survival times, thus increasing prevalence. Interestingly, it has been shown for HIV that drug treatment decreases incidence while at the same time increasing prevalence, because treatment not only prolongs life but decreases transmission by drastically reducing viral loads in infected patients (Fig. 5.1).

METHODS TO COUNT VIRAL INFECTIONS AND DISEASE

Accurate determination of incidence rates or prevalence of a virus depends on the ability to count infections. The values obtained depend on the methods used for counting. Changes in the case definition of a disease, or the diagnostic tools available to detect a viral infection can lead to sudden apparent changes in incidence or prevalence.

Passive surveillance is done when healthcare workers report cases of disease. Many viral diseases are designated as "reportable" and in theory, each and every case should be reported. The weak link in passive surveillance is the filing of the initial report, as in practice only a small fraction of cases of common, reportable viral diseases are actually reported (reporting frequency may be as low as 10%–15%). Even so, passive surveillance is useful for monitoring infectious disease trends.

Active case detection through investigation is another way to collect information on disease outbreaks. Active case detection seeks to classify an illness and determine the causative organism, to assess the extent of an outbreak and its economic and health impact, to stop the outbreak and to inform the public.

Serological surveys are an important tool for the viral epidemiologist. They are useful because many viral infections stimulate the production of antibodies (for example, immunoglobulin G) and detectable antiviral antibody titers are often maintained lifelong.

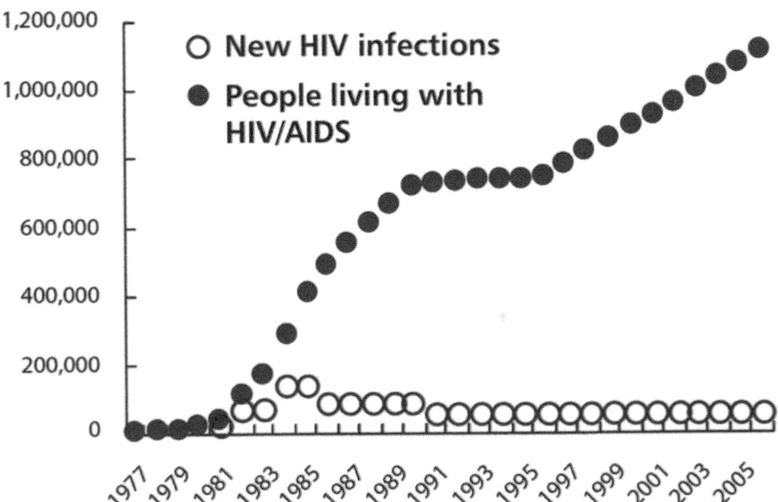

FIGURE 5.1 Incidence versus prevalence of HIV infections. The number of new HIV infections per year in the United States has remained steady but most infected individuals live years to decades with treatment, hence the yearly increase in prevalence. Source: *Campsmith, et al., 2008. CROI 2009. JAMA 300 (5), 520–529.*

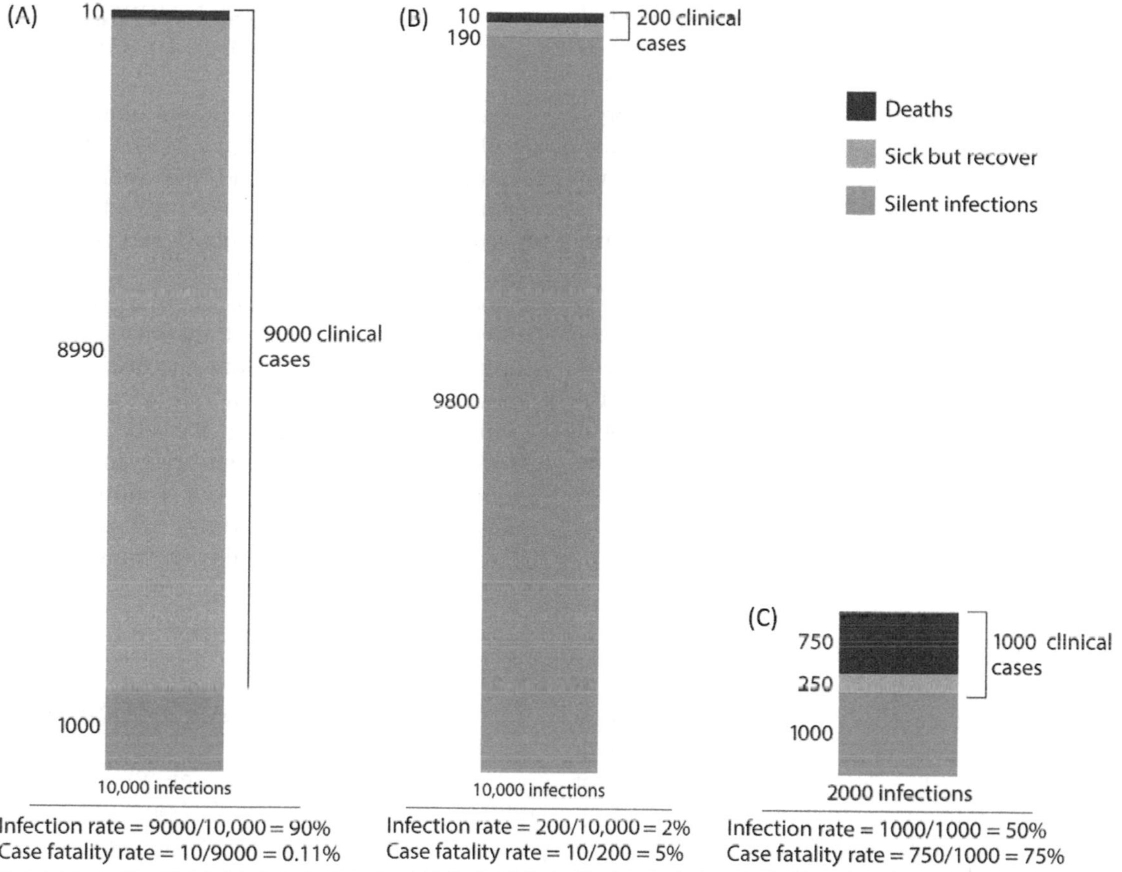

FIGURE 5.2 Infection ratio versus fatality ratio. The infection ratio is the number of clinical cases per case of infection. The case fatality ratio is the number of deaths per clinical cases. Three different scenarios are presented. The virus in panel A has high pathogenicity but low virulence. The virus in panel B has low pathogenicity. The virus in panel C has high virulence.

An important feature of serosurveys is that they provide a history of both clinically apparent and clinically inapparent infections. A question that can only be answered by serosurvey is whether or not those severely impacted by an infection represent the total number infections, or if severe disease is a rare outcome on a background of a many inapparent infections (Fig. 5.2).

INFECTION VERSUS DISEASE

Another important concept in infectious disease epidemiology is the difference between infection and disease. For example, one reason for the successful eradication of the smallpox virus was that over 95% of infections resulted in observable, distinct, disease. The virus was not able to hide or spread silently. In contrast, poliovirus has proven more difficult to eradicate. One reason is that there is less than one case of observable disease per 100 infections.

Serosurveys are needed to determine the case infection ratio, the number of *clinical* cases per 100 infections (the number of clinical cases/the number of infections). To evaluate and express the severity of clinical disease, another measure is used. This is the case fatality ratio or the number of deaths per 100 clinical cases. The case infection ratio of a virus might be 0.001 (1 in 10,000 infected individuals develops disease) but the case fatality rate for the same virus might be 1 (every clinical case results in death).

As regards serosurveys, it is also important to note that a serologic assay may not be able to differentiate between natural infection and immunization. Recently there has been increased interest in developing vaccines designed to allow for discrimination between infection and vaccination (DIVA vaccines).

Serosurveys can assist in determining the pathogenicity of a virus. *Pathogenicity* is a measure of the proportion of infections resulting in overt disease. Measles virus is highly pathogenic as over 95% of infected individuals will experience a disease episode. A related, but distinct term is virulence. Virulence is a measure of the *severity* of the disease and is often measured as the number of deaths per number of infections. A virus can have low pathogenicity (few infections lead to a disease episode) but high virulence (when disease does occur it is often fatal) or the opposite might be true (most infections lead to disease, but disease is almost always mild). It is important to remember that many viral infections are completely inapparent. Outbreaks of novel viral diseases can be very stressful for clinicians, public health workers, and the population at large because we may not have enough information to distinguish between pathogenicity and virulence. We cannot assume that every infection results in disease.

Another important concept, particularly as regard viruses, is that very similar viruses can vary greatly in pathogenicity, virulence, and transmissibility. This is because very small genetic differences between closely related virus strains can have huge biological effects. For example, influenza viruses can vary greatly in their pathogenicity, virulence, and transmissibility, based on just one or very few amino acid differences. We will examine some of the genetic/molecular bases of these differences in Chapter 23, Family *Orthomyxoviridae*.

DEVELOPING MODELS OF VIRUS TRANSMISSION

An important goal of infectious disease epidemiology is to provide useful models of virus transmission. Such models can be used to determine the types of interventions (for example, vaccination, quarantine, drug therapy) that will most effectively reduce transmission and limit the scope of an outbreak. Development of good models requires knowledge about the number of susceptible individuals in the population, the proportion of infectious individuals, and the rate of contact between the two groups. In the case of an acute virus, a patient may be infectious for days or weeks. The recovered patient usually has detectable antibody and no longer participates in the transmission chain. Of course vaccinated individuals are also unlikely to be links in a transmission chain, thus high vaccination rates can protect unvaccinated individuals.

In contrast, transmission of persistent viral infections can occur in the presence of antiviral antibody. Most herpes virus infections are persistent and individuals are infected for life. People and animals may transmit herpesviruses, with or without displaying symptoms. Patients latently infected with the chickenpox virus (human herpes virus 3 (HHV3) or varicella zoster virus) are only infectious if they develop the painful rash called "shingles" as the fluid vesicles contain large amounts of infectious virus. In contrast, HHV2, the virus that causes genital herpes, can be spread in the absence of a visible lesion. Other viruses that cause long-term persistent infections include HIV, hepatitis B virus, and HCV. These infections can be asymptomatic even when large numbers of virus particles are present in blood and body fluids. In summary, persistently infected individuals are capable of transmitting infection over many years or decades.

INCUBATION, LATENT, AND INFECTIOUS PERIODS

The time between acquiring an infection and the onset of illness is the *incubation period*. The time between acquiring an infection and the ability to transmit the agent is the *latent period*. The period during which the infected person can potentially transmit the infectious agent to others is the infectious period. If the incubation period is longer than the latent period, asymptomatic individuals unknowingly transmit virus. In a persistent infection such as HIV, a long incubation period provides months to years during which an active infection may go undetected. Following the ebolavirus outbreak of 2014–15, it came as a surprise that some survivors continued to shed virus (for example, in semen) for many weeks. Understanding the distinction between clinical illness and the ability to transmit virus is key to developing effective quarantine measures.

VIRUS TRANSMISSION

Understanding virus transmission is key to developing robust epidemiologic models. Some viruses are very host specific, infecting only a single species. For example, polio, measles mumps, rubella, and chickenpox viruses infect only humans. For viruses maintained in a single host population, transmission modes can vary. Mechanisms for transmission include direct transmission from one animal to another (respiratory, fecal-oral, sexual, blood, or from parent to offspring (vertical transmission), or via contaminated food, water, or fomites). The term vertical transmission is used to describe various modes of transmission from mother to offspring. These include: across the placenta, perinatally during birth, and via colostrum or milk.

Other viruses spread across species boundaries by many of the same of the routes described above, such as respiratory or via contaminated food, water, or fomites. Influenza viruses can be transmitted between humans and pigs at county fairs and petting zoos. Spread is not always from pigs to humans, spread from humans to pigs has been well documented. Avian influenza viruses can be transmitted to humans involved in the slaughter and processing of poultry. Rabies virus is notable in that it is transmitted via bites, through contaminated saliva. Veterinarians are at risk for contracting zoonotic viruses during exams or necropsies and research virologists can acquire viruses from infected laboratory animals or infected cultures.

Some viruses *must* infect different hosts to be maintained in nature. Examples include yellow fever virus (Box 5.1) and dengue virus. These are human pathogens but natural transmission is seldom directly from human

BOX 5.1

MOSQUITOS IN THE WATER BARRELS. HOW YELLOW FEVER REACHED THE WESTERN HEMISPHERE

Author: I. Tizard, Distinguished Professor, Texas A&M University.

The yellow fever virus originated in the rain forests of Central Africa where it was a disease of monkeys. It was transmitted between these monkeys by mosquitos of the genus *Aedes*, most notably *Aedes aegypti*. Provided the monkeys were left in peace, the virus could circulate among them. Occasionally local hunters would wander into the deep forest and some, the unlucky ones, would be bitten by an affected mosquito and develop severe disease. This was probably not a common occurrence, and, given the low human population density in the rain forests, the likelihood of transmission between two humans was very low and major disease outbreaks were rare.

This changed with the development of the transatlantic slave trade. Africans were taken from interior Africa, marched to the coast, and loaded onto slave ships. Some of these victims were probably infected with the yellow fever virus. The slave ships had to carry food and water for their cargo. Water barrels were filled at any convenient lake or river. Some of this water would have contained mosquito larvae. Thus eventually, the two major components required for disease transmission, multiple potential human victims and infected mosquitos came together as they were transported across the Atlantic.

The first yellow fever epidemic in the new world (as opposed to single cases in slaves) likely occurred in 1647 on the English colony of Barbados. The population here was sufficiently dense (about 200 persons per square mile) to sustain the epidemic. From Barbados the disease spread throughout the Caribbean, associated with the spread of the mosquito vector, *A. aegypti*. The virus reached Mexico, Florida, and northern South America within a year. The disease eventually established itself (became endemic) in much of Central America, the Caribbean islands, the southern United States, especially New Orleans, and occasionally reached as far north as the cities of Charleston, New York, and Philadelphia (cities that engaged in trade with the West Indies). It persisted in these urban areas because of the presence of *Aedes* mosquitos and readily available susceptible hosts.

Yellow fever is an acute hepatitis. As a result of liver damage, patients develop jaundice and thus appear yellow. Eventually it results in hemorrhagic disease, and bleeding into the gastrointestinal tract. Hence the Spanish name is "vomito negro" or black vomit. It is a lethal disease with 40% mortality. For example, in 1793, 44% of British soldiers based on Santo Domingo died within 3 months. The soldiers were based in the West Indies as part of the ongoing struggle between the British, Spanish, and French, for what were then exceedingly valuable island colonies, the source of most of the world's sugar at that time.

The French established a colony on the Western side of Haiti and this became a major sugar-producing site supported by a very large number of African slaves. When wars broke out between Britain and France, the Atlantic was dominated by the British Royal Navy who effectively cut Haiti off from France. As a result, a slave revolt led by Francois l'Overture succeeded in declaring Haiti's independence and it became the second independent state in the Western Hemisphere. In 1801, a brief truce between the British and French permitted Napoleon to attempt to reassert French sovereignty. He sent 20,000 soldiers under the command of his brother-in-law General LeClerc to retake the colony. This was rapidly achieved, but within a few months the French were dying of yellow fever at a rate of 50 per day. LeClerc himself died within a few months.

By 1803 only 3000 French soldiers were left alive, and the war with Britain had resumed. Napoleon recognized that he could no longer retain possession of colonies in the New World so rather than let the British capture them, he sold them to the United States. The Louisiana Purchase Treaty was signed in 1803. Given the importance of this to the United States, it is no exaggeration to say that yellow fever can be considered the most significant disease in American history.

For most of the 19th century, yellow fever continued to be a scourge throughout the Caribbean basin and sporadic outbreaks occurred regularly in New Orleans and around the US gulf coast. At this time, however, it was still not known how the disease was transmitted. But by the 1880s, there was a growing recognition that some diseases, malaria for example, could be transmitted by mosquitos. A Cuban physician, Dr. Carlos Findlay went so far as to propose that yellow fever was mosquito borne but was unable to prove this. His theory was correct but his timing was wrong! He left insufficient time for the virus to grow in the mosquito after it fed on a yellow fever victim before allowing it to bite a healthy human volunteer.

In 1898, the Spanish-American war broke out as a result of Spanish attempts to suppress a revolt in its colony of Cuba. The United States resolved to expel the Spaniards from Cuba and did so in a short campaign conducted during the dry season. Unfortunately once

BOX 5.1 *(cont'd)*

the fighting was over, US troops remained in the country during the rainy season and soon began to experience losses from yellow fever. Seriously alarmed, in 1900, the US Army established the "Yellow Fever Commission" under Major Walter Reed and sent it to Havana, Cuba. Their task was to determine the cause and seek to prevent the disease.

After first demonstrating that yellow fever was not a bacterial disease as some claimed, the commission began to consider seriously the theories of Carlos Findlay, namely that it was a mosquito borne disease. They paid soldiers to volunteer for studies. Their first experiment allowing mosquitos to feed on yellow fever victims and then immediately letting them feed on healthy volunteers did not work. Nobody developed yellow fever. One of the Commission members, however, James Carroll, allowed himself to be bitten by a mosquito that had fed on a yellow fever victim 12 days earlier. Carroll developed yellow fever but survived. Because Carroll had been in contact with a yellow fever case a few days previously, the experiment was repeated on two more volunteers who had not been in contact with such cases. They both developed yellow fever and one, Jesse Lazear, died. The Commission now conducted a series of elegant experiments that confirmed conclusively that yellow fever was transmitted by the *A. aegypti* mosquito.

The next step was to control or eliminate the disease. William Gorgas, the chief sanitary officer in Havana organized an aggressive mosquito erradication campaign. By eliminating standing water, the sites where mosquito larvae developed, the mosquito population was reduced to such an extent that the number of yellow fever cases in the city fell from 1400 in 1900 to none in 1902.

The French engineer Ferdinand de Lesseps built the Suez Canal in 1869. This encouraged him to attempt to build a Panama Canal. The French began digging in 1882 but gave up after several years as a result of several factors, the most important of which was yellow fever. Not knowing how yellow fever was transmitted, there was much standing water in the construction zone and mosquitos thrived. In the first year of construction 400 workers died, in the second, 1300. When construction was abandoned 7 years late it is estimated that more than 22,000 workers had died. de Lesseps and his company went bankrupt. The United States decided to complete the construction of the canal and this recommenced in 1907. Prior to this, however, William Gorgas, the eliminator of yellow fever from Havana was called in and pursued aggressive mosquito control policies in the canal zone. Removal of standing water and screening of workers housing worked well. The death rate from yellow fever was reduced to insignificance and the last case of yellow fever in Panama occurred in 1906.

In the years following, research into the cause of yellow fever demonstrated that it was caused by a virus. Eventually Max Theiler, working at the Rockefeller Foundation in New York, showed that it would infect mice when inoculated into the brain. By passing the virus many times through mice, he showed that the virus lost its virulence for monkeys. This attenuated 17D strain could not induce disease in human volunteers but did induce protective immunity. The 17D strain has proved to be an incredibly effective vaccine and has been used to protect millions of people. As a result, the virus is now confined to the remoter areas of rain forests in Amazonia and Central Africa where it causes occasional disease in unvaccinated visitors.

to human. Instead the natural transmission cycle is from infected mosquitos to humans (Fig. 5.3). Viruses transmitted by insects are collectively called arboviruses (from arthropod-borne). West Nile virus is another example of an arbovirus, but in this case the virus is largely maintained in a transmission cycle between mosquitos and birds. Humans can be infected, but usually are dead end hosts because we seldom transmit virus back to mosquitos (and human–human transmission is very rare). Transmission from an animal to insect vector requires that sufficient virus be present in the blood.

Once ingested by the insect, most arbovirus replicate in the insect to facilitate spread back to the human, animal, or avian host. Many are maintained for long periods in their insect hosts. For example, mosquitos can transmit virus to their offspring (transovarial transmission). Thus even in the absence of susceptible vertebrate hosts, the virus persists in nature. Virus transmission via insect vectors depends on a number of variables including the species of insect vector, the host preference of the blood-feeding vector, and environmental factors (temperature, rainfall) that may limit or facilitate vector spread. Vesicular stomatitis virus, a virus of livestock is thought to be maintained in infected midges or mosquitos but once an outbreak begins, animal to animal transmission occurs via virus-filled blisters.

Another notable mechanism of transmission of some zoonotic viruses is through dried feces or urine (Fig. 5.4). Rodents are natural hosts for a number of

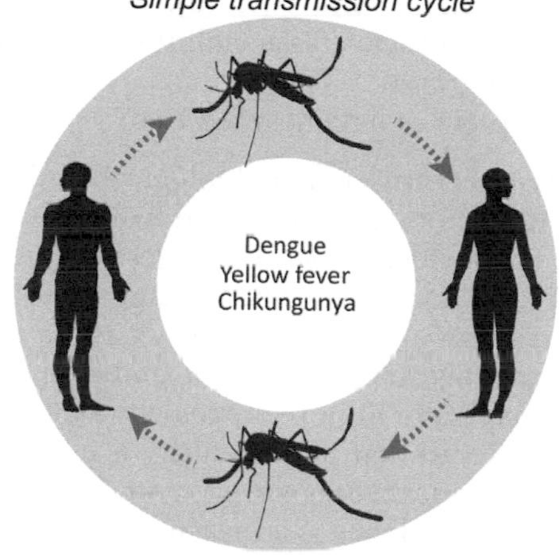

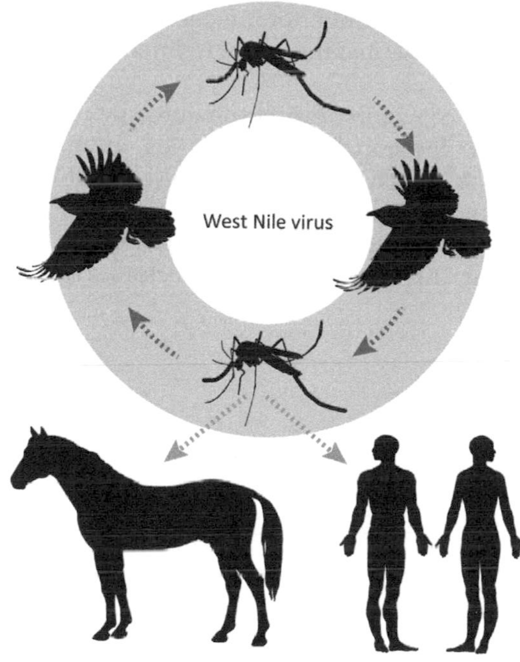

FIGURE 5.3 Transmission of viruses by insects. The top cycle depicts a simple transmission cycle involving humans and mosquitos. The bottom cycle shows human/horse infections via a complex transmission cycle. In this process the virus is normally maintained in birds and mosquitos. The virus may occasionally infect humans or horses but is usually not transmitted further.

viruses that are potential human pathogens. These viruses do not seem to harm their rodent hosts but can cause severe disease if humans are infected through inhaling contaminated dust. In the United States, Sin Nombre hantavirus (see Chapter 24, *Bunyaviridae*) is an example. It was first discovered as the cause of a small outbreak of severe pulmonary disease in the southwestern United States that occurred in 1993. The outbreak was linked to an explosion in the deer mouse population that year. Since that time over 600 cases of hantavirus pulmonary syndrome have been reported in the

FIGURE 5.4 Hantavirus transmission. The normal hosts of hantaviruses (Family *Bunyaviridae*) are rodents. They excrete virus in urine and feces. When rodent activity is high, structures may become contaminated. Humans can be infected by inhaling contaminated dust.

United States. They are caused by a group of distinct, but related hantaviruses whose natural hosts include a variety of rodents. A recent outbreak (2012) occurred among visitors to Yosemite National Park. It resulted in 10 cases with 3 deaths. Most of the victims stayed at a single location in the Park (the Signature Tent Cabins) and the outbreak was linked to high rodent activity in the area. Recommendations to avoid infection include appropriate rodent control and cleaning methods. Rodent infested areas should be sprayed and mopped with disinfectant as it is important to avoid generating dust that can be easily inhaled.

Virus transmissibility can be described quantitatively. Epidemiologists use the expression *basic reproductive rate* (R_0) to describe transmissibility. R_0 is defined as the average number of new infections initiated by a single individual in a completely susceptible population during the individual's infectious period. Basic reproductive rate is a function of both virus characteristics and contact patterns within a community. Measles virus is highly infectious with an R_0 of 12–18. R_0 is calculated by assuming that every contact is susceptible. However prior infection and/or vaccination reduces the number of susceptible individuals in the population. The effective or net reproductive number (R) is the actual average number of secondary cases and it equals the product of R_0 and the proportion of susceptible individuals in the population. R is smaller than R_0 because not all individuals in real populations are susceptible.

What is the value of calculating R? When R is greater than 1, we expect the incidence of infection to increase, but when R is less than 1, the infection burns itself out. When R equals 1, the number of infections is constant. Thus infection control programs can be modeled to determine the most effective methods available (for example, vaccination, quarantine, use of antiviral drugs) to control an outbreak. Of course we must also consider the human factor, the ability, and willingness of a population to adhere to recommended control methods and guidelines.

In summary, understanding the epidemiology of an outbreak requires knowledge of the particular virus, the host, the place (geography), and the time (season). Host factors include: age, nutrition, immunity, personal conduct, occupation, and interaction networks. Geography includes important factors such as temperature and humidity but also includes factors such as types of dwellings, water sources, and sanitary infrastructure. Many viral outbreaks are seasonal and the factors that influence seasonality include both host factors and geography.

In this chapter we have learned that:

- A variety of environmental, biological, and societal/behavioral factors impact virus transmission.
- Viruses (even closely related viruses) differ in their virulence, pathogenicity, and transmissibility, and these attributes are not linked. A virus may be highly transmissible but has low pathogenicity. A virus may have high pathogenicity but low virulence. Recall that pathogenicity is a measure of how many clinical cases are seen among infected individuals. Virulence is a measure of the severity of disease.
- Serosurveys are important for gathering information about the total number of infections within a group. Including those instances where infection was inapparent.
- A variety of factors impact the outcome of an infection. These include the age, sex, immune status, nutritional status, and overall health of the individual or animal.
- Modes of virus transmission include respiratory, fecal-oral, direct contact, insect vectors, vertical.

CHAPTER

6

Immunity and Resistance to Viruses

OUTLINE

After studying this chapter, you should be able to answer the following questions:

- What are the major differences between innate and specific immune responses?
- What kinds of unique molecules (proteins, nucleic acids, carbohydrates) are associated with pathogens?
- How do cells detect the presence of viral pathogens?
- What are the first responses to viral infection?
- How are interferons (IFNs) induced and what do they do?
- Describe at least one IFN-induced antiviral pathway.
- What is intrinsic viral immunity? Provide an example.

Immunity is the defense of the body against microbial invaders such as bacteria, viruses, parasitic protozoa, parasitic worms, and cancer cells. Immunity includes physical barriers to infection such as mucus, skin, and normal flora. Innate immunity involves cells and signaling pathways that are triggered by large groups of infectious agents. Specific immune responses use cells and antibodies that recognize *very specific* pathogens. There is a memory component to specific immunity that allows an animal to respond faster and more robustly the second (or third, fourth, . . .) time an infectious agent is encountered. In this chapter we review some basic information about innate and specific immunity (Fig. 6.1), with a focus on viral infections. If you are not familiar with the basic principles of immune responses, it would be helpful to read about the immune system in a general microbiology text. The following terms should be very familiar: antigen, antibody, lymphocyte, macrophage, T-cell, and B-cell. A brief description of the major cellular player in the immune response is provided in Fig. 6.2.

Viruses. DOI: http://dx.doi.org/10.1016/B978-0-12-803109-4.00006-4

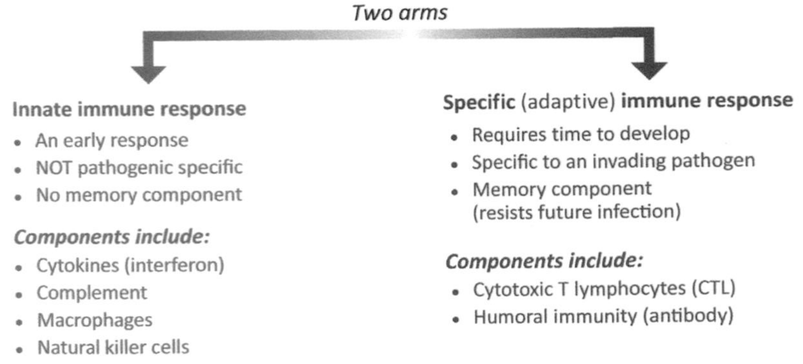

Two arms

Innate immune response
- An early response
- NOT pathogenic specific
- No memory component

Components include:
- Cytokines (interferon)
- Complement
- Macrophages
- Natural killer cells

Specific (adaptive) **immune response**
- Requires time to develop
- Specific to an invading pathogen
- Memory component (resists future infection)

Components include:
- Cytotoxic T lymphocytes (CTL)
- Humoral immunity (antibody)

FIGURE 6.1 Overview of immune responses. A general comparison of innate and specific immune responses to pathogens.

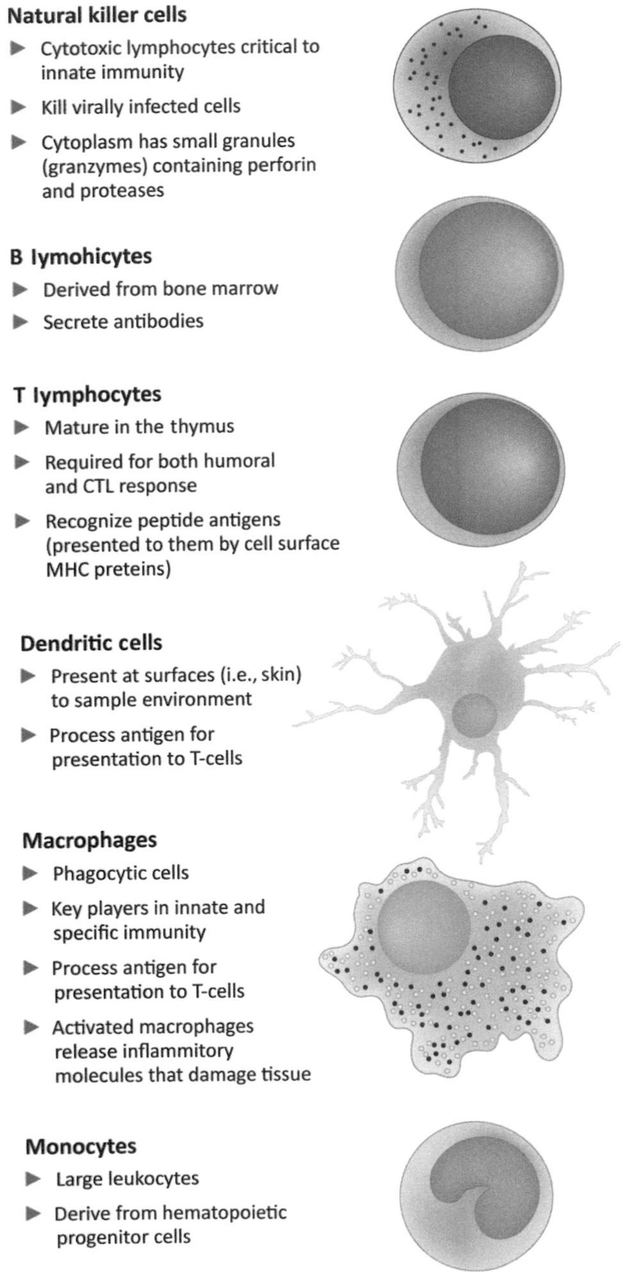

Natural killer cells
- ▶ Cytotoxic lymphocytes critical to innate immunity
- ▶ Kill virally infected cells
- ▶ Cytoplasm has small granules (granzymes) containing perforin and proteases

B lymohicytes
- ▶ Derived from bone marrow
- ▶ Secrete antibodies

T lymphocytes
- ▶ Mature in the thymus
- ▶ Required for both humoral and CTL response
- ▶ Recognize peptide antigens (presented to them by cell surface MHC preteins)

Dendritic cells
- ▶ Present at surfaces (i.e., skin) to sample environment
- ▶ Process antigen for presentation to T-cells

Macrophages
- ▶ Phagocytic cells
- ▶ Key players in innate and specific immunity
- ▶ Process antigen for presentation to T-cells
- ▶ Activated macrophages release inflammitory molecules that damage tissue

Monocytes
- ▶ Large leukocytes
- ▶ Derive from hematopoietic progenitor cells

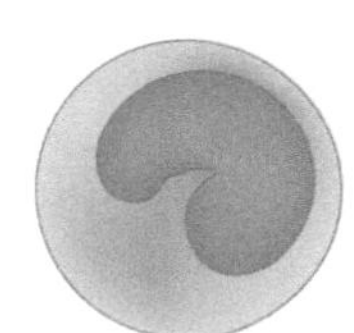

FIGURE 6.2 Cells of the immune system. Types of cells and their key roles in immunity to viruses are summarized.

INNATE IMMUNITY

The innate immune system is complex, involving multiple activators, signaling pathways, and responses. The signaling pathways connect activators to responses and these pathways intersect at many points. If signaling pathways are a roadmap, this chapter will introduce some highways and major intersections but will not consider the many smaller roads that may lead to the same destination. When reviewing this material try to keep the "big picture" in mind by considering the following questions:

- If the immune system recognizes molecules unique to pathogens, why do not mutations arise that allow the pathogens to escape? (Sometime they do!)
- If pathogens change in response to the body's defenses, what is the response of the host? (Develop new defenses!)
- Can host and infectious agent reach a balance? (Often, but not always.)

The foundation of innate immunity is the ability to detect and respond to foreign (pathogen-associated) macromolecules (Fig. 6.3). Thus when a virus breaches physical barriers and infects a cell, hard-wired, innate responses to infection are immediately triggered. The virus is recognized as an invader because the body recognizes and responds to specific groups of molecules, collectively called pathogen-associated molecular patterns (PAMPs). This phenomenon is called pattern-recognition and is accomplished by groups of proteins called pattern-recognition receptors (PRRs). In the case of viruses, several receptors recognize viral nucleic acids. When PRRs bind to a target they initiate signaling cascades that modulate cellular gene expression prompting the cell to synthesize many inflammatory and defensive molecules central to antiviral defenses. Three groups of PRRs are important for innate immunity to viruses:

- Toll-like receptors (TLRs),
- RIG (retinoic acid-inducible gene)-1-like receptors (RLRs),

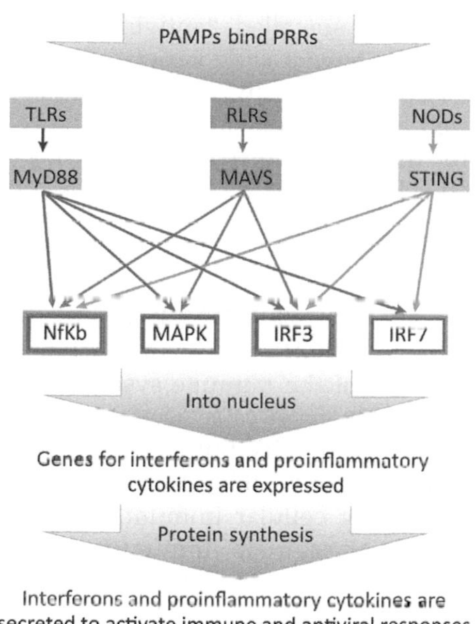

FIGURE 6.3 Signaling cascades are activated by PAMPs. PAMPs interact with classes of receptors (NODs, TLRs, RLRs) that in turn interact with major signaling molecules (MYD88, MAV, STING) that then activate specific transcription factors enabling genes encoding interferon and proinflammatory cytokines to be activated.

- NOD (nucleotide-binding oligomerization domain)-like receptors (NLRs).

Toll-Like Receptors

The word "toll" in German, means amazing or weird, and was used by geneticists to describe abnormal-looking fruit flies. The toll molecule was eventually found to be responsible for antifungal activities in *Drosophila*, thus when proteins with similar sequence, structure, and function were found in mammals, they were called TLRs). TLRs are transmembrane proteins. Some TLRs are found in the plasma membrane where they sense extracellular PAMPs. Intracellular TLRs are important in virus recognition as they detect intracellular viral nucleic acids. When triggered, TLRs stimulate the production of antiviral cytokines, particularly IFNs.

Most TLRs signal through a protein called MyD88. MyD88 in turn activates transcription factors such as nuclear factor kappa-β (NF-kβ), MAP kinase (MAPK), and IFN regulatory factor (IRF) 3 and IRF7. Activated transcription factors move from the cytosol into the nucleus where they bind DNA to turn on genes that code for IFNs and proinflammatory cytokines. Binding of PAMPs to TLRs not only triggers the innate immune response, but is also required to activate the adaptive immune system.

Retinoic Acid-Inducible Gene-Like Receptors

Two proteins, RIG-1 and melanoma differentiation-associated protein 5 (MDA5), are members of the RLR family. RLRs are found in the cytosol where they bind to certain types of viral nucleic acids. RLRs are helicases. RIG-1 senses RNAs with a 5′-triphosphate and base pairing at the 5′ end. Recall that cellular mRNAs have 5′-methyl guanosine caps, thus do not display a triphosphate at the 5′-end. However some RNA viruses lack "caps" and many have complementary sequences at their 5′- and 3′-ends. MDA5 recognizes the long dsRNA molecules that are formed during replication of the genomes of plus-strand RNA viruses.

When RIG-1 and MDA5 encounter their ligands, they are recruited to the mitochondrial signaling protein, MAVS, initiating a multistep signaling pathway leading to activation of the transcription factors MAPK, NF-kβ, and IRF3 with the subsequent induction of antiviral cytokines. Recall that these transcription factors are also key players in TLR signaling.

NLRs Are Receptors With NOD-Like Domains

The NOD domain is a nucleotide-binding oligomerization domain, thus NLRs bind to nucleic acids and form oligomers. NLRs are PRRs that detect pathogens in the cytosol. These DNA sensors include IFI16 (p204), the helicase DDX41, and cyclic GMP-AMP synthase. The endoplasmic reticulum resident protein, stimulator of IFN genes (STING), is the converging point of these DNA sensors. STING activates IRF3, IRF7, and NFkβ.

Inflammasomes

After PRRs interact with specific specific PAMPs, they act as scaffold proteins for assembly of multiprotein complexes called inflammasomes. Assembly of inflammasomes leads to the activation of caspaces (a type of protease) that cleave and activate proinflammatory cytokines such as IL-1β and IL-18.

Proinflammatory Molecules

Proinflammatory molecules and IFNs are released from cells as a result of PRR activation. These include cytokines such as interleukin-1 (IL-1), tumor necrosis factor alpha (TNF-α), and IL-6. Other proinflammatory molecules include nitric oxide (NO) (produced by the enzymes nitric oxide synthase 2 and cyclooxygenase-2) and proinflammatory lipids such as prostaglandins and leukotrienes. These molecules increase local blood flow, attract defensive cells (for example, neutrophils) and increase blood vessel permeability so that antimicrobial molecules and cells can reach the affected tissues.

Interferons

Signaling pathways activated by viral PAMPs result in the production of important antiviral molecules called IFNs. The name "interferon" refers to the fact that they were originally identified by their ability to *interfere* with many types of viral infection in cultured cells.

TYPE I IFNs

IFN-α and IFN-β are key players in the antiviral response. The importance of IFN-α is underscored by the presence of 13 IFN-α subtypes in humans. Type I IFNs are found in all mammals and similar molecules have been found in birds, reptiles, amphibians, and fish. Type I IFNs are secreted by most cell types. They are expressed in response to viral infection, are secreted from infected cells, and bind to IFN-α receptors on nearby cells thereby activating cell signaling cascades that result in the expression of about 300 or so IFN-stimulated genes (ISGs) (Fig. 6.4). Expression of ISGs leads to the so-called "antiviral state" of uninfected cells. ISGs have a wide variety of antiviral activities. The number and diverse antiviral activities of ISGs, as well as ability of viruses to circumvent them, underscore the truly ancient relationship between viruses and cells.

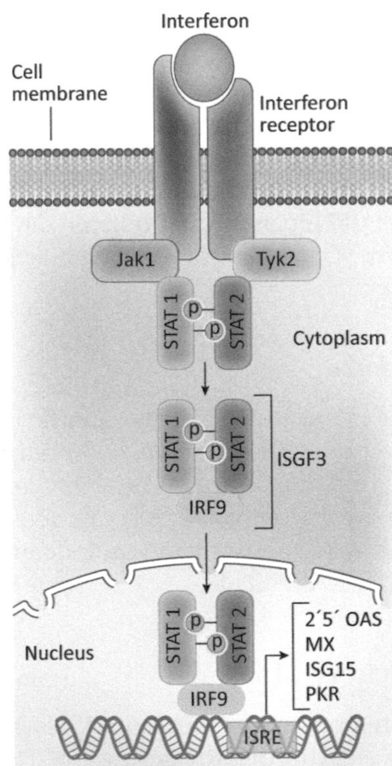

FIGURE 6.4 IFN signaling pathway. IFNs bind to receptors and signal through JAK STAT pathways. JAKS are found in association with IFN receptors. In the absence of a stimulus, JAK proteins are inactive. If IFN binds a receptor, the kinase domains of the JAKs phosphorylate the IFN receptor chains. This in turn activates STATs by exposing their nuclear localization signals. In the nucleus, STATs are transcriptional activators that drive expression of ISGs.

TYPE II IFN

IFN-γ is the only type II IFN. It signals through the IFN-γ receptor (IFNGR). IFN-γ production is largely restricted to cells of the immune system; however, the IFNGR is found on most cells. Therefore IFN-γ is produced by specialized cells, but can bind nearly all cell types to elicit a response. IFN-γ signaling plays a major role in establishing cellular immunity. One important role of IFN-γ is to regulate immunity and bridge the innate and specific immune response pathways. The patterns of gene expression induced by IFN-γ and IFN-α/β are similar. The genes induced by IFN-γ are a subset of the total induced by IFN-α/β. The IFN-γ transcriptional response generally is of lower magnitude than that of IFN-α/β. There is also cross-talk between Type I and Type II IFNs.

TYPE III IFNs

There are three types of IFN-λ (lambda) IFN-λ1, IFN-λ2, and IFN-λ3. These bind to high and low affinity receptors. The high affinity receptor (IFN-λ receptor 1) is expressed only in epithelial cells.

IFNs Bind to Receptors and Signal Through Janus Kinase Signal Transducers and Activators of Transcription Pathways

Janus kinases (JAKs) are expressed in all cells and they are key players in IFN signaling. They bind to various receptors on the inner side of the plasma membrane (PM). In the absence of a stimulus, the cytoplasmic domain of each IFN receptor chain is bound by a specific JAK protein in an inactive conformation. Upon IFN binding, the two JAK kinase domains are activated and phosphorylate the IFN receptor chains leading to the repositioning and binding of STAT (Signal Transducers and Activators of Transcription) proteins to the receptors. As a result, STATs are phosphorylated, released from the receptor, and dimerize. There are seven different STAT proteins but STAT1 and STAT2 are the most important when it comes to signaling through IFN receptors. STAT dimers have exposed nuclear localization signals. Once in the nucleus, STATs are transcriptional activators that drive expression of ISGs.

IFN-Stimulated Genes and the Antiviral State

There are over 300 ISGs and they code for proteins with a variety of activities. Some ISGs are regulators of IFN signaling. Others have specific or general antiviral effects. Cells responding to IFN binding are said to be in an "antiviral state" and they are refractory to infection by many viruses (not just the specific virus that that originally triggered the IFN response). Some types of viruses are more sensitive to the effects of IFNs than others. The following sections describe the activities of a few ISGs.

Protein Kinase RNA-Activated Pathway

The enzyme protein kinase RNA-activated (PKR) is produced in response to Type I IFNs. PKR remains in an inactive or latent state until, as its name suggests, it is activated (auto-phosphorylated) upon binding to viral dsRNA. Once active, PKR phosphorylates the eukaryotic translation initiation factor EIF2A. Phosphorylation of EIF2A inhibits mRNA translation, thereby preventing host and viral protein synthesis. Active PKR also induces cellular apoptosis, to prevent further viral spread.

2′-5′ Oligo A Synthase Pathway

Two ISGs are key to this pathway: ribonuclease L (RNaseL) and 2′-5′ oligo A synthase (OAS). RNaseL (for latent) is inactive when synthesized, but when activated destroys RNA within the cell, leading to autophagy and apoptosis. (Both of these processes are robust antiviral responses.) In order for RNAseL to become active, it must bind to polymers of 2′-5′-linked oligoadenylate (2′-5′ oligo A) and these are synthesized by OAS. However OAS is also produced in an inactive form and only becomes active upon binding to dsRNA. The requirement for a multistep pathway helps insure that RNaseL is not activated in the absence of viral infection. To summarize the pathway: (1) RNaseL and OAS are ISGs. (2) If a cell expressing these proteins is infected by an RNA virus, long dsRNA accumulates in the cytoplasm. (3) OAS is activated and synthesizes 2′-5′ oligo A. (4) RNaseL binds to 2′-5′oligo A and is activated. (5) Active RNaseL cleaves viral and cellular RNA.

Other Interferon Stimulated Genes

Tetherin is an ISG that inhibits a variety of enveloped viruses, including human immunodeficiency virus (HIV), Ebola virus, Lassa virus, vesicular stomatitis virus, and Nipah virus. As the name implies, tetherin inhibits release of enveloped viruses by tethering them to the cell surface. Tetherin is a small, glycosylated membrane protein (181 aa). It is found associated cholesterol-rich microdomains in the PM, preferential budding sites of several enveloped viruses. In polarized cells, tetherin is found specifically at the apical surface. The antiviral role of tetherin appears to be to physically crosslink virions to the PM. A number of viruses, including HIV and Ebola virus, have evolved mechanisms to counteract tetherin.

The *IFN-inducible transmembrane protein (IFITM) family of proteins* inhibits a long list of enveloped viruses including influenza A and B viruses, dengue virus, hepatitis C virus (HCV), and Ebola virus among others. IFITMs contain two membrane-associated domains separated by a conserved intracellular loop. They are found on cytoplasmic and endosomal membranes and block viral replication by preventing fusion of the viral envelope to the host membrane. Imaging studies suggest that IFITMs trap infecting virions leading to their destruction by lysosomes and autolysosomes.

How do IFITMs work to inhibit fusion of enveloped viruses? They may alter the physical properties of the membrane of the host cell. IFITMs seem to associate with proteins that reside in lipid rafts, a location favored by some enveloped viruses. The importance of IFITM-mediated virus restriction has been demonstrated in mice, where lack of IFITM3 increases susceptibility to influenza virus infection. A role for IFITM3 in human susceptibility to influenza virus infection has been suggested, based on evidence linking a minor human allele of IFITM3 with more severe disease.

Viperin (virus inhibitory protein, endoplasmic reticulum-associated, IFN-inducible, also known as RSAD2) is an ISG induced by Type I, Type II, and Type III IFNs. Viperin has broad antiviral activity against both RNA and DNA viruses including some herpesviruses, flaviviruses, retroviruses, orthomyxoviruses, paramyxoviruses, togaviruses, and rhabdoviruses. Viparin is expressed by a variety of animals, including mammals, fish, and reptiles. Viparin inhibits the release of some viruses but may have other means of inhibiting virus replication. At least one virus, human cytomegalovirus (CMV or human herpes virus 5 (HHV-5)) induces viparin expression to enhance viral infectivity.

Mx proteins are IFN inducible, large GTPases that belong to the *dynamin superfamily* of high molecular weight GTPases. (Dynamin in involved in membrane budding and is required for scission of nascent vesicles from parent membranes.) Mx proteins are expressed by most vertebrates and they commonly localize to the cytoplasm. An exception is mice, whose Mx proteins are nuclear and cytoplasmic. Mx proteins assemble to form large, highly ordered oligomers that associate with intracellular membranes. Humans encode two Mx proteins, MxA and MxB, however only MxA has antiviral activities. In mice the Mx1 protein confers significant resistance to influenza viruses. The antiviral activities of Mx proteins seem to arise from their ability to bind to viral nucleocapsids to inhibit early steps in virus replication. Mx proteins have a GTPase domain at the N-terminus, a central interactive domain in the middle, and an effector domain with leucine zipper motifs at the C-terminal end. Both the GTP binding and carboxy-terminal effector functions are required for its antiviral activities.

INTRINSIC IMMUNITY

Intrinsic immunity is a type of innate immunity. However the proteins that mediate intrinsic immunity usually restrict the replication of specific viruses. These viral restriction factors may not be IFN inducible and they do not activate signaling cascades. Their presence renders certain cells nonpermissive to certain viruses. Examples of intrinsic factors important in species-specific immunity to retroviruses are described below.

Restriction Factors

TRIM5α (tripartite motif-containing protein 5, alpha isoform) is a retrovirus restriction factor. It is a 493 amino acid E3 ubiquitin ligase expressed by cells of most primates. (Ubiquitin ligases direct ubiquitin conjugating enzymes to their target proteins. They assist in or directly catalyze transfer of uniquin to those targets. The human genome encodes numerous E3 ligases that interact with a diversity of substrates. Ubiquinated proteins are often targeted for destruction by the proteasome.) In humans TRIM5α is one of a cluster of eight TRIM genes on chromosome 11. There is evidence for positive selective pressure in the cluster and several TRIM genes have been identified as viral restriction factors. The species specificity of TRIMs is demonstrated by the following: HIV cannot replicate in African green monkeys or macaques and this restriction is conferred by TRIM5α. The restriction activity of TRIMs requires a direct interaction between the TRIM and a particular viral protein. Depending on the viral protein targeted, restriction can take place at different points in the virus replication pathway including entry, transcription, or release.

APOBEC3 or A3 proteins (apolipoprotein B mRNA editing enzyme catalytic polypeptide 3) catalyze the deamination of cytidine to uridine in single-stranded DNA. Primates encode 5–7 different A3 proteins (A3A to A3G), in a gene cluster on chromosome 20 (other mammals encode a single A3 protein). The antiretroviral activity of A3 proteins requires they be packaged into virions to be present during reverse transcription in the newly infected cell. Thus A3 proteins do not interfere with virus production in an infected cell, but the released progeny viruses will be restricted in their ability to establish productive infections of new cells because of the mutagenic properties of A3 proteins. Retroviruses can avoid the activity of A3 proteins by excluding them from virions. For example, the murine retrovirus murine leukemia virus (MLV) is not restricted by mouse A3 because it fails to interact with MLV proteins and is not packaged. Primate retroviruses, such as HIV-1, use a different method to exclude A3 proteins from virions. The best-studied examples are the interactions between A3G proteins and the VIF proteins encoded by primate lentiviruses (see Chapter 36: Family *Retroviridae*). HIV-1 VIF interacts with human A3G (hA3G), triggering its ubiquitination and proteosomal degradation, therefore HIV-1 replication is not blocked by hA3G (Chapter 37: Replication and Pathogenesis of Human Immunodeficiency Virus (HIV)). However the VIF proteins of most simian immunodeficiency viruses (SIVs) do not target hA3G for degradation. Therefore the SIVs are, for the most part, unable to replicate in human cells. (One of the adaptations allowing HIV to replicate to high levels in humans is its ability to inactivate hA3G.) There are a variety of SIVs, each adapted to a particular species of monkey and for the most part an SIV VIF inactivates the A3G of its "natural host," but may fail to inactivate A3G of other species of monkeys.

SAMHD1 (sterile alpha motif and histidine-aspartic domain (HD) containing protein 1) is a host restriction factor that limits retroviral replication in certain cell types. SAMHD1 is a deoxynucleoside triphosphate phosphohydrolase that degrades the intracellular pool of deoxynucleoside triphosphates available during early reverse transcription. It seems to be particularly effective in limiting retrovirus replication in monocyte-derived macrophages and dentritic cells. Retroviruses that can infect these cell types have evolved mechanisms to block SAMHD1 activity. For example, the VPX accessory protein of HIV-2 and SIVsm (Chapter 37: Replication and Pathogenesis of Human Immunodeficiency Virus (HIV)) targets SAMHD1 for proteosomal degradation.

Autophagy

The autophagosome is a vesicle with a double membrane (Box 6.1). In normal cells the autophagosome contains cellular components targeted for degradation. The contents of the autophagosome are degraded when it fuses to a lysosome. During viral

BOX 6.1

AUTOPHAGY

Autophagy is a complex and highly regulated process whereby cellular components (for example, protein aggregates or damaged organelles) or infectious agents (viruses) are engulfed by double-membrane vesicles in the cytosol. These vesicles, called autophagosomes, then fuse with lysosomes leading to degradation of their contents and providing building blocks for the cell. Not surprisingly, a general trigger for autophagy is nutrient deprivation.

The selective removal of intracellular infectious agents by autophagy is called xenophagy and the process has been shown to restrict virus titers. Autophagy also provides peptides for presentation to the adaptive immune system. In addition autophagy controls inflammasomes triggered by innate immunity signaling cascades, possibly preventing excessive immune activation. There are more than 30 autophagy-related gene products that control and drive these diverse processes.

Viral pathogens manipulate autophagy as an immune escape mechanism. Some viruses have been shown to interfere with the formation of autophagosomes while others inhibit their fusion with late endosomes or lysosomes. There are also examples of viruses coopting autophagosomal membranes for use as platforms for genome replication and to facilitate their release from infected cells. For example, picornaviruses and flaviviruses seem to rely on autophagosomal membranes for optimum replication and release from cells.

infection, autophagy can act as a surveillance mechanism that delivers viral antigens to endosomal/lysosomal compartments enriched in immune sensing molecules. Activated immune sensors can signal to activate autophagy. To evade this antiviral activity, many viruses actively block the autophagy pathway. Alternatively, some viruses subvert autophagy for their own benefit. Manipulated autophagy has been proposed to facilitate nearly every stage of the viral lifecycle in direct and indirect ways.

Natural Killer Cell Responses to Viral Infection

Natural killer (NK) cells are large granular, cytotoxic lymphocytes that, as part of the innate immune system, play a key role in host defense against viral infections. NK cells provide rapid responses to virally-infected cells, acting at around 3 days postinfection. NK cells may recognize viral antigens directly and may kill some virus-infected cells. The importance of NK cells in controlling virus infection is underscored by the ability of some viruses to alter NK function.

SPECIFIC IMMUNITY

Specific immunity refers to pathogen-specific immune responses that get better with time and experience (practice makes perfect). Two major types of specific immunity are antibody responses (humoral immunity) and cell-mediated immunity (CMI). These processes involve the activities of T and B lymphocytes. As summarized in Fig. 6.5, the general steps in activating specific immune responses are:

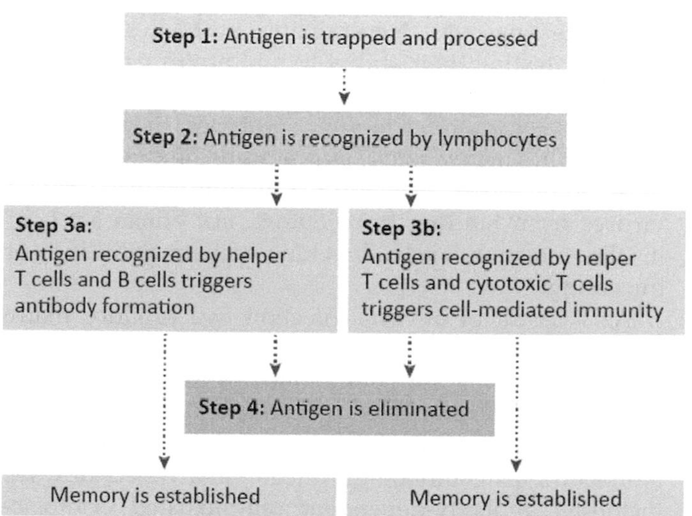

FIGURE 6.5 Overall steps in development of specific immunity.

- Antigen is trapped and processed.
- Antigen is recognized by T and B lymphocytes.
- Antigen is eliminated.
- Memory is established due to the presence of long-lived T and B lymphocytes.

Antibody Responses

Antibodies are produced by B lymphocytes in response to foreign antigens. Antibodies are found in the blood, lymph, and mucosal secretions. Their role is

Class I proteins

- Found on the surface of most mammalian cells
- Presented peptides derived from intracellular proteins
- Presented peptides represent a sampling of all the proteins synthesized in the cell (including viral proteins)

Class II proteins

- MHC Class II
  - B-cells
  - Macrophages
  - Dendritic cells
  - Activated T-cells
- Presented peptides are from extracellular proteins (that are taken up by cells)

FIGURE 6.6 MHC proteins bind and present both "self" and "foreign" peptides.

to bind to foreign antigens, either inactivating them directly, or tagging them for removal by complement or phagocytic cells (i.e., macrophages). From a protective standpoint, the most important antiviral antibodies are those that neutralize viruses, preventing them from infecting cells. Neutralizing antibodies are measured by testing their ability to inhibit virus replication in cultured cells or animals. It should be noted that when virologists refer to viral serotypes, they are often referring to neutralizing, not simply binding, antibodies. Some types of neutralizing antibodies prevent viruses from binding to receptors, but others act later in the infection cycle, blocking penetration, and/or uncoating.

As a result of a virus infection, we produce many antibodies that bind to, but do not neutralize viruses. Detecting binding antiviral antibodies is very important for viral diagnostics. However nonneutralizing antibodies also have a role in controlling a virus infection. Nonneutralizing antibodies function in a variety of ways including antibody-dependent cell-mediated cytotoxicity, complement-mediated cytotoxicity, and antibody-dependent cell-mediated phagocytosis. These mechanisms can result in lysis or phagocytosis of virions. The Fc portions of immunoglobulins are key to these mechanisms as they are recognized by immune cells (NK cells, macrophage, neutrophils, and eosinophils) and proteins (such as complement proteins) thereby targeting these effectors to virions or virus-infected cells.

Cell-Mediated Responses

Cell-mediated responses (CMI) are primarily directed at intracellular infectious agents, and as such are very important for control of virus replication. Two types of immune cells are involved in CMI. Cytotoxic T lymphocytes (CTL) recognize and kill virus-infected cells or can trigger apoptosis. CTL may also damage infected tissues and organs. (Another name for CTLs are $CD8^+$ T cells.) The other players in CMI are activated macrophages. Activated macrophage may degrade virus intracellularly. They are important in sites such as the lungs.

Antigen Presentation

Antigen presentation is central to specific immunity. Extracellular antigens can bind to professional antigen presenting cells (APCs) (macrophage, dendritic cells, and B cells). Viral (or other) antigens produced *inside* of cells are proteolytically processed and are presented on the surface of the cell. Antigens are presented by a set of cell surface proteins called major histocompatibility (MHC) proteins (Fig. 6.6). Their main function is to bind peptide fragments and display them on the cell surface for recognition by the appropriate T cells. All cells produce MHC class I molecules and these bind to fragments of cell proteins (self) or peptides derived from intracellular pathogens (nonself). APCs express MHC class II proteins.

An APC takes up extracellular antigen, processes it, and returns small peptides to the cell surface, bound to an MHC class II protein. Cells that *present* foreign antigens in the context of MHC can be recognized by antigen-sensitive cells such as $CD8^+$ and $CD4^+$ T cells. Cytotoxic $CD8^+$ T cells recognize peptides on MHC class I molecules and $CD4^+$ T cells recognize peptides binding to MHC class II molecules.

VIRUSES FIGHT BACK

The previous sections provide a glimpse into the myriad immune responses to infectious agents. The complex mix of immune responses has been driven by the evolution of infectious agents to counteract host defenses! In the previous sections we saw that how some retroviruses have evolved specific proteins

BOX 6.2

ESCAPE FROM NEUTRALIZING ANTIBODIES

Neutralizing antibodies are an important specific defense against viral invaders. Neutralizing antibodies not only to bind to a virus, they bind in a manner that blocks infection. A neutralizing antibody might block interactions with the receptor, or might bind to a viral capsid in a manner that inhibits uncoating of the genome. Only a small subset of the many antibodies that bind a virus are capable of neutralization. After an infection, it can take some time for the host to produce highly effective neutralizing antibodies but these persist to protect against future encounters with the agent.

It is not surprising that some viruses have developed mechanisms to avoid antibody-mediated neutralization. For example, one obvious target for neutralizing antibodies are viral attachment proteins. If attachment proteins are bound to antibody, they cannot interact with the cell receptor. However some viruses "hide" the attachment domains of their surface proteins. The key interacting amino acids may be within folds or clefts in the attachment protein and thereby remain inaccessible to antibody. This is the case with the SU protein of HIV and the HA protein of influenza virus. Many picornaviruses also hide their attachment domains in deep canyons or pits on the surface of their capsids. With their attachment domains hidden, other regions of attachment proteins become targets for antibody, some of it neutralizing. Often the targets for neutralizing antibodies are surface loops or projections. These regions are not only good targets for antibodies, they are regions that can tolerate a variety of amino acid replacements. Thus neutralizing antibodies developed in response to a natural infection or vaccination can be avoided by mutation of one or a few amino acids. In the case of HIV, a persistent infection, as protective neutralizing antibodies are produced, mutated viruses that are not recognized by those antibodies are favored. In the case of influenza virus, an acute infection, escape mutants arise as the virus moves through a large population. Thus next season's influenza virus may escape neutralization by last year's vaccine. Viruses can also escape from the effects of cellular responses by mutating T-cell epitopes.

HIV uses another mechanism to avoid neutralizing antibodies. Its surface protein, SU, is heavily glycosylated. About 50% of the total mass of HIV SU is carbohydrate. These carbohydrate chains are synthesized by the infecting cell, thus are not readily seen as "foreign." The carbohydrate surrounds the more antigenic peptide backbone, shielding it from antibody interactions.

to counteract specific retroviral restriction factors. However there are also general strategies, used by multiple viruses, to counteract immune responses.

Antigenic Variation

Some viruses (for example, influenza virus and HIV) defend against specific immune responses by mutation (Box 6.2). The mutated viruses display proteins that are not recognized by previously primed immune responses. For example, neutralizing antibodies may no longer neutralize the mutated viruses. Viruses may avoid preexisting immune responses by accumulating point mutations or by reassorting or recombining gene segments.

Suppression of Antigen Presentation

Many viruses interfere with the ability of a cell to present antigens, thereby avoiding detection. Recall that antigen presentation takes place in the context of MHC proteins. There are multiple steps in antigen presentation pathways and many are targeted by viruses (Fig. 6.7).

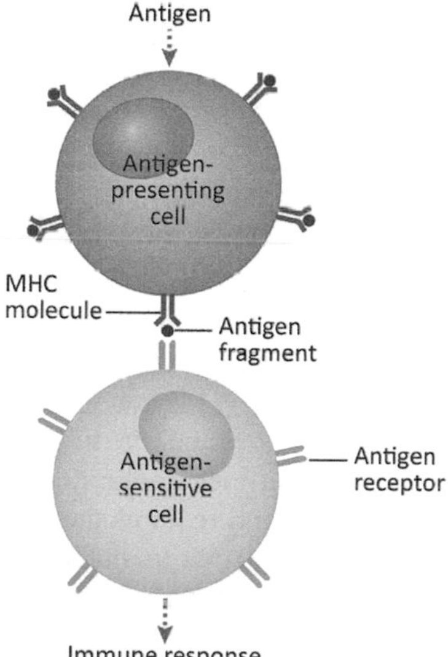

FIGURE 6.7 A general model of antigen presentation. In this figure the APC (top) might be a professional APC (dendritic cell, macrophage, or B-cell) or an infected fibroblast or epithelial cell. The antigen-sensitive cell (bottom) could be a T or B lymphocyte that becomes activates as a result of the interaction.

Suppression of IFN Responses

Some viruses induce strong IFN responses while other suppress IFN synthesis by a variety of mechanisms. A "global" mechanism to derail the antiviral effects of IFN is the ability of some viruses to inhibit cellular gene expression. Other viruses inactivate particular players in IFN signaling pathways.

Suppression or Coopting of Apoptosis and Autophagy

Apoptosis (programmed cell death) and autophagy (degradation of cellular components) are two responses that can inhibit virus replication. Apoptosis has an antiviral effect when cell death limits virus production. Autophagy can likewise interfere with virus replication, limiting the amount of virus produced by the infected cells. Thus it is not surprising that viruses have developed diverse mechanisms to suppress apoptosis and autophagy. Some viruses subvert these defenses for their own benefit.

Immunosuppression

The host cells for some viruses are lymphocytes. Destruction of lymphocytes leads to global immunosuppression that may persist long after the infecting virus has been eliminated.

Poxviruses Encode Molecules That Mimic Cytokines, IFNs, and Their Receptors

Poxviruses are large DNA viruses that encode proteins that mimic host defense molecules such as cytokines or their receptors. For example, poxviruses encode viroceptors that bind (and inactivate) host cytokines, preventing them from activating their true receptors. Virokines are viral cytokines that bind to receptors such that they can no longer response to the cell's cytokines.

Hiding From the Immune System

Some viruses replicate in tissue/organs that are not readily targeted by the immune system, for example, the brain. Others hide from immune recognition by entering latency. During latency genomes are maintained within cells but no (or very few) viral proteins are expressed. After years or decades of latency, the virus may emerge (replicate) at times when the immune system is compromised (by advancing age, stress, other infections). A few viruses hide from the immune system by infecting a fetus, thereby allowing the viral antigens to be seen as self-antigens. An example is bovine viral diarrhea virus. Persistently infected bovids may shed copious amounts of virus and are antibody negative.

Finally, micro-RNAs (miRNAs) are short (21–23 nt) host-encoded RNAs that bind to complementary sequences on mRNAs thereby blocking translation or inducing mRNA degradation. Some viruses have coopted cellular miRNAs to aid in their replication (Box 6.3).

In this chapter we have learned that

- Innate immune responses are rapid responses to groups of pathogens. They are initiated by detection of PAMPs such as viral nucleic acids. Groups of PRRs include TLRs, RLRs, and NLRs. Pattern-

BOX 6.3

MICRO-RNAs AND VIRAL INFECTION

Micro-RNAs (miRNAs) are short (21–23 nt) host-encoded RNAs that bind to complementary sequences on mRNAs thereby blocking translation or inducing mRNA degradation. Thus miRNAs play an important role in regulating gene expression. Some viruses have coopted cellular miRNAs to aid in their replication. HCV replicates in human hepatocytes and requires the abundant, hepatocyte specific miRNA miR-122 to stabilize viral RNA and enhance replication. Other viruses, notably the herpesviruses, encode their own miRNAs; they target both host and viral mRNAs. By targeting host mRNAs, herpesviral miRNAs interfere with processes such as immune response, cell cycle control, and intracellular trafficking.

In vertebrates, miRNAs do not appear to play a major role in restricting virus replication. This is in contrast to *plants*, *nematodes*, and *arthropods* that utilize small RNAs as a major mechanism to inhibit virus replication. These small *interfering* RNAs are the major players in antiviral responses in these groups of organisms. They function by targeting viral mRNAs for cleavage by a process called antiviral RNA interference. That is not to say that miRNAs have no role in antiviral defenses in vertebrates, but the major antiviral players are most certainly the proteins expressed by ISGs.

recognition initiates signaling pathways, many of which converge at common points.

- Innate responses to viruses include production of IFNs and proinflammatory cytokines.
- IFNs are released from virally infected cells and bind to receptors on other cells to initiate the antiviral state. The antiviral state results from synthesis of more than 300 ISG products such as PKR, OAS, tetherin, viperin, and Mx proteins.
- Some antiviral proteins (TRIM5α, APOBEC3G, and SAMHD1) are very species specific as regards host and virus. They serve as specific examples of host–virus coevolution.
- Specific immune responses depend on antibodies and CTL to target specific pathogens. Specific responses are primed and aided by innate responses.
- Specific responses depend on processing and presentation of antigens by professional APCs.
- Neutralizing antibodies against a virus are able to block infection.
- MHC proteins present peptide antigens on the surfaces of cells. MHC class I proteins are produced by all cells. MHC class II proteins are produced by APC.
- Viruses counteract host defenses of all types.
- Some viruses depend on host processes such as autophagy to support their replication.

CHAPTER

7

Viral Vaccines

OUTLINE

After reading this chapter, you should be able to answer the following questions:

- What is the purpose of vaccination?
- When considering vaccine development, what are "correlates of protection" and why is it helpful to determine them?
- What are the advantages of a killed/inactivated vaccine? What are the disadvantages?
- What are the advantages of a replication competent viral vaccine? What are the disadvantages?

The purpose of a vaccine is to provide protection against a pathogen (in our case a viral pathogen). Early attempts at "vaccine" development were based on the observation that persons surviving a disease, for example, smallpox, were protected against further occurrences of that particular disease. Today we know that immune protection against a specific viral infection is provided by "specialized" (also called "adaptive") responses that inhibit virus replication or kill infected cells. The goal of vaccination is to stimulate the body to develop a specialized, protective, immune response in the absence of disease. Immune responses target one or more viral proteins and there are a variety of ways that viral proteins can be delivered as vaccines. These include administration of:

- Weakened (so-called attenuated) viruses that replicate in the host without causing disease.
- Inactivated (killed) virions.
- Purified protein products.
- Nucleic acids (genes) to direct synthesis of desired proteins.

CLASSICAL VERSUS ENGINEERED VACCINES

Traditional or classical methods to produce a viral vaccine required growing large quantities of virus. Attenuated viruses were usually obtained by repeated passage in animals or cultured cells; unfortunately the results were impossible to predict and some viruses were never adequately attenuated even after prolonged passage (Fig. 7.1).

Today recombinant DNA technologies, along with our understanding of the molecular details of virus replication and pathogenesis, provide an array of opportunities to achieve vaccine goals. Recombinant DNA technologies can be used to make targeted mutations in a virus or to clone and express specific viral proteins. As shown in Fig. 7.2, examples recombinant DNA strategies for vaccine development include:

- Virulent viruses can be weakened by deletion or modifications of *specific* genes.
- Specific genes from a virulent virus (often genes for capsid or envelope proteins) can be transferred to a

Viruses. DOI: http://dx.doi.org/10.1016/B978-0-12-803109-4.00007-6

safer virus. The recombinant virus might be administered as a replicating virus or might be grown and inactivated.

- Specific genes from a virulent virus might be moved into bacteria, yeast, or cultured cells to produce large quantities of a particular viral protein.
- DNA molecules encoding specific proteins, or parts of proteins might be generated and administered directly to the patient.

In addition, the use of adjuvants (substances that modify the effects of other agents) provides even more flexibility, as there are a wide variety of adjuvants that can enhance or modify host immune responses.

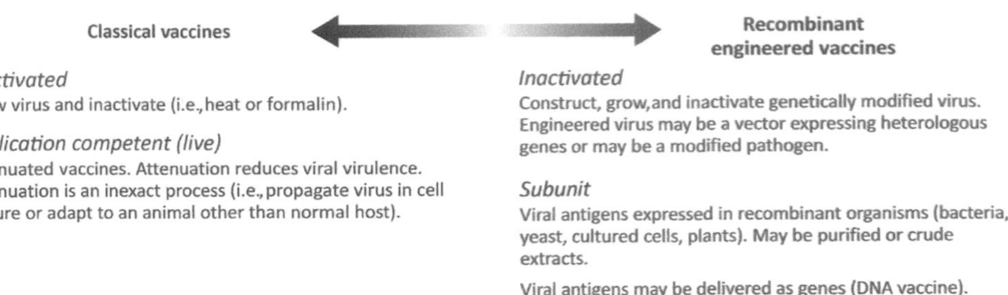

FIGURE 7.1 General types of viral vaccines.

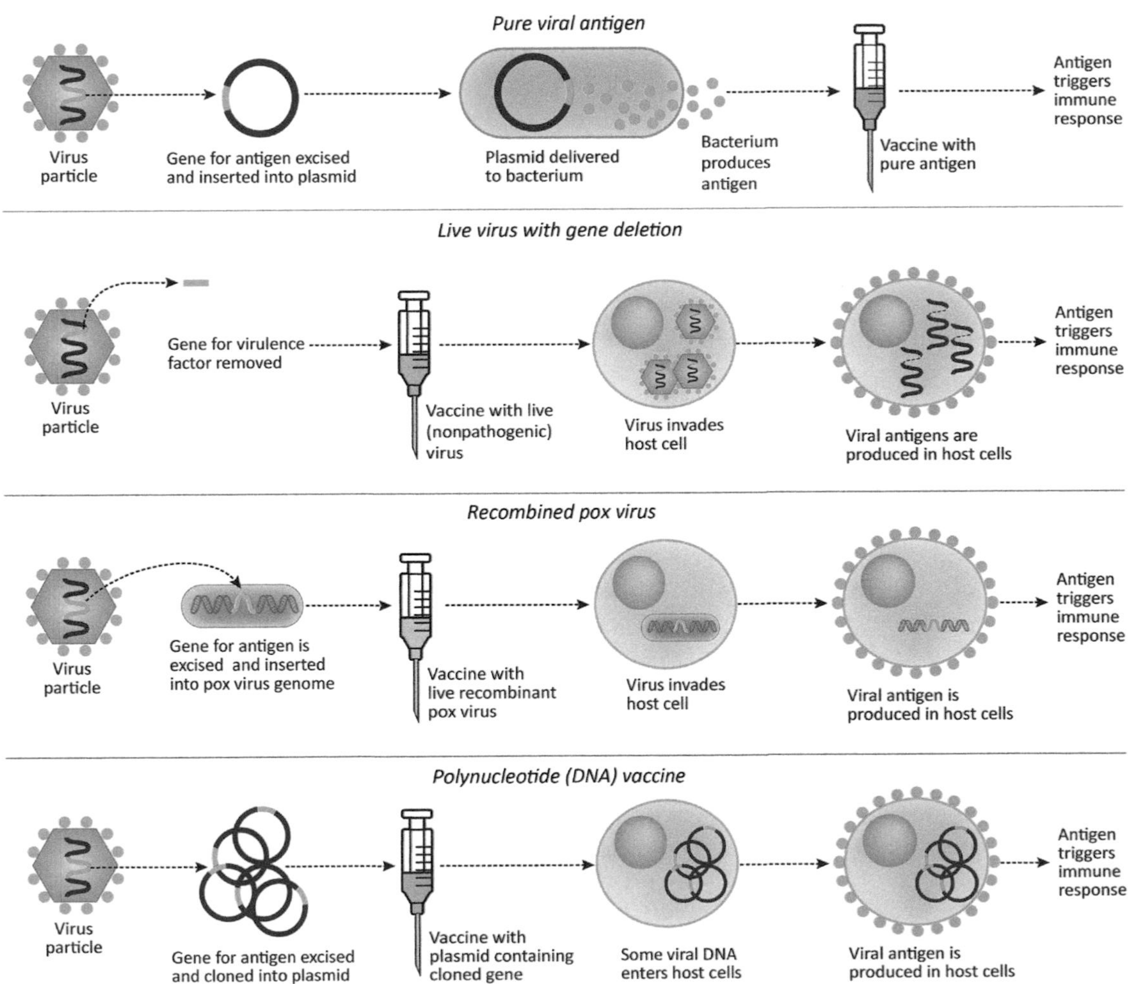

FIGURE 7.2 A variety of methods can be used to produce genetically engineered vaccines.

VACCINES PROVIDE A SOURCE OF ANTIGENS

The purpose of a vaccine is to provide a safe source of antigen, representative of a particular pathogen. ("Antigens" are simply the proteins of a pathogen that elicit immune responses.) Some viral vaccines contain specific protein antigens, but no virus capable of replication. Other viral vaccines contained weakened or attenuated viruses that *must* replicate in the host in order to induce a protective response. Recall that specific immune responses are pathogen-specific and get stronger with repeated exposures. Specific immune responses result from the interaction of T- and B-cells with antigens. Most often the viral antigens presented in a vaccine are proteins. To be effective a vaccine must provide antigens in a form that can be recognized by the immune system. A small, unfolded peptide might be sufficient, or complex protein structures (similar to a viral capsid) might be needed to elicit protection. Hence vaccines are not "one size fits all."

REPLICATING VERSUS INACTIVATED VACCINES

Whether a vaccine is derived by classical techniques or genetic engineering, a key consideration is whether or not the vaccine contains a source of infectious material (a virus that will replicate in the recipient). A summary of the relative advantages of replicating versus inactivated vaccines is provided in Fig. 7.3.

KILLED, INACTIVATED, AND SUBUNIT VACCINES

Often the goal of killed, inactivated, or subunit vaccines is to stimulate the production of neutralizing antibodies for the purpose of blocking infection. The term subunit vaccine is used to describe a vaccine that has one (or a few) viral antigens; the viral proteins of interest might be produced from cloned genetic material. Viral genes are introduced into bacteria, yeast, plants, or cultured cells where the desired protein is produced. The protein product is purified and administered. Another method to achieve the same ends is to administer DNA (cloned genes) directly into the recipient. Some of the DNA will enter cells and the desired proteins will be synthesized. In reality it is very difficult to administer enough DNA to allow for expression of ample quantities of the desired immunogen.

Examples of inactivated/subunit vaccines licensed for human use include inactivated poliovirus vaccines (such as the poliovirus vaccine developed by Jonas Salk), hepatitis B virus vaccine (HBV surface (S) protein produced in yeast cells) and papillomavirus (PV) vaccines (recombinant vaccines consisting of virus-like particles assembled from PV capsid proteins). Many inactivated vaccines are highly effective, but one consideration is that inactivated (killed) virus preparations, or subunit vaccines present a limited number antigens to the immune system. Recall that while a virus particle may be assembled from a few types of capsid/envelope proteins, there are *many* additional viral proteins produced in the infected cell (for example, the nonstructural proteins required for genome replication and/or subversion of immune responses). These nonstructural proteins are not present in inactivated virions. Of course the products of such genes can be produced by recombinant DNA techniques and administered as subunit vaccines.

Inactivated/subunit vaccines typically require administration of a "large" dose of antigen. Manufacture and purification of the proteins may be relatively expensive. Inactivated/subunit vaccines are typically administered by injection. A variety of adjuvants might be included in the vaccine to increase immunogenicity and to protect the vaccine components from being degraded too

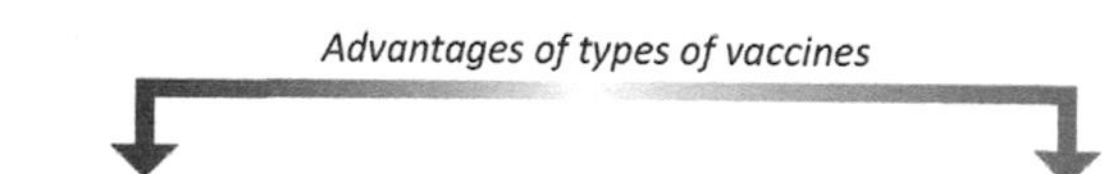

Living vaccines

- Fewer doses required
- Adjuvants unnecessary
- Smaller doses needed
- Can be given by natural route
- Stimulate humoral and cell-mediated responses
- Induce inteferon
- Protection lasts longer
- Less chance of hypersensitivity
- Relatively cheap to produce

Inactivated vaccines

- Do not replicate in recipient
- Stable on storage
- Unlikely to cause disease through residual virulence
- Unlikely to contain live contaminating organisms
- Will not spread to other people/animals
- Lower cost to develop

FIGURE 7.3 Relative advantages of killed and attenuated vaccines.

rapidly. Adjuvants are substances that modify the effect of other agents, such as the immunogens found in vaccines. One common use of adjuvants is to reduce the amount of antigen required to elicit an immune response. Adjuvants may also be used to direct the immune response in a particular manner, for example, by specifically activating cytotoxic T-cells versus antibody-secreting B cells.

In some cases, a viral pathogen is so virulent that any risk of improper inactivation is considered unacceptable. An example is rabies virus. Although traditional rabies vaccines used inactivated rabies viruses, modern vaccines are derived from a vaccinia virus vector expressing the rabies virus glycoprotein. The recombinant viruses are grown and then inactivated. Even if not properly inactivated, the recombinant vaccinia virus is relatively safe and certainly cannot cause rabies in the vaccine recipient.

VIRAL VACCINES THAT CONTAIN INFECTIOUS AGENTS

Viral vaccines containing infectious agents can be developed by classical techniques (attenuation by repeated passage in cultured cells) or by genetic engineering. The key point is that they contain viruses capable of *replicating* in the vaccine recipient. The "perfect" attenuated virus replicates enough to stimulate protective responses but not enough to cause harm. Production of an attenuated viral vaccine by classical methods requires that the virus be grown in cultured cells, or more rarely, in an animal. Some genetically engineered attenuated vaccines use a virus that has been specifically modified to reduce virulence. What sets these modern vaccines apart from classically attenuated viral vaccines is that *specific* genes are modified in a manner to prevent any possibility that the virus can revert to virulence.

Alternatively, a genetically engineered vaccine may contain a nonpathogenic virus that expresses proteins from a pathogen. For example, nonpathogenic poxviruses such as vaccinia virus, have been modified to produce the rabies virus envelope glycoprotein. The recombinant viruses replicate to a limited extent in vaccine recipients, but produce enough rabies glycoprotein to elicit a protective response. These vaccines are used for wild animals (raccoon, skunk, coyote) and are delivered in baits designed to be eaten. It should be noted that vaccine manufacturers often produce different types of vaccines for a particular agent. For example, the vaccine manufacturer Merial (Atlanta, GA, USA) produces a live recombinant vaccinia-virus-based rabies vaccine for use in wild animals but produces an inactivated recombinant vaccinia-virus-based rabies vaccines for pets and livestock. It should also be noted that a virus attenuated for one species may not be equally attenuated for all species. This is of particular importance in the practice of veterinary medicine.

ANTIGEN RECOGNITION

The purpose of vaccination is to induce protective, specialized immune responses to a pathogen. Specialized immunity is achieved by selective proliferation of T- and B-lymphocytes that recognize specific foreign antigens (these are often, but not always, proteins). Basic steps in the process were outlined in Chapter 6, Immunity and Resistance to Viruses. Antigen is "presented" to the appropriate lymphocyte populations (T- and B-cells) by "professional antigen presenting cells" such as macrophage or dendritic cells. Lymphocytes (via protein receptors on their surface) recognize (bind) antigen and are stimulated to proliferate. Development of a B-cell response results in production of antibodies that bind extracellular antigens. T-cells recognize antigens that have been processed and presented from within a cell. These processed antigens are short peptides bound to specific cellular proteins (MHC Class I or Class II proteins, see Chapter 6: Immunity and Resistance to Viruses) that carry the peptides to be displayed on the cell's surface.

The purpose of recognizing a foreign antigen is to allow its removal by the various players of the immune system. Binding of a pathogen by antibodies may render it available to phagocytes or may target it for destruction by complement-mediated lysis. In the case of viruses, the most protective antibodies are those that neutralize or block virus infection. These so-called neutralizing antibodies might: (1) prevent a virus from binding to a receptor; (2) prevent membrane fusion; or (3) prevent uncoating of the viral genome. Assays for the presence of neutralizing antibodies are relatively simple. If the virus under consideration replicates in cultured cells, it is mixed with antiserum that might contain neutralizing antibodies. (Typically, serial dilutions (2-fold or 10-fold) of the antiserum are mixed with a standardized virus sample.) The virus is then added to permissive cells and virus replication is monitored. The presence of neutralizing antibodies is indicated by a reduction in the number of cells that are successfully infected by the virus (a reduction in the number of plaques produced). There are many antiviral antibodies that bind, but do not neutralize, a virus. These might enhance phagocytosis or may promote complement-mediated lysis of the virus. In rare cases (for example, dengue virus, Chapter 15: Family *Flaviviridae*) binding antibodies may enhance or exacerbate infection and disease.

Cytotoxic T-lymphocytes recognize and kill virally infected cells thereby limiting the amount of virus produced. Recall that the T-cell response depends on recognition of peptides bound to MHC proteins and presented on the cell's surface. In the case a viral infection the cell will process (digest by proteolysis) some viral proteins as soon as they are synthesized, signaling to surveying T-cells that the cell is infected.

CORRELATES OF PROTECTION

An important consideration for vaccine development is to determine the type(s) of response that is (are) protective. Protection may require neutralizing antibodies, virus-specific cytotoxic T-cells, or both. Protection may require recognition of one or more viral proteins, or recognition of a specific epitope on a target protein. An obvious way to determine if a vaccine is protective is to compare the response of vaccinated versus nonvaccinated animals to a viral challenge. (We anticipate that the unvaccinated animals will get sick but the vaccinated animals will remain healthy.) This direct approach is the "gold standard" but uses a lot of animals and can be quite costly. Thus during the vaccine development process, it is often helpful to determine the specific type and magnitude of response that confers protection. Or in other words, what responses (both type of response and magnitude of response) *correlate* with the desired outcome: protection. For example, a specific titer of neutralizing antibody may serve to protect against infection by a particular viral pathogen. Various combinations of antigens, doses, delivery strategies, and adjuvants can then be administered, followed by measurement of neutralizing antibody titers. Only those preparations that induce adequate levels of neutralizing antibody need be tested further. Another advantage of understanding correlates of protection is that vaccine recipients can be tested to determine their status (protected or not protected). For example, in the case of rabies vaccines, protection is well correlated with high titers of antibody to the G protein. Among veterinarians or other "at risk" persons, antibody titers can be measured to provide information about the need for a booster dose (Box 7.1).

In the case of natural infections (for example, Ebola virus) we may examine survivors to determine the types of specific responses that correlate with survival. We can then design vaccines that mimic the natural, protective response (Box 7.2). But what happens when natural infection is not protective? When our immune system is unable to clear or control the virus after a natural infection? The best (worst?) example of this is infection with the human immunodeficiency virus (HIV). An untreated HIV infection is almost universally fatal, despite the fact that patients do mount an immune response. The inability for to control infection is due, in part, to the fact that HIV attacks the very immune cells we need to sustain a protective response. HIV also mutates rapidly and its surface proteins are designed to evade host responses. In short, there is little/no natural *protective* response, thus there are no well-defined correlates of protection to serve as the standard for vaccine design. (Worldwide, a very few persons have been identified who seem to be naturally resistant to HIV. They are an important resource in the fight against HIV!)

Vaccines to protect against influenza virus infection have been used for many decades, but we still face

BOX 7.1

RABIES VACCINE SUCCESS IN 1885

Louis Pasteur was a talented and innovative microbiologist. His work (among others) was key to understanding that diseases are caused by infectious agents that can be identified and grown in the laboratory. Pasteur initially developed and tested animal vaccines for the bacterial diseases anthrax (*Bacillus anthracis*) and fowl cholera (*Pasteurella multocida*). His studies of rabies were more challenging as he could not grow agent on an agar plate. Instead Pasteur used monkeys and rabbits to grow and attenuate the virus. He successfully used dried spinal cord material from infected rabbits to protect against rabies in dogs. He even determined that the vaccine could be administered to dogs *after* they were exposed to the virus. In 1885 Pasteur was asked to treat Joseph Meister, a 9 years old who had been bitten multiple times by a rabid dog. No one doubted that without treatment the boy would die from rabies. Pasteur treated Joseph Meister with injections of rabbit spinal cord material and the boy survived. This was such a celebrated and important event that people from around the world made contributions support Pasteur's work. Thus the Pasteur institute, dedicated to the study of biology, microorganisms, diseases, and vaccines, opened in 1888. To this day the institute continues to support cutting edge biomedical research.

BOX 7.2

HBV VACCINE: FIRST SUBUNIT VACCINE, FIRST RECOMBINANT VACCINE

Neither the discovery of the hepatitis B virus (HBV) nor the development of a protective vaccine followed predictable paths. Hepatitis is an infection of the liver that can be caused by a variety of infectious agents. By the mid-20th century, hepatitis B or serum hepatitis was recognized as a significant health problem, leading to increased risk of hepatocellular carcinoma. It was assumed that hepatitis B was caused by a viral infection, and well known that blood transfusions greatly increased the risk of infection. But efforts to culture the virus were unsuccessful for many years. A major breakthrough occurred in 1965 when Dr. Baruch Blumberg (who was not studying hepatitis B!) identified an antigen in a blood sample that was eventually identified as a HBV protein. Following this discovery, plasma from chronically infected patients was used a source of antigen for developing a blood test (1971), and eventually for developing a successful vaccine (licensed in the United States in 1981). Vaccine production was possible because some patients have large quantities of the hepatitis B surface (S) antigen that can be harvested from their plasma. Heat treatment of the product inactivates infectious virus, leaving the S antigen intact and highly immunogenic. The vaccine was very effective, but use of human plasma as starting material was not without risk and not without its detractors. But the success of the plasma-derived vaccine indicated that whole virus was not needed to induce protection. Rather, protection was provided by immune responses that specifically targeted the S antigen. This knowledge dovetailed beautifully with new recombinant DNA technologies (gene cloning and protein expression). By 1986 an HBV recombinant S antigen vaccine (a subunit vaccine) had been developed, tested, and approved. During HBV vaccine development, another important discovery was made: In order to induce a protective response, the S antigen had to be correctly folded and presented to immune system in a manner that mimicked the antigen produced during a natural infection. The recombinant vaccine product quickly replaced the human plasma-derived product in the United States, due to its superior safety.

vaccine challenges. Most human influenza virus infections are not fatal and infection with one particular strain of virus protects against future infection *with that strain*. However we remain susceptible to other strains of influenza virus, and there are many of these (see Chapter 23: Family *Orthomyxoviridae*) as influenza viruses mutate readily. Thus new vaccines must be manufactured each year, in an attempt to keep one step ahead of this viral pathogen. The "perfect" influenza virus vaccine would provide protection against many different strains. It remains to be seen when, or if, such a vaccine will be developed.

VACCINE DEVELOPMENT

Vaccine development can be complex. There are many factors that must be considered. Some of these are:

- Safety (Does the vaccine produce major/minor side effects? Is there risk of reversion?).
- Efficacy (How well does the vaccine work? How many vaccinated individuals are resistant to infection/disease?).
- Development costs (How much time/effort/money is needed to produce a vaccine candidate?).
- Manufacturing costs (How much money does it cost to produce each dose of vaccine?).
- Costs of transport and storage (Must the vaccine be refrigerated?).
- Ease of administration (An injectable vaccine requires adequate supply of sterile needles.).
- Number of doses required to achieve protection.
- Duration of immunity (How often do we need booster doses?).

A FINAL WORD ABOUT SAFETY

Vaccine safety is very important. In particular, the risk of vaccine-related adverse events (complications) versus risk of natural infection must be considered. If the risk (or perceived risk) of natural infection is extremely low, relatively minor side effects such as fever, headache, or a sore arm may be unacceptable. Complicating this is the occurrence of random events that happen to coincide with administration of millions of doses of vaccine. It is intellectually and emotionally difficult not to blame the vaccine if we become ill soon after having an injection, although there may be no connection at all. It has been said that vaccines are

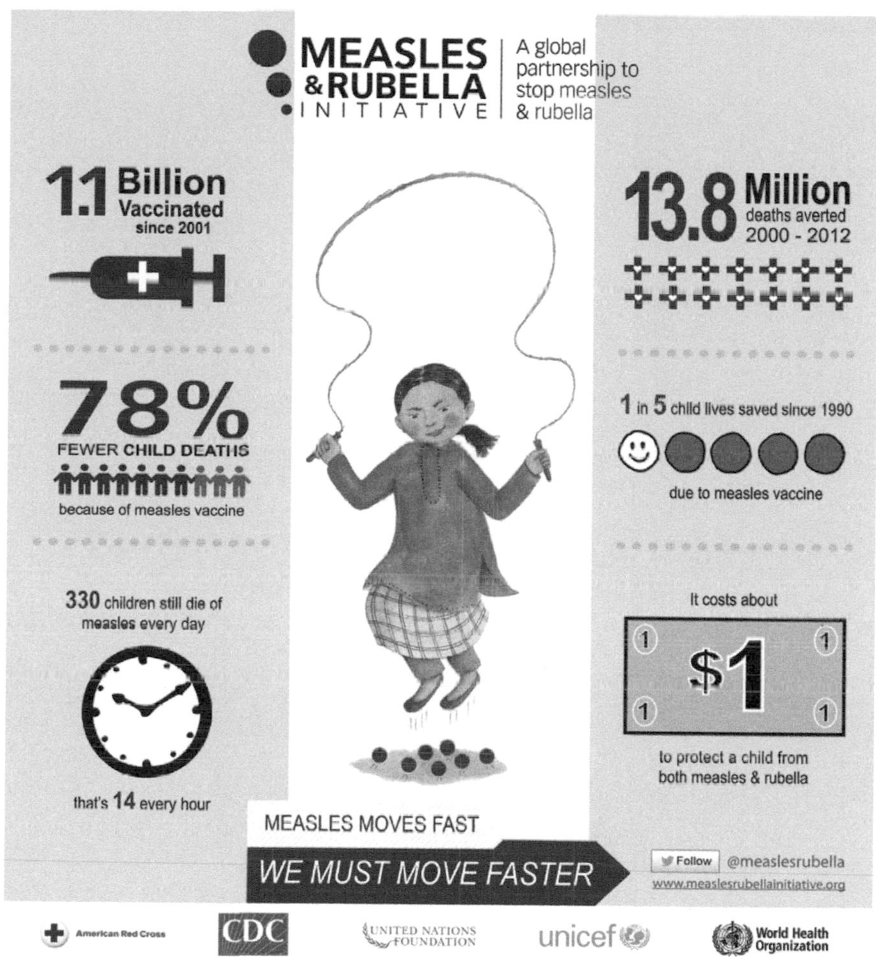

FIGURE 7.4 Entitled, "Measles & Rubella Initiative—A Global Partnership to Stop Measles & Rubella," this infographic created by the Centers for Disease Control and Prevention (CDC) brings to light a number of facts about the worldwide response to this highly contagious disease. *From CDC (2015).*

victims of their own success: *The more they successfully prevent disease, the easier it is to underestimate their value* (Fig. 7.4).

In this chapter we have learned that

- The purpose of vaccination is to protect against specific diseases. This is accomplished by administering antigens (usually proteins) that are processed and presented to T- and B-lymphocytes.
- Viruses encode a few to hundreds of proteins. Some of these are targeted successfully by the immune system to confer protection against repeated infection. An important part of vaccine development is to determine correlates of protection (the types immune responses are protective).
- The common feature of killed/inactivated/subunit vaccines is that they do not contain viruses that can replicate in the vaccine recipient. They may contain a few or many antigens. They can be produced by classical or recombinant DNA techniques. They are relatively inexpensive to develop, but can be costly to produce (high cost per dose). They are generally safe and can be given to immunocompromised or pregnant recipients.
- Some vaccines contain infectious viruses (capable of replicating in the recipient). They can be produced by classical methods (attenuate a virus by repeated passage) or by genetic engineering (altering specific viral genes). Classically attenuated viruses may revert to virulence or may be spread from a vaccinated individual to other contacts. In theory, genetic engineering can be used to generate very safe attenuated vaccines. Some live vaccines cannot be given to immunocompromised or pregnant individuals. A virus attenuated for one species may not be attenuated for all species.
- The greatest challenges for vaccine development are those diseases for which *natural infection* does not result in life-long immunity.

CHAPTER

8

Virus Evolution and Genetics

OUTLINE

After reading this chapter, you should be able to discuss the following:

- Did a single common ancestor give rise to all viruses?
- What are the two discrete processes that drive virus evolution?
- How is the mutation rate of a virus described?
- How is the evolution rate of a virus described?
- What are "nonessential genes" and are they always nonessential?
- What is "reverse genetics" and how is it used to probe the functions of virus genes?

In the first part of this chapter we discuss (in very general terms) when and how viruses originated and then consider the ongoing process of virus evolution. In the second part of this chapter we discuss the molecular genetics of some major groups of viruses.

VIRUS EVOLUTION

There is great diversity among viruses; they infect organisms from all Kingdoms of life and across all ecosystems. Animals, plants, single-celled eukaryotes, and prokaryotes are infected by a variety of viruses, both DNA and RNA. And some closely related viruses infect animals, plants, and bacteria. These facts suggest a very ancient origin for viruses.

There are three commonly proposed mechanisms for the origins of viruses:

- *Viruses descended from primitive precellular life forms.* This theory posits that viruses originated and evolved along with the primitive self-replicating molecules that were destined to become cells; the agents we recognize as viruses today were originally self-replicating molecules in the precellular world. If this is correct, it follows that cellular life forms were impacted by "viruses" from their earliest beginnings.
- *Viruses are "escaped" cellular genetic elements.* This theory posits that viruses evolved *after* cells. Their origins were cell-associated genetic elements that acquired protein coats, allowing for more efficient cell to cell transfer. The DNA genomes of some viruses do resemble plasmids, and retroviruses are related to the large group of nonviral retro-elements populating cellular genomes.
- *Retrograde evolution.* The theory of retrograde evolution states that viruses were once complex intracellular parasites that lost the ability for all independent metabolism; they retained only those genes required to manipulate the host cell and produce progeny virions.

It may be that each theory applies to a different viral lineage. For example, it is commonly thought that the first replicating molecules were RNA. Thus RNA viruses could be descended from these molecules.

Viruses. DOI: http://dx.doi.org/10.1016/B978-0-12-803109-4.00008-8

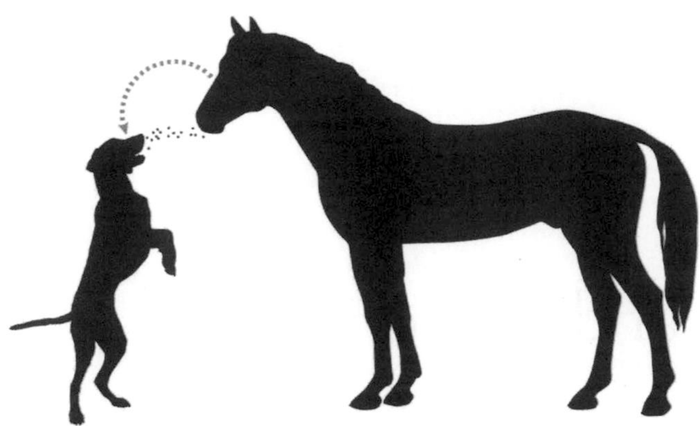

FIGURE 8.1 Virus moves to a new species: In 2004 equine influenza virus jumps to dogs.

As cellular life evolved, replicating RNA molecules would need to parasitize cells to access pools of biomolecules. The "escaped cellular element" theory could explain the origins of retroviruses, and the largest DNA viruses may have descended from more complex parasites. In contrast, some evolutionary biologists see evidence that large DNA virus are acquiring genes from their hosts.

Regardless of the origins of viruses, there is good evidence that they have shaped the evolution of single-celled and multicellular organisms for billions of years. Animal genomes are littered with pieces of RNA and DNA genomes that have very ancient origins and there are good examples of animals coopting viral genes. For example, genes necessary for the development of the mammalian placenta were obtained from retroviruses and some animal cells use captured viral gene fragments to defend against further infection!

While some virologists ponder the ancient origins of viruses, others examine the mechanisms that drive ongoing virus evolution. Examples of ongoing virus evolution include:

- Cross-species jumps that allow viruses to find new hosts. Influenza viruses (Chapter 23: Family *Orthomyxoviridae*) provide many examples of this type: Bird or swine influenza jumped into humans at the turn of the 20th century and horse influenza virus moved into dogs in 2004 (Fig. 8.1). Among the DNA viruses, a dog-adapted version of a feline parvovirus emerged in the 1970s and rapidly spread worldwide through the domestic dog population. Ongoing transmission in the new host is a measure of successful adaptation.
- Decreased virulence in a new host. Viruses that are well adapted to their hosts often cause little or no disease. When a virus jumps to new host, the virus may cause severe disease, but coevolution of virus and host often restores more balance. An instructive example is provided by studies of rabbit myxomavirus (Chapter 35: Family *Poxviridae*).
- Emergence of drug resistant viruses. As antiviral drugs have been developed, drug resistant viruses have emerged. The mutation and replication rates of human immunodeficiency virus (HIV) are so high that successful long-term treatment requires use of drug cocktails, containing three or four different active compounds (to minimize development of drug resistance).
- Immune escape mutants. HIV and influenza viruses are adept at escaping from host immune pressures. Point mutations in certain regions of influenza virus surface proteins [hemagglutinin (HA) and neuraminidase (NA)] are well tolerated by the virus but can thwart established immune responses. This is one reason that influenza viruses are closely monitored and vaccines are updated yearly (Chapter 23: Family *Orthomyxoviridae*).

Virus evolution is the outcome of two independent events (Fig. 8.2): The first event is mutation of the viral genome. For RNA viruses, this is often a frequent event during genome replication, as most RNA-dependent RNA polymerases have no proof reading functions (this is why some RNA viruses exist as a quasispecies as described in Chapter 10: Introduction to RNA Viruses. DNA damage may play a role in mutation of some DNA virus genomes. Virus genomes can also mutate as the result of recombination or even by capturing genes from the host or from other viruses!

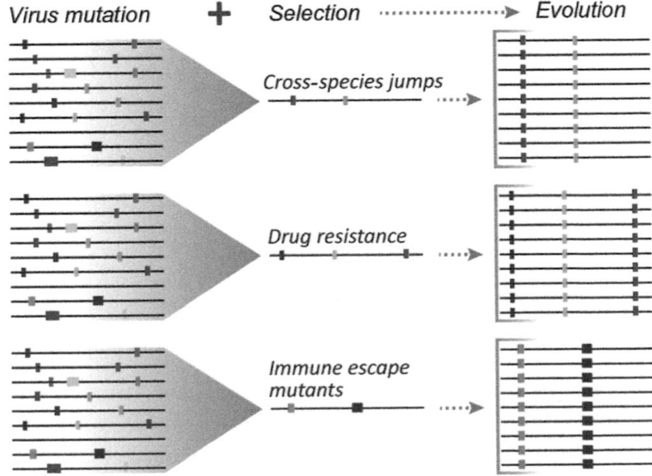

FIGURE 8.2 Evolution is the result of two independent processes: Mutation and selection.

There are several ways to express mutation rates. A common method is to express them as misincorporations per nucleotide synthesized. Another way is to express them as mutations/nucleotide/cell infection as this accounts for different modes of viral genome replication. Accepted estimates range from 10^{-6} to 10^{-4} mutations/nucleotide/cell infection for RNA viruses and range from 10^{-8} to 10^{-6} mutations/nucleotide/cell infection for DNA viruses. For some RNA viruses, these numbers translate into one or two mutations per genome replicated. Finally, viral mutation rates can be difficult to measure, as lethal mutations are quickly eliminated from the population.

The second event that drives virus evolution is *selection* of mutants that are more fit for a particular environment. For example, beneficial mutations may allow viruses to expand their host-range, evade immune responses, or tolerate antiviral drugs. Thus virus evolution requires initial mutations followed by selective pressures. A measurement of virus evolution that includes the parameter of natural selection is nucleotide substitutions per nucleotide site, *per year*. This measure reflects mutations that become *fixed* in the virus population. If a virus is highly adapted to a particular host and environment, this number may be very low. In contrast, the rate may be much higher if a virus is in the process of adapting to a new host or environment. Different lineages of the same virus may evolve at different rates depending on the selective pressures they encounter.

One common question about viruses is their "age." Is human immunodeficiency virus (HIV) a few decades old, a few hundred years old or many millions of years old? While we know that the virus we call HIV-1 *emerged* in the human population several decades ago, it is very closely related to viruses of nonhuman primates that are millions of years old. There are different methods for calculating rates of virus evolution (that will not be described here) and these can provide very different answers about the age of a virus. A relatively recent finding that impacts our view of virus age is that animal genomes harbor pieces of viral genomes (a variety of DNA and RNA viruses in addition to retroviruses) that are tens or hundreds of millions of years old. We can time the insertion events by looking at closely related species of animals. If nearly identical virus insertions are present in two different species, it is assumed that the viral gene insertion occurred prior to the split of those two species. Very surprising is the ease with which we recognize these so-called "endogenized" or "endogenous" viruses as *clearly related* to viruses widely circulating today. Research is finding that in some cases, the endogenized virus genes are expressed, and may serve to protect cells from infection with currently circulating viruses. In addition there are well characterized examples of co-evolution among organisms and viruses (Box 8.1).

BOX 8.1

COEVOLUTION OF VIRUS AND HOST

Animals have evolved a variety of strategies to control/limit/inhibit virus replication. In response, viruses have evolved mechanisms to "fight back." The ability to readily sequence both animal genomes and viruses is providing a detailed view into specific instances of evolution and coevolution of viruses and their hosts. The story we consider here is that of tetherin and its virally encoded antogonists.

Tetherin is a potent antiviral protein that inhibits a variety of enveloped viruses, including retroviruses. Tetherin expression is induced by interferon in response to viral infection (Chapter 6: Immunity and Resistance to Viruses). As the name implies, tetherin inhibits release of enveloped viruses from the cell surface (it "tethers" them to the cell). Tetherin appears to physically crosslink virions to the plasma membrane.

How then do viruses antagonize tetherin? The retroviruses simian immunodeficiency virus SIVcpz (from chimpanzees) and SIVgor (from gorillas) antagonize their hosts' tetherin with a protein called NEF [Chapter 37: Replication and Pathogenesis of Human Immunodeficiency Virus (HIV)]. SIV NEF binds to a specific amino acid in monkey tetherin to counteract its activity. However, human tetherin has a deletion where NEF would bind. Therefore SIVs replicate poorly in humans because human tetherin *is not* antagonized by SIV NEF.

So how then is HIV able to replicate to high levels in humans? HIV evolved a completely different tetherin antagonist! HIV encodes a protein called viral protein U (VPU) that counteracts tetherin by interacting at an entirely different site. HIV still encodes a NEF protein (it has a number of different functions that help HIV replicate) but it uses VPU to antagonize tetherin.

VIRUS GENETICS

Recombination

The principles and terminology used to describe the genetics of cellular organisms are applicable to viral genetics as well. In the sections above we used the terms mutation, selection, and recombination with the assumption they would be understood by students of biology. However recombination of virus genomes merits particular mention as it includes some unique mechanisms. Homologous recombination between DNA virus genomes occurs by the same molecular mechanisms described for cellular DNA. However RNA viruses also undergo genome "recombination" by a different molecular mechanism. RNA virus recombination is achieved through a process called "copy choice." Copy choice involves template switching during genome replication. The RNA replicase complex physically dissociates from one genome and reassociates with another. Regions of genome complementarity probably serve to position the replicase complex on the new template. The frequency of copy choice recombination is dependent upon the mechanisms used by a particular virus family for genome replication and transcription. As described in Chapter 17, Family *Coronaviridae*, coronaviruses synthesize their mRNA by a process of discontinuous transcription, whereby the transcription complex jumps from one position on the genome to another. It is thought that this mechanism results in frequent template switching during genome replication as well, resulting in frequent recombination events.

Another mechanism of "recombination" comes into play for segmented RNA viruses (i.e., orthomyxoviruses, reoviruses, and bunyaviruses). The orthomyxoviruses (influenza viruses) have segmented genomes (eight segments for influenza A viruses) and upon infection of a cell with two different viruses an assortment of progeny can be released (Fig. 23.8). This process is called *reassortment* as it involves shuffling of complete genome segments. Particular genome segments can also undergo copy choice recombination but reassortment is a more common event.

Genotypes and Phenotypes

The terms genotype and phenotype apply to viruses. The viral genotype is simply the sequence of its viral genome. But what kind of virus phenotypes can be measured? Quite an array as it turns out!

Virus phenotypes include: the ability to productively infect different cell types (host range); the ability to replicate at elevated temperatures; the ability to form different types of plaques (large, small, clear, or cloudy); the ability to transform cells; to replicate at an increased or decreased rate; to cause disease; to be neutralized by a particular antibody; to replicate in the presence of various drugs; to display increased or decreased mutation and/or recombination rates.

Nonessential Genes

Examining the genetics of viruses in *cultured cells* led to a surprising finding: Not all viral genes appeared to be essential for replication. In fact, some of these so-called nonessential genes were initially identified because they were lost during passage in cultured cells, and the mutated viruses outcompeted their wild-type parents.

It was counterintuitive to some virologists to think that viruses, with their very ancient origins, maintained nonessential genes for millions of years. They hypothesized that while some genes are not essential in cultured cells, they must be advantageous under more exacting conditions, such as replication and spread in a natural population of susceptible hosts. And indeed, this appears to be the case. Over and over again, genes identified as nonessential in cultured cells are indeed required to maintain the virus in an animal population. For example, the herpesvirus thymidine kinase gene is nonessential for virus replication in cell culture but is essential for productive infection of natural hosts.

The loss of nonessential genes in cultured cells is virus evolution in action. Given an environment without physical barriers to infection, antiviral responses, or limited nutrients viruses discard "useless" genes. The benefit is a smaller genome that it can be replicated at a faster rate, using fewer resources. Thus viruses have a set of essential genes, required under all conditions (for example, capsid proteins, polymerases, replication origin binding proteins) and other genes that may only be required in a specific ecological niche. A gene may not be *absolutely* required in a particular niche, but its presence may confer a strong advantage. Thus not only is host environment important, so too are the presence or absence of viral competitors!

Forward and Reverse Genetics

The classical study of virus genetics (also called "forward" genetics) involved identifying a novel phenotype and then identifying the associated mutations. To accomplish this, virus stocks were subject to mutagenesis and large numbers of mutants were analyzed individually. A variety of methods were used to sort out and classify mutants. The approach worked well for bacteriophages (bacterial viruses) and it was also applied to animal viruses when methods were developed to grow and plaque purify them using cultured cells. However, it can be challenging to design strategies for selecting a desired phenotype, and brute screening can take years.

Today, studies of viral genetics depend largely on a general methodology called *reverse genetics*. The term reverse genetics is used to describe activities that *involve making specific mutations* in a viral genome and *then observing* the phenotype. Mutations can range from a single nucleotide substitution to partial or complete gene deletions. Sometimes the order or arrangement of genes is altered. The goal of these changes is to link gene function to altered phenotypes, thereby generating models that describe gene function. Most significant human/animal viral pathogens can be studied using reverse genetic approaches. Perhaps one of the most impressive (and controversial) feats in reverse genetics was the resurrection of the deadly 1918 pandemic strain of influenza virus, using genome sequences derived from decades-old, formalin-fixed lung autopsy materials and frozen tissues from victims buried in permafrost.

Reverse genetic analysis involves both *design* of a mutation and the ability to *incorporate* the mutation into a virus. Generating gene mutations uses a variety of methods available through modern molecular biology. (If anyone doubts the utility of "basic research" consider the powerful toolkit it has provided for the study of human and animal pathogens, genetic disorders, and cancer.) In the study of viruses, principles governing both the design of mutations and their incorporation into viruses benefit from having detailed knowledge of viral genome structure and replication strategy. As will be seen in the coming chapters, these vary widely, but there are some common principles.

Reverse Genetics of Positive-Strand RNA Viruses

The *genomes* of positive sense RNA viruses are infectious in the absence of any viral proteins (Chapter 10: Introduction to RNA Viruses). The infectious "heart" of a positive-strand RNA virus is its genome. Therefore the genomes of many positive-strand RNA virus can be synthesized and introduced into a cell with the aim of producing virions. The general process is outlined in Fig. 8.3.

Reverse Genetics of Negative-Strand RNA Viruses

The *genomes* of negative sense, ambisense and double-stranded RNA viruses are not infectious in the absence of viral proteins (Chapter 10: Introduction to RNA Viruses). In a natural infection, the genome is associated with viral proteins, key among them an RNA-dependent RNA polymerase. Therefore one must introduce, or express within a cell, both a viral genome and a subset of viral proteins before the replication cycle to produce virions can begin.

Reverse Genetics of Retroviruses

The genomes of retroviruses are mRNA (Chapter 36: Family *Retroviridae*). However they do not replicate in the manner of positive-strand RNA viruses, as their RNA genomes are not translated to initiate a productive infection. Instead, early in the infection process, retroviruses synthesize of a DNA copy of their RNA genome. Therefore, reverse genetics of retroviruses uses plasmids containing the DNA copies of retroviral genomes. The retroviral genome is expressed (transcribed) in the cell nucleus to produce the set of mRNAs needed to drive virion production.

Reverse Genetics of DNA Viruses

The genomes of many DNA viruses are infectious and they can be synthesized, manipulated, and introduced back into cells as plasmids, with the resulting production of virions. However, some DNA viruses also require a set of proteins to initiate an infection. Herpesvirus (Chapter 34: Family *Herpesviridae*) and poxvirus (Chapter 35: Family *Poxviridae*) virions contain proteins needed to initiate a productive infection cycle. Introducing mutations into these viruses depends on homologous recombination within an infected cell to exchange specific genome segments.

In this chapter we learned that:

- Two separate processes, mutation and selection are the drivers of virus evolution.
- Virus *mutation rates* are described in several ways, such as mutations per nucleotide synthesized, mutations per genome synthesized or mutations/nucleotide/cell infection. For some RNA viruses, these numbers translate into one or two mutations per genome replicated. Estimates may try to account for the length of the virus replication cycle, the number of genomes produced, or the method of genome synthesis.
- The *evolution rate* of a virus is often described as nucleotide substitutions per nucleotide site, *per year*.
- Viruses have core sets of essential genes that are always required for genome replication. Examples include capsid proteins and polymerases. Other genes, sometimes called nonessential, are dispensable, and may be lost, under certain specific

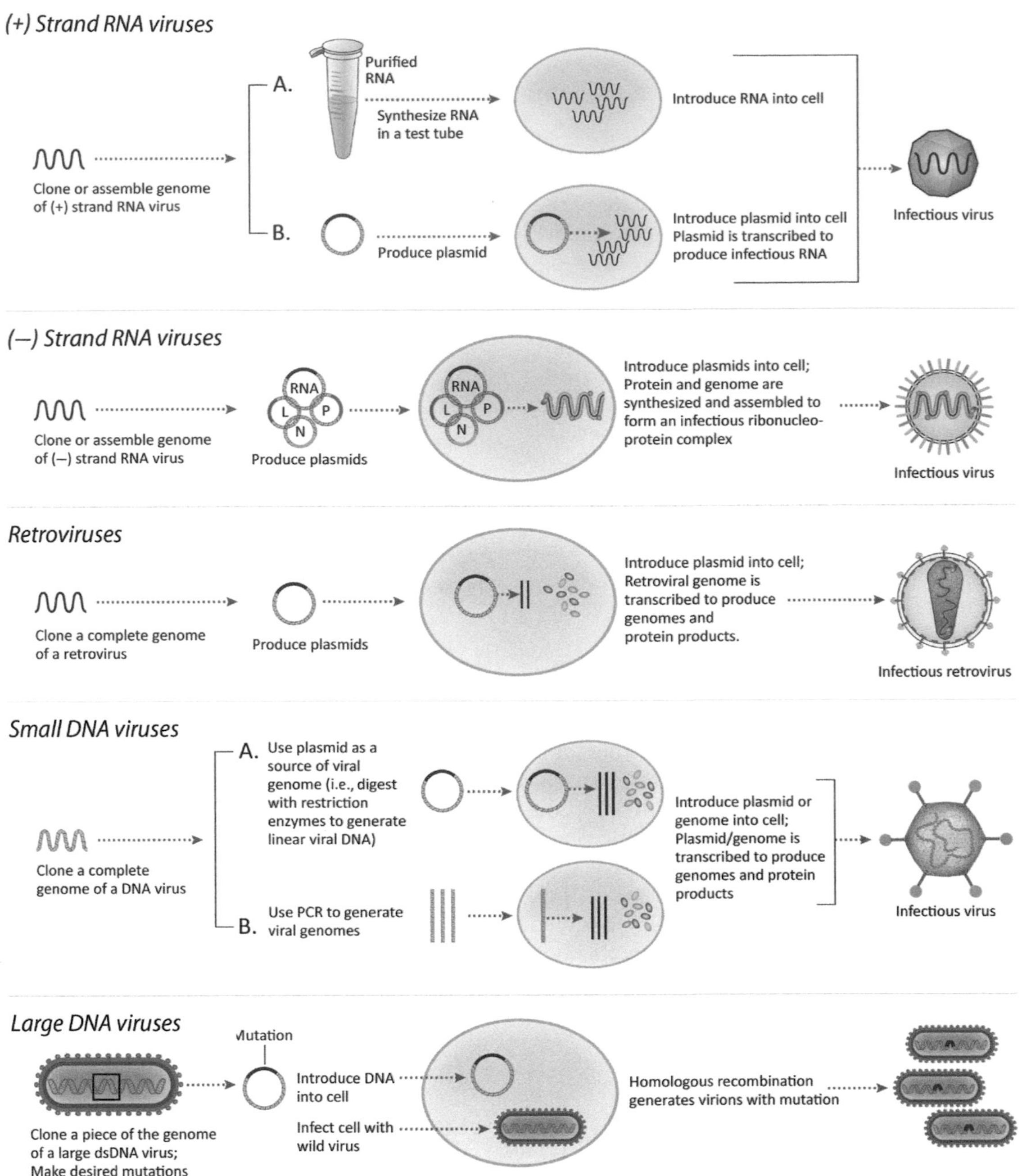

FIGURE 8.3 Reverse genetics systems for some common DNA and RNA viruses.

some conditions (for example, during replication in cultured cells).

- The process of reverse genetics is used to determine the function of viral genes. It involves cloning a viral genome, designing and making specific mutations, and reintroducing the genome into cells to obtain mutated virions whose phenotype can be determined.

CHAPTER

9

Viral Pathogenesis

OUTLINE

After reading this chapter, you should be able to answer the following questions:

- What is disease?
- What is the meaning of the term "pathogenesis"?
- What is the meaning of the term "virulence"?
- Explain how modern genetic methods have improved development of animal models to study viral diseases.
- Provide an example of a disease process that benefits the viral pathogen.
- Provide an example of a disease process that does not benefit the pathogen.
- What are the molecular differences between the attenuated version of a virus (used as a vaccine) and its highly virulent "parent"?
- What events turn an asymptomatic persistent infection into a lethal disease?
- How do some patients survive infections with highly lethal viruses? Conversely, what makes a usually innocuous virus lethal in a small percentage of infections?
- Is a particular disease caused by direct virus damage or by immune responses (or a combination of both)?

How do viruses cause disease? Sounds like a simple question until we begin to examine the many cases whereby viruses replicate *without* causing disease. What makes one virus a killer while another does no harm?

Pathogens are disease-causing agents and pathogenesis (meaning "origins of disease") is the study of the disease *process*. In early days of infectious disease research pathologists examined organs and tissues (macroscopically and microscopically) to look for damage. Modern studies extend this approach to ask questions about the *underlying molecular events* that cause the observed damage. In other cases researchers seek to ferret out the causes of disease in the absence of obvious lesions or tissue damage.

Examples of the types of questions asked by those with an interest in viral pathogenesis include:

The cytopathic effects observed in cultured cells are caused directly by virus infection: They are direct effects of virus replication. Cultured cells can be damaged in a variety of ways including virus inhibition of cell transcription and/or translation, changes in membrane permeability, alterations of cytoskeleton or trafficking pathways, cell–cell fusion, or induction of apoptosis.

In an animal, disease may arise from virally induced cell death but may also be caused immune by responses directed against infected cells. While necessary and essential, the immune system can damage the host in the process of clearing an infection (Box 9.1).

Infectious disease always begins with infection, so we begin this chapter by looking at the infection process. (Keep in mind however that infection is not synonymous with disease.) The molecular steps required for productive infection of a cell were

Viruses. DOI: http://dx.doi.org/10.1016/B978-0-12-803109-4.00009-X

BOX 9.1

ACUTE FEBRILE ILLNESS

Many different viruses cause "acute febrile illness" with symptoms including fever, lethargy, depression, prostration, and/or anorexia. These general symptoms may or may not be accompanied by more specific signs and symptoms such as diarrhea, rash, liver damage, or neurologic deficits.

Why do so many virus infections cause fever (particularly if replication is localized)? The answer is that the body's response to localized infection is often global, due to secretion of IFNs and other cytokines from infected cells. Many cytokines associated with a viral infection are pyrogens (fever inducing molecules) that affect the hypothalamus. Cytokines that act as pyrogens include interleukin (IL)-1α, IL-1β, IL-6, IL-8, tumor necrosis factor-β, and IFNs-α, -β, and -γ.

discussed in earlier chapters; however, a cultured cell is not an intact organism so here we briefly consider the barriers that must be overcome to allow a virus to successfully reach target cells or tissues in the process of a natural infection.

NATURAL BARRIERS TO INFECTION

Intact skin is a major barrier to viral infection. The outmost layers of the skin are made up of dead cells that cannot support virus replication. Viruses that directly infect the skin (for example, papillomaviruses) require breaks or lesions that allow access to lower layers of dividing cells (Chapter 32: Family *Papillomaviridae*). Insect or animal bites provide another route past the skin barrier. Human activities provide additional avenues for infection. These include medical, dental and veterinary procedures, cosmetic procedures (for example, tattoos), and i.v. drug use. Mucosal surfaces of the eye, respiratory, gastrointestinal, and genitourinary tracts, although protected by a mucus layer and/or digestive enzymes, provide easier access to viral infection. Thus many viruses are transmitted by respiratory, fecal-oral, or sexual transmission (Fig. 9.1).

PRIMARY REPLICATION

Primary replication refers to the initial site of virus replication. This is often, though not always, an epithelial cell. Epithelial cell infection may remain localized in skin (for example, warts), upper respiratory tract, or gastrointestinal tract. Viruses in the respiratory or digestive tract also encounter lymphoid tissues that may serve as primary sites of replication.

Microfold or M cells are specialized epithelial cells in the gastrointestinal tract (Fig. 9.2). They lack microvilli and are found primarily at sites of gut-associated lymphoid tissues, including Peyer's patches. While most epithelial cells provide a strong barrier to gut contents, the role of M cells is to sample those very contents. They do this by their high capacity for transcytosis. They take up material at their apical surfaces (luminal side of the gut) and release it basolaterally to immune cells. In a perfect system, the M cells alert immune cells that respond effectively. But some viruses target M cells to gain access to lymphoid tissues, thus facilitating replication and spread.

MOVEMENT TO SECONDARY REPLICATION SITES

Some virus infections are very localized with replication confined to the infection site. Examples include papillomaviruses and the poxvirus, molluscum contagiosum virus. However many viruses spread from the initial site of infection to other organs. How do these viruses reach other sites in the body? There are three general mechanisms: Blood, lymphatics, and nerve cells. The presence of virus in the blood is called viremia. Viruses in the blood gain access to many tissues and organs. Some viruses are "free" in the blood, present as extracellular virions in plasma. Other viruses in the "blood" are in fact, cell-associated. Some viral infections produce a very short-term or transient viremia; other viruses are continuously found in the blood. Viruses that infect T-cell, B-cells, monocytes, or other circulating lymphocytes may be carried to numerous lymphoid tissues. A prime example is HIV, carried in an infectious form by dendritic cells to tissues rich in T-cell targets. Blood monocytes are also targets for HIV infection. These persistently infected cells transport virus out of the blood and into target organs. Thus HIV invades the host by targeting these

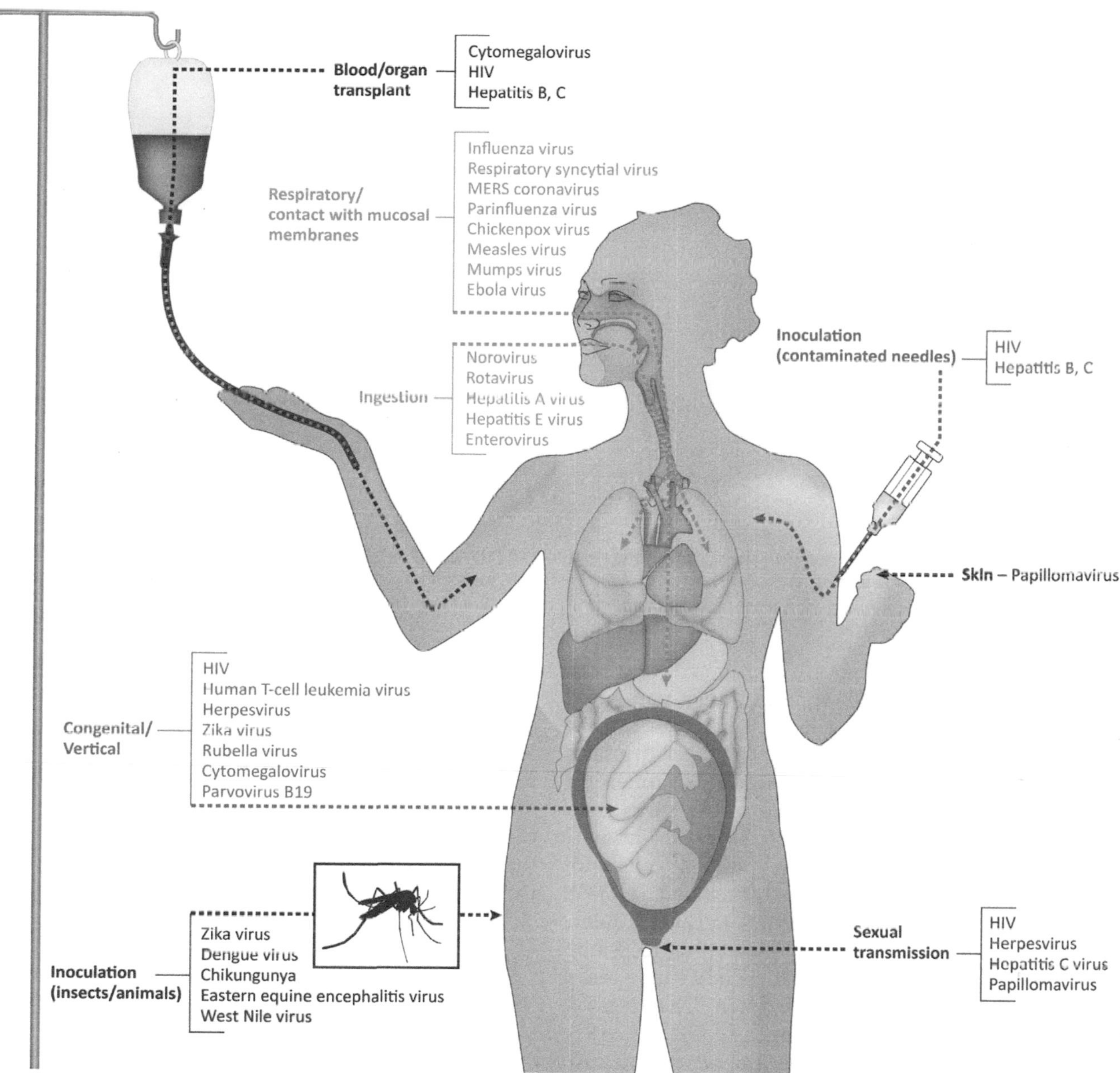

FIGURE 9.1 Routes of invasion. There are a variety of ways that viruses can access permissive cells. Some viruses use multiple routes. Many viruses that are normally transmitted by other routes can be acquired via transfusion or organ transplant.

"protective" cells (Chapter 37: Replication and Pathogenesis of Human Immunodeficiency Virus (HIV)).

Some viruses such as rabies, mammalian bornavirus, and some herpesviruses use neurons as their pathway through the body. Neurons can provide protected routes of travel and direct access into the central nervous system (CNS). Rabies virus enters neurons near the initial site of infection (for example a bite wound) and travels along neurons at rates of up to 50–100 nm/day. While moving within a neuron, rabies virus is protected from immune surveillance. Rabies virus also crosses synaptic junctions very efficiently.

How do viruses move within neurons? As mentioned in Chapter 3, Virus Interactions With the Cell, viruses use elements of the cytoskeleton to move into, across, and out of cells. Movement within neurons is an instructive example, given the long distances that must be covered. Rabies virus is transported the length of axons by interacting with *dynein molecular motors* that move cargo from synapses to the cell body in a process called retrograde transport. *Kinesins* are motors that move cargo from the cell body to nerve endings, a process called anterograde transport. Herpesviruses and rabies virus make use of both retrograde and anterograde transport systems.

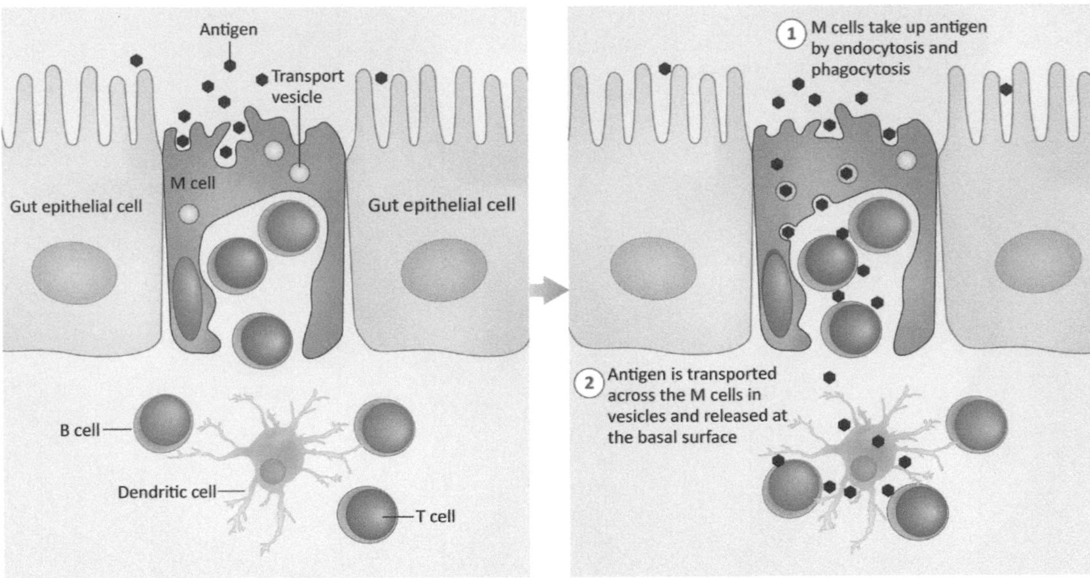

FIGURE 9.2 Microfold or M cells are specialized cells in the gastrointestinal tract. Their primary purpose is to sample the environment of the gut. Materials in the gut, including pathogens, cross M cells by transcytosis and are delivered to immune cells. Some viruses exploit M cells to gain entry into the host.

Some viruses use a combination of transit pathways Fig. 9.3. One example is measles virus which infects via the respiratory route. Virus released after primary infection causes a viremia and spreads to numerous lymphoid organs. Another example is the virus that causes chickenpox (HHV3 or vericella-zoster virus). This virus replicates in respiratory epithelial cells and is released into the blood. From the blood, HHV3 moves into the skin where it replicates to cause a rash. From the skin the virus moves into neurons to establish latency in the CNS. Decades later HHV3 may reactivate from latency, traveling back down the nerves to the skin to productively replicate once again.

Some viruses do not infect peripheral nerves but access the CNS by crossing the so-called blood–brain barrier (BBB). The brain is protected from most viruses by virtue of the BBB, a layer of tightly joined endothelial cells (Fig. 9.4). There are three major pathways across the BBB. One mechanism is via infected lymphocytes. Other viruses cross the BBB when endothelial cells are damaged, for example, by excessive levels of cytokines. The third way to cross the BBB is by infecting the endothelial cells themselves. Viruses that sometimes cross the BBB include poliovirus (PV), measles virus, West Nile virus, equine encephalitis viruses, and HIV. While crossing the BBB has serious consequences for the infected host, none of the viruses listed here requires access to the CNS, as they all productively infect nonneuronal tissues. In contrast, rabies virus must reach the CNS to replicate sufficiently to be transmitted to a new host.

Sometimes very closely related viruses differ in their ability to reach or invade certain target tissues or organs. Pathogens may reach sites that their avirulent relatives fail to reach. Invasiveness is the capacity of a virus to enter a tissue or organ. Viruses that can invade the CNS are more likely to cause neurologic disease than those that cannot. Some viruses cross the placenta to invade a fetus, while others cannot.

GENESIS OF DISEASE

Disease is a disorder of structure or function of an organism. Disease can result from death, damage, or dysfunction of specific cell types. Damage may be directly due to virus infection (akin to cytopathic effects in cultured cells) or may be caused by immune responses to infection. Many diseases are probably a combination of both.

General types of diseases caused by viruses include:

- respiratory,
- intestinal,
- skin lesions/rashes,
- single organs (liver, kidney, heart, etc.),
- multiorgan/systemic,
- hemorrhagic,
- neurologic,
- immunosuppressive,
- benign and malignant tumors.

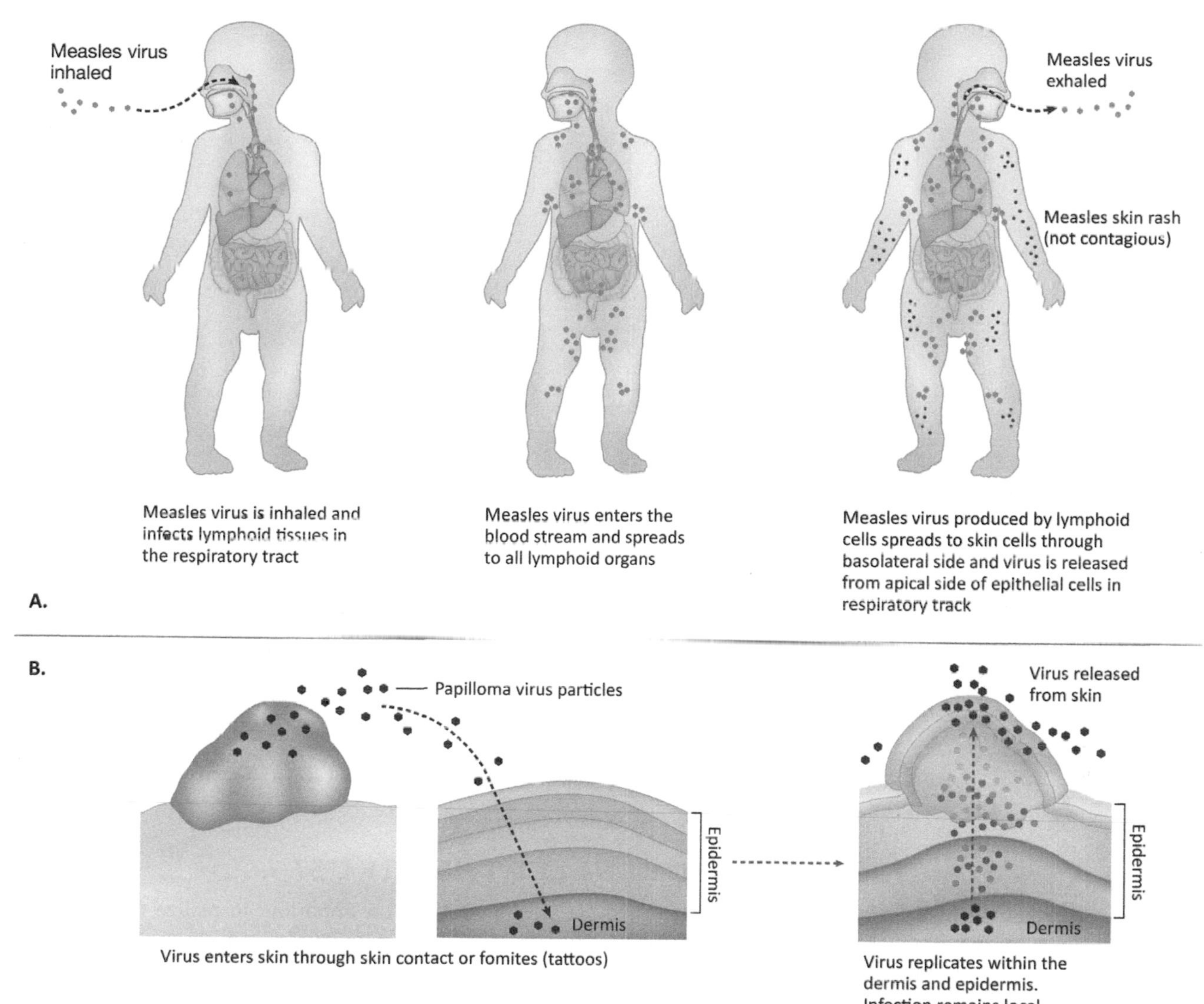

FIGURE 9.3 Generalized versus localized infection. Panel A: Measles virus gains access the body via the respiratory route. Virus released after primary infection causes a viremia and spreads to numerous lymphoid organs. Additional replication seeds the skin, causing a rash. However virus is transmitted by release from respiratory epithelial cells. Panel B: Papillomaviruses cause warts, infecting skin and remaining highly localized. They are transmitted by direct contact with infected skin or fomites.

There are many different mechanisms by which viruses cause disease. Infection of the respiratory tract can lead to inflammation, generating the mucus typical of such infections. Inflammation may result when cells are directly killed by a virus and/or because of host responses to infection. Inflammation in the lung can result in pneumonia with reduced capacity to absorb oxygen. Damage to intestinal epithelial cells may inhibit absorption of nutrients and/or water into the gastrointestinal tract. The short-term effect can be diarrhea; a long-term effect may be weight-loss and wasting (Box 9.2). Direct and indirect damage to the immune system eventually leads to immunosuppression of the non-treated HIV patient. Direct or indirect damage to any organ (for example, liver, kidney, brain, heart) may adversely impact normal function (Box 9.3).

IMMUNOPATHOGENESIS

The purpose of the immune system is to clear pathogens from the infected host (Fig. 9.5). Unfortunately this necessary activity can sometimes cause disease. Disease is a tolerable, even desirable, outcome if it results in clearance of a pathogen and lifetime protection. However in rare cases, the immune system causes life-threatening damage (with or without virus clearance). For example, many viruses trigger the release of cytokines such as interferons (IFNs). IFNs trigger cell-signaling cascades that result in synthesis of antiviral proteins. However IFNs also cause of the fever, aches, and lethargy that accompany many viral infections. In extreme cases cytokines can disturb permeability of blood vessels resulting in fluid

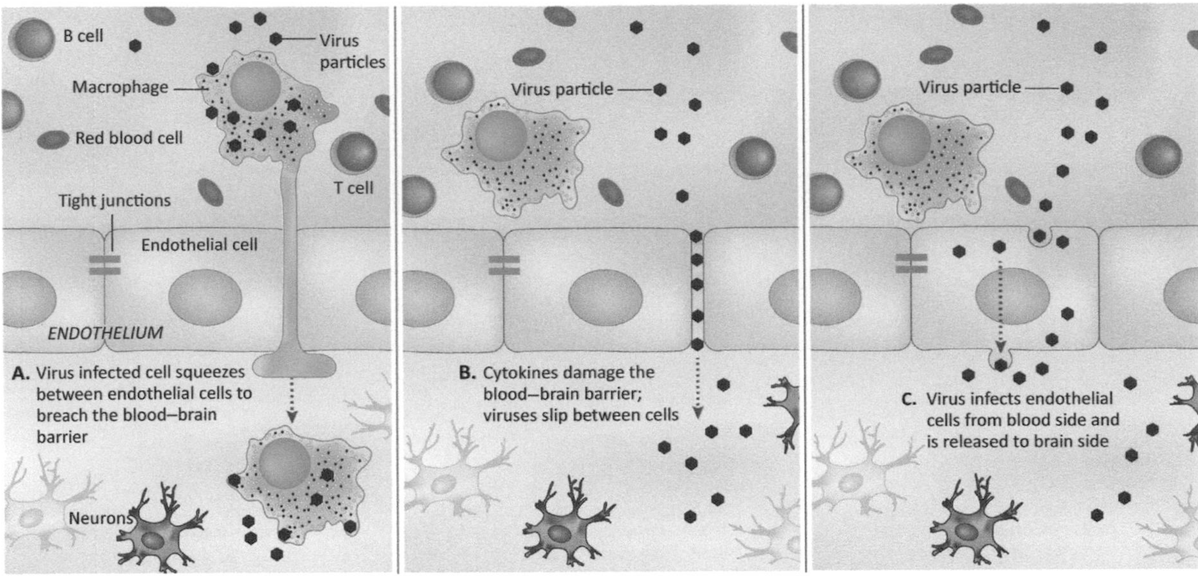

FIGURE 9.4 Endothelial cells lining blood vessels are key to maintaining the BBB. However some viruses cross the BBB, gaining access to the brain and spinal cord. Panel A: HIV infects T-lymphocytes and macrophages; these cells can carry HIV across the BBB. Panel B: Other viruses induce high levels of inflammatory molecules and cytokines that can damage the BBB. Panel C: Some viruses infect endothelial cells and can be released into the CNS.

BOX 9.2

DOES DISEASE BENEFIT THE VIRUS?

There are many examples whereby a specific disease process benefits the virus. Often the benefit is transmission to a new host. For example, human noroviruses (Family *Caliciviridae*) are intestinal pathogens and disease symptoms may include explosive vomiting and diarrhea. These highly unpleasant symptoms most certainly help disperse noroviruses to surfaces on which they remain infectious for extended periods of time. Respiratory viruses benefit from the coughing and sneezing that accompany those infections. A more extreme example is rabies virus. Infected dogs become aggressive, attacking people, or other animals without provocation. On the other extreme, rabies can render an animal lethargic and vulnerable to an unsuspecting predator (or to a kindly human!). The virus is the only winner in this scenario.

However, it is also important to realize that while a disease may kill or debilitate a host, those outcomes do not always benefit the pathogen. For example, polio virus (PV) infections most often result in asymptomatic replication in the gastrointestinal tract. PV (Chapter 11: Family *Picornaviridae*) replicates to high titer in the gastrointestinal tract and is shed into the environment and may go on to infect new hosts. On the other hand paralytic polio, a disabling disease, occurs only if PV invades the CNS. It is not at all obvious how this benefits the virus. In fact, the ability to infect the CNS has proven highly detrimental to PV, which is being pushed to extinction due to worldwide monitoring and vaccine programs!

imbalance, hemorrhage, and shock. Many aspects of viral hemorrhagic fever are caused by excessively high levels of cytokines. Thus overzealous inflammatory responses may cause tissue damage beyond that caused directly by the virus.

VIRULENCE

Virulence is the *relative* ability of an infectious agent to cause disease. Thus virulent viruses have a greater propensity to cause disease (to be pathogens) in a

BOX 9.3

SMALL MOUSE AND SMALL VIRUS YIELD BIG RESULTS

Lymphocytic choriomeningitis virus or LCMV, a rodent virus, is a member of the family *Arenaviridae*. Arenaviruses are enveloped, ambisense RNA viruses that encode only four proteins. Most arenaviruses have rodent hosts but some infect humans as well. LCMV was first isolated in 1933 during an encephalitis outbreak in the United States. Shortly thereafter LCMV was linked to disease in mouse colonies. The natural host for LCMV is the house mouse.

Very early studies of LCMV disease in mice pointed directly to immune responses for causation, a novel finding for the time. In the LCMV system, infection of adult mice causes acute disease followed by virus clearance (or death, depending on the route of inoculation) while infection of mice in utero, or shortly after birth, results in a long-term persistent infection. Furthermore, immunosuppression of adult mice, by various means, turns a lethal infection into a persistent one.

Studies to investigate the cellular and molecular mechanisms of LCMV immunopathogenesis went hand in hand with advances in understanding adaptive immunity. LCMV had several advantages as a virus model: LCMV infection is noncytopathic and immune responses are the only cause of death of infected cells. The mouse model was highly tractable and inbred strains of mice provided a means of examining genetics of both host and virus. The small size of the virus made it relatively easy to link virus genotype and phenotype. Those advantages continued as reverse genetics systems for LCMV were developed and genetically engineered mice became a commonplace tool. The following model for LCMV immunopathogenesis is supported by an impressive number of studies:

- Virus replication is noncytopathic but in immunocompetent mice, virus-specific cytotoxic T lymphocytes (CTLs) kill infected cells. This causes an acute disease episode, but clears the infection.
- Infection of mice, before complete development of their immune systems, results in tolerance (specifically lack of LCMV-specific CTL) and long-term persistent infection.
- During persistent infection, antibodies to LCMV are produced but they are not able to clear the infection. Instead, the formation and deposition of virus-antibody immune complexes causes glomerulonephritis and arteritis. Thus disease that eventually results from persistent infection is also immune mediated.

Other important lessons learned from LCMV include:

- A variety of specific genetic factors, both host and viral, can alter the outcome of infection. Discrete and specific molecular interactions can explain those outcomes.
- Noncytopathic LCMV infection of differentiated cells can and does compromise cell function resulting in diseases such as a growth hormone deficiency syndrome.
- Persistent, noncytopathic LCVM infection measurably alters behavior and learning in mice.

greater proportion of infected hosts. Virulence determinants or factors are those genes and proteins that play key roles in disease development. Virulence determinants can range from surface/capsid proteins that determine cell/tissue tropism to small adapter proteins that alter cell-signaling cascades. Differences in virulence may even be determined by cis-acting viral sequences that control the relative abundance of specific viral transcripts and their protein products. In the case of RNA viruses, the degree of fidelity of the RNA-dependent RNA polymerase can be a virulence determinant, as described in Chapter 10, Introduction to RNA Viruses.

Virulence of a virus also depends upon genetics, age, and overall health of the host. And it would not be surprising to find that the microbiome (the mixture of commensal bacteria and viruses in the host) can have an impact on virulence. In experimental situations, the route of inoculation or dose administered can have a major effect on virulence. Thus the study of virulence determinants requires an understanding of the complex interactions between virus and host (and perhaps environment).

Understanding the virulence determinants of a virus brings us full circle back to pathogenesis: The process by which a virus infects certain tissues, inhibits or induces various immune responses, or alters basic cellular processes. Thus the study of viral pathogenesis requires an understanding of virus molecular biology, host cell biology, and immune responses.

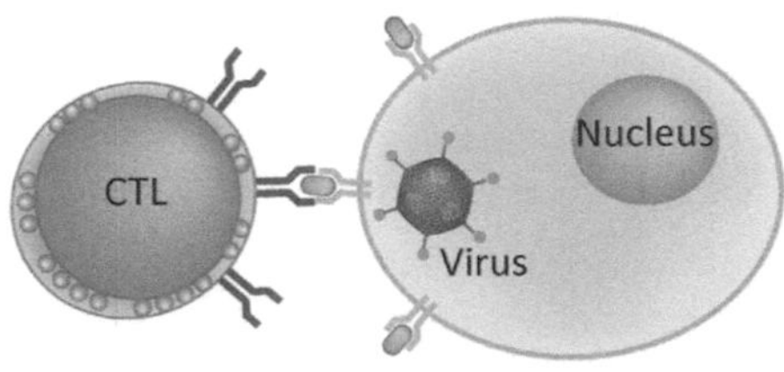

Cytotoxic T lymphocyte (CTL) binds cognate antigen on virus infected cell; antigen recognition activates CTL

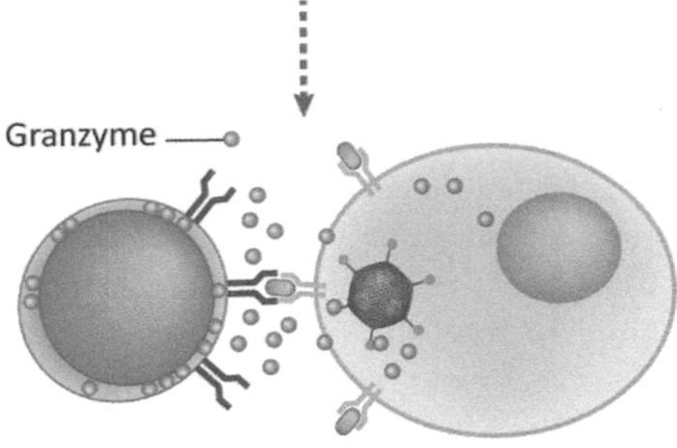

CTL releases granzymes that enter target cell to activate caspaces

Apoptosis

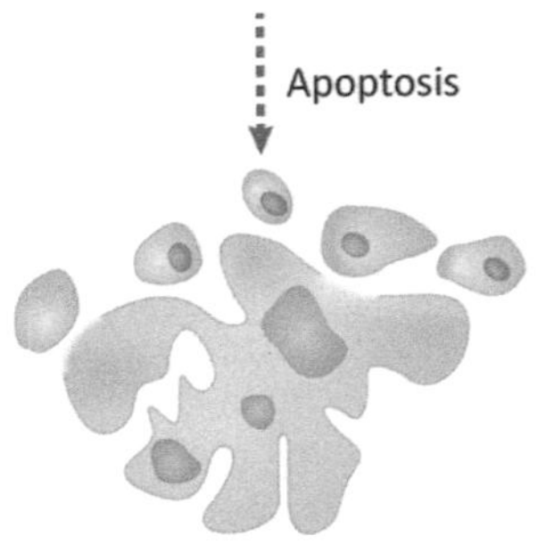

FIGURE 9.5 Killing of an infected cell after recognition by a cytotoxic T-cell.

Why is it important to study pathogenesis? An obvious answer is that understanding pathogenesis is important for treatment. It allows the clinician to move past the empirical toward more directed treatments. It also provides a blueprint for understanding the likely course of an infection and may also provide a basis for administering supportive therapies that do not directly target the virus.

APPROACHES TO STUDY VIRAL PATHOGENESIS

Understanding the pathways from initial infection, to dissemination, to replication in target tissues is a complex process. Disease development may take days to weeks, or years to decades. Studying natural infections can provide some insight into pathogenesis, but unraveling the molecular details often requires use of well-defined model systems.

Diseases caused by human viruses are often studied in animal models. Developing such models may include adaption (by repeated passage) of a human virus to growth in animals. Or human viruses may be used to inoculate animals by unnatural routes, or at very high doses. An alternative is to study the pathogenesis of an animal virus in its natural host to provide insight into pathogenesis of a related human virus. For example, the overall mechanisms by which bovine papillomaviruses cause warts in cows are instructive in understanding the development of human warts by human papillomavirus. Sometimes models must be developed to study animal viruses. Viruses that infect large animals, companion animals or endangered species may be adapted to growth in smaller/more abundant laboratory animals (for example, rodents). Developing a model system can be difficult, expensive, and time consuming. In the past it was largely an empirical process, but that has changed with the availability of modern methods to manipulate the genomes of animals and their viruses. Approaches to animal model development include:

- Adapting a virus to a new host by repeated passage.
- Genetic manipulation of a laboratory animal to support virus replication. For example, inserting the receptor for a human virus into the mouse genome.
- Identify specific genes that render a laboratory animal resistant to infection with a human virus, and remove them by genetic engineering.
- Use mice as hosts for *tissues or cells* from other organisms. For example, immunodeficient mice can be transplanted with human cells or tissues (tiny pieces!). Mice have been successfully "humanized" as models to examine HIV and dengue virus pathogenesis.
- Genetically manipulate the virus to allow for replication in a model animal.
- Add one or more viral genes to an animal's genome to study their effects in the absence of virus replication.

All model systems have their limits! Evaluating the data generated by use of a particular model requires an understanding of the strengths and limitations of the model. Indeed, comprehensive descriptions of viral pathogenesis often include results gained from several different model systems. Models for HIV pathogenesis include use of monkeys infected with simian immunodeficiency viruses (SIVs), chimpanzees infected with genetically engineered HIV/SIV chimeric viruses, humanized mice, and mice genetically engineered to express one or more HIV genes (in the absence of virus replication).

In this chapter we learned that:

- Disease is disorder of structure or function in an animal. Colds are respiratory diseases that result from virus replication and/or inflammatory responses in the epithelial cells of the upper respiratory tract. Diarrhea is caused by fluid and electrolyte imbalance in the digestive tract. Many severe viral diseases result from death of infected cells in the CNS.
- Pathogens are infectious agents that cause disease and pathogenesis is the study of the disease process. Pathogenesis can be straightforward (rapid, virus-mediated destruction of a particular cell population) or may be a lengthy and complex process (development of immunodeficiency or a cancer many years after initial infection).
- Virulence is the relative ability of an infectious agent to cause disease. A highly virulent virus may cause debilitating disease while an avirulent virus is "without" virulence.
- The study of viral pathogenesis requires well-characterized animal models. Modern genetic methods have improved development of such models. Genes can be added to, or removed from model organisms. Viruses can be genetically altered as well.
- Disease may be beneficial to a virus, aiding in spread to new hosts, for example. Alternatively disease may harm the host with no obvious benefit to the virus (paralytic polio). Finally, causing disease may be detrimental to a virus, if in response humans develop and aggressively utilize effective vaccines.

CHAPTER

10

Introduction to RNA Viruses

OUTLINE

After studying this chapter, you should be able to:

- Define "positive-strand RNA virus."
- Describe "negative-strand RNA virus" and "ambisense RNA virus."
- Understand the activities of RNA-dependent RNA polymerase (RdRp).
- Describe three general methods for priming RNA synthesis.
- Describe some functions of the noncoding regions (NCRs) of an RNA virus genome.

DEFINITION AND BASIC PROPERTIES OF RNA VIRUSES

RNA viruses replicate their genomes using virally encoded RNA-dependent RNA polymerase (RdRp). The RNA genome is the template for synthesis of additional RNA strands (a molecule of RNA is the template and molecules of RNA are produced). During replication of RNA viruses, there at least three types of RNA that must be synthesized: the genome, a copy of the genome (copy genome), and mRNAs. Some RNA viruses also synthesize copies of subgenomic mRNAs. RdRp is the key player for all of these processes (Fig. 10.1).

RdRps of RNA viruses probably arose from a common ancestor. The RdRp, in association with other proteins required for viral genome synthesis is often called the *replicase complex*. The replicase complex consists of the set of proteins required to produce infectious genomes. In addition to the RdRp, the replicase complex may contain RNA-helicases (to unwind base-paired regions of the RNA genome) and ATPases (to supply energy for the polymerization process). The number of proteins in the replicase complex differs among virus families. There may also be a requirement for host cell proteins.

The biochemical requirements for genome synthesis may or may not be identical to those required for synthesis of mRNAs. If the two processes differ, the term *transcription complex* is sometimes used to describe the particular set of proteins required for *viral mRNA synthesis*. Some RdRps have methylase activities that synthesize mRNA "caps" and RdRp may "stutter" at poly U tracts to generate polyadenylated mRNAs.

Viruses. DOI: http://dx.doi.org/10.1016/B978-0-12-803109-4.00010-6

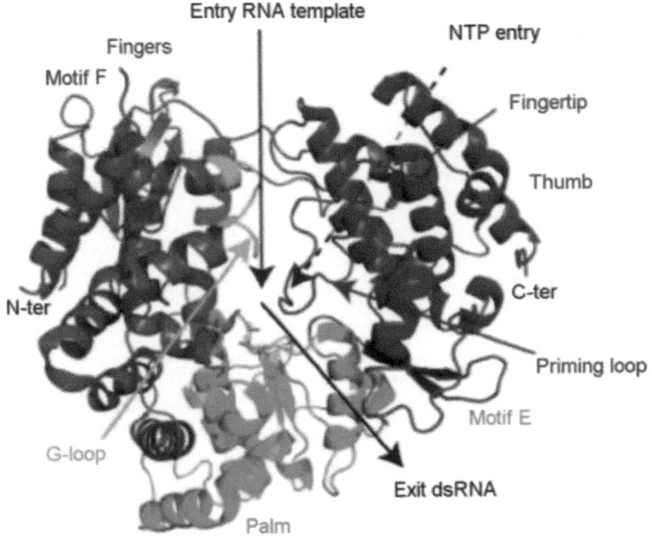

FIGURE 10.1 Ribbon diagram of flavivirus RdRp. Fingers, palm, and thumb subdomains are colored in blue, green, and red, respectively. The ssRNA template entry and the dsRNA exit are shown by black arrows. A dotted arrow points to the NTP entry tunnel at the back of the RdRp. Motifs A, C, E, F, the G-loop, and the priming loop are colored in orange, yellow, gray, magenta, cyan, and purple, respectively. The Asp residues of catalytic motifs A and C (Asp-533, Asp-663, and Asp-664) are represented as stick models. N-ter and C-ter indicate the termini of the RdRp domain. *From Bollati et al. (2009).*

TABLE 10.1 Positive-Strand RNA Viruses of Animals[a]

Virus family	Virion morphology	Genome type
Arteriviridae	Enveloped, helical nucleocapsid	Unsegmented
Astroviridae	Nonenveloped, icosahedral	Unsegmented
Caliciviridae	Nonenveloped, icosahedral	Unsegmented
Coronaviridae	Enveloped, helical nucleocapsid	Unsegmented
Flaviviridae	Enveloped, icosahedral nucleocapsid	Unsegmented
Hepeviridae	Nonenveloped, icosahedral	Unsegmented
Nodaviridae	Nonenveloped, icosahedral	Bisegmented
Picornaviridae	Nonenveloped, icosahedral	Unsegmented
Togaviridae	Enveloped, icosahedral nucleocapsid	Unsegmented

[a]*Only those families with members that infect vertebrates are included in this list.*

SUBGROUPS OF RNA VIRUSES

Positive Strand RNA Viruses

RNA viruses can be subdivided into groups based on type of RNA that serves as the genome. Positive or plus (+)-strand RNA viruses have genomes that are functional mRNAs (Table 10.1). Upon penetration into the host cell, ribosomes assemble on the genome to synthesize viral proteins. Genomes of positive-strand RNA viruses are single-stranded molecules of RNA and may be capped and polyadenylated. During the replication cycle of positive-strand RNA viruses, among the first proteins to be synthesized are those needed to synthesize additional genomes and mRNAs. Thus the infecting genome has two functions: It *is* an mRNA and also serves as the template for synthesis of additional viral RNAs. A functional definition of a positive-strand virus is that purified or chemically synthesized genomes are infectious (Fig. 10.2 and Box 10.1).

Positive-strand RNA viruses often use large complexes of cellular membranes for genome replication. They actively modify host cell membranes to construct viral replication scaffolds. Positive-strand RNA viruses of animals also use a common strategy to express RdRp. RdRp is a nonstructural protein, meaning that it is not found within the assembled virion. Instead it is translated directly from the infecting genome shortly after penetration. RdRp and other viral proteins needed for viral RNA synthesis are encoded as a *polyprotein* that is cleaved by virally encoded proteases. In the case of the picornaviruses and the flaviviruses, all viral proteins (structural and nonstructural) are synthesized as part of a single long polyprotein. Other positive-strand RNA viruses (i.e., togaviruses, coronaviruses, arteriviruses) synthesize an RdRp-containing polyprotein from genome-length mRNA, but use subgenomic mRNAs to encode structural and other proteins.

Negative Sense and Ambisense RNA Viruses

There are three additional groups of RNA viruses whose genomes are not mRNAs. They are the negative or minus (−)-strand RNA viruses, the closely related ambisense RNA viruses, and double-stranded RNA viruses (Table 10.2 and Box 10.2). For each of these groups of viruses, the first synthetic event after genome penetration is transcription (Fig. 10.3). This is accomplished by viral proteins (including viral RdRp) that enter cell with the genome. Transcription/replication complexes usually contain between two and four proteins. They associate with the genome through interactions with RNA-binding nucleocapsid (N) or capsid proteins. Therefore, naked (purified away from protein) genomic RNA is not infectious, cannot be translated, and will eventually be degraded if transcription is blocked. Before genome replication can proceed, viral mRNAs must be transcribed and translated. Because the virion of a negative-strand

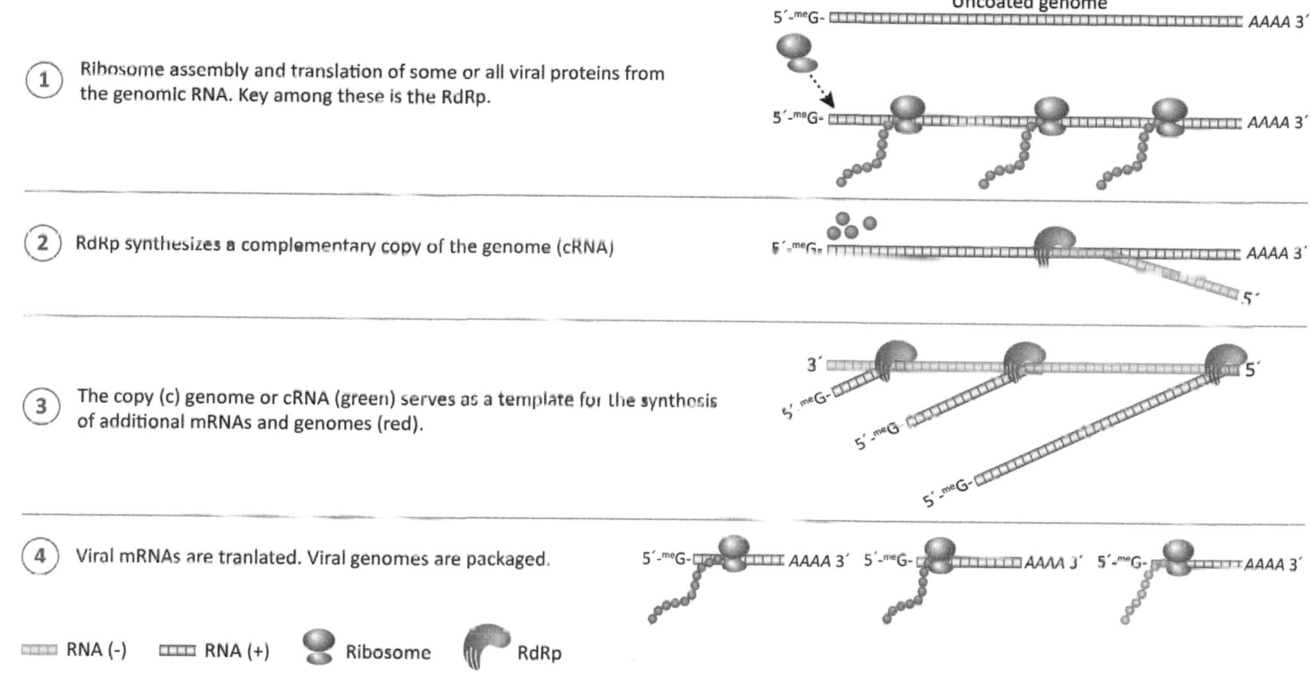

FIGURE 10.2 Schematic representation of replication of positive-strand RNA virus genomes. The genome of a positive-strand RNA virus is an mRNA that is translated, upon entry into the cells, to produce proteins needed for transcription and genome replication (for example, RdRp). After initial rounds of translation, the genome serves as the template for synthesis of copy RNA. The copy RNA is the template for synthesis of additional genomes and subgenomic mRNAs.

BOX 10.1

POSITIVE-STRAND RNA VIRUSES

- Purified genomes (or chemically synthesized genomes) are infectious if introduced into a permissive cell.
- The genome serves as an mRNA.
- The first synthetic event in the replication cycle is protein synthesis.
- Genome replication is cytoplasmic.
- Genomes of positive-strand RNA viruses fold into complex structures. These RNA structural elements have key roles in genome replication, transcription, and translation.
- Families include: *Picornaviridae, Flaviviridae, Togaviridae, Hepeviridae, Coronaviridae, Arteriviridae, Toroviridae, among many others.*

RNA virus contains RdRp, it is possible to synthesize viral mRNAs in a test tube (Fig. 10.4). If purified virions are gently lysed under appropriate buffer conditions, with the addition of NTPs, mRNAs will be transcribed in the test tube. However, genome RNA will not be synthesized under these conditions (Table 10.1). Finally, the genomes of negative-strand RNA viruses are often shown in diagrams with the 3′ end to the left, opposite to the usual convention of illustrating a strand of nucleic acid.

Negative-strand RNA viruses use the genome sense strand as the template for synthesis of all mRNAs. In contrast, viruses that use an ambisense coding strategy transcribe some mRNAs from the *copy* genome. There are virus families in which some members are considered negative-strand RNA viruses while others use an ambisense strategy. Thus these two strategies are closely related. Some ambisense viruses *package copy genomes* that can be used as templates for transcription, such that the full complement of viral genes can be transcribed soon after infection. It should be noted that packaged copy genomes are not mRNAs and are not translated.

dsRNA Viruses

Family *Reoviridae* is a large family of dsRNA viruses (Table 10.3). Reoviruses also have segmented genomes, packaging 11–12 segments of dsRNA. Reoviruses are nonenveloped and particles consist of two or three concentric icosahedral capsid layers. The genome segments are found within the innermost, $T=1$ icosahedral shell. A unique feature of the reovirus replication cycle is that the genome segments are transcribed from *within* the capsid. The mRNA products leave the capsid through pores at the vertices of the capsid (see Chapter 26: Family *Reoviridae*).

TABLE 10.2 Negative-Strand RNA Viruses of Animals[a]

Virus family	Virion morphology	Genome type
Arenaviridae	Enveloped, helical nuclecapsid	Bisegmented (some use ambisense coding strategy)
Bornaviridae[b]	Enveloped, helical nuclecapsid	Unsegmented
Bunyaviridae	Enveloped, helical nuclecapsid	Trisegmented (some use ambisense coding strategy)
Filoviridae[b]	Enveloped, helical nuclecapsid	Unsegmented
Nymaviridae[b]	Enveloped, helical nucleocapsid	Unsegmented
Orthmyxoviridae	Enveloped, helical nuclecapsid	Segmented
Paramyxoviridae[b]	Enveloped, helical nuclecapsid	Unsegmented
Pneumoviridae[b]	Enveloped, helical nuclecapsid	Unsegmented
Rhabdoviridae[b]	Enveloped, helical nuclecapsid	Unsegmented

[a]*Only those families with members that infect vertebrates are included in this list.*
[b]*Order* Mononegavirales.

STRUCTURAL FEATURES OF RNA VIRUS GENOMES

The genomes of RNA viruses have some common general features. Obviously there are one or more open reading frames that encode the viral proteins. But there are also regions of RNA that do not code for protein. These non-coding regions (NCRs) or untranslated regions (UTRs) are often highly conserved within a virus family, indicating that they have important functions. The preferred usage can vary by virus family but the terms NCR and UTR are largely interchangeable. NCRs may have specific, critical nucleotide sequences but in some cases they are regions of the genome that fold into conserved structures, and *structure* may be more critical than a specific sequence.

All RNA genomes have NCRs at their 5′ and 3′ ends. NCRs vary in size, from very long (several hundred nt) in the case of picornaviral 5′ NCRs to just a short base-paired hairpin in the case of flavivirus 3′ NCRs. Complementary RNA sequences are sometimes found at the 5′ and 3′ ends of RNA genomes, allowing them to circularize.

The 5′ and 3′ NCRs are required, and are often sufficient, to direct genome replication. Virally encoded proteins, such as the RdRp recognize specific sequences and/or structures at the ends of the genome. To simplify studies of the genome replication process, researchers often generate "minigenomes" that lack some virally encoded proteins. Of course a source of RdRp must be supplied. RdRp may be encoded in the minigenome or may be supplied *in trans* (by using a

BOX 10.2

NEGATIVE AND AMBISENSE-STRAND RNA VIRUSES

- Purified genomes (or chemically synthesized genomes) are NOT infectious.
- RdRp is packaged within the virion.
- Genomes serve as templates for transcription and the first synthetic event in the replication cycle is mRNA synthesis.
- Many replicate in the cytoplasm, a few replicate in the nucleus.
- Viral genomes are often tightly associated with a nucleocapsid (N) protein.
- Families of negative-strand RNA viruses include *Orthomyxoviridae*, *Paramyxoviridae*, *Rhabdoviridae*, *Bornaviridae*, and *Filoviridae*. The *Bunyaviridae* includes some members that use a negative sense coding strategy and other that use an ambisense coding strategy.

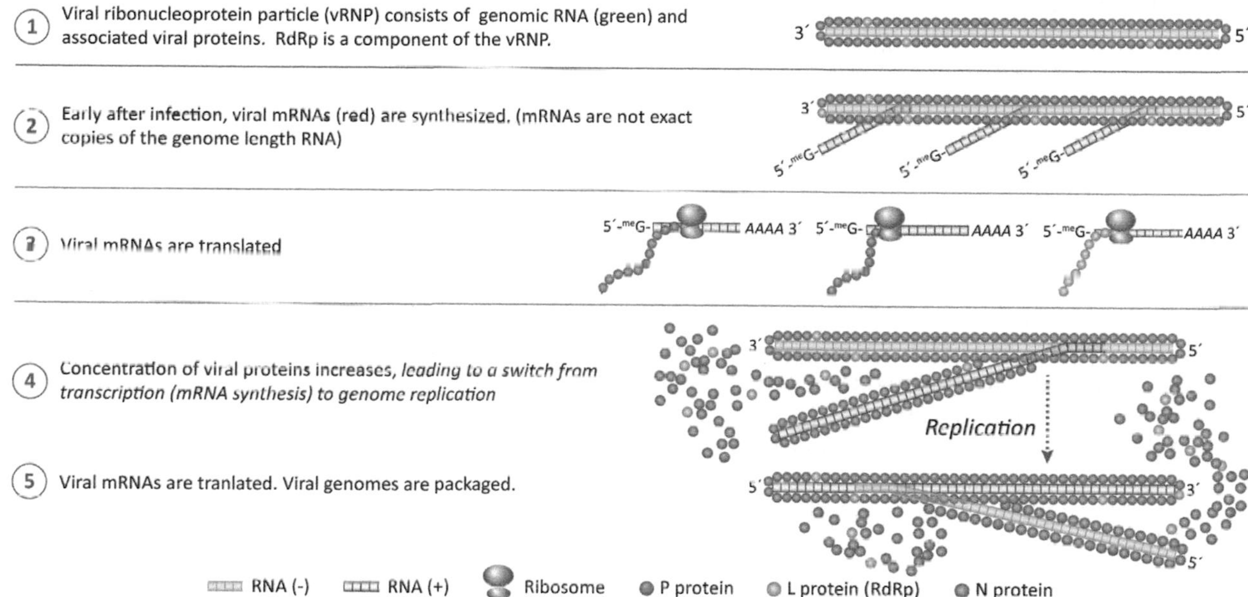

FIGURE 10.3 Schematic representation of replication of genomes of minus-strand RNA viruses. Genomes are associated with RNA-binding proteins and RdRp. Upon entry into the cell, the active *transcription complex* synthesizes mRNAs. This process can also occur in a test tube (see Fig. 10.4). Translation of mRNAs produces proteins required for genome replication. (Thus if protein synthesis is blocked in the infected cell, mRNAs continue to by synthesized but genome replication does not occur.) Newly synthesized proteins provide the switch from transcription to genome replication.

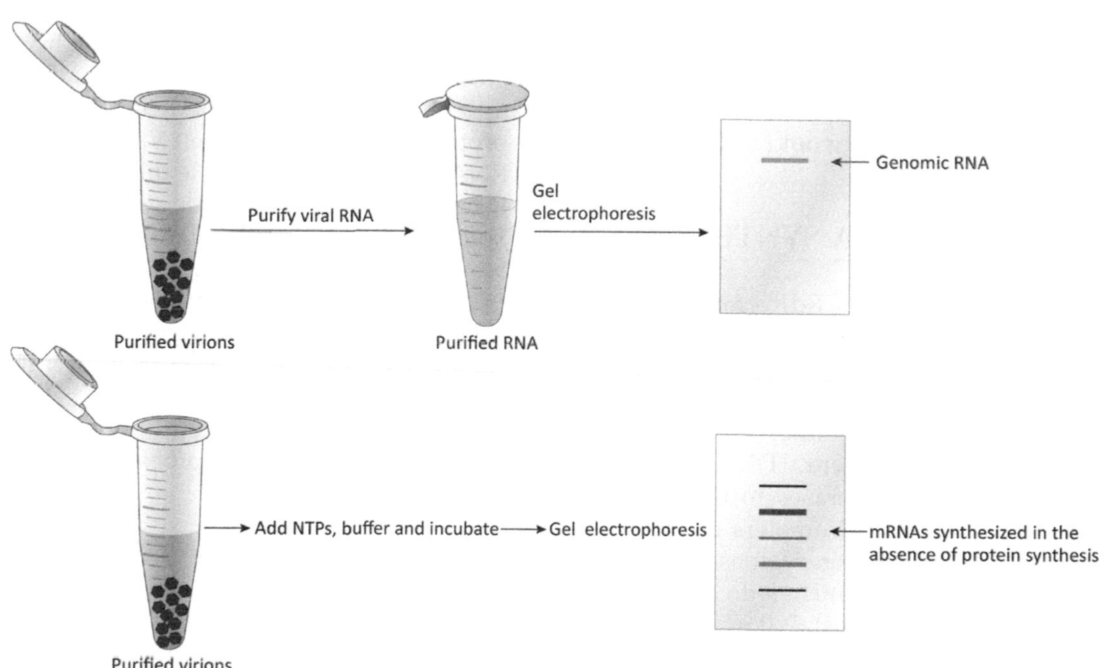

FIGURE 10.4 The genomes of viruses in the order *Mononegavirales* are unsegmented, negative-strand RNA. If RNA is extracted from purified virions and analyzed by gel electrophoresis, a single RNA is seen (top panel). However, virions contain RdRp and if they are gently lysed, in a buffer solution containing NTPs, mRNAs are synthesized. The lower panel shows the products of mRNA synthesis as a set of subgenomic mRNAs of different sizes. Note that genome synthesis does not occur under these conditions.

TABLE 10.3 Double-Stranded RNA Viruses of Animals[a]

Virus family	Virion morphology	Genome type
Birnaviridae	Nonenveloped, icosahedral	Bisegmented
Picobirnaviridae	Nonenveloped, icosahedral	Bisegmented
Reoviridae	Nonenveloped, icosahedral with multilayered capsids	Segmented (10–12)

[a]*Only those families with members that infect vertebrates are included in this list.*

cell line stably expressing the viral RdRp, for example). The sequences required to direct RNA replication are often fairly simple and can be linked to virtually any RNA sequence to drive its replication.

Many RNA genomes also have promoters to direct synthesis of subgenomic mRNAs. These promoter sequences can be rather short but provide a means to direct the RdRp to internal sites on the genome. There may also be specific RNA sequences that signal polyadenylation. There are a variety of different strategies that RNA viruses use to regulate transcription and genome replication, but all involve RNA sequences found in the genome.

The RNA genomes of some viruses are highly structured and extensively base paired. An example is the *internal ribosome entry sites* (IRES) of picornaviruses (Chapter 11: Family *Picornaviridae*). The IRES serves as a platform for ribosome assembly. Picornaviruses are among the RNA viruses that encode RNA-helicases to unwind highly structured regions of the genome, such as the IRES. Without the help of a helicase, the RdRp would not be able to disrupt the base-paired RNA in the IRES and genome replication would stall.

PRIMING VIRAL RNA SYNTHESIS

While all RNA viruses use an RdRp for replication and transcription, there are a variety of strategies used for priming RNA synthesis, regulating RNA synthesis, and capping and polyadenylating mRNAs.

In the eucaryotic cell, RNA synthesis (from a DNA template) is primer independent. DNA sequences (promoters) direct RNA polymerases to the "correct" position on the DNA genome. Promoters can be quite long and complex and promoter regions themselves are not transcribed. What mechanisms do viral RdRps use to prime viral RNA synthesis? It is particularly important, in the case of genome synthesis, that genetic information not be lost or modified; however, mRNAs are often capped and polyadenylated. Are the methods for priming viral mRNA synthesis the same or different from the methods of priming genome replication? The RNA viruses seem to have experimented widely. Various mechanisms by which RdRps initiate RNA synthesis are described below.

- De novo or primer-independent transcription and replication employs the RdRp, the RNA template, an initiation NTP, and a second NTP. The initial NTP can be considered the primer. The initial complex is not highly stable.
- De novo initiation is sometimes used to generate short RNA oligonucleotides that are subsequently used as primers in a mechanism known as *prime and realign*. The short internally initiated oligos slip back to extend from 5′ end of the template. Bunyaviruses and arenaviruses use this process. As a result the 5′ ends of the genome and copy genome contain nontemplated nucleotides.
- RNA structures such as hairpins can be used for priming in a process where the 3′ end of the template RNA loops back upon itself to serve as the primer. This process is "template priming" as part of the template (the genome) functions as the primer.
- Some RNA viruses (i.e., picornaviruses) use a protein to prime RNA replication. A tyrosine hydroxyl group on the primer protein (VPg) is linked to GTP to serve as the first nucleotide in the new strand. The protein interactions between VPg, RdRp, and the RNA template form a stable initiation complex (Chapter 11: Family *Picornaviridae*).
- Some RNA viruses (for example, influenza viruses) "steal" short (10–15 nt) capped oligonucleotides from cellular mRNAs to prime viral mRNA synthesis (a process called cap-snatching) (Chapter 23: Family *Orthomyxoviridae*). This priming process cannot be used for genome synthesis and indeed, the genome segments of influenza viruses are primed by a de novo process.

MECHANISMS TO GENERATE CAPPED mRNAs

RNA viruses use different mechanisms to cap their mRNAs. In the case of the influenza viruses the fragment of cellular mRNA used for priming viral mRNA synthesis also provides the methyl-G cap (Chapter 23: Family *Orthomyxoviridae*). The RdRp proteins of many RNA viruses have methylase activities, thus the RdRp also synthesizes the cap. Some RNA viruses (for example, piconaviruses) do not use capped mRNAs.

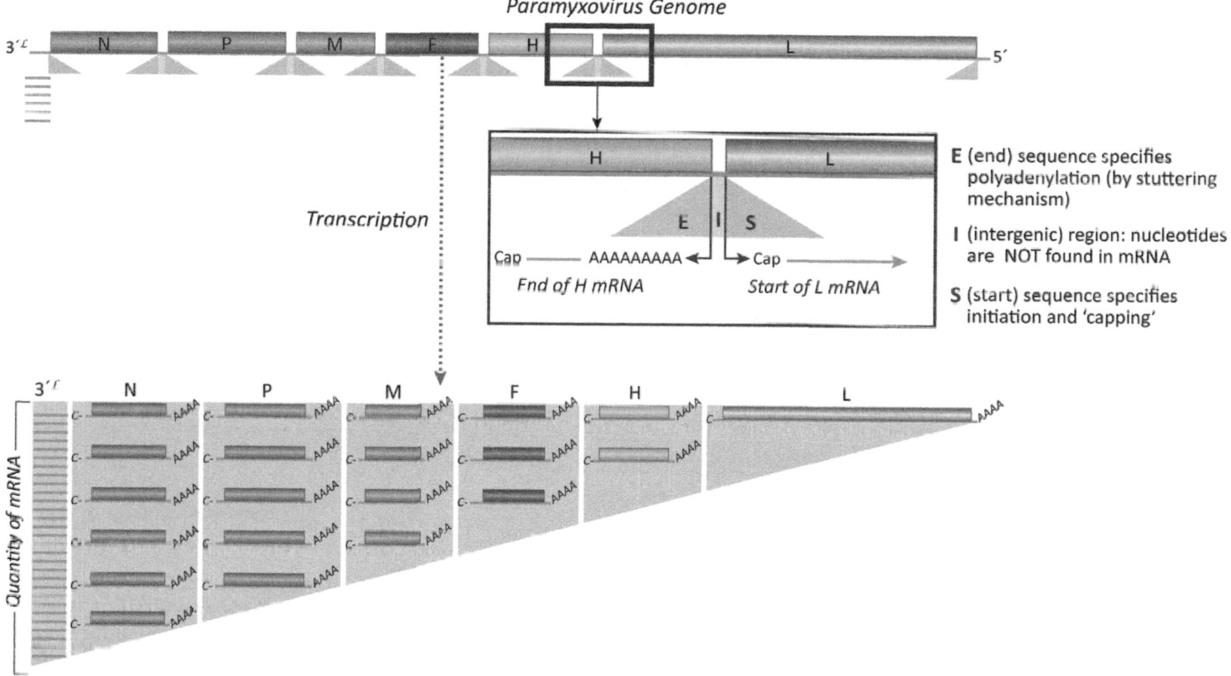

FIGURE 10.5 A strategy to regulate mRNA synthesis. This figure shows the organization of a paramyxovirus genome (paramyxoviruses are members of the order *Mononegavirales*; negative-strand RNA viruses with unsegmented genomes). Each protein-coding region is flanked by regulatory sequences that control capping and polyadenylation. The order of the genes on the genome regulates the relative quantities of mRNAs synthesized. The RdRp always initiates mRNA synthesis at the very 3′ end of the genome. After polyadenylating an upstream mRNA, RdRp must remain associated with the genome to initiate synthesis of the downstream mRNA. If the RdRp dissociates from the genome, it cycles back to the 3′ end to reinitiate transcription. Because RdRp *does* often dissociate from the genome during transcription, the downstream genes are produced in lower quantities. Note that the N protein (a structural protein required in large amounts) is positioned at the 3′ end of the genome while the L protein (the RdRp), an enzyme is positioned at the 5′ end of the genome.

MECHANISMS TO GENERATE POLYADENYLATED mRNAs

RNA viruses also use a few different mechanisms to polyadenylate their mRNAs. For example, the picornaviruses use poly A tracts encoded in the genome. Among the negative-strand RNA viruses, those in the order *Mononegavirales* use a stuttering mechanism to synthesize long poly A tracts from short poly U tracts (Fig. 10.5). Other RNA viruses do not polyadenylated their mRNAs. The flaviviruses have a short hairpin at the 3′ of their genome-length mRNA.

MECHANISMS TO REGULATE SYNTHESIS OF GENOMES AND TRANSCRIPTS

Even with fairly simple genomes, RNA viruses must, and do, regulate the amounts of genome, copy genome, and mRNAs that are synthesized during an infection. It would be "wasteful" if a positive-strand RNA virus had to make a new copy genome for synthesis of every genome. It is much more efficient to synthesize many genomes from each copy genome. *Having different structures at the 5′ and 3′ ends of the genome and copy RNA would be one way to accomplish this.* We will see in a later chapter that togaviruses (Chapter 16: Family *Togaviridae*) use slightly different versions of the replicase complex to synthesize genomes versus copy genomes and that the replicase proteins are "regulated" by proteolytic cleavage. Internal promoters for mRNA synthesis can vary in sequence, controlling the relative affinity of the transcription complex for each mRNA.

A large group of unsegmented negative-strand RNA viruses (Order *Mononegavirales*) synthesize mRNAs sequentially, from the 3′ end to the 5′ end of the infecting genome (Chapters 19–22: Family *Rhabdoviridae*, Families *Paramyxoviridae* and *Pneumoviridae*, Family *Filoviridae*, and Family *Bornaviridae*) (Fig. 10.5). Those genes closest to the 3′ end of the genome are more abundantly expressed, as the transcription complex always initiates synthesis at the 3′ end of the genome.

RNA VIRUSES AND QUASISPECIES

An important feature of RNA viruses is that many exist in nature as *quasispecies*. The term quasispecies is

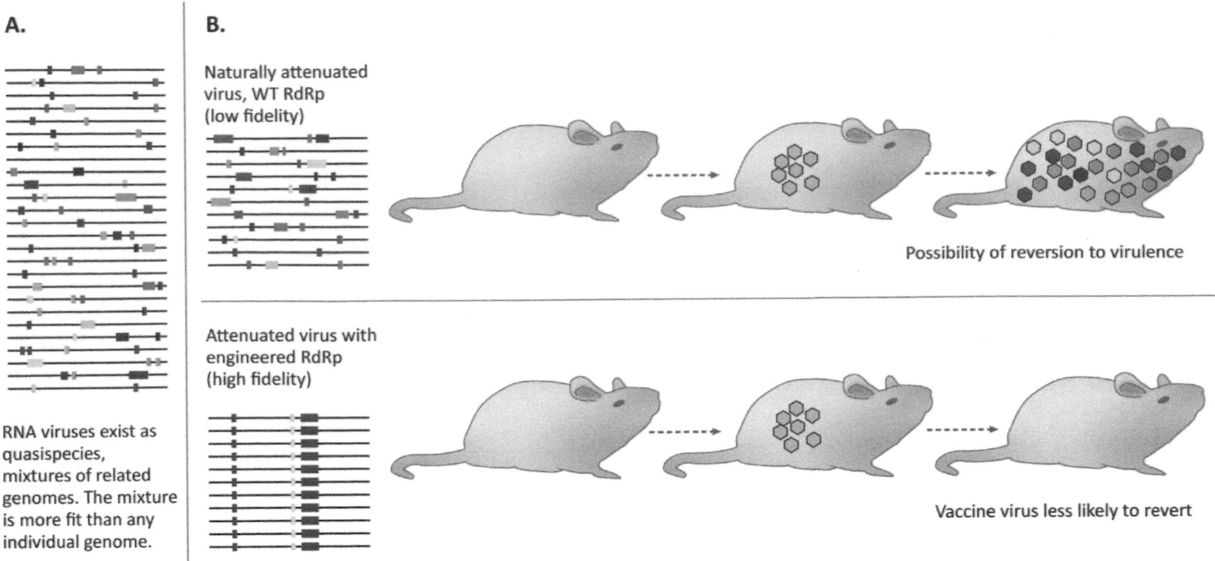

FIGURE 10.6 (A) Positive-strand RNA viruses exist as quasispecies, complex mixtures of related genomes. The mixture is more fit than any individual genome; fitness is maintained by generation of new variants in response to selective pressures. (B) Potential for safer vaccines. If the fidelity of RdRp is increased the population remains more homogeneous. Therefore an attenuated virus with a high-fidelity RdRp is more likely to remain attenuated.

BOX 10.3

RNA VIRUS QUASISPECIES AND FITNESS

The RdRps of RNA viruses control RNA synthesis error rates. Molecular/biochemical studies have revealed that RdRp mutations can have a measurable impact on fidelity and this in turn measurably impacts virus fitness. A well-studied example is polio virus (PV) a picornavirus. Vignuzzi et al. (2006) characterized a PV mutant with a high-fidelity polymerase. While the mutant replicated as well as the wild-type virus (produced equivalent numbers of virions), there was measurably less diversity in the population. The less diverse population performed poorly, when compared to the wild-type virus, when exposed to *adverse* growth conditions. (Adverse conditions included exposure to an antiviral drug in cultured cells and inoculation into mice.) The more diverse population was more adaptable and fit under these circumstances!

Based on this result, it might be tempting to assume that a PV mutant with a less faithful RdRp (generating a more highly mutated, thus diverse population) would have increased fitness compared with wild-type PV. However this is not the case, as demonstrated in a study published by Korboukh et al. (2014). These researchers identified a PV with a single amino acid substitution in the RdRp that produced a "mutator phenotype." The progeny population was 2–3 fold more diverse than the wild-type population. This mutant population could not compete with the wild-type population and was driven to extinction in direct competition studies. Thus increasing the wild-type mutation rate was also deleterious. There were simply too many mutations in each genome produced; the progeny were not viable.

While these studies might seem esoteric, they actually provide practical and valuable insights. For example, they have led to proposals that drugs to increase RdRp error rates (as opposed to inhibiting these enzymes) will increase RNA virus mutation rates to a catastrophic level. On the other hand, attenuated vaccine strains might be engineered to have more faithful RdRps thereby decreasing their ability to revert to virulence (Fig. 10.6).

used to describe a group of closely related, but nonidentical genomes (Fig. 10.6). (Quasispecies are not unique to RNA viruses, some retroviruses, HIV for example, exist as quasispecies).

Poliovirus (PV) is a good example of a virus that forms a quasispecies. If one examines genome sequences from a mouse experimentally infected with PV serotype 1, we find that the genomes are not identical, although they are all clearly related to one another. So, which one of these genomes is the "best" or the "fittest"? To the surprise of many virologists, it turns out that the population (quasispecies) may be more fit than any individual genome. Or put another way, we cannot find any single genome in the population that replicates better than the group as a whole and in fact, most individual genomes replicate more poorly than the group. Why this occurs is not always clear, but an animal is a very complex ecosystem. Different members of the quasispecies may be better adapted to different niches in the animal.

How does a quasispecies form? PV genomes can be chemically synthesized and/or cloned to generate a single genome sequence. But as the cloned virus replicates, mutations accumulate (generating a quasispecies). Measurable levels of mutation occur because the *fidelity* of PV RdRp is low. RdRps do not have proofreading activities (as do many DNA polymerases). If a mistake occurs, there are only two possibilities: RNA synthesis can stop, or RNA synthesis can continue beyond the mistake to generate a point mutation. The fidelity of RdRps has been studied and they generate transition mutations at a frequency of 10^{-3}–10^{-5} (one transition mutation for every 10^3–10^5 nt synthesized) and transversion mutations at a frequency of 10^{-6}–10^{-7} (one transversion mutation for every 10^6–10^7 nt synthesized). A rate of one mutation per 10^5 nucleotides synthesized ensures that during an infection, many progeny will contain a mutation. The significance of the quasispecies model for RNA viruses is that there is no single "fittest" or "best" genome sequence. *The population is fitter than the individual* (Box 10.3).

In this chapter we have learned:

- The definition of an RNA virus.
- The subgroups within the RNA virus group (positive strand, negative strand, ambisense, and double strand) and the differences in their replication strategies.
- The types of noncoding regions commonly found in the genomes of RNA viruses.
- The different mechanisms used to prime RNA synthesis.
- That some RNA viruses exist as quasispecies, mixed populations that are more fit than any individual member.

References

Bollati, M., Alvarez, K., Assenberg, R., Baronti, C., Canard, B., Cook, S., et al., 2009. Antiviral Res. 87, 125–128.

Korboukh, V.K., Lee, C.A., Acevedo, A., Vignuzzi, M., Xiao, Y., Arnold, J.J., et al., 2014. RNA virus population diversity, an optimum for maximal fitness and virulence. J. Biol. Chem. 289, 29531–29544. Available from: http://dx.doi.org/10.1074/jbc.M114.592303, Epub 2014 Sep 11.

Vignuzzi, M., Stone, J.K., Arnold, J.J., Cameron, C.E., Andino, R., 2006. Quasispecies diversity determines pathogenesis through copperative interactions in a viral population. Nature. 439, 344–348. Available from: http://dx.doi.org/10.1038/nature04388, Epub 2005, Dec 4.

CHAPTER

11

Family *Picornaviridae*

OUTLINE

After reading this chapter, you should be able to answer the following questions:

- What are the key features of all members of the family *Picornaviridae*?
- What strategies do picornaviruses use for protein expression?
- What is the primer for genome replication?
- What strategy is used by picornaviruses for ribosome assembly/translation initiation?
- At what site in the cell do picornaviruses replicate?
- What are some of the major human and animal diseases caused by picornaviruses?

Members of the family *Picornaviridae* are small (pico) RNA viruses. Family *Picornaviridae* is one of five named families in the order *Picornavirales*, a large group of unenveloped positive-strand RNA viruses of plants, insects, and animals. Virions are unenveloped with $T = 3$ icosahedral capsids (Fig. 11.1 and Box 11.1).

GENOME ORGANIZATION

Picornaviral genomes are a single molecule of positive-sense, single-strand RNA (Fig. 11.2). Genome sizes range from ~7500 to 9000 nucleotides (nt). The 5′ terminus of all picornaviral genomes is covalently linked to a small protein called VPg (Viral Protein, genome linked). The picornaviral genome has a long untranslated region (UTR) (called the 5′ UTR) that ranges from 600 to 1200 nt in length. This region of folds into a stem-loop structure that functions as an internal ribosome entry site (IRES). Recall that the IRES serves as a scaffold for ribosome assembly (Fig. 3.13) thus picornaviral mRNAs can be translated even though they lack a 5′-methyl-G cap. There is also an UTR at the 3′ end of picornaviral genomes (40–330 nt). The very 3′ end of picornaviral genomes is a stretch of polyadenine (polyA) that ranges from 35 to 100 nt. Another important structural region is found within the coding region of the PV genome: A 61 nt stem-loop structure required for genome replication and called the *cis-acting replication element*.

Between the 5′ and 3′ UTRs of all picornaviruses is a single open reading frame (ORF) encoding a polyprotein of ~2300 amino acids. Structural (capsid) proteins are encoded at the 5′end of the ORF while the RdRp (also called $3D^{pol}$) is encoded at the 3′ end of the ORF. In between are other nonstructural proteins required for replication, including proteases, helicases, and ATPases. The protein expression strategy of the picornaviruses is to generate all mature proteins from

Viruses. DOI: http://dx.doi.org/10.1016/B978-0-12-803109-4.00011-8

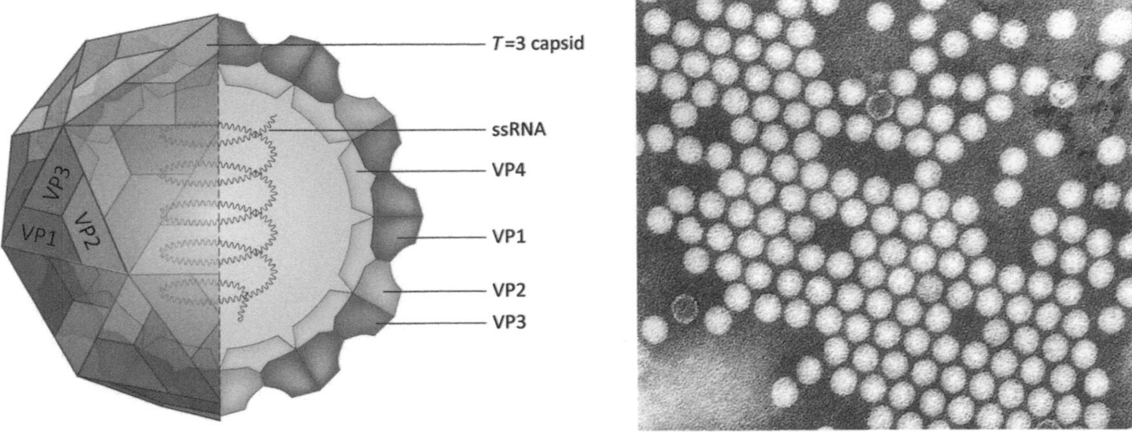

FIGURE 11.1 Virion structure of a picornavirus (left) and negative stain electron micrograph of PV (right). *From CDC/Dr. Fred Murphy, Sylvia Whitfield.*

BOX 11.1

GENERAL CHARACTERISTICS

Picornavirus genomes are a single strand of positive-sense RNA ranging in size from 7.5 to 9.0 kb. Genomes are covalently attached to a viral protein at their 5′ end. Genomes are polyadenylated at the 3′ end. A long UTR at the 5′ end of the genome folds into a complex stem-loop structure to form an IRES. Genome replication is cytoplasmic. A protein primer (VPg) is used to primer RNA synthesis. Genomes contain a single long ORF that encodes the full complement of viral proteins. Proteins are produced by proteolytic processing of a long precursor.

Virions are unenveloped with $T = 3$ icosahedral symmetry, ~30 nm diameter. Capsids are built from 180 capsid proteins (60 copies each of VP1, VP2, VP3). A fourth structural protein, VP4, is located inside of the capsid.

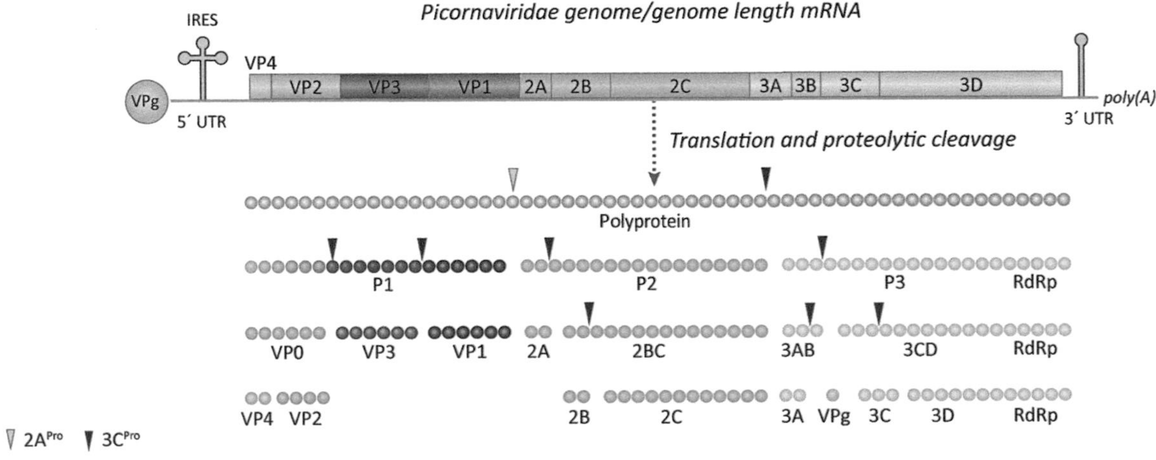

FIGURE 11.2 Schematic diagram of a picornaviral genome showing a single long ORF flanked by 5′ and 3′ UTRs. The 5′ end of the genomic RNA is covalently linked to a small viral protein, VPg. A highly base-paired structure in the 5′ UTR, called the internal ribosome entry site or IRES, is the site of ribosome assembly.

a single polyprotein precursor, by a series of stepwise proteolytic cleavages. The total number of protein products ranges from 11 to 15.

VIRION STRUCTURE

Virions are unenveloped with $T = 3$ icosahedral symmetry. Virions are about 30 nm in diameter. Virions are assembled from 180 capsid proteins: 60 copies each of VP1, VP2, VP3 assemble to form the outer shell (picornavirus capsids are sometimes referred to as pseudo $T = 3$ because they are not assembled from 180 copies of a single capsid protein). A fourth structural protein, VP4 is inside the capsid. The structural proteins are produced as part of a long precursor polyprotein. It is likely that the three structural proteins assemble into a protomer as the precursor is cleaved. VP2 is actually part of a precursor (VP0) that undergoes a maturation cleavage after capsid formation to generate VP2 and VP4.

PICORNAVIRUS REPLICATION (POLIOVIRUS MODEL)

For many centuries paralytic poliomyelitis, although endemic worldwide, was a relatively rare disease. However this changed in the early 20th century with major epidemics that began in Europe. By 1910, epidemics of paralytic polio were occurring throughout the developed world, including the United States, and the infectious agent, poliovirus (PV) had been isolated. By the 1940s and 1950s, polio was killing or paralyzing over half a million people per year. In the United States, president Franklin Roosevelt (a polio survivor) pushed for creation of the National Foundation for Infantile Paralysis. Today the Foundation is known as the March of Dimes. The Foundation supported basic research on PV and supported development of vaccines. Fundamental aspects of picornaviral structure, genome organization, protein expression, and assembly were first described for PV. The replication cycle of PV, summarized here (Fig. 11.3), serves as a model for many other picornaviruses (Box 11.2).

Binding and Penetration

PV virions bind to a cell protein called the poliovirus receptor (PVR). The PVR is an integral membrane protein and member of the immunoglobulin superfamily of proteins. The PVR molecule has three immunoglobulin-like domains that extend from the cell surface. (The receptors for other picornaviruses include a wide variety of cell surface molecules including CD55 of the complement cascade, ICAM-1 (a cellular adhesion molecule), integrins, and heparin

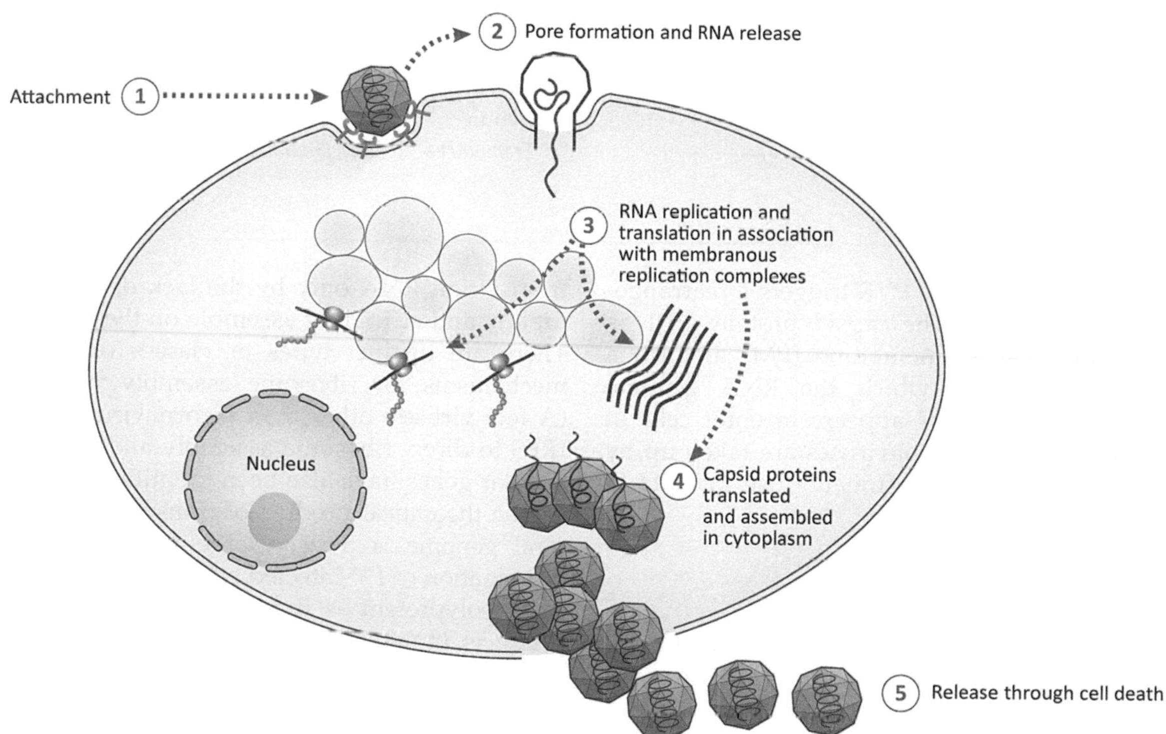

FIGURE 11.3 Overview of the picornavirus replication cycle showing replication in the cytoplasm. RNA synthesis is membrane-associated. Virions assemble in the cytoplasm. Replication is often cytopathic.

BOX 11.2

TAXONOMY

Order
Picornavirales

Family
Picornaviridae

Genus	Species
Genus *Apthovirus*	Species include *Foot and mouth disease virus, equine rhinitis A virus*
Genus *Aquamavirus*	(One species)
Genus *Avihepatovirus*	Species include *Duck hepatitis A virus*
Genus *Avisivirus*	*Avisivirus A, B and C*
Genus *Cardiovirus*	Species include *Encephalomyocarditis virus, Theilovirus*
Genus *Cosavirus*	*Cosavirus A, B, D, E, F*
Genus *Dicipivirus*	*Cadicivirus A*
Genus *Enterovirus*	Species include *Human enterovirus A, B, C, D, Human rhinoviruses A, B, C, Bovine enterovirus, Porcine enterovirus B*
Genus *Erbovirus*	*Erbovirus A (Equine rhinitis B virus)*
Genus *Gallivirus*	*Gallivirus A*
Genus *Hepatovirus*	*Hepatitis A - I*
Genus *Hunnivirus*	*Hunnivirus A*
Genus *Kobuvirus*	Species include *Aichi virus, Bovine kobuvirus*
Genus *Megrivirus*	*Megrivirus A*
Genus *Mischivirus*	*Mischivirus A, B, C*
Genus *Mosavirus*	*Mosavirus A*
Genus *Oscivirus*	*Oscivirus A*
Genus *Parechovirus*	Species include *Human parechovirus, Ljungan virus*
Genus *Pasivirus*	*Pasivirus A*
Genus *Passerivirus*	*Passerivirus A*
Genus *Rosavirus*	*Rosavirus A*
Genus *Salivirus*	*Salivirus A*
Genus *Sapelovirus*	Species include *Porcine, Simian, Avian sapeloviruses*
Genus *Senecavirus*	Species include *Seneca Valley virus*
Genus *Teschovirus*	Species include *Porcine teschovirus*
Genun *Tremovirus*	Species include *Avian encephalomyelitis virus*

sulfate.) Binding of PV to the PVR triggers a rearrangement of capsid proteins. The capsid protein, VP1, is inserted into the plasma membrane (PM) to form a protein channel through which the RNA genome enters the cell (Fig. 3.4). PV appears to enter cells at the PM although other picornaviruses are taken up by endocytosis (these viruses presumably use low pH to trigger capsid rearrangements).

Translation

The picornavirus genome is an mRNA and the full complement of viral proteins is synthesized from the poly(A)-containing genome length mRNA. VPg, a small peptide covalently linked to the 5′ end of the genome, is quickly removed from the infecting genome and the mRNA is translated. Thus, PV mRNA differs from virion RNA only by the lack of VPg. Ribosomes rapidly and efficiently assemble on the IRES (Fig. 11.4). There are distinct types or classes of IRES and the mechanisms for ribosome assembly vary somewhat. (A few viruses, other than picornaviruses, also use an IRES to direct ribosome assembly and IRES-containing cellular genes have also been identified.)

The theoretical product of translation of the picornaviral genome is a very large polyprotein. However, examination of PV-infected cells reveals little to no complete polyprotein as it is rapidly cleaved into smaller products. In the case of PV, the first polyprotein cleavage is carried out *in cis* by the viral protease ($2A^{Pro}$) to release the P1 polyprotein (Fig. 11.2). PV encodes a second protease ($3C^{Pro}$) that performs additional cleavages.

Very soon after PV infection, translation of host proteins is inhibited. This happens because a PV protease

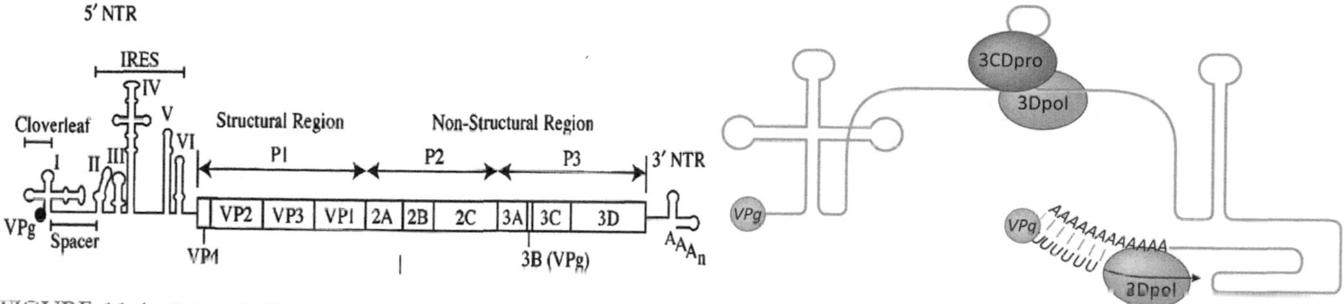

FIGURE 11.4 Internal ribosome entry site. Regions at the 5′ end of the picornavirus genome fold into complex base-paired structures that form sites for cap-independent ribosome assembly.

FIGURE 11.5 Minus-strand RNA synthesis. Schematic representation showing picornaviral RdRp ($3D^{Pol}$) using a protein primer to initiate RNA synthesis. VPg is urydylated by $3D^{Pol}$ (as part of protein complex, not shown) to produce polyurydylated VPg, which binds the 3′ end of the genome strand to serve as a primer.

($3C^{pro}$) cleaves the cell "cap-binding protein" a critical component of the ribosome initiation complex. Inhibition of cap-dependent translation allows the virus exclusive access to the cell's translation machinery. In addition, inhibiting cellular protein synthesis interferes with the cell's ability to mount an antiviral defense. (Picornaviruses vary in their ability to shut off host cell translation, and among those that do so, the mechanisms vary.)

Genome Replication

In order for the PV genome to be replicated, it must be free of ribosomes. Both virally encoded and host proteins are required for genome replication, which occurs in the cytoplasm and is associated with cell membranes. Picornaviral RdRp ($3D^{pol}$) uses a protein primer to initiate RNA synthesis. VPg is a short (22–24 amino acids) polypeptide found covalently linked to the 5′ ends of newly synthesized viral RNAs. To serve as a primer, VPg is first urydylated by $3D^{Pol}$. UMP is linked to VPg via the hydroxyl group on a tyrosine near the N-terminus of VPg. Additional U residues are then added. This generates the urydylated VPg that interacts with the polyA tail on the genome strand, to initiate synthesis of the "negative" or copy RNA (cRNA) strand (Fig. 11.5). The cRNA is not an mRNA and is not translated. It functions as the template for synthesis of additional genomes. Uridylated VPg primes the synthesis of all picornaviral RNAs, thus all copies of viral RNAs are covalently linked to a molecule of VPg.

The synthesis of viral RNAs heavily favors genome strands. The ratio of genome to cRNA is 30–70:1 in the infected cell. Thus each cRNA is the template for many genomes. The synthesis of genome strands is favored because formation of replication complexes is much more efficient on the 3′ end of the cRNA than on the 3′ end of the viral genomic RNA.

Assembly

The process of capsid assembly begins when local concentrations of proteins are sufficiently high. VP1, VP3, and VP0 (VP2 + VP4) form protomers and five protomers assemble to form a pentamer. Twelve pentamers assemble to form an empty capsid. RNA is then inserted into the empty capsid.

Maturation

PV virions are released upon cell death. After release VP0 is cleaved to produce VP2 + VP4. VP4 is located inside the capsid, associated with RNA. Now the virion is ready to infect a new cell.

PICORNAVIRUSES AND DISEASE

To date there are 26 Genera in the family *Picornaviridae*. Of importance to human health are several members of the genus *Enterovirus*. Primary sites of replication of members of the genus *Enterovirus* are epithelial cells and lymphoid tissues of upper respiratory tract (rhinoviruses) and gastrointestinal tracts. In the gastrointestinal tract, enterovirus infections are very often asymptomatic.

Genus *Enterovirus*

PV is a member of the genus *Enterovirus* and is transmitted by the fecal-oral route, for the most part causing inapparent infections. However, PV can cause a viremia and may cross the blood–brain barrier resulting in the clinical disease we call paralytic poliomyelitis. PV can also infect the spinal cord to cause aseptic meningitis. (See also Box 11.3, history and pathogenesis of PV.)

BOX 11.3

PV HISTORY, VACCINES, AND ERADICATION

For many centuries, paralytic polio, although endemic worldwide, was relatively rare. However this changed in the early 20th century with major epidemics that began in Europe. By 1910 epidemics of paralytic polio were occurring throughout the developed world (including the United States) and the infectious agent, PV had been isolated. By the 1940s and 1950s, polio was killing or paralyzing over half a million people per year.

Early studies of PV were hampered by the belief that it replicated only in neurons (made sense, the disease was neurological). Vaccines were first introduced in 1935 but failed because it was not known that there were three distinct serotypes of PV. By 1948–49 it was understood that vaccine preparations must protect against all three serotypes. Another important breakthrough occurred in 1948 when Drs. Enders, Weller, and Robbins demonstrated that PV replicates in intestinal cells. This seemingly simple finding explained the epidemiology of the disease and lead to an understanding that the PV was an enteric virus, spread by the fecal-oral route. (The discovery also made culture of PV much easier, and facilitated vaccine development. For their discovery, Drs. Enders, Weller, and Robbins were awarded the Nobel Prize in Physiology or Medicine in 1954.)

How/why did a rare paralytic disease of children and infants suddenly explode into a worldwide epidemic? Did the virus change, or was it something else? Evidence points to improved sanitation as a major cause of the PV epidemic. Historically, infants were exposed to PV very early in life but they were also protected from neurologic disease by maternal antibody. Most adults were immune to PV through repeated exposure as children. However, as clean drinking water became the standard in the developed world, infants were no longer exposed to PV. By the early 20th century millions of children and adults had never been exposed to the virus thus were not immune. The existence of a large susceptible population set the stage for notable outbreaks of paralytic disease.

How does an enteric virus cause paralytic disease? PV is usually transmitted by the fecal-oral route, replicating in epithelial cells in the gastrointestinal tract and lymphoid tissue such as the Peyer's patches. When virus is limited to the gastrointestinal tract, there is no paralytic disease even though large amounts of virus are produced and excreted. However PV can move out of lymphoid tissues to produce a viremia. Rarely (0.5%–2% of infections) the virus gains access to the brain or spinal cord. Within the CNS, PV has a tropism for motor neurons and their destruction results in muscle weakness and/or paralysis.

Two different types of vaccines were developed to protect against poliomyelitis. A comparison of these vaccines is quite instructive. Dr. Jonas Salk led the effort to create a killed vaccine. In brief, he developed methods to culture all three PV serotypes, purified the virions, and inactivated them (with formalin) to produce an injectable vaccine. The inactivated, injectable vaccine induces production of neutralizing antibodies to PV capsid proteins. Most of the antibody response is found as circulating IgG, which protects against CNS disease quite efficiently. Salk's vaccine did not protect against infection however, orally ingested PV could replicate in the gastrointestinal tract.

Dr. Albert Sabin led the effort to create an oral vaccine containing live-attenuated (weakened) viruses. The attenuated viruses replicate in the gastrointestinal tract, stimulating development of immune responses that protect against infection with wild-type viruses. The weakened viruses in the oral vaccine are unable to reach the CNS.

Both vaccines have relative advantages and disadvantages:

Salk's vaccine (inactivated PV or IPV) must be injected. As IPV cannot replicate, it is very safe. Vaccination does not lead to infection and the vaccine itself cannot be transmitted. Most people get vaccinated when they are children. Current recommendations call for four doses of IPV.

Sabin's vaccine is given orally (today it is called oral poliovirus vaccine or OPV), thus is easier and cheaper to administer. However OPV contains live virus and should not be given to immunocompromised individuals. OPV replicates in the gastrointestinal tract thus can be transmitted to contacts of vaccinated individuals. In the mid-20th century this was considered an advantage by some, as it expanded the number of protected individuals in the population. However, replication meant that the virus could (and did) change. So-called "revertant" viruses would (rarely) cause paralytic disease in vaccinated individuals or their contacts.

The use of killed and attenuated PV vaccines has changed over the years. The Salk vaccine was the first PV vaccine licensed in the United States. However early on, a poorly prepared batch of vaccine (containing replication competent virus) resulted in many cases of vaccine-associated paralytic disease, diminishing initial enthusiasm for the Salk vaccine. While the live-attenuated Sabin vaccine did (rarely) cause vaccine-associated cases of paralytic polio, for many years its advantages were thought to

BOX 11.3 *(cont'd)*

outweigh the risk. However, as the incidence of vaccination increased and cases of natural disease plunged, risks of the attenuated vaccine became unacceptable and OPV was banned in the United States by the year 2000.

OPV continues to be used in parts of the world where health care infrastructure does not allow for use of IPV. Use of OPV is a key part of the World Health Organization's drive to eradicate PV. Indeed, use of OPV has reduced cases of paralytic poliomyelitis by 99%. The original OPV formula included three attenuated strains, PV1, PV2, and PV3. Recently it has been suggested that OPV be reformulated to eliminate PV2. All cases of PV2-associated paralytic polio since 1999 have been vaccine-associated and PV2 was declared eradicated in 2015. Thus there is measurable risk, but no known benefit, to including PV2 in current OPV formulations.

The goal of worldwide elimination of polio is close, but has not yet been achieved due to political instability, conflict, hard-to-reach populations, and poor health care infrastructure. The complete elimination of polio will require finding political and social solutions to reach as yet unprotected populations. Even then the PV story will not be completely closed, as there will be need for continued discussions about the eventual destruction of vaccine strains and laboratory stocks.

There are hundreds of enteroviruses in addition to PV and some of these cause human disease. Hand, foot, and mouth disease is a generally mild disease of infants and children younger than 5 years old. Symptoms of hand, foot, and mouth disease [not to be confused with foot and mouth disease (FMDV) of cattle, sheep, pigs, and horses caused by a different picornavirus] include fever, mouth sores, and a skin rash. Large outbreaks are uncommon in the United States; however, in some areas of Asia large outbreaks are common.

In rare cases, enteroviruses other than PV reach and replicate in, target organs such as the CNS, heart, vascular endothelium, and others. Infection of these organs can result in diseases ranging from paralysis to mild acute aseptic meningitis syndromes to pericarditis to nonspecific febrile illness. For example, in 2015 enterovirus C105 was isolated from a hospitalized child with flaccid paralysis. Despite their name, enteroviruses do not cause epidemics of diarrheal disease.

Within the genus *Enterovirus* are three species of rhinovirus, which are generally transmitted by aerosols, replicate in the upper respiratory tract, and cause the common cold (mild upper respiratory tract disease). The rhinoviruses are not acid stable, prefer to replicate at 33°C, and there are hundreds of serotypes. The large number of serotypes is the reason that there is no vaccine for the common cold. (There are also viruses other than picornaviruses that cause the "common cold.")

Genus *Hepatovirus*

Hepatitis A is a liver disease that results from infection with the hepatitis A virus (HAV). Disease can range from a mild illness lasting a few weeks to a severe illness lasting several months. Hepatitis A is contracted when a person ingests food or water contaminated with fecal matter containing HAV. HAV is the only member of the genus *Hepatovirus*. HAV is resistant to high temperatures and low pH. In cultured cells the virus replicates slowly and does not cause a cytopathic effect. (This is in contrast to members of the genus *Enterovirus* that replicate to high titers in cultured cells and usually do cause cytopathic effects.) There is only a single serotype of HAV (although there are different genotypes). HAV infects humans and other primates. A safe and effective vaccine is available and currently all children in the United States are routinely vaccinated, as are travelers to regions where HAV is prevalent. Vaccination is also recommended for persons at high risk for HAV infection.

HAV is stable in the environment (as might be expected of a virus that is transmitted by the fecal-oral route). Experiments have shown that HAV can remain infectious for weeks or months in water or soil sediments. HAV is concentrated from seawater by bivalves (clams, oysters, etc.) so they are a common source of infection. Persons with any type of immune disorder (old age is enough) are at high risk for contracting HAV from uncooked muscles, clams, or oysters. HAV is also highly resistant to drying, thus infectious virus can also contaminate uncooked fruits and vegetables, such as lettuce, sprouts, strawberries, or green onions. HAV is also resistant to inactivation by detergents and solvents.

Genus *Apthovirus*: Foot and Mouth Disease Virus

Foot and mouth disease (FMDV) is a serious disease of cattle, sheep, pigs, and horses. The virus is found worldwide (except in the United States and parts of Europe). While mortality is low, morbidity is very high. FMDV is a "vesicular" disease, causing painful fluid-filled vesicles in the mouths and on the hooves of cattle and swine. FMDV is spread by contact, respiratory, and fecal-oral routes. It is *very* easily transmitted, moving over long distances (miles) in the air. The virus is also easily transmitted via fomites (cars, trucks, shoes) thus poses a high risk to food production biosecurity. FMDV is also very stable in the environment and infectious virus has even been detected in smoked meat products.

FMDV is endemic in many parts of the world and vaccines are available. However there are multiple serotypes and it is not feasible to vaccinate against them all. Countries that are free of FMDV take great precautions to retain their "FMDV-Free" status. Such countries, including the United States, exclude importation of meat from countries where the virus is endemic, or where vaccination is practiced. Rare outbreaks in FMDV-free countries disrupt normal agricultural activities causing huge economic losses and destruction of large numbers of animals.

An informative example is the FMDV outbreak that occurred in England in 2001. The first cases were reported in February 2001. The likely source of the outbreak was uncooked waste food fed to pigs. While only ~2000 cases of disease were confirmed, the outbreak was contained by slaughter and disposal of over 10 *million* sheep and cattle. This amounted to the slaughter of 80,000–90,000 animals per week. Veterinarians from the United States traveled to Great Britain to help in the control effort. (Because of the high infectivity and stability of the virus, once veterinarians had visited an infected farm, they could not visit another holding until they, their vehicles and instruments were thoroughly decontaminated.) During the outbreak, all movement of livestock was halted. Also, with the intention of controlling the spread of the disease, public rights of way across farmland were closed, severely disrupting tourism in rural England. Economic losses due to curtailing tourism may have been greater than the agricultural losses.

In this chapter we have learned that:

- Members of the family *Picornaviridae* are small, unenveloped, positive-strand RNA viruses with $T = 3$ icosahedral symmetry.
- Picornavirus replication is cytoplasmic. Genome replication requires host cell membrane complexes.
- Genomes are covalently linked to a viral protein (VPg) that serves as the primer for RNA synthesis.
- The genome length RNA serves the single mRNA for expression of all viral proteins.
- Translation initiation is dependent on a stem-loop structure called an IRES found near the 5′ end of the genome.
- The family *Picornaviridae* is large and contains pathogens of humans and animals. The genus *Enterovirus* includes isolates that replicate in upper respiratory tract (rhinoviruses) and viruses that replicate in the gastrointestinal tract (enteroviruses).

CHAPTER

12

Family *Caliciviridae*

OUTLINE

After reading this chapter, you should be able to answer the following questions.

- What features are shared by members of the family *Caliciviridae*?
- What are the characteristic features of genome replication and translation initiation?
- At what site in the cell do caliciviruses replicate?
- In what ways are caliciviruses similar/different from picornaviruses?
- What are the major human diseases caused by caliciviruses?
- Why are human caliciviruses difficult to study?

Caliciviruses are small, unenveloped positive-strand RNA viruses with $T = 3$ icosahedral symmetry and a particle size of 27–40 nm in diameter (Fig. 12.1 and Box 12.1). The name calicivirus is derived from the Latin word "calyx" meaning cup or goblet, as many strains have visible cup-shaped depressions on the capsid surface. Caliciviruses have been isolated from humans, cattle, pigs, cats, rabbits, chickens, reptiles, dolphins, and amphibians (Box 12.2). They are not very well studied because many, including human noroviruses, are difficult to grow in cultured cells. Feline calicivirus (FCV) and murine noroviruses are often used as surrogates for human noroviruses. Novel caliciviruses continue to be isolated, at least in part, in an effort to find better model systems. An enteric calicivirus recently isolated from rhesus monkeys (Tulane virus) grows in cultured cells and may prove to be a useful model for human noroviruses.

There are currently five genera in the family *Caliciviridae* (Box 12.2). The genera *Norovirus* and *Sapovirus* contain human isolates that cause enteric disease. The genus *Vesivirus* includes FCV a common upper respiratory tract pathogen of cats. *Lagosviruses* and *Neboviruses* infect rabbits and bovids, respectively. Rabbit hemorrhagic disease virus (RHDV) emerged in China in 1984 and has since spread worldwide.

GENOME ORGANIZATION

Calicivirus genomes are unsegmented positive-sense RNA ranging in size from 7.3 to 8.3 kb. Genome replication is cytoplasmic. Caliciviruses prime RNA synthesis with a protein primer (VPg). VPg is also required for translation. Genomes contain two or three open reading frames (ORFs) and are polyadenylated (Fig. 12.2). Nonstructural (NS) proteins are synthesized from genome-length mRNA to produce a polyprotein processed by viral proteases. The major capsid protein of noroviruses and vesiviruses is expressed from ORF2 using subgenomic mRNA. The gene for the major capsid protein of lagosviruses and sapoviruses is found in the same reading frame as the nonstructural proteins, and

Viruses. DOI: http://dx.doi.org/10.1016/B978-0-12-803109-4.00012-X

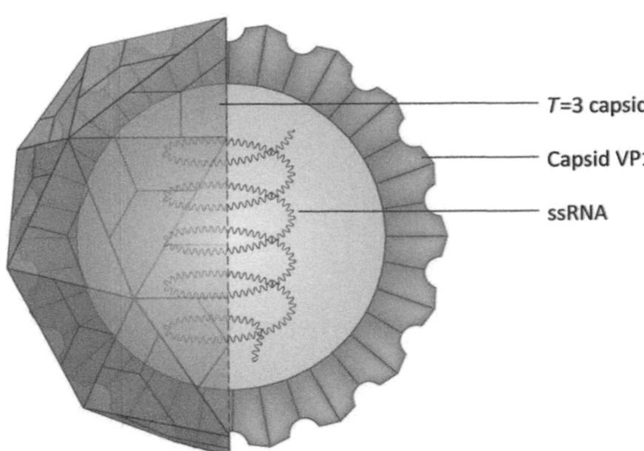

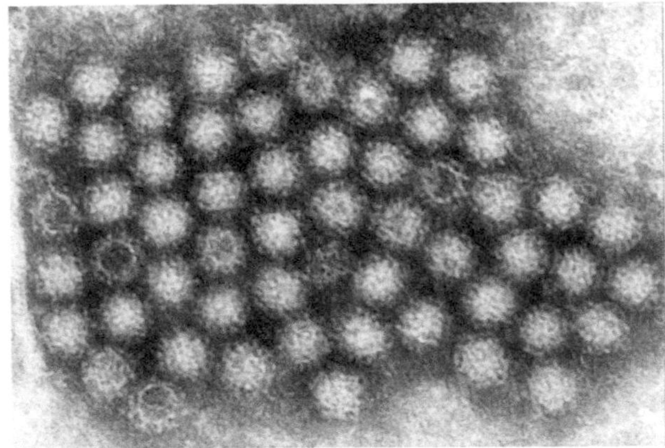

FIGURE 12.1 Virion structure (left). Negative stain electron micrograph of human norovirus (right). Content Providers: CDC/ Charles D. Humphrey.

BOX 12.1

GENERAL CHARACTERISTICS

Caliciviral genomes are unsegmented positive-sense RNA ranging in size from 7.3 to 9.0 kb. Genome replication is cytoplasmic. Caliciviruses prime RNA synthesis with a protein primer (VPg). Genomes contain two or three ORFs and are polyadenylated. Nonstructural (NS) proteins are synthesized from genome-length mRNA generating a polyprotein which is processed by viral proteases. The capsid protein is expressed from a subgenomic mRNA. VPg is required for translation.

Virions are unenveloped with $T=3$ icosahedral symmetry, ~35–39 nm diameter. Capsids are built from 180 copies of the capsid (C) protein.

BOX 12.2

TAXONOMY

Family *Caliciviridae*
Genus *Lagovirus*
Rabbit hemorrhagic disease virus
Genus *Nebovirus*
Newbury-1 virus (bovine origin)
Genus *Norovirus*
Norwalk and Norwalk-like viruses
Genus *Sapovirus*
Sapporovirus
Genus *Vesivirus*
Feline calicivirus, vesicular examthema of swine virus

could be produced by cleavage of a long polyprotein (Fig. 12.3). However transcription of a subgenomic mRNA is also used to produce these capsid proteins. ORF3 encodes the minor capsid protein VP2; it is produced from the subgenomic mRNA by a ribosomal termination–reinitiation mechanism (Fig. 3.14). The 5′ and 3′ nontranslated regions of calicivirus genomes are very short; however, recent evidence suggests that highly base-paired regions of secondary structure form using sequences that extend into coding regions.

VIRION STRUCTURE

Calicivirus capsids assemble from 90 dimers of VP1 and a few molecules of VP2 to produce virions of

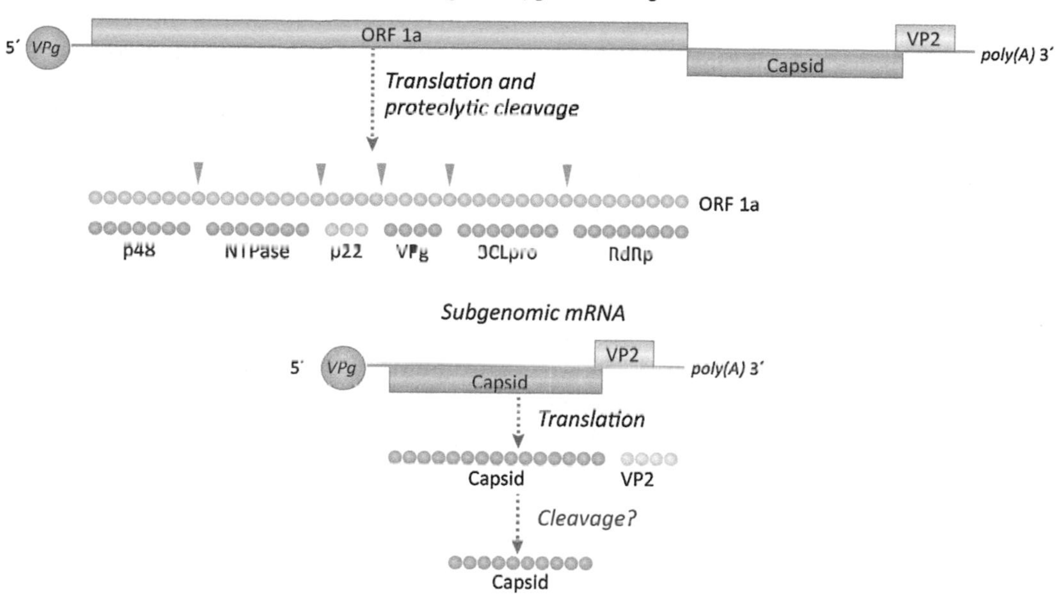

FIGURE 12.2 Example of calicivirus genome organization and protein expression strategy.

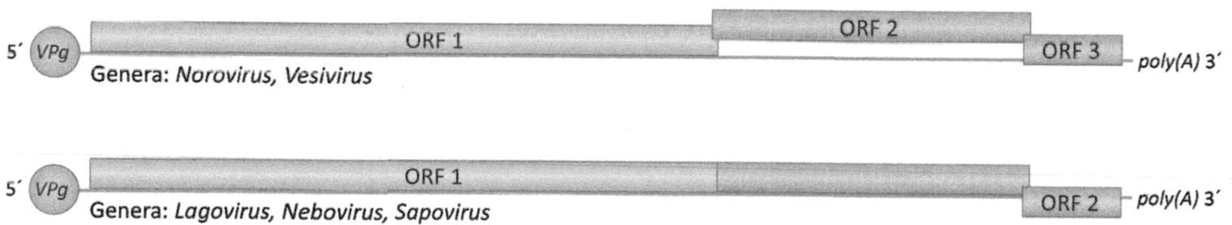

FIGURE 12.3 Genome comparison. The genome organization of the sapoviruses, lagoviruses, and neboviruses differ, as the capsid protein (VP1) is continuous with that of the NS proteins and can be released by proteolytic cleavage. All caliciviruses encode a second, minor structural protein (VP2) from an ORF near the 3′ end of the genome. Murine noroviruses contain four ORFs.

27–40 nm. Recombinant VP1 (expressed in baculovirus, for example) can efficiently self-assemble to form virus-like particles (VLP). The process does not require RNA or the minor capsid protein VP2.

GENERAL REPLICATION CYCLE

Many medically important caliciviruses are difficult to grow in cultured cells, in particular the human noroviruses. As new caliciviruses are discovered, new details will likely emerge. Below is a general description of calicivirus replication.

Attachment/Penetration

Noroviruses interact with histo blood group antigens, carbohydrates on the cell surface. Other caliciviruses interact with protein receptors. Different modes penetration into the cell have been documented. Some caliciviruses penetrate at the plasma membrane while others penetrate after endocytosis.

Genome Replication

As with other positive-strand RNA viruses, translation of the infecting genome immediately follows penetration, producing a long precursor polyprotein that is proteolytically processed to produce nonstructural (NS) proteins that include a protease and the RdRp. Following translation of NS proteins, RNA synthesis can begin. RNA synthesis takes place on host cell membranes and is primed by VPg such that caliciviral RNAs have VPg covalently linked to the 5′ end. A genome-length copy RNA serves as template for synthesis of both genomes and subgenomic mRNAs.

Translation

The calicivirus VPg protein (13–15 kDa) is much larger than the ~22 amino acid VPg of picornaviruses and it seems to play a larger role in replication. In addition to its role in genome replication, calicivirus VPg is essential for translation of viral RNA (Fig. 3.13). Removal of VPg from caliciviral RNA decreases infectivity as well as the ability of the RNA to support translation in cell-free systems. Conversely, addition of recombinant VPg to cell-free extracts inhibits both cap- and IRES-dependent translation suggesting that caliciviral VPg competes for initiation factors, effectively removing them from translation systems. VPg proteins from various caliciviruses have been shown to interact with eukaryotic initiation factors (eIF3, eIF4E, eIF4G), further supporting a role in translation.

Caliciviruses use a variety of protein expression strategies. In the case of vesiviruses and noroviruses, genome-length mRNA is used to express an ORF1 polyprotein that is proteolytically processed to produce the NS proteins NS1–NS7 (including NTPase, VPg, protease, and RdRp). A subgenomic mRNA is produced and contains one or two ORFs. The major capsid protein (VP1) is encoded by ORF2. The genome organization of the sapoviruses, lagoviruses, and neboviruses is a bit different as the capsid protein (VP1) is continuous with that of the NS proteins. However VP1 is also produced from a subgenomic mRNA. All caliciviruses encode a second, minor structural protein (VP2) from an ORF near the 3′ end of the genome. Murine noroviruses contain four ORFs.

Two different strategies are used to produce the minor capsid protein VP2 from the bicistronic mRNA. Noroviruses use leaky scanning (Fig. 3.14) for expression of VP2. In the case of FCV (vesivirus) and RHDV (lagosvirus), VP2 is expressed through RNA termination–reinitiation or stop–start (Fig. 3.14).

Assembly/Release

Calicivirus virions have $T = 3$ icosahedral symmetry. Capsids assemble from 180 copies of VP1 and a few copies of VP2. If the VP1 gene is cloned and expressed in cells, the recombinant VP1 protein assembles to form VLP. This experimental result indicates that neither VP2 nor viral RNA is absolutely required for formation of capsids. Calicivirus-infected cells undergo lysis to release progeny.

DISEASES

Acute Gasterointestinal Disease

In the United States, the Centers for Disease Control and Prevention (CDC) estimates that about half of all outbreaks of food-related illnesses are caused by noroviruses. However, noroviruses are very contagious and are quite stable on contaminated surfaces; therefore the majority of norovirus infections are not a *direct* result of ingesting contaminated food or water, but rather result from contact with contaminated surfaces. The CDC estimates that each year in the United States noroviruses cause 19–21 million cases of acute gastroenteritis. Common symptoms are stomach pain, nausea, vomiting, and diarrhea lasting from 24 to 48 hours. Virus can be shed in feces for several days. While most norovirus infections are self-limiting, serious illness may occur in young children and the elderly. Unfortunately, norovirus infections are not uncommon in health care and long-term care settings. Several large norovirus outbreaks have been recorded on cruise ships (cruise ship disease). The CDC reports that over 90% of diarrheal disease outbreaks on cruise ships are caused by noroviruses. It is difficult to control these outbreaks due to crowded conditions and shared dining areas. Also, noroviruses are very resistant to many common disinfectants so may survive routine cleaning.

Sapporoviruses are also associated with acute gasteroenteritis of humans.

Feline Calicivirus

FCV causes variable and sometimes complex disease syndromes in cats. FCVs are genetically diverse, and different strains can vary widely in their virulence. FCV can be isolated from about half of house cats presenting with upper respiratory tract infections. In the case of FCV infection, ulceration of the mouth is a common finding. A few strains of FCV cause more severe systemic (multiorgan) disease. FCV is transmitted via saliva, feces, urine, and respiratory secretions and is very stable in the environment. It can be transmitted through the air, orally, and by fomites. Some cats become persistently infected, serving as an ongoing source of infection. Within persistently infected cats, it has been documented that mutant viruses, with changes in the capsid protein, arise. This allows the virus to escape from immune responses and may also lead to the generation of strains with increased virulence. Vaccines are available for FCV, but due to genetic variation, protection is not complete. FCV is one of the few caliciviruses that grows well in cultured cells.

Rabbit Calicivirus

Rabbit hemorrhagic disease virus (RHDV, genus *Lagosvirus*, emerged in China in 1984. It is a highly infectious and often fatal disease that affects adult wild and domestic rabbits of the species *Oryctolagus cuniculus* (European or common rabbits, including all breeds of domesticated rabbits). Other species of rabbits, such as wild rabbits found in the United States, are resistant. RHDV is considered a major rabbit pathogen in some countries where it threatens rabbits raised for food and kept as pets. In some areas of Europe the virus has reduced wild rabbit populations to levels that threaten endangered predators such as the Iberian lynx. In stark contrast, RHDV has been introduced into Australia and New Zealand as a means of controlling overwhelming populations of nonnative European rabbits. The virus is transmitted by a variety of mechanisms; either by direct contact between rabbits or indirectly through contaminated fomites. Blood feeding insects may also be vectors. Like other caliciviruses, RHDV is very stable in the environment. The disease is characterized by acute necrotizing hepatitis and disseminated intravascular coagulation. Disease progression can be rapid, sometimes resulting in sudden death in the absence of clinical symptoms.

In this chapter we have learned that:

- Members of the family *Caliciviridae* are small, unenveloped, positive-strand RNA viruses.
- Calicivirus replication is cytoplasmic. Genome replication requires host cell membrane complexes and is primed by a virally encoded protein, VPg.
- In addition to its role as a primer for RNA synthesis, VPg is required for translation of genome length and subgenomic mRNAs.
- Among family members there is some variability in protein expression strategies but genome-length RNA is translated to produce a large polyprotein that is processed by viral proteases. A subgenomic mRNA encodes one or two capsid proteins.
- Human caliciviruses (noroviruses and sapoviruses) cannot be cultured in the laboratory so surrogates are used to probe virus replication and environmental properties of caliciviruses.
- Norwalk virus and Norwalk-like viruses (noroviruses) cause acute gastroenteritis in humans. These agents are transmitted by the fecal–oral route. Virions are stable in the environment and are highly infectious. Outbreaks of cruise ships have involved hundreds of passengers.

CHAPTER

13

Family *Hepeviridae*

OUTLINE

After studying this chapter, you should be able to:

- Describe the shared characteristics of members of the family *Hepeviridae*.
- Describe the general strategies used by hepeviruses for transcription, translation, and genome replication.
- List modes of hepatitis E virus (HEV) transmission.

Members of the family *Hepeviridae* are small, unenveloped RNA viruses (Fig. 13.1 and Box 13.1). The family name was derived from the name of the clinical disease "hepatitis E," caused by the first identified member of the family. Hepatitis E disease (an infection of the liver) is usually acute and self-limiting. The notable exception is an approximately 20% case fatality rate among pregnant women. Infection with HEV has been associated with consumption of raw or undercooked meats (particularly pork), raw shellfish, and contamination of drinking or irrigation water with animal manure. The host range (pigs, chickens, rabbits, rats, mongoose, deer, and possibly cattle and sheep) of HEV is expanding as more animals are being surveyed. In developed countries HEV infection is sporadic and zoonotic and is associated with particular genetic groups (genogroups 3 and 4) of the virus. In contrast, in underdeveloped countries HEV circulates among humans and the most common genogroups are 1 and 2. Endemic areas include Asia, Africa, Southern Europe, and Mexico. In endemic regions, large waterborne epidemics with thousands of people affected have been recorded.

GENOME ORGANIZATION

Hepevirus genomes are unsegmented, capped, and polyadenylated positive-strand RNA of about 7.2 kb. The genome contains three open-reading frames (ORFs) (Fig. 13.2). The ORF1 polyprotein is translated from genome-length RNA to encode the nonstructural (NS) proteins. The RNA-dependent RNA polymerase (RdRp) is located at the carboxyl terminus of the polyprotein, which is presumed to be cleaved by viral proteases. A second ORF encodes the capsid protein. HEV ORF3 encodes a protein of 123 aa that may be a minor capsid protein. The capsid protein and ORF3 protein are expressed from a subgenomic mRNA. The genome contains very short 5′ and 3′ nontranslated regions.

VIRION MORPHOLOGY

Virions have icosahedral symmetry ($T=1$) with a diameter of ~32–34 nm. The inability to grow HEV has precluded detailed structural analysis of the virion. However when expressed alone, capsid proteins assemble into virus like particles (VLPs). Structures of VLPs are shown in Fig. 13.3.

GENERAL REPLICATION CYCLE OF HEV

HEV is a hepatotropic virus and a useful cell culture system is lacking. However some aspects of HEV

Viruses. DOI: http://dx.doi.org/10.1016/B978-0-12-803109-4.00013-1

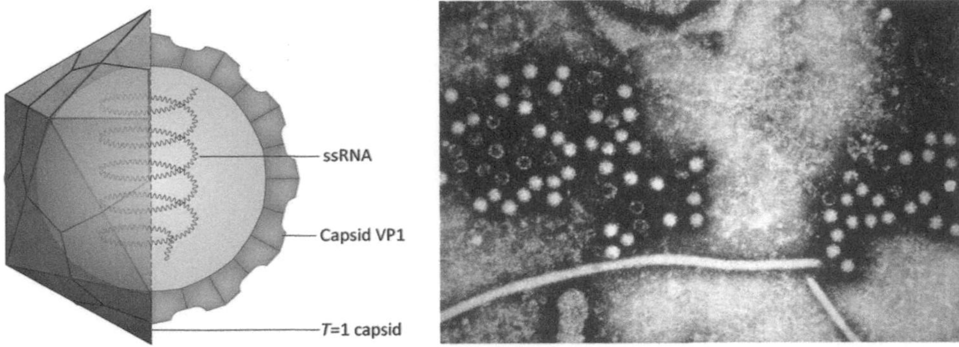

FIGURE 13.1 Capsid structure (left). Negative strain, electron micrograph of HEV (right). *Content provider: CDC. Public Health Image Library Image #5605.*

BOX 13.1

DESCRIPTION

Members of the family *Hepeviridae* are positive-strand RNA viruses. Genomes (~7.2 kb) have 5′ methyl G caps and 3′ poly (A) tails. Genomes contain three ORFS. There are two mRNAs (genome length and a subgenomic mRNA). Genome-length mRNA is translated to produce proteins required for RNA replication while the subgenomic mRNA is used to produce the capsid protein. Virions are unenveloped with $T = 3$ icosahedral symmetry and a size of ~32–34 nm. Some of the capsid protein is glycosylated, an unusual finding for an unenveloped virus.

replication have been determined. HEV replicates in the cytosol. Its RNA genome is translated to produce a polyprotein (~1693 amino acids) that is likely proteolytically processed to produce several NS proteins, including the RdRp. Examination of protein motifs suggests that a methylatransferase, helicase, and cysteine protease are present. Replication of the genome produces a copy RNA that serves as the template for synthesis of genome-length and subgenomic mRNAs.

The single subgenomic mRNA is bicistronic, translated to produce the ~660 aa capsid protein (from ORF2) and a nonstructural protein (from ORF3). When the capsid gene is cloned, and the protein expressed in cultured cells, the resulting products include a 74 kDa

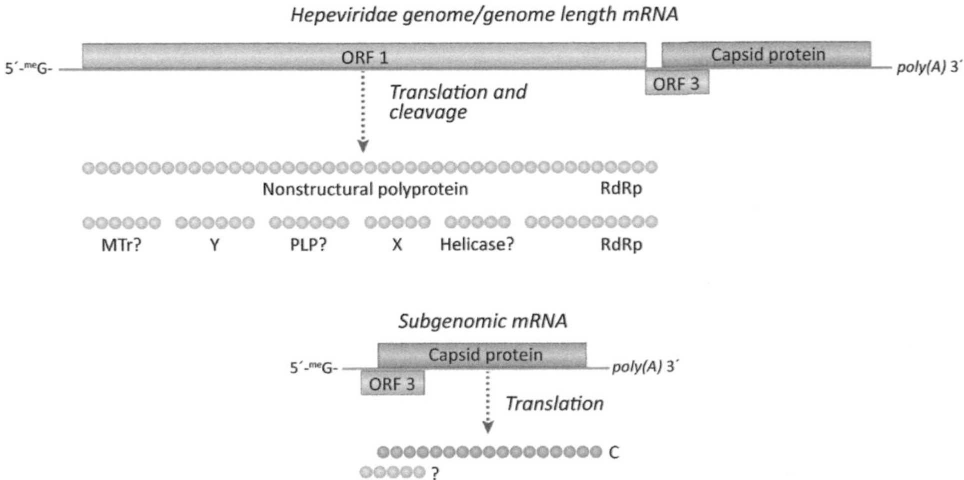

FIGURE 13.2 Hepevirus genome organization. Functions of protein products are not well characterized.

FIGURE 13.3 Cryo-EM model of virion sized HEV capsid. From: NCBI Molecular Modeling data base MMDB ID: 84140

protein (unglycosylated) and an 88 kDa *glycosylated* protein. The capsid protein contains an N-terminal signal sequence that directs the protein into the ER where is acquires N-linked glycosylation. Mutation of the N-linked glycosylation sites prevents formation of infectious virions suggesting that glycosylation is important. Because there are no robust systems for producing infectious virus in culture, details about capsid assembly/virion morphogenesis are currently lacking. There is evidence for a role for lipids in these processes; however, hepeviruses do not appear to be "typical" enveloped viruses, so what roles do glycosylated capsid proteins and lipids play in their assembly, morphogenesis, and infectivity?

DISEASE

HEV is often transmitted via food and environmental routes, although HEV-infected mothers can transmit the virus to the fetus. There is one known serotype of HEV; however, genome sequencing provides evidence for four distinct genotypes. Genotypes 1 and 2 have only been isolated from humans. Genotypes 3 and 4 infect a variety of animals and are zoonotic, causing sporadic outbreaks among humans (Box 13.2).

In highly endemic areas of the world (Fig. 13.4), large human outbreaks are associated with genotypes 1 and 2. Highly endemic areas include developing countries with inadequate water supplies and poor sanitation. Large epidemics have been reported in Asia, the Middle East, Africa, and Central America. According to estimates by the World Health Organization, an estimated 20 million HEV infections occur worldwide each year, leading to an estimated 3.3 million symptomatic cases of hepatitis E and over 50,000 HEV-related deaths. Symptomatic infection appears to be most common in young adults aged 15–40 years. Studies from developing countries have shown excess mortality in pregnant women (ranging from 20% to 25% in the third trimester). High mortality among pregnant women seems to be restricted to genotypes 1 and 2. The reason for increased mortality is unclear and continues to be debated. A vaccine to prevent HEV infection has been developed and is licensed in China, but is not available elsewhere.

In areas with low endemicity, sporadic outbreaks of HEV are associated with genotypes 3 and 4. These sporadic outbreaks are associated with zoonotic transmission (from animals to humans). HEV infection is very common among swine (based on serosurveys) but does not appear to cause any disease. In the United States, sporadic HEV is most often associated with consumption of uncooked or undercooked pork or deer meat.

In this chapter we learned that:

- Members of the family *Hepeviridae* are positive-strand RNA viruses. Virions appear to be unenveloped but the capsid protein is glycosylated and there may be some role for lipids in virion morphogenesis.
- The HEV genome is capped and polyadenylated and contains three ORFs. The NS proteins are translated from genome-length mRNA. A subgenomic mRNA is used to produce the capsid protein.

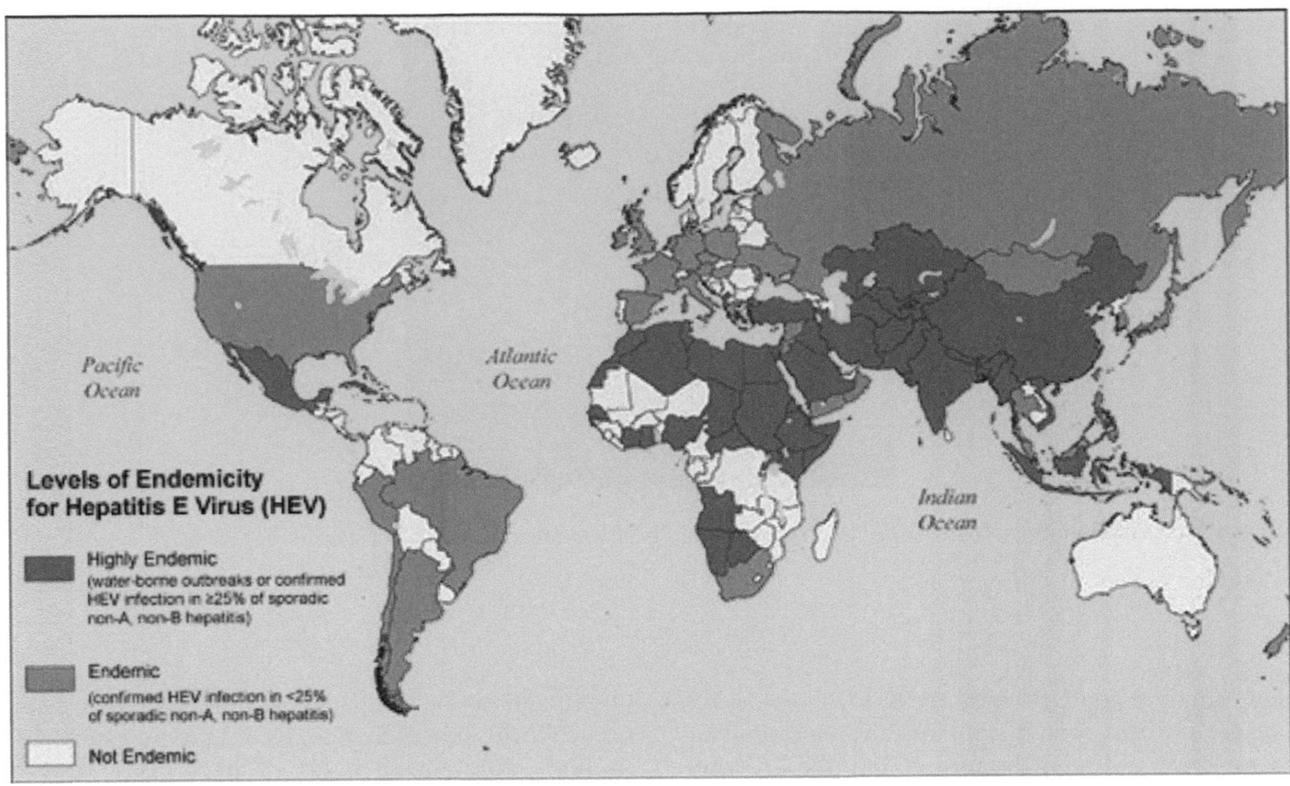

FIGURE 13.4 Levels of endemicity of HEV. Light brown regions HEV is not endemic, orange regions HEV is endemic (sporadic cases), dark brown regions HEV is highly endemic (waterborne outbreaks occur). *From CDC.*

BOX 13.2

TAXONOMY

Family *Hepeviridae*
Genus *Orthohepevirus* (four species)

Species *Orthohepevirus A* (HEV) contains multiple isolates and genogroups

Species *Orthohepevirus B* (avian HEV)
Species *Orthohepevirus C* (rodent viruses)
Species *Orthohepevirus D* (bat isolate)

Genus *Piscihepevirus* (fish, one species).

- In developed countries transmission is largely zoonotic and sporadic. However in underdeveloped countries, waterborne epidemics may affect thousands.
- Disease is usually acute and self-limiting but severe disease can occur in pregnant women.
- Many aspects of HEV replication and pathogenesis are unknown, due to lack of tractable experimental systems.

CHAPTER

14

Family *Astroviridae*

OUTLINE

After reading this chapter, you should be able to discuss the following:

- What are the major structural and replicative features of astroviruses?
- What disease is most frequently associated with astrovirus infection?
- How are human astroviruses (HAstVs) most often transmitted?
- How is the astrovirus capsid protein processed to produce the proteins found in the mature virion?

Astroviruses (from the Greek, *astron* meaning star) were discovered in 1975, in association with an outbreak of diarrhea in humans. Since that time they have been isolated from many other mammals including pigs, cats, minks, dogs, rats, bats, calves, sheep, and deer as well as from marine mammals such as sea lions and dolphins. Astroviruses have also been isolated from birds; they can cause significant disease in turkeys, ducks, and chickens. Astroviruses are unsegmented, positive-sense RNA viruses with ~7–9 kb genomes. These small, unenveloped viruses have spikes that project about 41 nm from the surface of the capsid, giving them a star-like appearance (Fig. 14.1 and Box 14.1). HAstVs cause gastroenteritis in children and adults. Symptoms last 3–4 days and include diarrhea, nausea, vomiting, fever, malaise, and abdominal pain. For the most part, disease is self-limiting.

GENOME ORGANIZATION

Astrovirus genomes are unsegmented positive-strand RNA; genomes are not capped but have a 3′ poly(A) tail. Genomes have three long overlapping reading frames (ORFs) that encode polyproteins (Fig. 14.2). There are short untranslated regions at the 5′ and 3′ ends of the genome. Similar to other positive-strand RNA viruses (for example togaviruses and coronaviruses), two ORFs covering the 5′ half of the genome encode nonstructural proteins (NSPs) that include proteases, membrane-associated proteins, an NTP-binding protein, and the RNA-dependent RNA polymerase (RdRp). ORFs for NSPs are overlapping and synthesis of the longer polyprotein product likely requires a *ribosomal frame-shift* between ORF1a and ORF1b (Fig. 14.2). Astrovirus RNA is not capped, and based on the presence of a protein sequence similar to calicivirus VPg (in ORF1b), it is postulated that RNA synthesis is initiated using a protein primer. Astrovirus replication is cytoplasmic.

VIRION MORPHOLOGY

Astrovirus particles have $T = 3$ icosahedral symmetry. The capsid protein is encoded from ORF2, expressed from a subgenomic mRNA. The capsid precursor undergoes multiple cleavages. The full-length precursor (VP90) is cleaved by cellular proteases (caspases) to generate the VP70 product. If capsase inhibitors are added to infected cells, release of virions is blocked. However VP70-containing capsids are likely noninfectious until VP70 is further processed by trypsin-like proteases to generate mature virions containing three polypeptides (VP25, VP27, and VP34). Addition of trypsin to cultured cells produces

Viruses. DOI: http://dx.doi.org/10.1016/B978-0-12-803109-4.00014-3

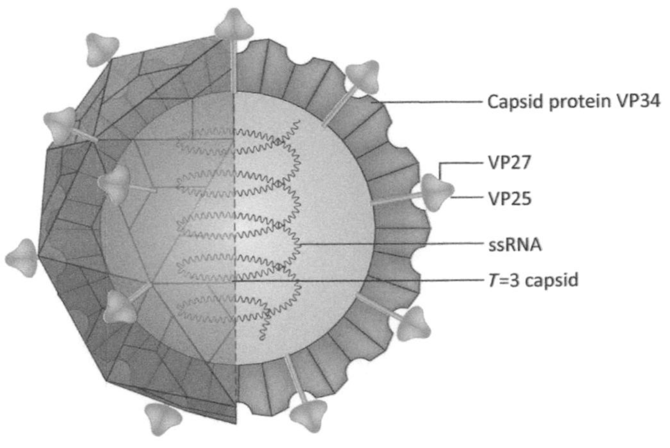

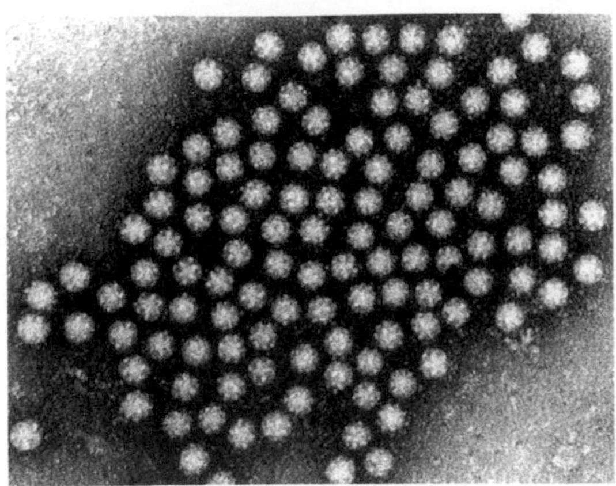

FIGURE 14.1 Virion structure (left) and electron micrograph (right).

BOX 14.1

GENERAL CHARACTERISTICS

Astrovirus genomes are single stranded, unsegmented positive-strand RNA ~6.8–7.9 kb. The genome is not capped but has a poly A tail. Genomes have three overlapping ORFs that encode polyproteins. Nonstructural proteins (NSPs) are cleaved by viral proteases. A third ORF encodes the capsid precursor. The capsid precursor is cleaved by host proteases. Two mRNAs are present in infected cells (full-length and a single subgenomic mRNA). Replication is cytoplasmic.

Virions (~40–45 nm in diameter) are unenveloped. Capsids have $T=3$ icosahedral symmetry with spike-like projections at the vertices.

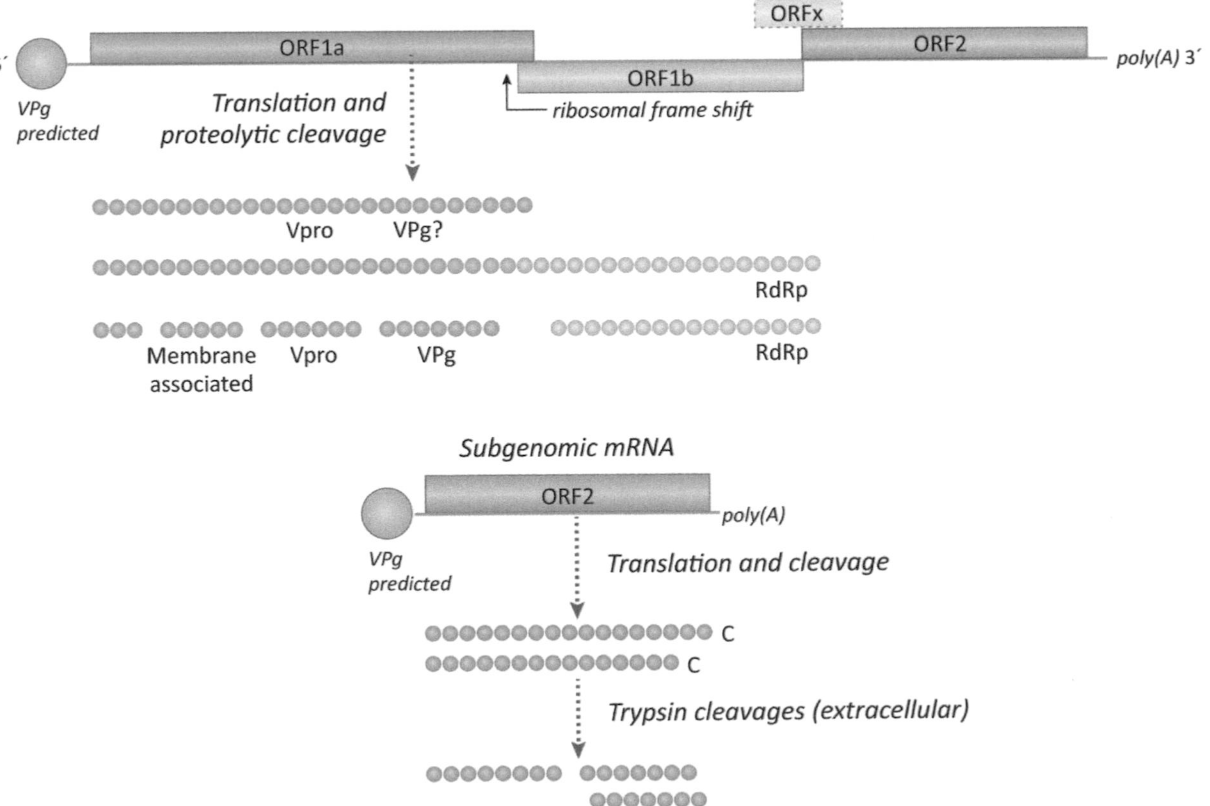

FIGURE 14.2 Genome organization.

infectious particles and similar enzymes are present in the intestine during a natural infection. The capsid core is formed by VP34 while VP25 and VP27 form the spikes on the virion surface. Binding sites for neutralizing antibodies map to VP25 and VP27.

Due to the lack of robust cell culture systems, many details of astrovirus replication have not been confirmed. However the overall replication cycle of astroviruses is predicted to be quite similar to that of other positive-strand RNA viruses. Uptake of virions is thought to be by endocytosis and the uncoated genomic RNA would be translated to produce the viral RNA replication machinery.

DISEASES CAUSED BY ASTROVIRUSES

HAstVs are thought to be the second or third most common cause of viral diarrhea in young children. They have also been isolated from sporadic outbreaks of acute gastroenteritis in adults. A few studies have associated astroviruses with chronic diarrhea in immunocompromised children and adults. HAstVs are found worldwide. The main mode of human astrovirus transmission is by contaminated food (including bivalve mollusks) and water, although direct person-to-person transmission has also been documented (Fig. 14.3).

There are multiple serotypes of human astrovirus and the main target cells are enterocytes (epithelial cells of the intestinal tract). Astrovirus infection does not notably alter intestinal architecture and does not induce inflammation. It has been proposed that pathogenesis may be caused by apoptotic death of infected epithelial cells. Symptomatic infections are most common in children younger than 2 years of age, and it is estimated that 5%−9% of cases of viral diarrhea in young children are caused by astroviruses. In the US population the presence of antiastrovirus antibodies is very high, indicating that most infections are asymptomatic or very mild. Outbreaks of astrovirus-associated diarrhea have been reported among elderly patients and military recruits. Food-borne outbreaks, affecting thousands of individuals, have occurred in Japan. In temperate climates astrovirus infection is highest during winter months while in tropical regions prevalence is highest during the rainy season (Box 14.2).

Rarely, astroviruses have been isolated from organs other than the gastrointestinal tract. They have been isolated from a few children with CNS disease although disease causation has not been confirmed.

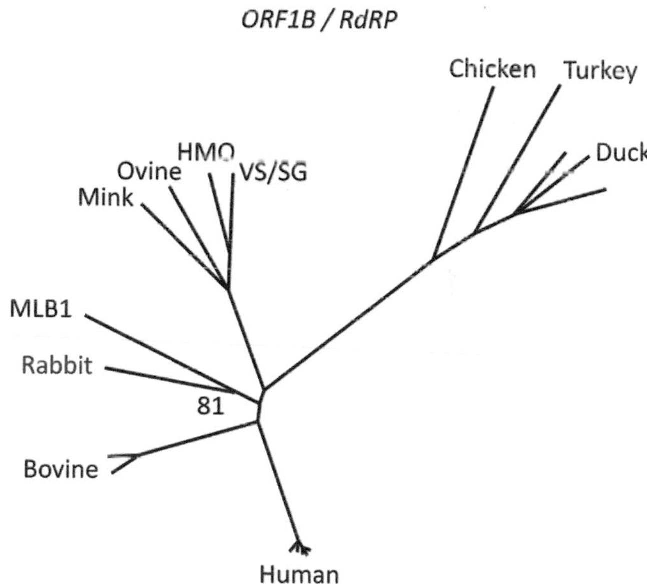

FIGURE 14.3 Astrovirus phyologeny based on RdRp sequences (Stenglein et al., 2012).

BOX 14.2

TAXONOMY

Family *Astroviridae*

Genus *Avastrovirus* (three numbered species from birds)

- Astrovirus 1 (turkey)
- Astrovirus 2 (chicken)
- Astrovirus 3 (duck)

Genus *Mamastrovirus* (19 numbered species from mammals; note that many were identified by sequencing studies and have not been cultured)

- Mamastrovirus 1, 6, 8, 9 (humans and human stool)
- Mamastrovirus 2 (feline)
- Mamastrovirus 3 (porcine)
- Mamastroviruses 4, 11 (sea lion)
- Mamastrovirus 5 (canine stool)
- Mamastrovirus 7 (bottlenose dolphin)
- Mamastrovirus 10 (mink)
- Mamastroviruses 12, 14, 15, 16, 17, 18, 19 (bat)
- Mamastrovirus 13 (sheep).

However, there is a good example of CNS-associated astrovirus infection in an animal model. Shaking mink syndrome is a neurologic disorder of farmed minks. Outbreaks have occurred in Denmark, Sweden, and Finland. Examination of diseased mink revealed brain lesions (nonsuppurative encephalomyelitis) and experimental infection of brain homogenates into healthy mink recapitulated the disease, a result highly suggestive of an infectious agent. Attempts to culture an infectious agent were unsuccessful but the agent was finally identified using metagenomics. Nucleic acids sequences were obtained from brain material of diseased and healthy mink. Comparisons revealed an astrovirus genome associated only with diseased mink. The CNS-associated astrovirus shares about 80% nucleotide identity with an enteric mink astrovirus.

In this chapter we learned that:

- Astroviruses are unenveloped, positive-strand RNA viruses. Their name derives from their star-shaped virions.
- Astroviruses were first identified in association with outbreaks of gastroenteritis.
- HAstVs are most often transmitted by the fecal oral route, through contaminated food and water.
- The astrovirus capsid protein is processed by host proteases. One cleavage is mediated by intracellular caspases and others by extracellular trypsin-like proteases.

References

Stenglein, M.D., Velazquez, E., Greenacre, C., Wilkes, R.P., Ruby, J.G., Lankton, J.S., et al., 2012. Complete genome sequence of an astrovirus identified in a domestic rabbit (*Oryctolagus cuniculus*) with gastroenteritis. Virol. J. 9, 216. Available from: http://dx.doi.org/10.1186/1743-422X-9-216.

CHAPTER

15

Family *Flaviviridae*

OUTLINE

After reading this chapter, you should be able to answer the following questions:

- What structural features are shared by all members of the family *Flaviviridae*?
- What are the characteristic features of genome replication and translation initiation?
- At what site in the cell do members of the family *Flaviviridae* replicate?
- What are some of the major human and animal diseases caused by members of the family *Flaviviridae*?
- What is the basis for heptotropism of Hepatitis C virus (HCV)?

The *Flaviviridae* family is a large family of unsegmented positive-strand RNA viruses (Box 15.1). Virions are enveloped but do not contain the prominent spikes often seen on enveloped virions (Fig. 15.1). The family is quite large and contains many important human pathogens including Yellow fever virus (YFV), Dengue fever virus (DFV), Zika virus, and HCV. There are four genera within the family (Box 15.2):

Genus *Flavivirus*: Most members of the genus *Flavivirus* are transmitted by insect vectors; thus they are also often referred to as *arboviruses* (arthropod-borne means "transmitted by insects"). There are close to 70 identified members in the genus *Flavivirus*, and several are significant human pathogens (YFV, DFV, encephalitis viruses such as West Nile virus (WNV) and Zika virus).

Genus *Hepacivirus*: The type strain of the genus *Hepacivirus* is HCV, a blood-borne hepatitis virus. HCV constitutes a large group of related viruses that can be subdivided based on genome sequence. Even within an infected individual, viral genomes are not identical. Instead "swarms" of genetically related viruses are found. Thus HCV exists as a quasispecies. It is likely that genetic variability contributes to the ability of HCV to cause chronic infection, persisting for many decades in the infected host. There are also members of the genus *Hepacivirus* that infect other mammals; they have been recovered from dogs and horses where they do not appear to cause major disease outbreaks, but may cause sporadic or subclinical hepatitis. Both animal and human hepaciviruses are difficult to culture, but because they are important pathogens, details of their replication cycles have been elucidated using a variety of model systems.

Genus *Pestivirus*: There are three pestiviruses of veterinary significance: bovine viral diarrhea virus (BVDV), border disease virus (BDV) of sheep and goats, and classical swine fever virus (CSFV). These three pestiviruses are routinely transmitted by contact, but also by vertical transmission to the fetus. Vertical transmission may result in

Viruses. DOI: http://dx.doi.org/10.1016/B978-0-12-803109-4.00015-5

BOX 15.1

GENERAL CHARACTERISTICS

Flavivirus genomes are positive-strand RNA, unsegmented, and range in size from 9.5–12.5 kb. Some have 5′ capped genomes. Hairpin structures are found at the 3′ end of nonpolyadylated genomes. The genome serves as the single mRNA. Viral proteins (structural and nonstructural) are synthesized as part of a large polyprotein precursor. Translation is interesting as some of the polyprotein is targeted to ER while other regions remain cytosolic. Replication is cytoplasmic.

A single capsid (C) protein forms the capsid. Enveloped virions contain two or three membrane-associated glycoproteins. Virions have a smooth, golf-ball-type appearance due to the arrangement of the surface glycoproteins that lie flat on the envelope.

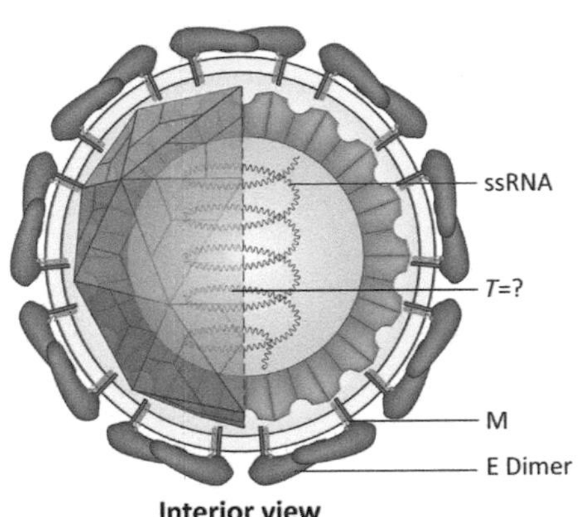

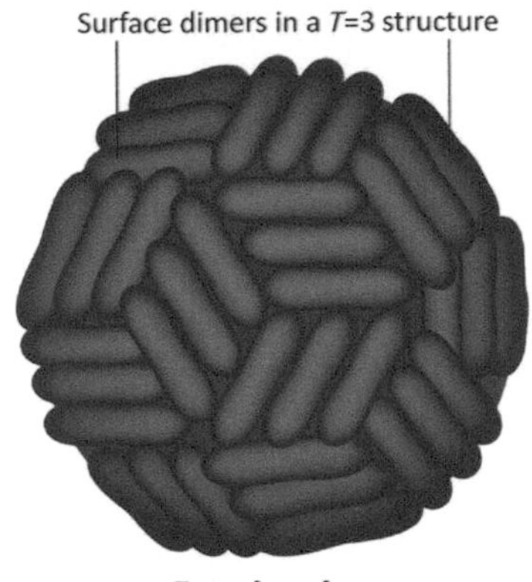

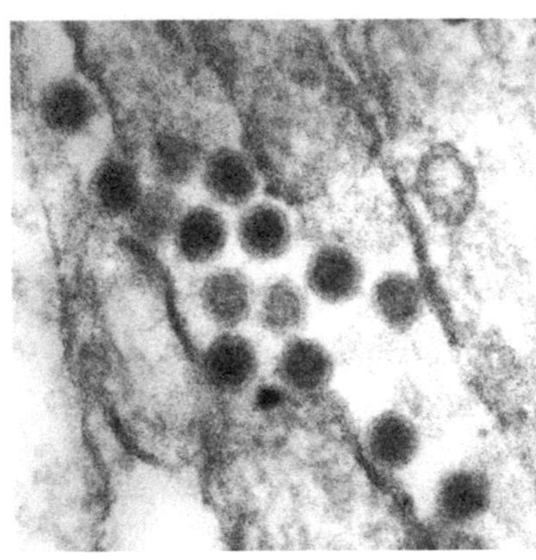

FIGURE 15.1 Virion structure in cross section (right). Surface view of the virion (middle). Negative stain of St. Louis encephalitis virus within a mosquito salivary gland tissue sample (left). *From CDC. Dr. Fred Murphy; Sylvia Whitfield. Public Health Image Library image ID# 10228.*

BOX 15.2

TAXONOMY

Family *Flaviviridae*

Genus *Flavivirus*

Arthropod vectored, over 70 members known

Genus *Hepacivirus*

Hepatitis C virus

Genus *Pestivirus*

Bovine viral diarrhea virus (BVDV)
Border disease virus (sheep and goats)
Classical swine fever virus (CSFV).

persistently infected (PI) offspring that are viremic, antibody negative, and constantly shed virus. Thus these viruses can be difficult to eradicate from livestock herds (CSFV has been eradicated from the United States).

Genus *Pegivirus*: Pegiviruses cause persistent infections and have been detected in humans and animals. Relatively little is known about their role in disease.

GENOME STRUCTURE/ORGANIZATION

Members of the family *Flaviviridae* have positive sense, single-stranded RNA genomes. Genome sizes range from ~9000 to 12,500 nt (Box 15.1). Members of the genus *Flavivirus* have a methyl-G-cap protecting the 5′ end of their genomes. However, pestiviruses and hepacivirus genomes are not capped. The genome 3′ ends of all members of the family *Flaviviridae* have a

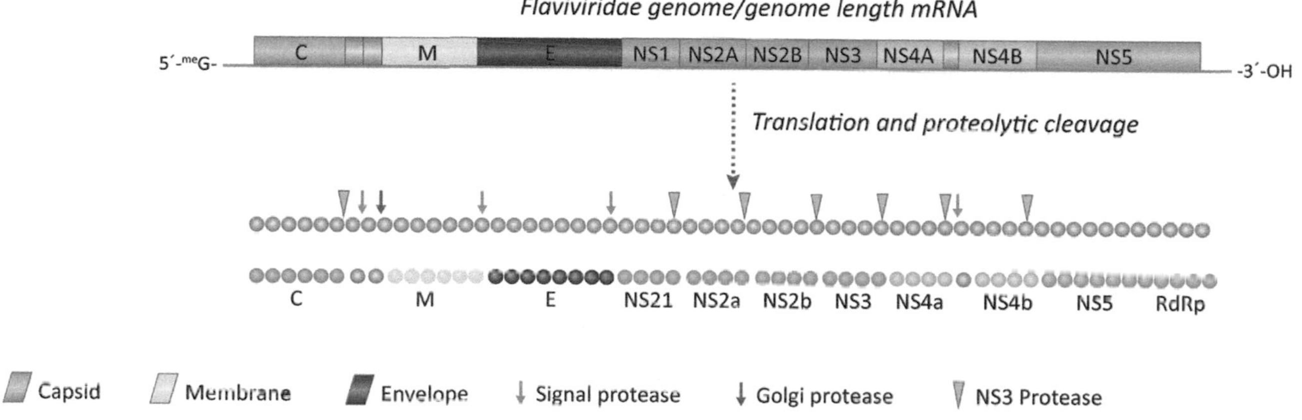

FIGURE 15.2 Genome and gene expression strategy (genus *Flavivirus*). The genome contains a single long ORF. No subgenomic mRNAs are expressed. NS5 is the RdRp.

hairpin structure that stabilizes the genome and is probably required for genome replication; genomes are not polyadenylated. Genomes contain 5′ and 3′ noncoding regions (NCRs) that precede and follow the single long open-reading frame (ORF) (Fig. 15.2). 5′ NCRs are ~300–400 nt. The 5′ NCR of HCV contains an internal ribosome entry site (IRES). Pestiviruses also have stem/loop structures in their 5′ NCR.

Between the 5′ and 3′ NCRs of all members of the family *Flaviviridae* is a single long ORF encoding a polyprotein. A single capsid (C) and two or three envelope proteins are encoded at the 5′-end of the ORF while the RNA-dependent RNA polymerase (RdRp, the NSP5B protein) is encoded at the 3′ end of the ORF.

VIRION STRUCTURE

Virions are enveloped with a diameter of about 30 nm. A single capsid protein (C) assembles to form the capsid. Two or three glycoproteins (Fig. 15.3) are associated with the envelope. Flaviviruses have a unique morphology for enveloped viruses as the glycoproteins lie flat on the surface of the mature virion, giving it a smooth golf-ball-like appearance (Fig. 15.1). The structure of the envelope glycoproteins is pH sensitive. They form spikes at low pH and the spike form mediates membrane fusion.

GENERAL OVERVIEW OF REPLICATION

After attachment (various receptors that differ by virus), virions are endocytosed and fusion between viral envelope and cell membrane occurs at low pH (Fig. 15.4). The low pH environment triggers viral envelope protein rearrangement so that the envelope (E) proteins change conformation from a "flat" to an unfolded (spike) structure, driving their insertion into the endosomal membrane. Membrane fusion releases the infecting genome, which is efficiently translated.

The process of translation initiation varies by genus. The hepaciviruses and pestiviruses have an IRES in the 5′ NCR while the members of the genus *Flavivirus* use cap-directed translation. The protein expression strategy of the members of the family *Flaviviridae* is to generate the majority of their proteins from a single polyprotein precursor, by a series of stepwise proteolytic cleavages. This is very similar to the protein expression strategy of the picornaviruses (Chapter 11: Family *Picornaviridae*). Some products of the polyprotein remain cytosolic (for example, the capsid protein and several nonstructural proteins, NSPs) while others (for example, the envelope proteins) cross the ER membrane during synthesis. Cytosolic regions of the precursor polyprotein are cleaved by virally encoded proteases while cleavages in the ER are accomplished with host cell proteases. Viral proteins residing in the ER lumen traffic through the ER and Golgi where they are glycosylated.

After translation the infecting genome is replicated. Genome replication occurs on cell membrane scaffolds and requires the viral NSPs. Viral genomes associate with capsid proteins in the cytoplasm. Capsid proteins also interact with portions of the viral glycoproteins displayed on the cytoplasmic face of the ER. Viral assembly occurs by budding of the capsid into the lumen of the ER. Virions are released when the ER vesicles fuse with the plasma membrane, releasing their cargo outside the cell, a process called *exocytosis*. It is interesting to note that immature virions have spike-like E proteins. These change conformation to lie flat on the surface of the mature virion at neutral pH. This shows that pH controls the conformation of the E proteins, and that E protein conformation is reversible.

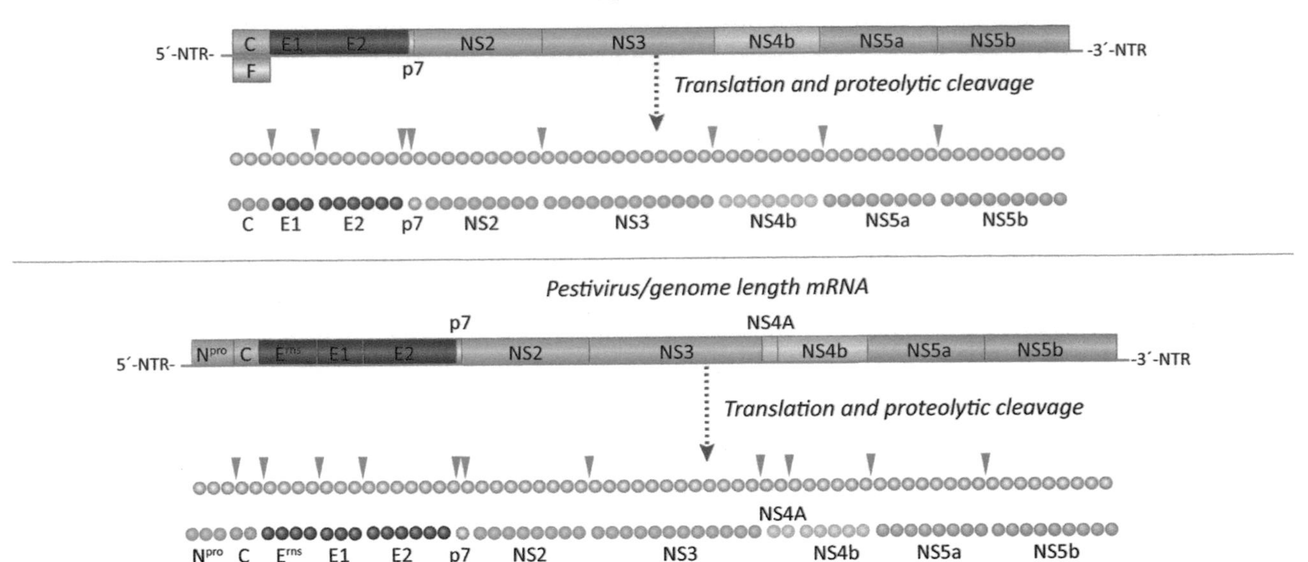

FIGURE 15.3 Comparison of hepacivirus and pestivirus genomes. Note that HCV encodes two envelope (E) proteins; pestiviruses encode 3 E proteins. Pestiviruses encode a protease (N^{Pro}) upstream of the stuctural proteins.

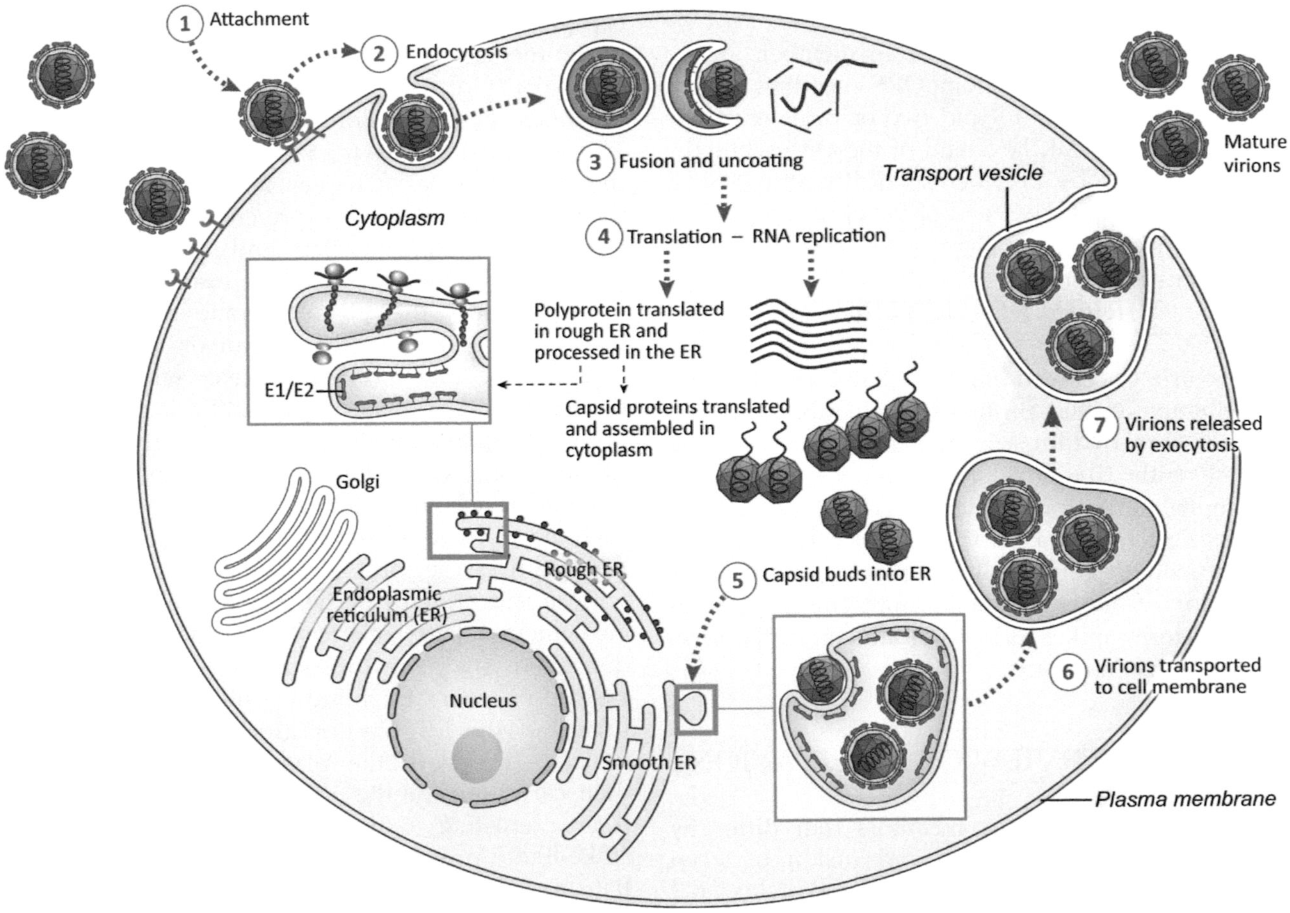

FIGURE 15.4 Flavivirus replication cycle.

PROTEIN PRODUCTS

The overall genome organization of members of the family *Flaviviridae* is conserved such that the structural proteins (capsid and envelope) are encoded at the 5′ end of the single long ORF. The RdRp (NSP5B) is encoded at the very 3′ end of the ORF. However the pestiviruses are notably distinct in that they encode a protease domain (N^{pro}) upstream of C. N^{pro} cleaves the polyprotein to generate the correct N-terminus of the C protein (Fig. 15.3).

All members of the family *Flaviviridae* encode a single capsid (C) protein. In addition there are two or three envelope glycoproteins. Members of the genus *Flavivirus* encode prM and E, hepaciviruses encode E1 and E2 while pestiviruses encode Erns, E1, and E2. Erns is an unusual glycoprotein that is found in both membrane anchored and secreted forms. Erns is a ribonuclease (RNase) with a preference for single-stranded RNA!

The NSPs of members of the family *Flaviviridae* include the RdRp (NS5B), proteases, helicase, and others involved in genome replication and subversion of immune responses. For our purposes the hepacivirus NSPs provide a well-studied example (Table 15.1).

DISEASES

Genus *Hepacivirus*: Hepatitis C Virus

About 3% of the world's population is infected with HCV. While blood-screening has virtually eliminated blood transfusion-associated HCV infection, virus continues to be transmitted by other routes such as illegal injection drug use. Transmission is aided by the silent nature of many HCV infections. HCV has always been, and continues to be, a challenging virus to study as most isolates are impossible to culture and there is no inexpensive small animal model (Box 15.3).

The HCV pandemic is relatively recent, with worldwide increases in HCV infections paralleling the increase in manufacture and use of syringes for medical purposes (and illegal injection drug use). In some cases medical injections resulted in large numbers of infected individuals because needles and syringes were reused without proper sterilization. Such was the case in Egypt. In contrast, in the United States an epidemic of illegal injection drug use from the 1950s to the 1980s is believed to have caused many HCV infections. It has been estimated that there were probably less than 500,000 persons with chronic HCV infection in the United States in the early 1950s; by the mid-1990s an estimated 5 million infections had occurred. In the United States today, serosurveys reveal a high rate of HCV infection among people born between 1945 and 1964. Rates of new infections have decreased since 1980s, largely due to development of blood-screening tests (Box 15.4).

Humans are the only known reservoir of HCV. Acute HCV infection is usually asymptomatic with only a few patients presenting with typical symptoms of acute viral hepatitis. An inapparent acute infection becomes an inapparent chronic infection in ~70% of cases. However chronic infection is accompanied by an increased likelihood of developing serious medical problems such as fatty deposits in the liver, liver disease, and hepatocellular carcinoma. The large numbers of chronically infected individuals have encouraged development of novel antiviral drugs to combat HCV infection. However the genetic diversity of HCV means that not all strains are susceptible to all drugs. Anti-HCV drugs include protease inhibitors and drugs that inhibit the RdRp (NS5B) and NS5A. A current barrier to treatment is the high cost of many new and highly effective drugs. (One of the first treatments for HCV infection was a combination of recombinant interferon alpha and ribavirin. However that treatment is not well tolerated, is only partially successful and is not recommended when other treatments are available.)

TABLE 15.1 NSPs of HCV and Their Functions

NSP	Function
P7	A 63 amino acid peptide that oligomerizes in membranes to form a functional viroporin, a cation-selective ion channel that allows leakage of protons from acidic vesicular compartments. Required for virion production.
NS2	Cysteine protease
NS3	N-terminal domain is a serine protease and the C-terminal domain is an RNA helicase-NTPase.
NS4A	A cofactor for the NS3 serine protease.
NS4B	Recruits and rearranges ER membranes forming a membranous structure that serves as the site of viral genome replication.
NS5A	A zinc-binding metalloprotein that binds the viral RNA and various host factors
NS5B	RdRp

BOX 15.3

DISCOVERY OF HCV

Michael Houghton, Qui-Lim Choo, George Kuo, and Daniel W. Bradley codiscovered Hepatitis C virus in 1989 (Choo et al., 1989). Dr. Houghton reviewed the details of this discovery in a 2009 manuscript (Houghton, 2009). Dr. Houghton's paper is an accessible and well-written manuscript that may be of interest to anyone who wishes to truly understand how many scientific "breakthroughs" are achieved. For the less adventuresome, a brief summary of this important discovery is presented here.

Hepatitis is liver disease and can be caused be several different viruses. By the 1970s it was understood that there were many cases of posttransfusion hepatitis that were not caused by either the hepatitis A virus or the hepatitis B virus. The so-called non-A, non-B hepatitis (NANBH) agent appeared to be a virus, but one that could not be replicated in cultured cells. The NANBH agent was transmissible to chimpanzees, a characteristic that eventually led to the solution of this puzzle of a virus.

As described in a breakthrough study published in the journal *Science* in 1989, Dr. Houghton's group cloned random pieces of nucleic acid present in plasma of an experimentally infected chimpanzee. They started with a highly infectious plasma sample and centrifuged it at very high speed to recover any/all virus particles or nucleic acids that might be present. They then used the enzyme reverse transcriptase to generate copy DNA (cDNA) from any nucleic acid present in the sample. These short pieces of cDNA were cloned into a bacteriophage that had been engineered to produce protein products from the cloned cDNA. (A bacteriophage is a virus that infects bacteria). The recombinant bacteriophages were grown on lawns of susceptible *E. coli* cells to produce millions of plaques. The proteins and nucleic acids from these plaques were transferred to membranes, followed by incubation with immune sera from infected chimpanzees. The researchers were looking for plaques that would bind antibodies present in sera from infected chimpanzees but not present in sera from uninfected controls. The researchers screened over a million plaques before identifying a single clone (called 5-1-1). This particular bacteriophage encoded a small peptide that bound to antibodies present *only* in chimps infected with the NANBH agent.

The gene fragment was sequenced and was recognized to be a piece of a virus. This first piece of viral genome was then used to identify additional overlapping pieces using nucleic acid hybridization technology. In this manner the researchers pieced together almost the entire genome of the agent we know today as HCV. It was soon possible to inject the genome into chimpanzees and recover infectious HCV to verify their findings. Obviously the inability to culture the virus and the need to use chimpanzees as experimental animals made for a very challenging system. However the high toll of HCV (millions of blood-borne infections) and the need to develop a blood-screening test provided ample incentive to study this intractable virus.

It is highly unlikely that the studies described above would be performed today. Instead we likely would use high throughput nucleic acid sequencing techniques and powerful computers to generate and analyze all nucleic acids (host, viral, and bacterial) present in any sample. Nucleic acids present in experimentally infected and uninfected animals would be compared, with hopes of finding unique viral sequences in infected animals.

BOX 15.4

HCV: WHY THE LIVER?

HCV is very highly tropic for the human liver, replicating only in hepatocytes. Most isolates of HCV will not even replicate in cultured hepatocyte cell lines. What is so special about the liver?

First entry of HCV requires several proteins, some of which are preferentially expressed on hepatocytes. Adaptation of hepatocytes to growth in culture does not preserve liver architecture or the expression of the full complement of necessary receptor proteins. Once HCV gains entry into a cell its replication also requires the presence of a micro-RNA (miRNA) that is only expressed in liver cells. miRNAs are small (~22 nt) noncoding RNAs that are important regulators of protein synthesis. Their regulatory activity is based on their ability to base-pair with complementary sequences in mRNAs. The miRNA required by HCV is called miR-122. miR-122 is a liver-specific miRNA. In the adult human liver miR-122 constitutes ~70% of total miRNA; it is clearly very important for normal functioning of liver cells.

In the case of HCV replication, miR-122 base pairs to two complementary sites at the 5′ end of the viral genome. A protein called Argonaute-2 binds to miR-122 generating a double-stranded RNA/protein complex. Experiments have shown that this complex *protects* the 5′ end of the viral genome from degradation by a ubiquitous host cell exo-RNase that normally functions to degrade decapped miRNAs. Recall that the HCV genome is not capped. Thus in the absence of miR-122 HCV RNAs are not stable.

Genus *Flavivirus*

There are many human and animal pathogens within the genus *Flavivirus*. YFV, DFV, and WNV are transmitted by mosquitoes. YFV and DFV are major human pathogens while WNV is primarily an avian virus that only occasionally infects mammals. YFV and DFV are transmitted between human and mosquito hosts in a "simple transmission cycle." YFV infects both human and nonhuman primates (Fig. 15.5) while DFV is strictly a human pathogen (Fig. 15.6). The mosquito hosts of YFV and DFV are members of the genus *Aedes*. These viruses actively replicate in their mosquito hosts, a requirement for their eventual retransmission to humans.

Yellow Fever Virus

YFV has two transmission cycles (Fig. 15.5), the so-called urban transmission cycle and a sylvatic (forest) transmission cycle. Nonhuman primate hosts maintain YFV in the sylvatic cycle, maintaining a virus reservoir that can move back into urban environments. YFV can cause severe illness and death; up to 50% of severely affected persons will die without treatment. Worldwide, the numbers of YF cases are ~200,000 per year with about ~30,000 deaths. Vaccination is the single most important preventive measure against yellow fever. There is a single serotype of YFV and the vaccine is safe, affordable, effective, and appears to provide long-term protection. The YFV vaccine is a live

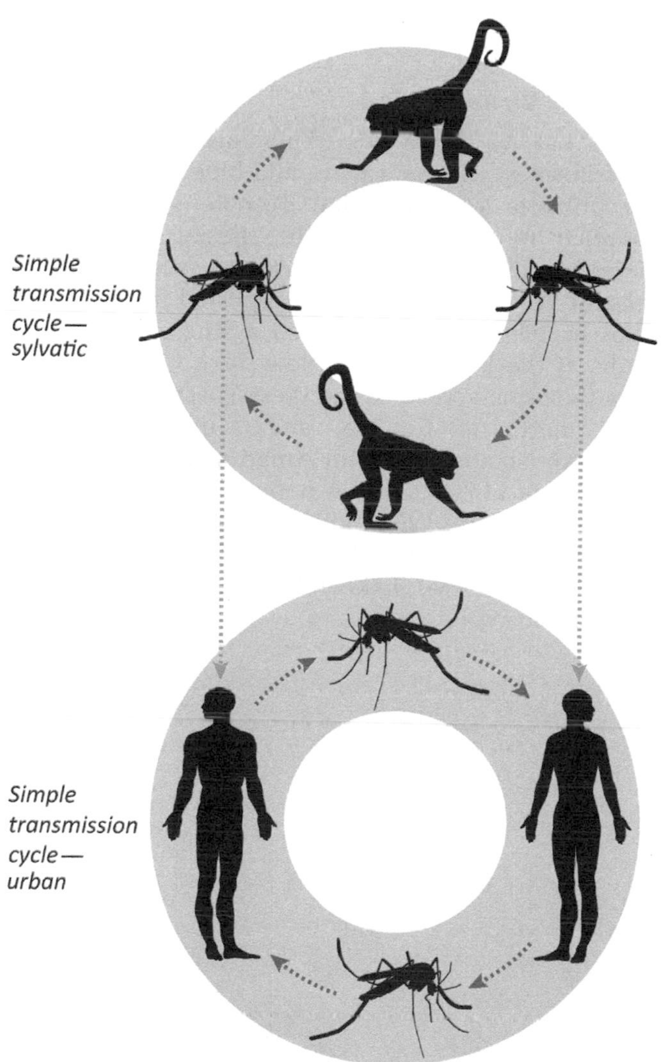

FIGURE 15.5 YFV transmission cycles. Human and nonhuman primates are the only vertebrate hosts for YFV. Vaccination can break human to human transmission cycles but the sylvatic (forest) cycle remains intact allowing for possible reintroduction into humans.

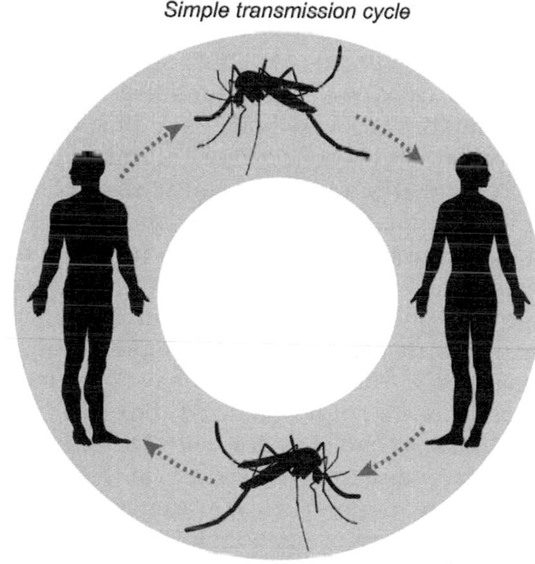

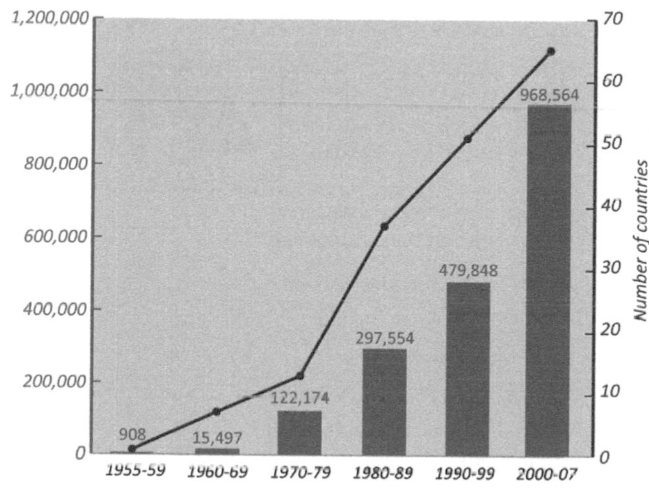

FIGURE 15.6 Dengue fever transmission cycle and worldwide incidence.

attenuated strain called 17D. The same vaccine has been used since the 1940s.

Dengue Fever Virus

DFV is transmitted between humans and mosquitos in a simple transmission cycle. Before World War II, DFV was restricted to two small areas in Southeast Asia. However during the war, cargo ships carried infected *Aedes* mosquitos throughout the world. Since the 1960s the number of countries reporting DFV and the numbers of cases of DFV have skyrocketed, from less than 1000 cases reported from 1955 to 1959 to close to a million reported cases between 2000 and 2007 (Fig. 15.6). Humans are the only mammalian hosts for DFV. Mosquito control in urban areas, including use of screening on houses and eliminating mosquito habitat (standing water), is currently the only way of slowing the spread of DFV. Primary infection with DFV is usually a self-limited febrile illness. A notable symptom is severe bone pain, thus the disease is also called "breakbone fever." Primary infection with DFV probably protects against reinfection with the same serotype. However there are four serotypes of DFV and infection with one serotype does not protect against the others. In fact, infection with a second serotype may increase risk of complications such as dengue hemorrhagic fever and dengue hemorrhagic shock syndrome. These conditions have high mortality rates if patients are not hospitalized (Box 15.4).

Why does a second infection with DFV increase the risk of severe disease? As shown in Fig. 15.7, the immune system responds to the primary infection by producing a mixture of neutralizing and binding antibodies. The neutralizing antibodies are serotype-specific, while binding antibodies are not. In fact, binding antibodies can increase infection of macrophages. Recall that macrophages are professional phagocytic cells that display Fc receptors on their surface. Fc is the constant fragment of the IgG molecule (the tail of the Y-shaped molecule). The Fc portion of IgG binds to the Fc receptors on macrophages to facilitate antigen uptake. However, in the case of infection with a second DFV serotype, nonneutralizing or poorly neutralizing antibodies bind to virions and these antibody bound virions can more easily infect macrophages. As macrophages are a source of proinflammatory cytokines, it is postulated that increased levels of infection and death lead to dengue hemorrhagic fever and dengue shock syndrome. There are currently no vaccines for DFV (although some are in development). A major problem for vaccine development is the need to generate adequate levels of neutralizing antibody against all four serotypes in order to avoid the complication of vaccine-associated immune enhancement of disease.

West Nile Virus

The natural hosts for WNV are birds (largely Passeriformes). Crows, robins, and blue jays are highly susceptible to lethal infection. To date, scientists have identified more than 138 bird species that can be infected, and more than 43 mosquito species that can transmit WNV. WNV was first reported in the United States in 1999. It was identified as the cause of bird deaths at the Bronx zoo in New York City. Within 5 years the virus spread to the West Coast of the United States, north to Canada, and southward to the Caribbean Islands and Latin America. WNV continues to adversely affect North American bird populations. In December 2013 WNV killed more than two dozen bald eagles in Utah. The deaths of ~20,000 water birds around the Great Salt Lake in November 2013 was also attributed to WNV.

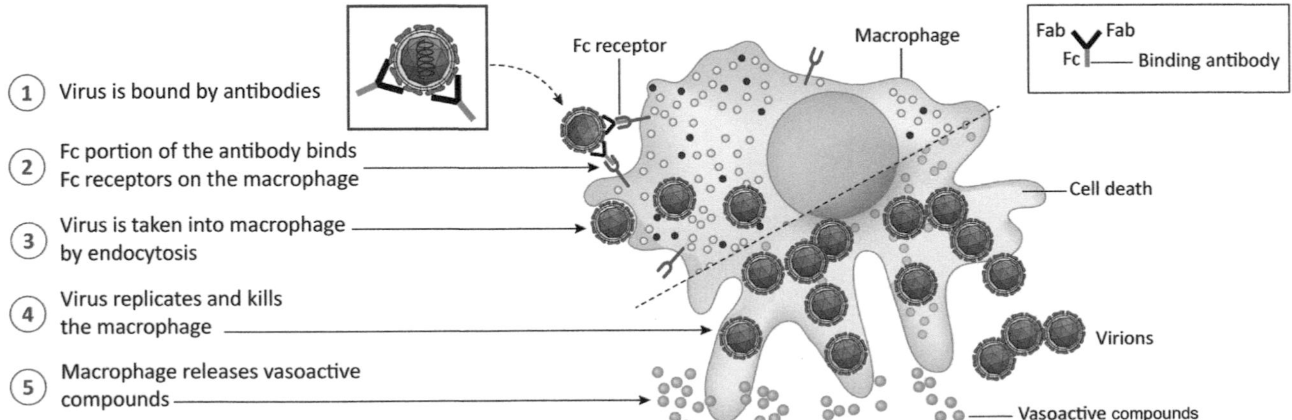

FIGURE 15.7 A model for immune enhancement of DFV infection. In this model, nonneutralizing or poorly neutralizing antibodies bind to virions and enhance infection of macrophages.

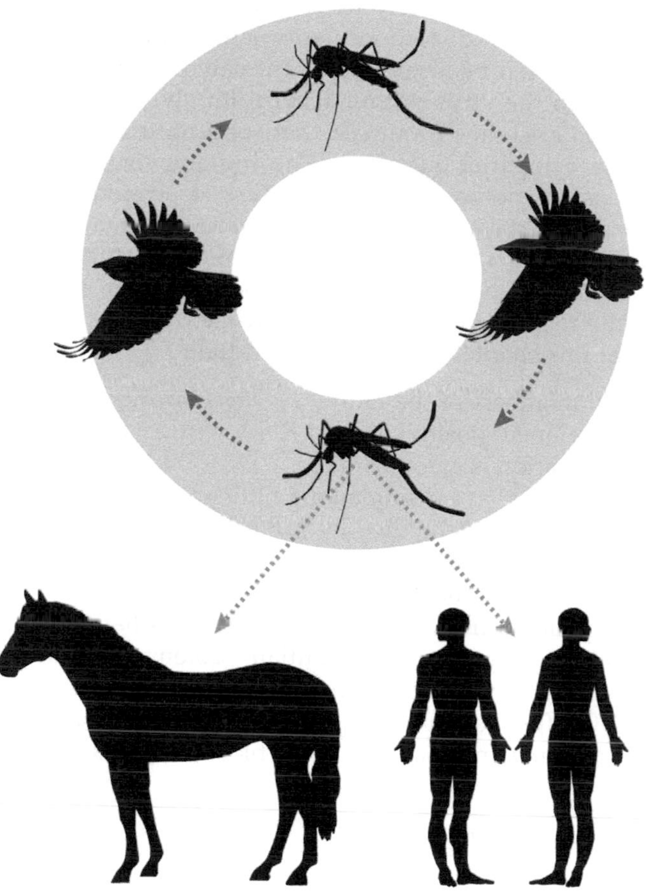

FIGURE 15.8 WNV transmission. The normal vertebrate hosts for WNV are birds. However mosquitoes may transmit virus to other mammals such as horses and humans which are dead end hosts.

Humans are not natural hosts for WNV (Fig. 15.8) and approximately 80% of WNV infections in humans are subclinical. The most common clinical manifestation is West Nile fever. Symptoms include fever, headaches, fatigue, muscle pain or aches, malaise, nausea, anorexia, vomiting, myalgia, and rash. More serious are WNV infections that involve the central nervous system (CNS). These are rare, occurring in only ~1% of infected individuals. CNS complications include encephalitis, meningitis, meningoencephalitis, and WNV poliomyelitis (flaccid paralysis due to spinal cord inflammation). Elderly, the very young or immunosuppressed persons are most susceptible. Currently, no human vaccine against WNV infection is available. Rates of WNV infection can be reduced by adequate mosquito control, personal protective measures such as mosquito repellent, window screens, long sleeves, and pants. Vaccines approved for use in horses are available and should be used prior to mosquito season.

Zika Virus: An Emerging Threat

Zika virus infects humans and nonhuman primates and is spread by *Aedes* sp. mosquitoes. The most common symptoms of Zika disease are fever, rash, joint pain, conjunctivitis, malaise, and headache. The illness is usually mild with symptoms lasting for several days to a week. Many infections are probably asymptomatic. Unfortunately, Zika virus infection during pregnancy can cause a serious birth defect called microcephaly. The head of a microcephalic baby is smaller than normal; the brain is often small and not properly developed. There is only one known serotype of Zika virus and once a person has been infected, he or she is probably protected from future infections.

Zika virus was first discovered in Uganda 1947 in monkeys and was detected in humans in 1952. Prior to 2015, cases were reported in tropical Africa, Southeast Asia, and the Pacific Islands but because the symptoms of Zika are similar to those of many other diseases, many cases may not have been recognized. The first large outbreak of disease caused by Zika was reported from the Island of Yap (Federated States of Micronesia) in 2007. In 2015 the first confirmed human case of Zika virus infection occurred in Brazil. By early 2016, the World Health Organization (WHO) declared Zika virus a worldwide public health emergency. A driving factor in declaring Zika virus a public health emergency was the association between Zika virus infection and microcephaly.

Zika virus is primarily transmitted to people through the bite of an infected *Aedes* mosquito. However, sexual transmission of Zika virus has been documented, and given the link to birth defects it is imperative that we learn more about this mode of transmission.

Currently there is no vaccine for Zika virus. Prevention includes protection against mosquito bites. In areas where Zika virus is prevalent, pregnant women should be tested for possible exposure, should practice safe sex (use of condoms) and avoid mosquito bites. People who have been traveling to Zika virus endemic areas should be aware that they may be infected, even if they have not experienced symptoms. Men should adopt safe sex practices and couples planning a pregnancy should seek out medical advice.

Genus *Pestivirus*

Three pestiviruses cause major diseases of livestock: BVDV (cattle), BDV (sheep), and CSFV (swine). These viruses sometimes cause inapparent, persistent, lifelong infections making it difficult to eliminate them from the environment.

BVDV and the results of persistent infection (PI) will be briefly described here. BVDV affects cattle and the clinical signs of BVDV are highly variable, ranging from few to no signs, to severe disease that quickly kills the animal. BVDV is transmitted by contact but can also be transmitted vertically from an infected cow to her calf. Congenital infections may result in resorption of the conceptus, abortion, stillbirth, or live-birth depending on the time of infection. Infected fetuses that survive may be *immunotolerant* (they do not make antiviral antibodies) and PI. Although the virus replicates, there is little disease and the PI animal is source of transmission!

A few PI animals develop a lethal disease, called mucosal disease. Mucosal disease is characterized by fever, leukopenia, bloody diarrhea, loss of appetite, dehydration, erosive lesions of the nares and mouth, and death within a few days of onset of symptoms.

What has changed? What causes mucosal disease? Mucoal disease is caused by *mutated viruses that emerge in the PI animal*. These viruses replicate much more robustly (in both cultured cells and in the infected animal) causing cell death. They are referred to as cytopathic (cp) BVDV. A variety of mutations can change BVDV to a cytopathic type. The mutations are often gene insertions and the inserted genes are of host origin (Fig. 15.9). The common denominator appears to be an increase in the rate of cleavage of the precursor polyprotein, from slow to very rapid. To summarize, BVDV is often of low virulence and can cause persistent infection of immunotolerant calves. This helps to maintain the virus within a herd. Rarely, recombinant viruses arise in PI animals that replicate robustly and cause cytopathic infection. The mutant viruses cause extensive tissue damage leading to rapid death. Therefore these mutant viruses are not maintained long term in a herd. The change from noncytopathic to cytopathic replication may be as simple as an increase in the relative rate of protein cleavage.

In this chapter we have learned that:

- All members of the family *Flaviviridae* are enveloped, positive-strand RNA viruses and their replication is cytoplasmic.
- Genomes have a single long ORF and individual proteins are generated by proteolytic cleavage.
- Some members of the family *Flaviviridae* have capped genomes and express proteins using cap-dependent translation initiation. Others have uncapped genomes and initiate protein synthesis in an IRES-dependent manner.
- HCV (genus *Hepacivirus*) is a significant human pathogen and major cause of blood-borne hepatitis.

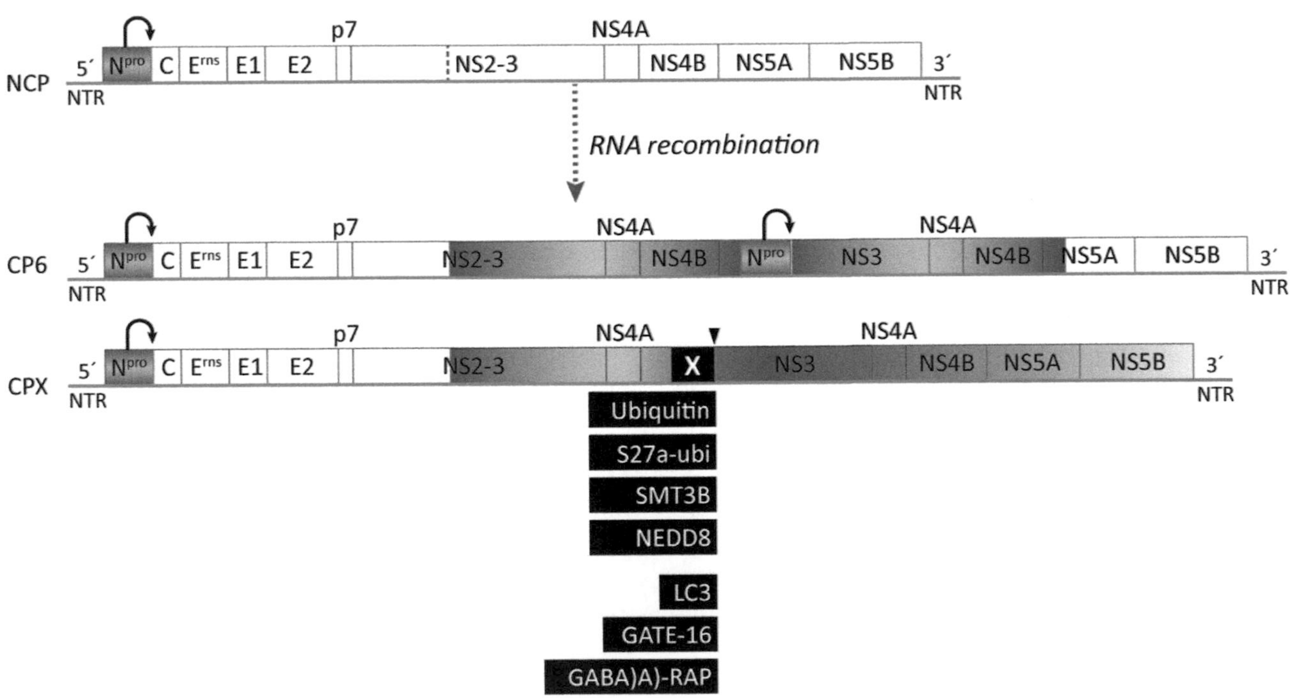

FIGURE 15.9 Examples of genome insertions that increase replication of the pestivirus BVDV to generate the cytopathic biotype. BVDV is generally noncytopathic and causes persistent infections. During the life-time of a PI bovid, the virus may accumulate gene duplications or insertions of host genes to generate a cytopathic biotype with high replicative ability. Increased replication is linked to enhance proteolysis of the precursor polyprotein.

Most infections are asymptomatic and chronic. HCV infection is a major cause of liver damage, liver failure, and hepatocellular carcinoma.
- Members of the genus *Flavivirus* include human and animal pathogens that are transmitted by insects. They replicate in both animal and insect hosts.
- Members of the genus *Pestivirus* include important pathogens of food animals. They are spread by contact and may cause persistent infections

References

Choo, Q.L., Kuo, G., Weiner, A.J., Overby, L.R., Bradley, D.W., Houghton, M., 1989. Isolation of a cDNA clone derived from a blood–borne non-A, non-B viral hepatitis genome. Science. 244, 359–362.

Houghton, M., 2009. The long and winding road leading to the identification of the hepatitis C virus. J. Hepatol. 51, 939–948.

CHAPTER

16

Family *Togaviridae*

OUTLINE

After studying this chapter, you should be able to:

- Describe the shared characteristics of members of the family *Togaviridae*.
- Describe the general strategies used by togaviruses for transcription, translation, and genome replication.
- List some human and animal diseases caused by togaviruses.
- Describe transmission of alphaviruses.

Togaviruses are enveloped viruses with unsegmented positive-strand RNA genomes (Figs. 16.1 and 16.2). The family contains two genera (Box 16.1). Genus *Alphavirus* contains many human pathogens. The alphaviruses are arboviruses, transmitted by insects. Included among alphaviruses are pathogens such as Eastern equine encephalitis virus (EEEV), Western equine encephalitis virus (WEEV), Venezuelan equine encephalitis virus (VEEV), Sindbis virus (SINV), Ross River virus (RRV), Semliki Forest virus (SFV), and Chikungunya virus. Chikungunya is emerging as a worldwide human pathogen. The ability to infect mosquitoes is key to the natural transmission cycle of alphaviruses. Mosquitoes are infected by feeding on a viremic vertebrate host and the virus moves to, and replicates within the salivary glands. Only after virus has replicated within the salivary glands can it be transmitted to a new vertebrate host.

The second genus in the family, *Rubivirus*, contains a single virus, Rubella, or German measles virus. Rubella virus is a human pathogen that is transmitted by the aerosol route. Rubella virus has the potential to cause severe congenital defects if the virus is contracted during pregnancy.

GENOME STRUCTURE

Togavirus genomes are a single piece of positive-strand RNA, 10–12 kb in length (Box 16.2). Togavirus genomes have a methylguanosine cap at the 5′ end and a poly(A) stretch at the 3′ end. There are short 5′ and 3′ untranslated regions. A short untranslated region upstream of the open-reading frame (ORF) encoding the structural polyprotein serves a promoter synthesis of a subgenomic mRNA. Nonstructural proteins (nsp) are encoded at the 5′ end of the genome. Nsps are synthesized as a polyprotein that is cleaved by viral proteases. The RNA-dependent RNA polymerase (RdRp) is nsp4. Genes for the three structural proteins, C, E1, and E2 (and sometimes an E3 protein), are found at the 3′ end of the genome. They are

Viruses. DOI: http://dx.doi.org/10.1016/B978-0-12-803109-4.00016-7

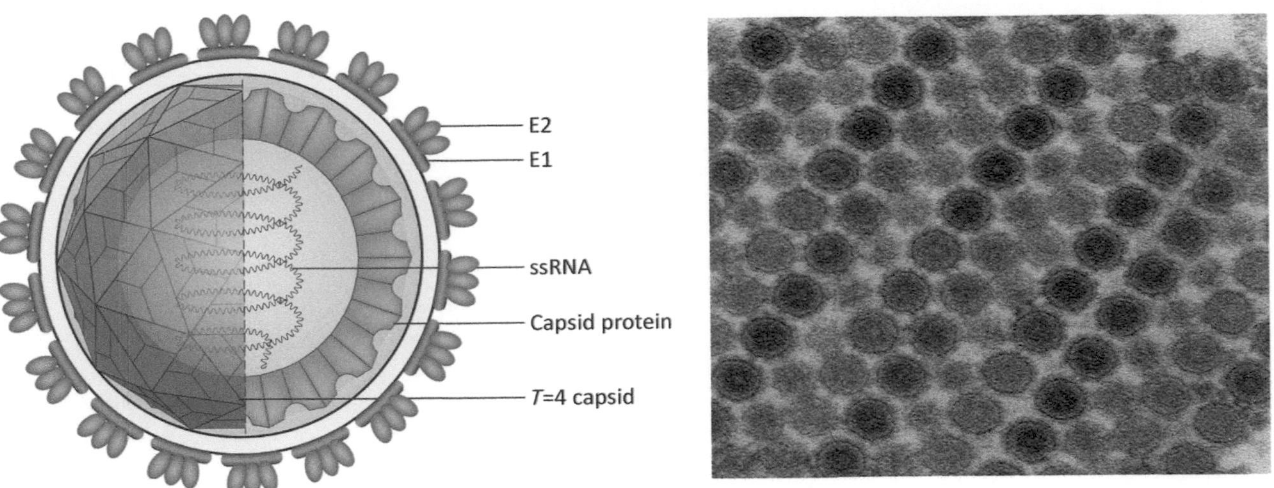

FIGURE 16.1 Virion structure (left) and transmission electron micrograph showing Chikungunya virions (right). *From CDC and Cynthia Goldsmith. CDC Public Health Image Library, Image #17369.*

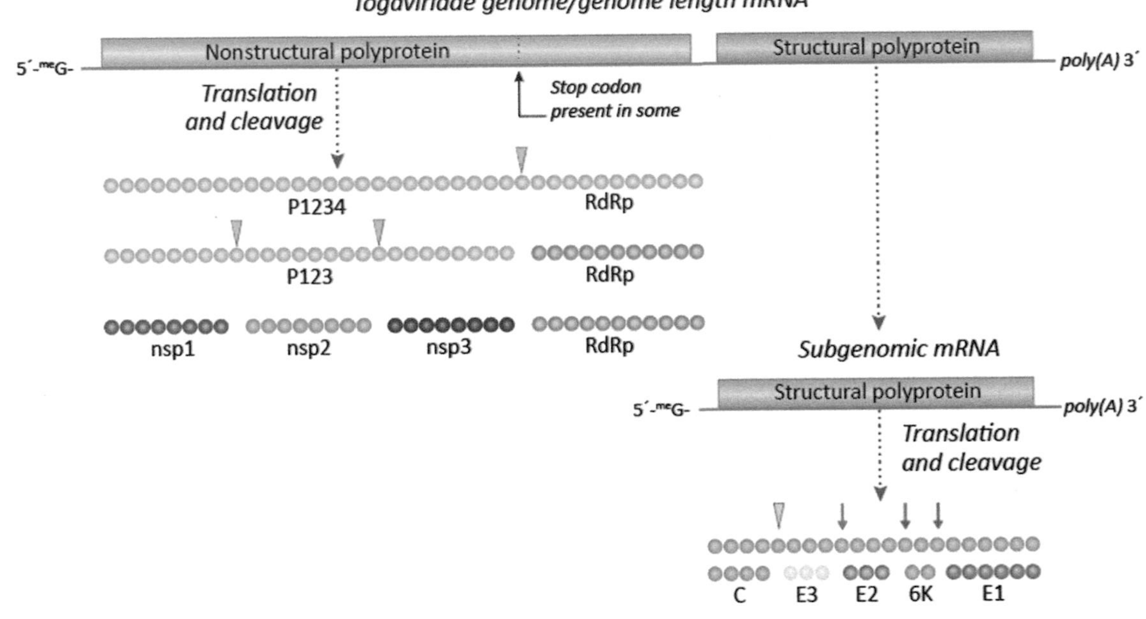

FIGURE 16.2 Genome organization and protein expression strategy.

synthesized as a polyprotein precursor from a subgenomic mRNA (Fig. 16.2).

VIRION STRUCTURE

Togavirus capsids are assembled from 240 copies of a single capsid (C) protein to form a $T = 4$ icosahedron. The capsid is surrounded by a lipid bilayer-containing protein spikes assembled from the envelope glycoproteins, E1 and E2. E1 and E2 associate to form heterodimers and each spike is a trimer of heterodimers. The envelope glycoproteins associate with capsid proteins resulting in a very regular structure for the envelope (Fig. 16.1). Virions are ~70 nm in diameter.

TOGAVIRUS REPLICATION

The alphaviruses replicate in both vertebrate and insect cells. In vertebrate cell lines alphaviruses replicate very rapidly, with the release of progeny virus

BOX 16.1

TAXONOMY

Family *Togaviridae*

Genus *Alphavirus*

Thirty recognized species that cluster into two groups. Old World alphaviruses include: Ross River virus, Semliki virus, Sindbis virus, and Chikungunya virus. The New World alphaviruses include EEEV and VEEV. Western equine encephalitis is a recombinant between a New World and an Old World alphavirus.

Genus *Rubivirus*

One species: Rubella virus.

BOX 16.2

GENERAL CHARACTERISTICS

Flavivirus genomes are a single strand of positive-sense RNA ranging from 10 to 12 kb in length. Genomes are have 5′ methyl-G caps and have poly(A) tails at the 3′ end. Genomes contain two ORFs. A polyprotein encoding the nsps is found at the 5′ end of the genome. The structural proteins (capsid and envelope glycoproteins) are synthesized from a polyprotein translated from a subgenomic mRNA.

Enveloped virions (~70 nm diameter) have $T=4$ icosahedral symmetry as the $T=4$ capsid is closely associated with 240 copies of the E1 and E2 proteins.

typically within 4–6 hours after infection. Infection causes extensive cytopathic effects (CPE) in vertebrate cells. In contrast, replication in insect cells is noncytopathic, producing a persistent infection.

Attachment and Penetration

The receptors for alphaviruses are most likely ubiquitous (common) cell surface molecules allowing the viruses to attach to both insect and vertebrate cells. Virions are endocytosed and fusion between the viral envelope and the endosomal membrane is triggered in the low pH environment.

Translation

After penetration and uncoating, genome-length RNA is translated to produce a polyprotein that is cleaved by virally encoded proteases to produce the nsps (including RdRp) required for genome replication. Most togavirus genomes have a single stop codon between nsp3 and nsp4. Therefore ribosomes must read-through or suppress the stop codon in order to synthesize the long polyprotein that contains nsp4 (Fig. 3.14).

Togavirus structural proteins (C, E1, and E2) are translated from a subgenomic mRNA that is synthesized during genome replication. The capsid protein, C, is cytosolic and is cleaved from the polyprotein precursor by a viral protease. E1 and E2 are transported across the endoplasmic reticulum (ER) membrane during synthesis. The polyprotein is cleaved by resident ER proteases. In vertebrate cells, E proteins are glycosylated and transported to the plasma membrane. The cleavages generating C, E1, and E2 also generate two small protein products with a shared amino terminus. They are called 6K (for 6 kDa MW) and transframe (TF). TF is formed by ribosomal frameshifting. Both proteins have a transmembrane region and they are both found, in low quantities, in purified virions. Mutations in 6K are associated with greatly decreased virion production and/or deformed virions (Box 16.3).

Genome Replication

Genome replication is cytoplasmic. Early after infection copies of the genome (cRNA) are synthesized. Full-length cRNA serves as the template for synthesis of additional genomes as well as subgenomic mRNAs. In cases where the process has been studied in detail, it appears that an uncleaved nsp precursor (called P123) plus a molecule of cleaved nsP4 make up the complex that synthesizes the negative-strand cRNA. However P123 is eventually cleaved to produce nsP1 + nsP2 + nsP3. When this occurs, the complex comprised of nsp1 + nsp2 + nsp3 + nsp4 synthesizes genomes and subgenomic mRNA (Fig. 16.3). In

BOX 16.3

PROTEIN DISCOVERY THROUGH BIOINFORMATICS (DISCOVERY OF A NEW ALPHAVIRUS PROTEIN)

SFV is one of the most extensively studied alphaviruses because it is relatively benign and grows quite well in cultured cells. It is a common model for studies of alphavirus replication and structure. The 3′ ORF of SFV encodes a precursor polyprotein that is cleaved to generate structural proteins C, E3, E2, 6K, and E1. The 6K protein contains a transmembrane domain and is present at low amounts in purified virions. The 6K protein is proposed to be a viroporin, possibly involved in virus budding.

Researchers studying 6K routinely noted an 8K protein along with it. It was thought to be a modified form of 6K protein, although this was never confirmed. The 8K product was finally identified in 2008, through use of bioinformatics tools, biochemistry, and genetics (Firth et al., 2008). Researchers analyzing viral genomes for potential ribosomal frameshifting sites found a previously unrecognized site in the genomes of most alphaviruses. The conserved sequence was located within the 6K coding region and had all of the structural hallmarks of a ribosomal frameshifting site; its use predicted an 8K protein. Using SFV as their model virus, the investigators biochemically analyzed the 8K protein (using mass spectroscopy with peptide sequencing as well as immunoprecipitation) and demonstrated it was the product of a ribosomal frameshift in 6K. They named the new protein TF. Finally the investigators mutated SFV by introducing a stop codon just downstream of the predicted frameshifting site. They were able to grow the mutant virus, but the 8K product was missing! Since the 2008 publication, other groups have identified TF protein in virions of other alphaviruses, for example, Sindbis and Chikungunya.

The take-home message of this story is not that there is another alphavirus protein product to memorize, it is the following: After Firth, Chung, Fleeton, and Atkins developed and used a *bioinformatics approach* to identify a *potential* new frameshifting site, they did additional work to *confirm* the presence of the protein using a well-characterized model alphavirus. They used *two different biochemical methods* to characterize the protein and then followed up with a *genetic method* for additional confirmation. Their work set the stage for others to confirm the findings using pathogens of global importance.

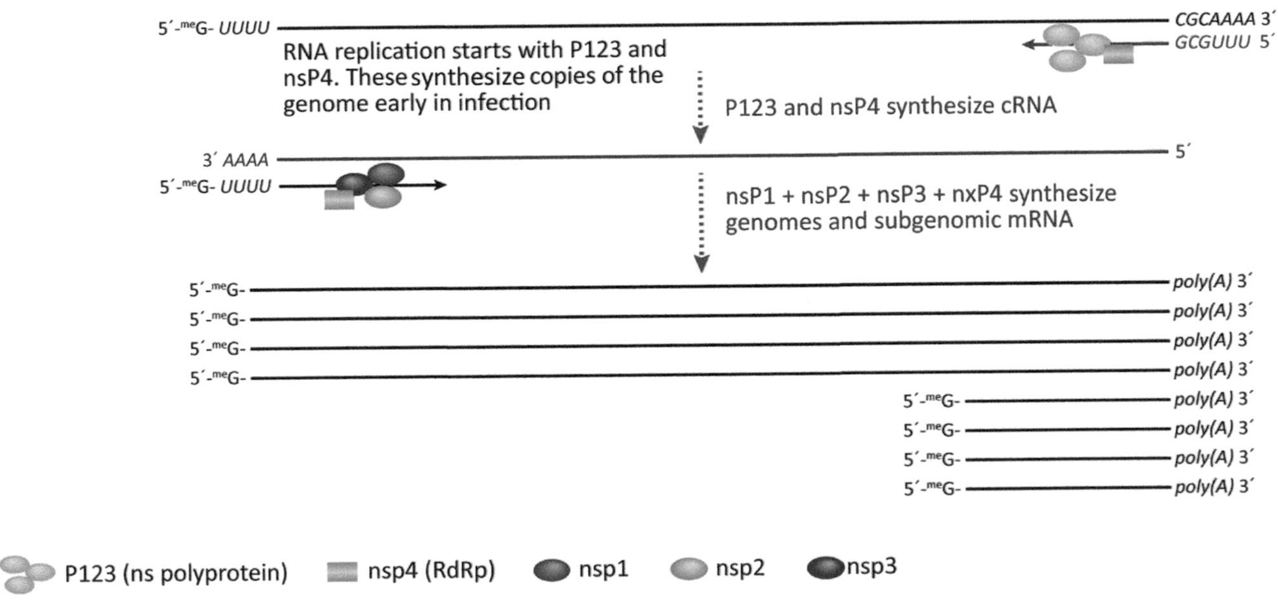

FIGURE 16.3 Togavirus model for regulation of synthesis of cRNA and genomic RNA. Proteolytic processing of the nonstructure polyprotein precursor controls synthesis of cRNA (unprocessed P123 plus nsP4) versus synthesis of genomic RNA and mRNA (nsP1, nsP2, nsP3, and nsP4).

TABLE 16.1 Togavirus nsps

Nonstructural protein	Function
nsP1	Enzymatic activities include: Methyltransferase and guanylyltransferase. These serve to cap togavirus mRNAs. An amphipathic helix and palmitoylation anchor the nsP1 to the cellular membranes.
nsP2	Enzymatic activities include: Nucleotide triphosphatase (NTPase) helicase, RNA triphosphatase, and protease. May act as a transcription factor to recruit the RNA synthetic complex to the subgenomic promoter. Involved in the shutoff of host macromolecular synthesis.
nsP3	Phosphoprotein important for minus-strand (cRNA) synthesis. Exact function not known. Binds to cellular amphiphysin-1 and -2, indicating a role in induction of membrane curvature.
nsP4	Enzymatic activities include: RdRp. Promoter binding and recognition.

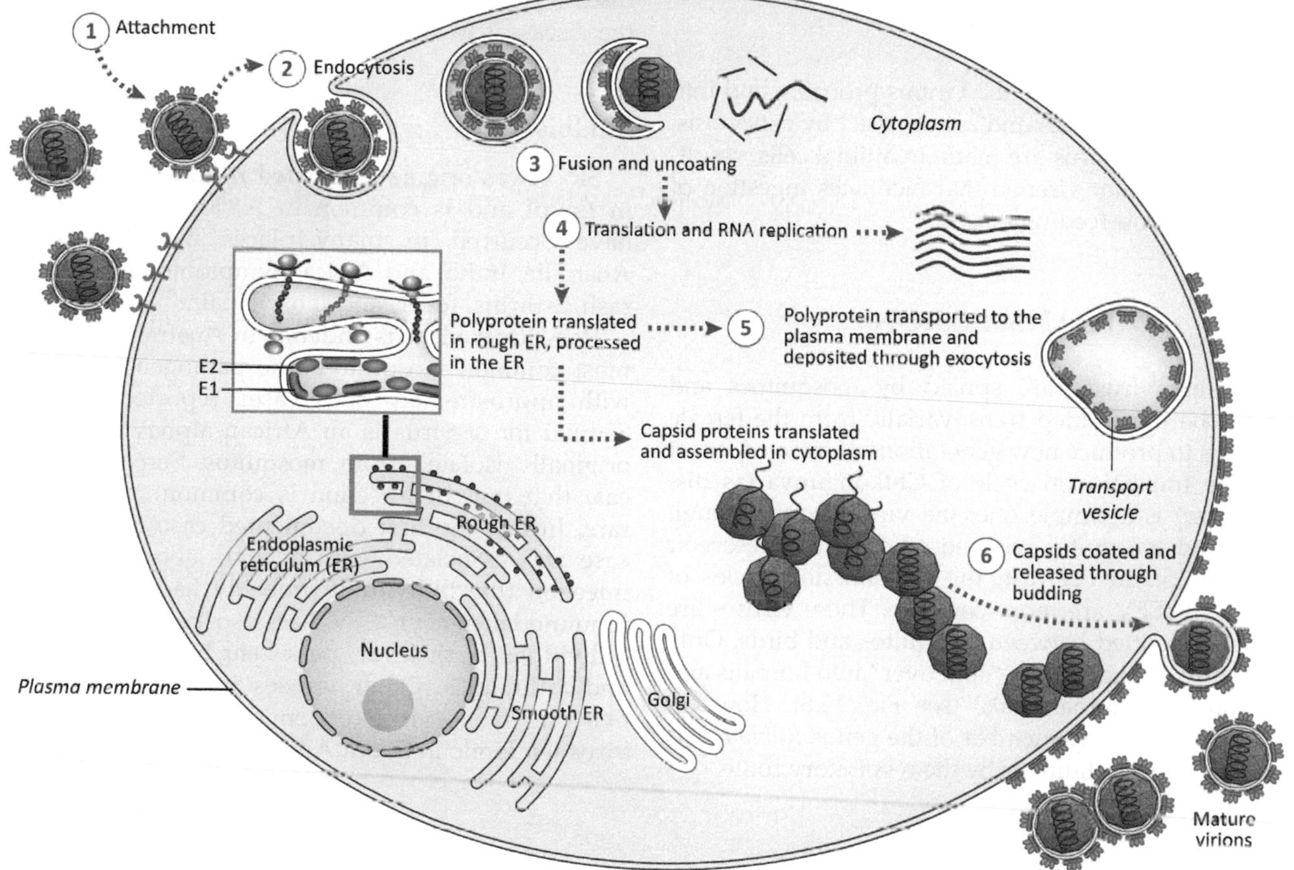

FIGURE 16.4 Overview of the togavirus replication cycle (vertebrate cell). Virion release from insect cells occurs by exocytosis.

summary, the uncleaved precursor P123 binds to the 3′ end of the genome strand but the nsP1 + nsP2 + nsP3 complex interacts with the 3′ end of the cRNA to efficiently synthesize new positive-sense RNAs. This is a clever mechanism to control the relative amounts of cRNA (just a little) and genome RNA (a lot) without encoding different sets of replicase proteins. The details of this process were determined by generating viruses with mutations at P123 eliminate cleavage sites. The RNA products produced by the mutated viruses were then analyzed. Togavirus nsps and activities are summarized in Table 16.1.

Assembly and Release

In vertebrate cells, capsid proteins and genomes assemble in the cytoplasm (Fig. 16.4). Capsids associate with E proteins at the plasma membrane where budding occurs. In the case of animal cells infection is usually cytopathic; however, in insect cells the infection can be

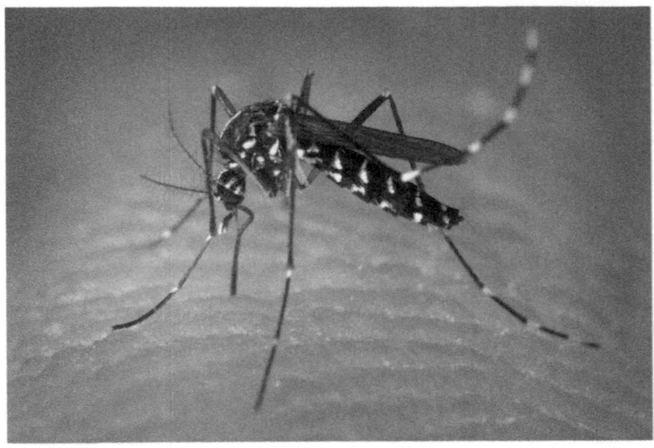

FIGURE 16.5 Asian tiger mosquito, *Aedes albopictus*.

noncytopathic. In this case the virions probably bud into intracytoplasmic vesicles and are released by exocytosis. Large amounts of virus are made in animal cells, resulting in the high titer viremia that facilitates ingestion of the virus by blood-feeding insects.

TRANSMISSION

Most alphaviruses are spread by mosquitoes and virus can be transmitted transovarially from the female to her eggs to produce new generations of infected insect hosts. The transmission cycle of Chikungunya (as discussed later) is a simple one; the virus cycles through humans and mosquitos with no other animal reservoir (see Fig. 15.5). In contrast, the transmission cycles of EEEV and WEEV are more complex. These viruses are usually transmitted between mosquitos and birds. Only infrequently do the viruses "spill over" into humans and horses (and other mammals) (see Fig. 15.8). However rubella virus, the only member of the genus *Rubivirus*, is transmitted among humans by the respiratory route.

ALPHAVIRUSES ASSOCIATED WITH RASH AND ARTHRITIS

Chikungunya Virus

In the Kimakonde language of Mozambique Africa chikungunya means "to walk bent over," thus is a very good name for this human virus that causes crippling arthritic disease with rapid onset. The virus is widespread; from 2004 to 2007, a large epidemic affected islands in the Indian Ocean and India and spread to Southeast Asia and Europe. In 2013, Chikungunya virus was found on islands in the Caribbean. Most cases of chikungunya in the United States are acquired during travel to Caribbean, but there has been limited local spread within the United States. The symptoms of Chikungunya virus infection are similar to those of Dengue virus, thus in areas where both viruses are common, these illnesses can be confused.

Chikungunya is transmitted by mosquitoes in a simple transmission cycle. The mosquito vectors are *Aedes* sp. Common symptoms of chikungunya virus infection are fever and joint pain, but there may also be headache, muscle pain, joint swelling, or rash. Musculoskeletal symptoms are recurrent or persistent in approximately 40% of adults. The disease is usually self-limiting and rarely life threatening but infants infected at birth are susceptible to severe central nervous system disease. There is no vaccine and the best way to avoid the disease is to avoid contact with mosquitos (Fig. 16.5).

Sindbis Virus

SINV was originally isolated from *Culex* mosquitoes in Egypt and is common in Africa. Small outbreaks have occurred in many places including Europe, Australia, India, and Africa. Symptoms include itching rash, arthritis, fever, and muscle pain.

Ross River virus is endemic in Australia. RRV is the most common mosquito-borne pathogen in Australia, with approximately 5000 cases reported each year. Semliki forest virus is an African alphavirus that was originally isolated from mosquitos. Serosurveys indicate that human infection is common, but disease is rare. In fact the first documented case of human disease was associated with a fatal, laboratory-acquired infection (the individual probably had an underlying immunodeficiency). SFV has also been isolated from individuals with fever, persistent headache, myalgias, and arthralgias. Other viruses in this group include O'nyong'nyong virus (endemic in Africa) and Mayaro virus (endemic in South America).

ALPHAVIRUSES OF VETERINARY IMPORTANCE

EEEV is a New World alphavirus that is maintained in a bird to mosquito cycle in freshwater hardwood swamps. The mosquito, *Culiseta melanura*, feeds almost exclusively on avian hosts. Transmission to mammals requires mosquito species capable of creating a "bridge" between infected birds and uninfected mammals. Some *Aedes*, *Coquillettidia*, and *Culex* species serve as bridges to mammals. Humans and horses are considered "dead end" hosts for EEEV as the concentrations of virus in the bloodstream are not high enough for transmission back to mosquitos. Survival

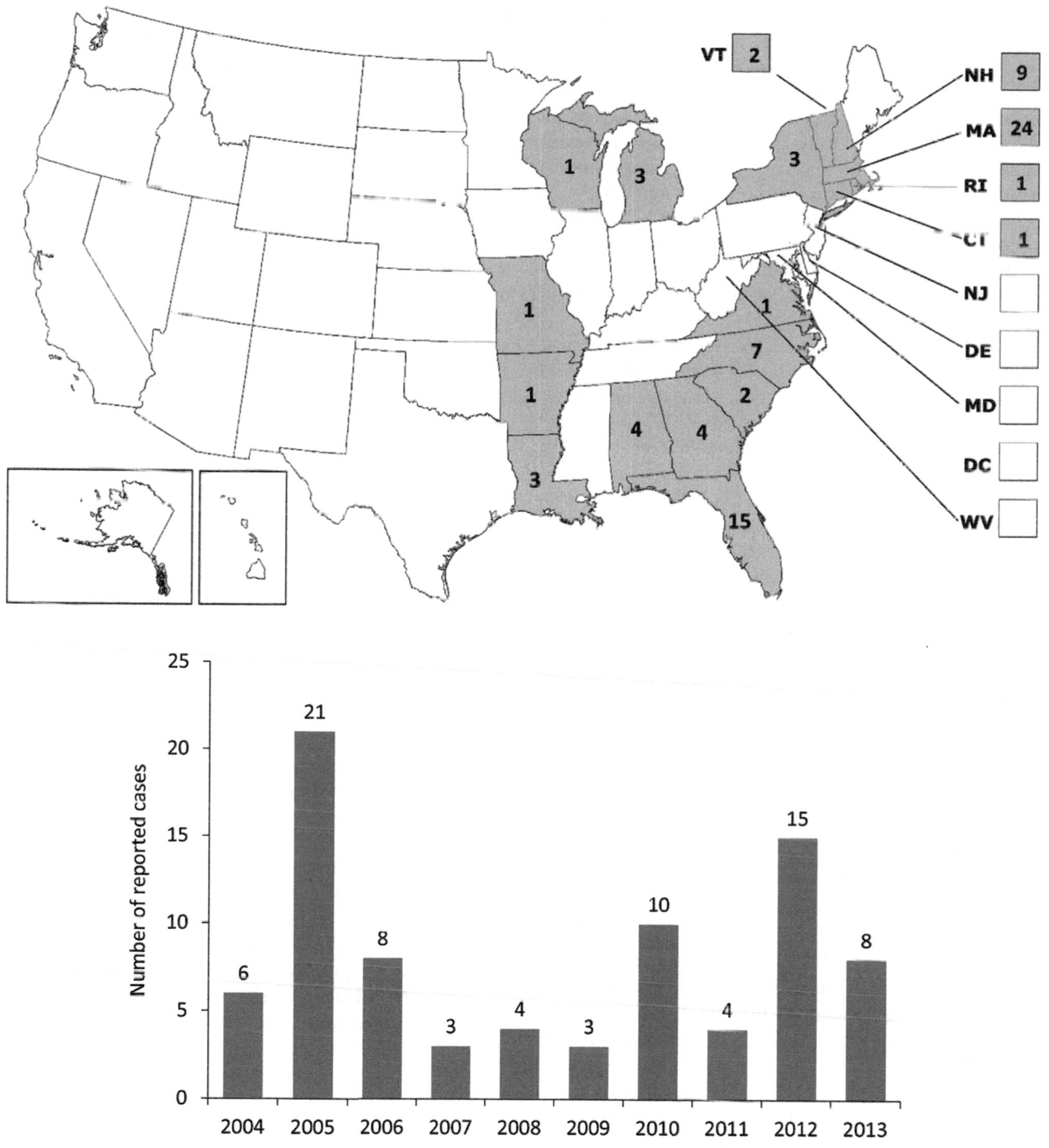

FIGURE 16.6 Top: Reported cases of neuroinvasive EEEV by state for years 2004–13. EEEV cases have been reported in Alabama (4), Arkansas (1), Connecticut (1), Florida (15), Georgia (4), Louisiana (3), Massachusetts (24), Michigan (3), Missouri (1), New Hampshire (9), New York (3), North Carolina (7), Rhode Island (1), South Carolina (2), Vermont (2), Virginia (1), and Wisconsin (1). Bottom: In the United States, the number of EEEV neuroinvasive disease cases reported each year varies. From 2004 through 2013, an average of eight cases were reported annually (range 3–21). *From ArboNET, Arboviral Diseases Branch, Centers for Disease Control and Prevention. https://www.cdc.gov/EasternEquineEncephalitis/resources/EEEV by-year_2004-2013.pdf.*

rates of infected horses are 70%–80% and a vaccine is available. Among humans there are on average, eight cases of EEEV-associated encephalitis reported yearly; the disease is fatal in about one-third of cases of encephalitis (Fig. 16.6). However, many humans infected with EEEV remain asymptomatic.

WEEV is uncommon. WEEV is transmitted by *Culex* and *Culiseta* mosquitos. WEE is genetically very interesting as it is a *recombinant virus* generated from two other alphaviruses: One a Sindbis-like (Old Word) virus and the other an EEEV-like (New World) virus. WEEV is found primarily in the United States in states west of the Mississippi River (hence the name) and is also found in South America. The usual cycle of WEEV transmission is between birds and mosquitoes. Infection of horses or people is most often subclinical. The overall mortality of WEE in humans is low (approximately 4%) and is associated mostly with infection in the elderly.

VEEV. Some types of VEEV (so-called *enzootic* subtypes) cycle among *mosquitoes and rodents* and they remain in localized areas. These sometimes "jump" to horses. Other strains of VEEV (so-called *epizootic* subtypes) primarily infect horses as their mammalian hosts (*they cycle between horses and mosquitoes*). These strains can cause quite severe disease in horses. Infected horses may shed virus in nasal, eye, and mouth secretions, as well as in urine and milk. Therefore among horses, VEEV is not restricted to spread by mosquitos but can be transmitted by the respiratory route. Healthy adults (humans) often experience flu-like symptoms but the virus may cause a fatal encephalitis in very young, very old, or immunosuppressed people.

Rubella Virus

Rubella or German measles is caused by the Rubella virus, the only member of the genus *Rubivirus*. Rubella virus shares overall genome organization and morphology with the alphaviruses but is not transmitted by insects. Humans are the only host. Rubella has an incubation period of 2–3 weeks and virus spreads primarily via the aerosol route. The virus is also found in urine, feces, and on the skin. Infections are acute and there is no prolonged shedding or carrier state. In children disease is mild and of short duration (1–3 days) and may go unnoticed. Persons infected as adults can have a more serious and prolonged disease course.

Rubella virus does the most damage in pregnant women. If infection occurs before the third trimester, the virus can severely damage the developing fetus. The earlier in the pregnancy a woman is infected, the more damage the virus causes. Miscarriages occur about 20% of the time, and children born with congenital rubella syndrome (CRS) face serious lifelong health issues. Children born with CRS can shed virus for months, transmitting it to other infants and pregnant women. Infants born with CRS can present with deafness, eye abnormalities, and congenital heart disease. Other problems may include low birth weight, microcephaly, and transient spleen, liver, or bone marrow problems. Children may also have developmental delays and intellectual disability. As of 2015 the World Health Organization estimates 110,000 babies are born with CRS every year. An effective rubella vaccine is available. In the United States it is usually is administered in a cocktail with measles and mumps vaccines (MMR vaccine). Rubella vaccination is very important for nonimmune women who may become pregnant. Widespread vaccination during the past decade has virtually eliminated rubella and CRS in many developed countries.

In this chapter we have learned that:

- Members of the family *Togaviridae* have enveloped virions with $T = 4$ icosahedral symmetry. Genomes are unsegmented positive-strand RNA (capped and polyadenylated).
- Togavirus nsps, including RdRp, are translated as a polyprotein from capped genome-length mRNA. Togavirus structural proteins include a capsid protein (C) and two or three envelope glycoproteins. Structural proteins are translated as a precursor polyprotein from a subgenomic mRNA.
- Most alphaviruses are transmitted by mosquitos. Chikungunya is an emerging human pathogen. EEEV and WEEV primarily cycle between avian and mosquito hosts but sometimes "spill over" into humans and horses. Some strains of VEEV cause quite severe disease in horses and can be transmitted by respiratory secretions.
- Rubella virus is the only member of the genus *Rubivirus*. It generally causes only mild disease in children but infection of nonimmune pregnant women can result in babies with serious birth defects. Rubella vaccination is very important for nonimmune women who may become pregnant.

References

Firth, A.E., Chung, B.Y., Fleeton, M.N., Atkins, J.F., 2008. Discovery of frame- shifting in alphavirus 6K resolves a 20-year enigma. Virol. J. 5, 108.

CHAPTER

17

Family *Coronaviridae*

OUTLINE

After reading this chapter, you should be able to answer the following questions:

- What are the major structural and replicative features of coronaviruses?
- Which coronavirus protein is responsible for attachment and fusion?
- By what mechanism are coronavirus mRNAs synthesized?
- Why do coronaviruses undergo high rates of RNA recombination?
- What human diseases are caused by coronaviruses?
- How does the virus that causes feline infectious peritonitis (FIP) virus differ from the virus that causes feline enteritis?

Members of the family *Coronaviridae* are large, enveloped, single-stranded RNA viruses. They are the largest known RNA viruses, with genomes ranging from 25 to 32 kb and a virion of 118–136 nm in diameter. Virions are roughly spherical and are notable for the large spike (S) glycoprotein that extends 16–21 nm from the virus envelope (Fig. 17.1 and Box 17.1). The family contains two subfamilies, the *Coronavirinae* and the *Torovirinae* (Box 17.2). The toroviruses are notable for having a helical, doughnut-shaped nucleocapsid (Fig. 17.2).

Members of the subfamily *Coronavirinae* are widespread among mammals, often causing only mild respiratory or enteric infections. Over 60 coronaviruses (CoVs) have been isolated from bats (BtCoV) and most of these are in the genus betacoronavirus. Bats serve as large (and highly mobile) CoV reservoirs; many bat species have their own unique BtCoV, suggesting a very long history of coevolution. Until 2002 CoVs were considered only minor pathogens of humans. However an outbreak of severe acute respiratory syndrome (SARS) that began in 2002 was linked to infection with a new CoV (SARS-CoV). The outbreak increased interest in CoV replication, distribution, evolution, transmission, and pathogenesis. In 2014 another coronavirus (distinct from SARS-CoV) was isolated in connection with an outbreak of severe respiratory disease in the Middle East. This virus, called Middle East respiratory syndrome coronavirus (MERS-CoV), continues to cause sporadic cases of severe respiratory illness.

Several important animal diseases are caused by CoVs. Infectious bronchitis virus (IBV) of chickens was the first CoV identified (in the 1930s). A pig coronavirus caused the deaths of millions of piglets in the United States in 2014. Later in this chapter, we consider the unique pathogenesis of the feline CoV

Viruses. DOI: http://dx.doi.org/10.1016/B978-0-12-803109-4.00017-9

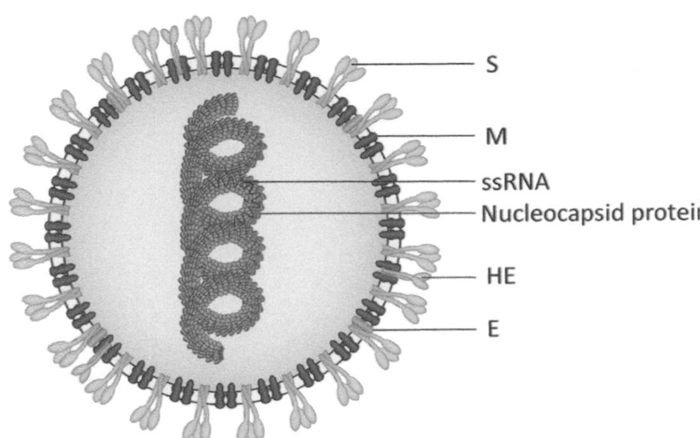

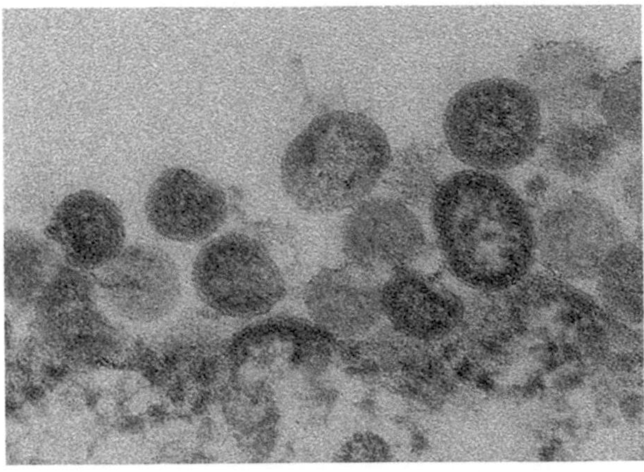

FIGURE 17.1 Virion structure (subfamily *Coronavirinae*) (right). TEM image of MERS-CoV virions in culture (left). *From CDC/Cynthia Goldsmith, Azaibi Tamin, Ph.D. Public Health Image Library Image #17280.*

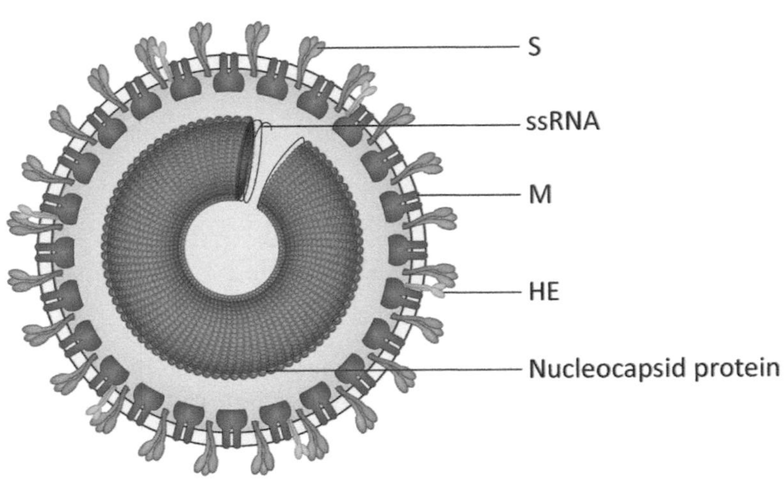

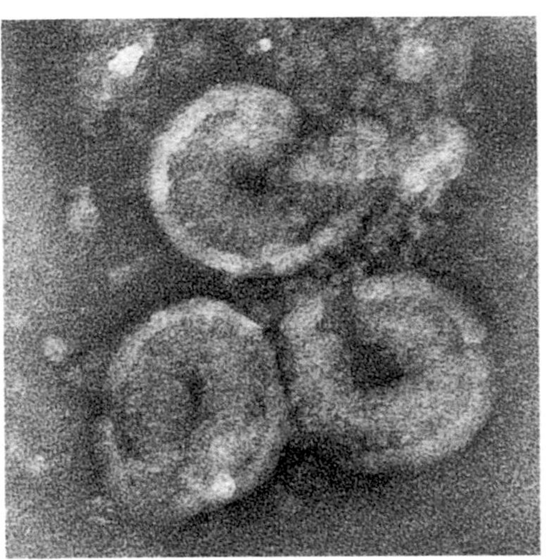

FIGURE 17.2 Torovirus structure (left) and transmission electron micrograph of torovirus particles (right). *From Graham Beards.*

BOX 17.1

GENERAL CHARACTERISTICS

Genomes are monopartite, single-strand RNA, positive sense, capped, and polyadenylated, 25–32 kb.

Virions are enveloped with prominent spikes. Virion size is 118–140 nm. Within the envelope is a flexible (subfamily *Coronavirinae*) or a doughnut-shaped (subfamily *Torovirinae*) nucleocapsid.

Transcription and genome replication are cytoplasmic. Genome-length RNA serves as mRNA for a long polyprotein precursor [encoding several nonstructural proteins (nsps) including RdRp]. A set of 3′ coterminal mRNAs encode the structural proteins and some nsps. Transcription is discontinuous, leading to high rates of template switching.

(FeCoV) causative agent of FIP, a deadly disease of domestic cats. A rodent coronavirus, mouse hepatitis virus (MHV), has served for many years as a useful model system for investigating CoV replication and pathogenesis.

CORONAVIRUS GENOME ORGANIZATION

The 25–32 kb positive-strand RNA genome contains 7–10 open reading frames (ORFs). Almost two-thirds

BOX 17.2

TAXONOMY

Family *Coronaviridae*
Subfamily *Coronavirinae*

Genus *Alphacoronavirus*
Genus *Betacoronavirus*
Genus *Deltacoronavirus*
Genus *Gammacoronavirus*

Subfamily *Torovirinae*

Genus *Torovirus*
Toroviruses have been isolated from mammals with gastroenteritis, but rarely from humans. Virions are enveloped, about 120–140 nm in diameter. They have surface spikes and a doughnut-shaped nucleocapsid.

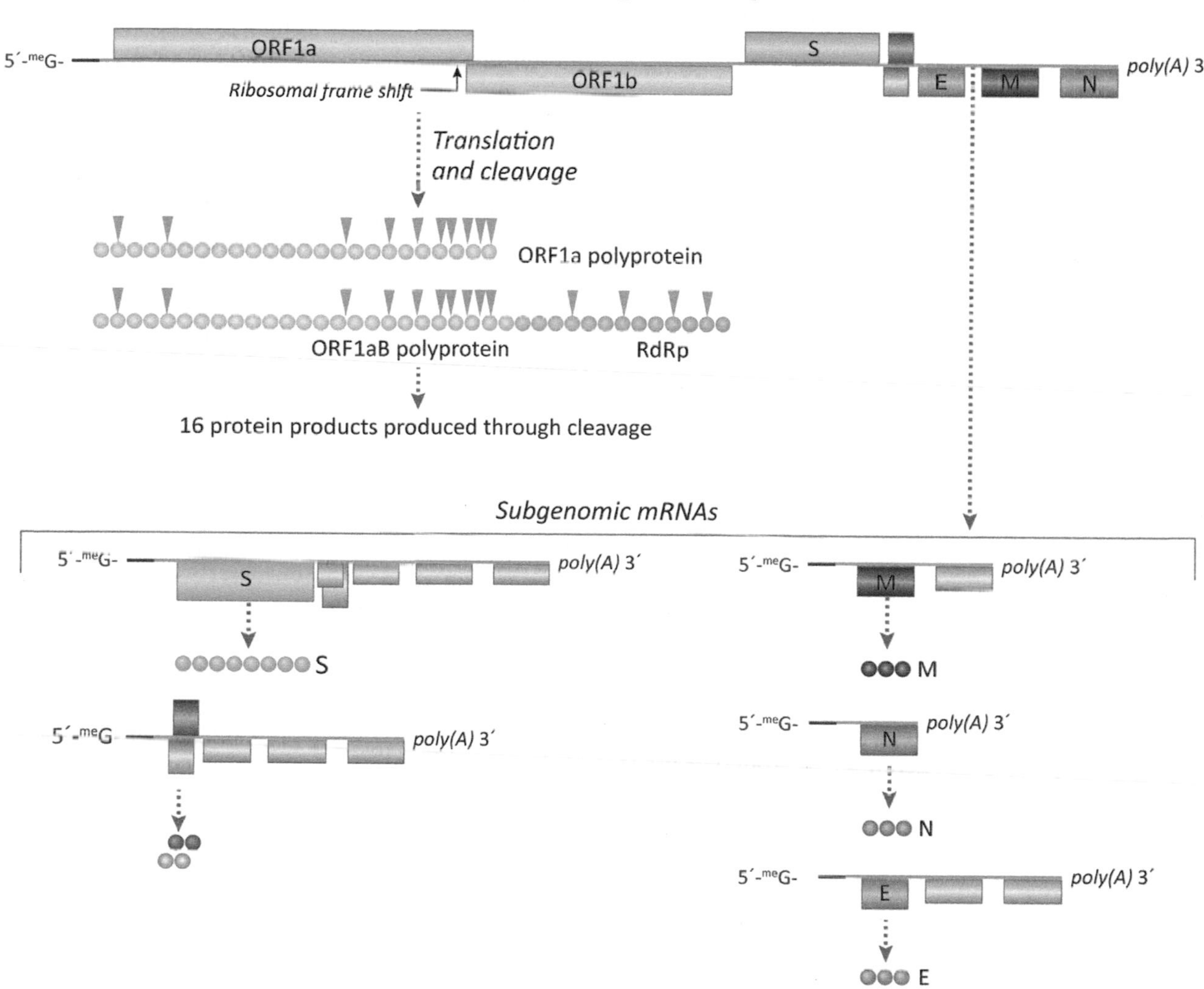

FIGURE 17.3 Coronavirus genome organization and gene expression strategy.

of the genome encodes nonstructural proteins (nsps) that are required for transcription and genome replication. Among these is nsp12, the large 930 amino acid RNA-dependent RNA polymerase (RdRp). Nsp12 forms a multiprotein complex with other CoV nsps. CoV nsps are synthesized as long precursor polypeptides, cleaved by virally encoded proteases (Fig. 17.3). The REP1a polyprotein is cleaved to produce 11

TABLE 17.1 Coronavirus nsps That form the Replicase/Transcriptase Complex*

Nonstructural protein	Function
nsp1	Interferon antagonist (not present in all coronaviruses)
nsp2	Not known
nsp3	Papain-like protease domains and several other protein interaction domains. May tether the RNA genome to the replicase/transcriptase complex.
nsp4	Transmembrane scaffold. Involved in membrane remodeling.
nsp5	Main protease (M^{pro}) (also called 3C-like protease)
nsp6	Transmembrane scaffold. Involved in membrane remodeling.
nsp7	Forms a large complex with nsp8.
nsp8	Forms a large complex with nsp7. The complex may function as a processivity clamp for the RdRp.
nsp9	Single-stranded RNA-binding protein
nsp10	Zinc-binding cofactor for 2′-*O*-methyltransferase (nsp16)
nsp12	RdRp
nsp13	RNA 5′ triphosphatase (cap synthesis), RNA helicase
nsp14	N7-methyltransferase, Exo N 3′–5′ exonuclease (provides a proofreading function for the coronavirus RdRp)
nsp15	Nendo U endonuclease (cleaves single- and double-stranded RNA downstream of uridylate residues, producing 2–3 cyclic phosphates).
nsp16	2′-*O*-methyltransferase (cap synthesis)

*Products of polyproteins 1a and 1b.

smaller products (nsp1–nsp11; listed in Table 17.1). *Rep1b* follows *Rep1a* in the genome, separated from it by a frame-shifting site that is slightly upstream from the *Rep1a* stop codon. About 20%–25% of the time, ribosomes reaching the frame-shift site will slip into the minus-1 reading frame and a longer polyprotein, called REP1b is synthesized. REP1b codes for five additional proteins, nsps 12–16, among which are the RdRp (nsp12), a protein with helicase and phosphatase activities (nsp13), a protein with exonuclease and methyl transferase activities (nsp14), an endoclease (nsp15) and a second methyltransferase (nsp16). The guanine-N7-methyltransferase, nsp13, and the 2′ O-methyltransferase, nsp16, synthesize the CoV 5′-cap. In addition to the final cleavage products, long-lived intermediates may have activities different from those of the fully cleaved products.

Seven to 10 additional ORFs lie downstream of the replicase-associated genes. The largest of these encodes the spike (S) protein. The order (S, E, M, N) of the structural proteins on the coronavirus genome is well conserved (their specific structural roles will be discussed later). Dispersed among the structural protein genes are a variable number of additional small ORFs. These lie in-between, or overlap, the structural protein genes and they are not well conserved among CoVs. These are called *accessory proteins* (by convention they are named using the number of the shortest mRNA on which they are encoded, Box 17.3). Mutation studies show that at least some of the accessory proteins are not required for coronavirus replication in cell cultures. However, mutating the accessory proteins does have a profound effect on the ability of the CoVs to replicate in their hosts and impacts viral pathogenesis.

As expected for an RNA virus, the CoV genome contains regulatory elements. Cis-acting regulatory elements are required for replication, transcription, and genome packaging. At the 5′ end of the genome, elements that participate in viral RNA synthesis extend past the 5′ untranslated region (UTR) and into the coding region of *rep1a*. This region folds up into a set of seven stem-loops. Folding of the CoV 5′ RNA is dynamic, allowing for changing conformations that control critical aspects of RNA and protein synthesis. Functional analyses of cis-acting regulatory regions have shown that stability of RNA stem-loops can be critical for viral fitness.

At the 3′ end of the CoV genome, additional cis-acting elements are found in the 3′ UTR. The 3′ cis-acting elements are conserved (and are interchangeable) among betacoronaviruses. As with the 5′ UTR, this region is probably dynamic, with different conformations controlling different steps in RNA replication/synthesis.

Other important cis-acting sequences are those that direct ribosomal frame shifting near the end of the *rep1a* gene. Finally, genome packaging signals have been localized to internal regions of the HCoV-SARS genome.

VIRION STRUCTURE

Coronavirus particles are enveloped with prominent spikes. Virions are spherical and range in size from 118 to 140 nm. Within the envelope is a flexible (subfamily *Coronavirinae*) or a doughnut-shaped (subfamily *Torovirinae*, see Box 17.4) nucleocapsid that consists of genomic RNA associated with the nucleoprotein (N). The spike (S) protein is the major glycoprotein that extends from the surface of the virion. Other

BOX 17.3

CORONAVIRUSES ACCESSORIZE

Coronaviruses encode a heterogeneous group of proteins, many of which are not absolutely necessary for replication in cultured cells, collectively called accessory proteins. The numbers of accessory proteins encoded by CoVs differ among even closely related coronaviruses. The genes encoding accessory proteins are interspersed among the structural genes. They are numbered according to the mRNA from which they are produced. SARS-CoV expresses eight accessory proteins (3a, 3b, 6, 7a, 7b, 8a, 8b, and 9b). They are all small peptides, ranging in size from about 45 to 180 amino acids. A few are found in purified virions and several are membrane-associated. Several have been identified in infected cells from autopsy patients, good evidence that they are expressed during a natural infection. For the most part, their functions are not well characterized but a few activities have been determined: The SARS-CoV 3a protein forms an ion channel in the plasma membrane and may modulate virus releases. SARS-CoV 3b interferes with the interferon pathway, and protein 6 may be an antagonist of type I IFN. Both SARS and MERS-CoV infections are characterized by a delay in type I IFN production and poor development of adaptive immune responses. It is likely that accessory proteins are involved in their pathogenesis.

BOX 17.4

TOROVIRUSES

The family *Coronaviridae* contains two subfamilies, the *Coronavirinae* and the *Torovirinae*. As expected, toroviruses share many characteristics with the other members of the family *Coronaviridae*, including basic virion architecture, genome organization, and mechanism for synthesizing a 3′-nested set of mRNAs. Torovirus particles are round, pleomorphic, enveloped viruses about 120–140 nm in diameter with obvious surface spike proteins. What are the unique features of toroviruses that have resulted in their being assigned to a distinct subfamily? Their most distinctive morphologic feature is a doughnut-shaped nucleocapsid (Fig. 17.2). All toroviruses identified to date have a glycoprotein with HE activities (Fig. 17.4). HE mediates binding to sialic acid residues but is also a so-called receptor-destroying enzyme because it cleaves (thus destroys) sialic acid. One theory is that the activities of HE allow toroviruses to penetrate the mucus layer in the respiratory and digestive tracts. Toroviruses infect vertebrates and many have been isolated from mammals with gastroenteritis. Toroviruses have been isolated from humans, cattle, pigs, sheep, goats, and horses. Toroviruses have not been extensively studies as most are difficult to grow in cultured cells.

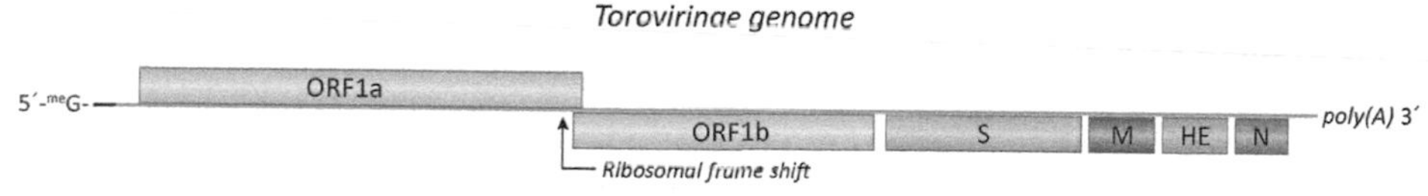

FIGURE 17.4 Torovirus genome organization.

membrane-associated proteins include membrane (M) and envelope (E).

CORONAVIRUS STRUCTURAL PROTEINS

All CoVs encode four structural proteins: Three membrane-associated proteins (S, M, and E) and a single nucleocapsid (N) protein. However some betacoronaviruses have an additional membrane protein with hemagglutinating and esterase activities, hence it is called HE. The order of the structural genes on the CoV genome is (HE) S, M, E, N (Fig. 17.3). The order of structural genes on the torovirus genome is S, M, HE, and N (Fig. 17.4).

The spike protein (S) forms a prominent projection from the virus envelope and gives CoVs their characteristic appearance. S is glycosylated and is the attachment and fusion protein. Among the CoVs there are some differences in the manner in which S is processed (cleaved into S1 and S2 fragments). S is always ER associated and modified by N-linked glycosylation.

In many beta- and gammacoronaviruses S is partially or completely cleaved by host furin-like proteases in the ER (prior to assembly of new virions). The extent of proteolysis correlates with the number of highly basic residues at the S1/S2 cleavage site. The S1 (N-terminal) and S2 (C-terminal) products remain noncovalently associated. In contrast, HCoV-SARS S is *not* cleaved during assembly or release. Instead it is cleaved in an acidified endosome during entry/penetration. It appears that there are two critical cleavage events that act in concert to mediate fusion: A cleavage at the S1/S2 boundary and a second cleavage within S2 (called the S2′ cleavage site). Some CoV S proteins are not cleaved, but the terms S1 and S2 are still used to refer to the corresponding N- and C-terminal domains of the protein. It is also likely that, in some cases, processing of CoV S in cultured cells is not identical to the process in the infected host.

Comparisons of S1 sequences show that they diverge extensively and are not highly conserved, even within multiple isolates of a single type of coronavirus. We can reasonably speculate that sequence divergence is, at least in part, a result of host immune responses. In contrast to S1, the S2 product is highly conserved across the subfamily *Coronavirinae*.

The membrane protein (M) is the most abundant protein in the virion. M contains three hydrophobic domains, thus is tightly associated with the virus envelope. M plays a major role in promoting membrane curvature. It has a short ectodomain (extracellular domain) that is modified by glycosylation. M also interacts with the N and E proteins. Expression of HCoV-SARS M (in the absence of other viral proteins) results in self-assembly and release of membrane-enveloped vesicles. Virus-like particles are released when M is coexpressed with either N or E.

The envelope (E) protein is found in very small amounts in the virion (about 20 molecules per virion), although larger amounts of E are present in infected cells. Different CoVs have variable requirements for E during particle formation, ranging from required for particle formation to not essential. In fact, virus titers close to 1×10^6 pfu per mL have been reported for HCoV-SARS lacking the E protein. Studies show that E assembles in membranes to form ion channels, thus E is a viroporin. Viroporins influence the electrochemical balance in subcellular compartments.

The nucleocapsid (N) protein is the only protein found in the ribonucleoprotein particle. N forms homodimers and homooligomers and binds genomic RNA, packaging it into a long flexible nucleocapsid. In the infected cell, N localizes to the cytoplasm, and for some CoVs N is also found in the nucleolus. N interacts with other CoV structural proteins, thus has a role in assembly and budding. N also colocalizes with replicase–transcriptase components and is required for RNA synthesis. Other roles for N include modulating cell-cycle (promoting cell-cycle arrest) and inhibiting host cell translation.

An additional structural protein is found in most betacoronaviruses, including the human coronavirus HCoV-HKU1. The hemagglutinin-esterase (HE) is 48 kDa glycoprotein that projects outwards 5–10 nm from the virion. HE binds sialic acid units on glycoproteins and glycolipids. The esterase activity removes acetyl groups from O-acetylated sialic acid, thus may be a receptor-destroying enzyme.

CORONAVIRUS REPLICATION CYCLE

The overall scheme of CoV replication is similar to that of other positive-strand RNA viruses but the process for synthesis of subgenomic mRNAs is unique.

Attachment

CoV S is the receptor-binding protein. The receptors for some CoVs have been identified. Two human CoVs, HCoV-SARS, and HCoV-NL63 bind different regions of angiotensin converting enzyme 2 (ACE2). ACE2 is a cell-surface, zinc-binding carboxypeptidase important for regulation of cardiac function and blood pressure. ACE2 is expressed in epithelial cells of the lung and small intestine, as well as many other organs. HCoV-MERS binds to dipeptidyl peptidase 4 (DPP4). Several other CoVs (FeCoV, FIPV, CCoV, and TGEV) bind to aminopeptidase N. (We might well ask if it is significant that many CoV receptors are proteases.) However, not all CoVs bind to protein receptors. Bovine CV and the HCoV-OC43 bind to sialic acid units found on glycoproteins and glycolipids.

Penetration

CoVs are enveloped, so penetration occurs as a consequence of a membrane fusion event. Just as cleavage of the S protein varies among CoVs, so do the sites and requirements of the fusion process. CoV S proteins contain a hydrophobic "fusion peptide" that becomes exposed upon a large-scale rearrangement of S. The location and details of S rearrangement are variable, but can be triggered by factors such as proteolytic cleavage of S and/or acid pH. Processing of HCoV-SARS has been well studied and will serve as our example. HCoV-SARS S is found as an uncleaved product on extracellular virions. It is cleaved only after attachment and endocytosis. There are two critical cleavage events that precede (and are required for) fusion. One cleavage is at the S1/S2 boundary and the other is a cleavage within S2 (S2′). Evidence supports cleavage at the S1/S2 boundary by cathepsin L; however, if one exposes HCoV-SARS to extracellular proteases (such as trypsin or elastase) in cell cultures, infection is *greatly enhanced*, suggesting that the endosomal proteases are not highly efficient. This may be relevant to HCoV-SARS-infection and pathogenesis: In the human respiratory tract, a transmembrane serine protease (TMPRSS2) is expressed in pneumocytes *and binds to ACE2 (the receptor for HCoV-SARS)*. The interaction between ACE2 and TMPRSS2 would conveniently put uncleaved S in close proximity to a host cell protease upon initial binding.

Amplification

The first synthetic event in the CoV replication cycle is translation of the viral genome by host cell ribosomes. REP1a and REP1b are translated from genomic RNA and these polyproteins are necessary for virus replication to move forward. Some of the REP1a products (nsp3, nsp4, nsp6) have transmembrane domains and these serve to anchor the replication–transcription complex to cell membranes, a prerequisite for synthesis of additional viral RNAs. The interaction also causes remodeling of host cell membranes to form structures dedicated to virus RNA synthesis.

RNA Synthesis

In vitro, the CoV RdRp requires a primer, specifically a short RNA oligonucleotide. It so happens that the CoVs encode two different proteins with RdRp activity. The nsp8 gene product is thought to be a primase capable of synthesizing short oligonucleotides. Nsp12 is the elongating polymerase. Other viral nsps in the replication–transcription complex are involved in cap synthesis (Table 17.1). CoVs also encode two ribonucleases. NendoU (nsp15) is a Nidovirales endonuclease that cleaves both single- and double-stranded RNA, cutting downstream of uridylate residues. ExoN (nsp14) is a 3′–5′ exonuclease. CoVs with mutations in ExoN have an *enhanced mutation rate* suggesting a role for ExoN in proofreading during RNA synthesis. Quite an unexpected finding for an RNA virus!

In addition to genome-length RNA, a set of subgenomic (sg) mRNAs is found in the infected cell. The sg mRNAs are used for the expression of the structural and accessory proteins. All are capped and polyadenylated and they share a common 3′-end forming a so-called "nested set" of mRNAs. A closer look at the sg mRNAs reveals that each contains an identical leader sequence of 70–100 nt at the 5′-end. The leader sequences found on all sg mRNAs are identical; however, this sequence is found only *once* in the genome, near the 5′-end. Leader sequences are fused to downstream sequences (sometimes called the body RNAs) at short, 8–9 nt motifs called the transcription regulating sequence (TRS). A TRS is found upstream of the ORFs-encoding structural proteins (these are called TRS body or TRS-B). A TRS is also present just downstream of the leader sequence in the 5′ UTR. These findings provide clues to the unique strategy used for CoV mRNA synthesis: CoV sg mRNAs are generated by a process of discontinuous transcription illustrated (simplified) in Fig. 17.5.

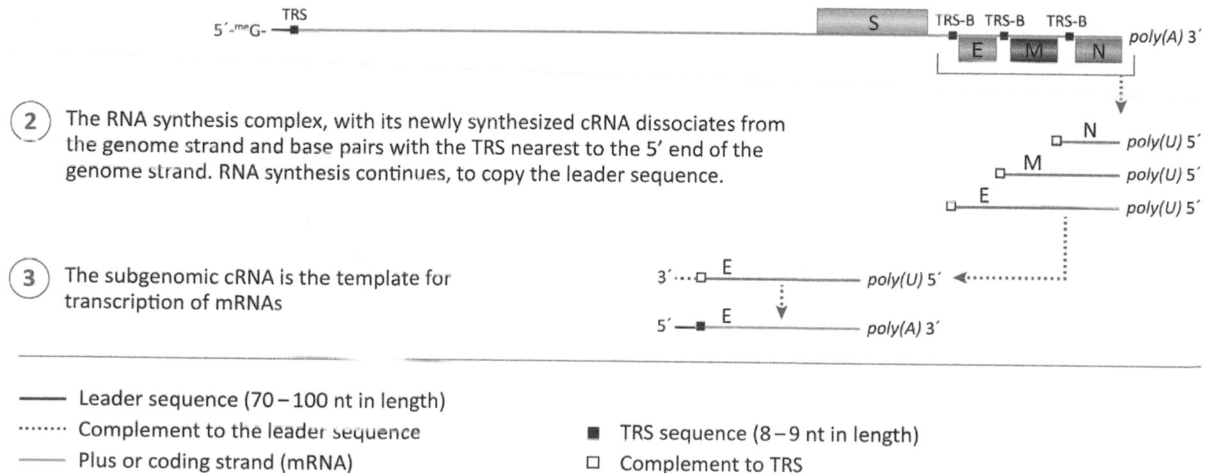

FIGURE 17.5 Coronavirus discontinuous transcription.

Discontinuous transcription very likely contributes to the high level of CoV genome recombination (Box 17.5). This type of RNA virus genome recombination is called a *copy-choice* mechanism (Fig. 17.6). Quantification of CoV RNAs presents in infected cells shows that cRNAs (negative-sense RNAs with 5′ oligo (U)) are present at levels 0.1–0.01 times lower than "positive-sense" genomes and mRNAs. This finding suggests that recombinants are more likely to occur *during synthesis of negative strands.*

BOX 17.5

CORONAVIRUS RECOMBINATION

Discontinuous transcription likely contributes to the readily observed examples of CoV genome recombination. If the replicase–transcriptase complex *must* dissociate/reassociate with template to generate subgenomic cRNAs, it follows that the same process sometimes occurs during synthesis of genome-length cRNA. Studies in the MHV experimental system show that genome recombination is relatively common event. Examination of feline and canine CoVs (CCoVs) provide real world examples of this phenomenon. FeCoVs are common in cats and usually cause mild enteric illness. On occasion enteric FeCoVs mutate within an infected cat and gain the ability to infect macrophages, an event which is accompanied by production of severe systemic disease called FIP. However, among cats with FIP, some researchers found a few "novel" viruses. Sequencing these novel viruses revealed that they contained genes from both feline and CCoVs. As CCoV is a relatively common dog virus (and it is common to have both cats and dogs in a household), it was hypothesized that the novel FIP-associated virus arose in cats that became infected with *both* FeCoV and CCoV; in the coinfected cats, macrophage-infecting recombinants emerged, causing the clinical disease, FIP (Terada et al., 2014). The FIP-associated recombinant viruses do not appear to be transmissible from one cat to another.

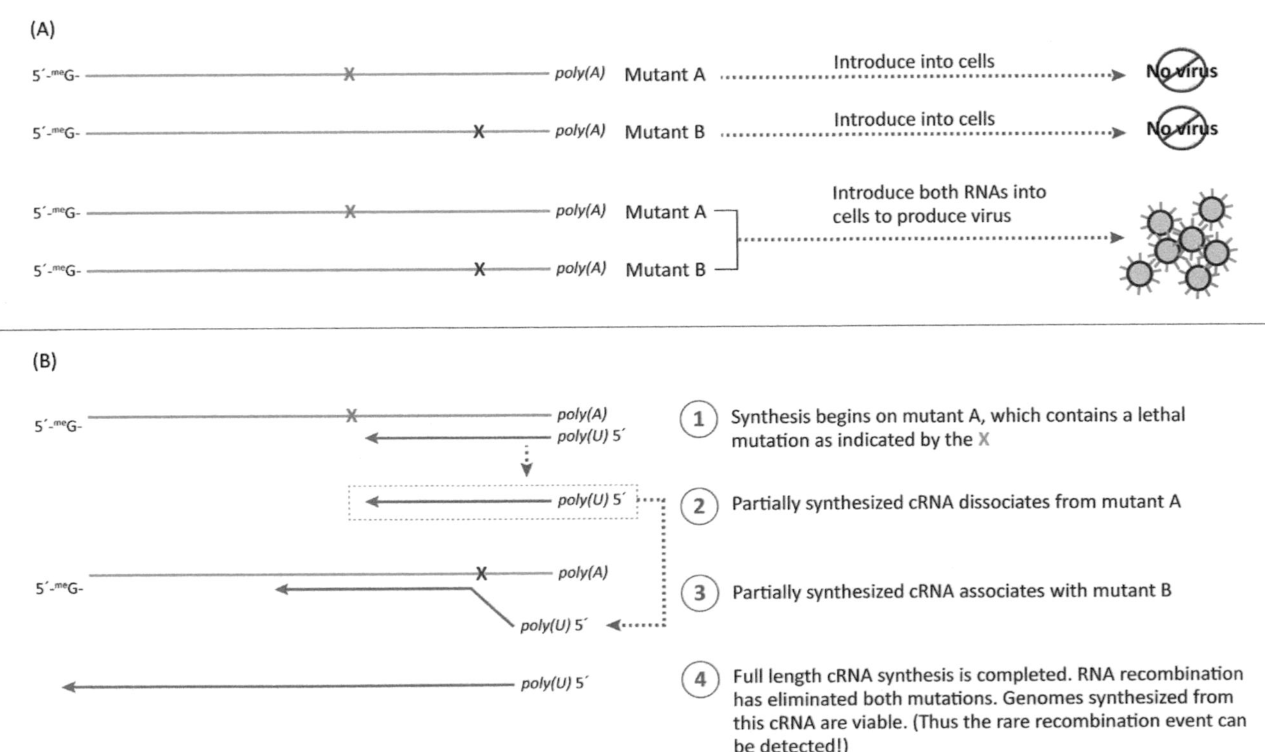

FIGURE 17.6 Coronavirus RNA recombination. (A). If individual mutant coronaviruses are used to infect cells there is no virus growth. However, if mutants are used together to infect cells, virus growth is restored suggesting that genome recombination has occurred. (B). This panel illustrates the mechanism of RNA copy choice 'recombination'.

Assembly and Release

Virion assembly takes place on membranes. Genomic RNA is bound by N protein, associates with M protein and buds into ER/Golgi membranes. M packs tightly into membranes and is thought to cause the membrane curvature that drives budding. S and E are also membrane proteins and are acquired during the budding process.

The ion channel activity of E is that of a *viroporin*; it alters cell secretory pathways to promote virus release. A function of E may be to increase the pH of the transport vesicles. Virus particles contained within membrane bound vesicles are released from cells by exocytosis.

DISEASES CAUSED BY CORONAVIRUSES

The first coronaviruses isolated were from poultry with respiratory disease (infectious bronchitis) in the 1930s. IBV remains a worldwide problem, particularly in high-density commercial production facilities. Until 2002 human coronaviruses (HCoVs) were associated only with mild respiratory tract disease, with estimates that they caused 15%–25% of all "common colds." That changed in 2002 when a human coronavirus was identified as the cause of an apparently new disease called SARS. The SARS outbreak was controlled, but in 2014 another novel CoV was isolated from patients hospitalized with severe respiratory disease in Saudi Arabia. As most infected patients lived in or had traveled to Middle East countries, the new disease was named MERS and the coronavirus responsible is called HCoV-MERS. Most animal and human CoVs are transmitted by the fecal-oral route and their initial replication site is in epithelial cells where virus production causes produces local respiratory symptoms or diarrhea. However, on occasion, CoVs cause severe to fatal disease. Some examples are presented here.

Severe Acute Respiratory Syndrome

SARS facts:

- SARS-CoV emerged in the human population in China in 2002.
- Epidemiologic studies and genetic analysis indicated the virus most likely jumped from bats into farm-raised Himalayan palm civets (*Paguma larvata*) and then into humans.
- Human to human transmission was by respiratory and fecal routes.
- Approximately 8000 cases were reported worldwide.
- Twenty-six countries were affected.
- Seven hundred and seventy-four deaths occurred (~10% case mortality rate).
- Economic losses in Hong Kong were ~5.9 billion ($US).
- In July 2003, WHO reported that the last known human chain of transmission was broken.
- Bats and birds are natural reservoirs of SARS-like viruses.
- Laboratory-associated infections occurred in China in 2004.

SARS-infection causes a triphasic pattern of disease. The first phase is nonspecific with fever, cough, sore throat, and myalgia. Breathing difficulties (dyspnea) show up 7–14 days after appearance of the first symptoms. The second phase of the disease includes shortness of breath, fever, onset of hypoxia, and often diarrhea. In the most serious cases, patients progress to a third phase with development of acute respiratory distress requiring hospitalization and mechanical respiration. Three viral proteins have been implicated in HCoV-SARS pathogenesis. An accessory protein (encoded from orf3a) interferes with cell signaling pathways, another accessory protein (encoded from orf6) interferes with interferon signaling and the E glycoprotein is a strong inducer of proinflammatory cytokines. All of these proteins are dispensable for virus replication in cell cultures, but in mouse models of disease their deletion reduces disease.

Middle East Respiratory Syndrome

MERS facts:

- MERS begins with coughing, fever, and breathing problems but may progress to pneumonia and kidney failure.
- Over 1600 human cases and the outbreak is ongoing.
- Case fatality rate >30%.
- Cases are sporadic and cannot be linked to a single source (based on genome sequencing).
- Countries most affected include those in the Arabian Peninsula (Bahrain, Iran, Jordan, Kuwait, Lebanon, Oman, Qatar, Saudi Arabia, United Arab Emirates, and Yemen).
- Casual transmission from person to person very rare.
- Most person to person transmission occurs in a hospital setting.
- Many healthy camels in the Arabian Peninsula have antibodies specific to CoV-MERS (indicating past infection) but infections often occur among people with no known contact with camels.

- A virus very similar to CoV-MERS has been found in some bats (sequence analysis suggests that the virus moved from bats to camels).
- Many questions about the epidemiology of CoV-MERS remain unanswered.

Sporadic cases of MERS continue to be reported. Strict infection control procedures in hospitalized patients limit person to person spread in that setting and transmission among casual or household contacts is rare.

While MERS-CoV and SARS-CoV are both betacoronaviruses, they derive from different sources and have unique characteristics. SARS-CoV has a tropism for ciliated respiratory epithelial cells and its receptor is ACE2. In contrast MERS-CoV has a tropism for nonciliated respiratory epithelial cells and its receptor is DPP4. Continued comparison of these two viruses will provide new insights into CoV transmission and pathogenesis. The propensity for CoV recombination, and their ubiquitous presence in bats, makes a case for close surveillance and study of these potential human pathogens.

Feline Coronavirus and Feline Infectious Peritonitis

FeCoV is the most common virus found in cat fecal samples and is spread through the fecal-oral route. Infection is usually subclinical or associated with a transient, mild diarrhea. Immunity to FeCoV is neither solid nor long-lasting. As antibody levels decrease, cats may be reinfected and once again experience mild diarrhea. Most kittens infected with FeCoV clear the virus, but about 15% become chronic shedders. Chronic shedders are at highest risk for developing FIP, a systemic (multiorgan), lethal disease. FIP develops when enteric FeCoV mutates to become capable of infecting monocytes and macrophages. This results in systemic viral infection, as the virus is no longer confined to the digestive tract. The virus associated with monocytes/macrophages is called the FIP *biotype*.

What kinds of mutations change the cell tropism of enteric FeCoV and result in development of FIP? Mutation occurs independently within each cat, with every FIP biotype virus having unique genetic features. Specific viral genes (for example, the orf3C protein) are important for the change in cell tropism. After changing cell tropism, the virus continues to mutate, becoming better adapted to peritoneal macrophages. This occurs despite preexisting host immune responses. Early signs of FIP are nonspecific and include anorexia, weight loss, inactivity, and dehydration. FIP can occur in any age cat, but cats less than 1 year of age or greater than 10 years of age are more susceptible. Constant virus replication in macrophages leads to B-cell activation and production of nonprotective (nonneutralizing) antibodies. In fact, immune complexes are damaging as they activate the complement system and lead to immune-mediated vasculitis. The classic lesions associated with the severest form of FIP are aggregates of macrophages, neutrophils, and lymphocytes that form in very small veins. These aggregates are called pyogranulomas and they are associated with development of edema and accumulation of large volumes of protein-rich fluids.

In this chapter we have learned that:

- Members of the family *Coronaviridae* are large, positive-strand RNA viruses.
- They are enveloped, with a helical nucleocapsids.
- A long spike (S) protein forms extends 16–21 nm from the surface of virions. S is the attachment and fusion protein.
- mRNAs are synthesized by a process of discontinuous transcription, a process that leads to high rates of RNA recombination.
- There are two subfamilies, *Coronavirinae* and *Torovirinae*. Most HCoVs cause mild respiratory or enteric disease. Notable exceptions are CoV-SARS and HCoV-MERS that can cause severe respiratory disease. The toroviruses are less well studied but have been isolated from a variety of mammals (intestinal and respiratory tracts).
- Members of the family *Coronaviridae* are notable for the mechanism of transcription, which is discontinuous (the RdRp moves from one location on the template RNA to a distant location). This process leads to high rates of RNA recombination during genome replication.

References

Terada, Y., et al., 2014. Emergence of pathogenic coronaviruses in cats by homologous recombination between feline and canine coronaviruses. PLoS One. 9 (9), e106534. doi:10.1371/journal.pone.0106534.

CHAPTER

18

Family *Arteriviridae*

OUTLINE

After reading this chapter, you should be able to discuss the following:

- What are the major structural and replicative features of arteriviruses?
- By what mechanism are arterivirus mRNAs synthesized?
- What diseases are caused by arteriviruses?
- Equine arteritis virus (EAV) can cause a long-term persistent infection in stallions. What evidence supports a requirement for the hormone testosterone in this process?

The family *Arteriviridae* is one of four virus families in the order *Nidovirales*. Arteriviruses are enveloped viruses with unsegmented plus-strand RNA genomes (Fig. 18.1). At 12.7–15.7 kb, their genomes are considerably smaller than the coronaviruses but they share many characteristics with them, including overall genome organization and use of discontinuous transcription to synthesize subgenomic RNAs (Fig. 18.2). They lack the notable spike proteins of the coronaviruses. Members of the family include equine arteritis virus (EAV), porcine reproductive and respiratory syndrome virus (PRRSV), lactate dehydrogenase elevating virus of mice, and simian hemorrhagic fever virus (SHFV). In the late 1987 PRRSV emerged as a serious pathogen of domesticated pigs in the United States; it is now found worldwide and some strains are quite virulent.

GENOME ORGANIZATION

Arterivirus genomes are organized in the same overall manner as other virus families in the order *Nidovirales*. The ~13–16 kb positive-strand RNA genome contains 10–13 open-reading frames (ORFs). Approximately three-fourths of the genome encodes the nonstructural proteins (nsp) required for transcription and genome replication. The nsp are encoded by two overlapping ORFs. ORF1a contains three to four protease domains and three transmembrane domains. The size of ORF1a is quite variable; for example, PRRSV ORF1a is ~800 bases longer than EAV ORF1a. ORF1b is much more conserved among the arteriviruses. It is expressed by a ribosomal frameshift and encodes the RdRp, a helicase, and the NendoU endoribonuclease. Arterivirus NendoU is related to the coronavirus protein of the same name and hydrolyzes single- and double-stranded RNA. The exact role of NendoU in the arterivirus replication cycle remains unknown.

Structural proteins are encoded downstream of ORFs1b. EAV and PRRSV encode eight proteins from short overlapping ORFs. The nucleoprotein (N) is encoded at the very 3′ end of the genome. The remaining structural proteins are all found associated with lipid envelope. The SHFV genome encodes additional ORFs. These seem to have arisen by a gene duplication of ORFs 2a, 2b, 3, and 4.

Viruses. DOI: http://dx.doi.org/10.1016/B978-0-12-803109-4.00018-0

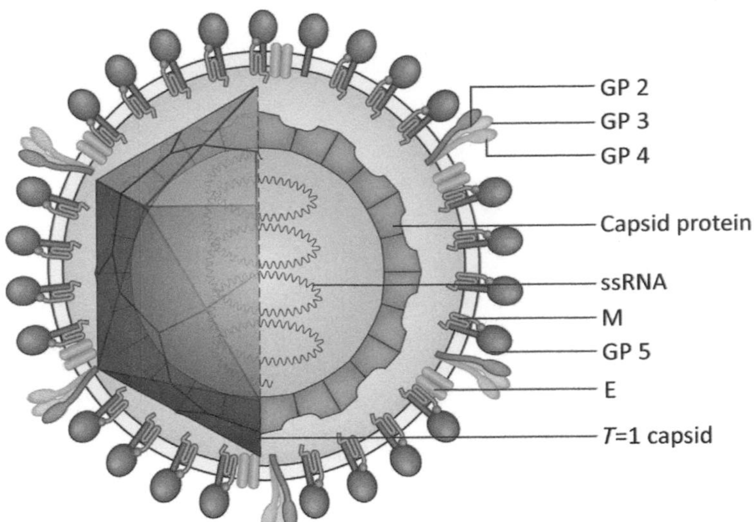

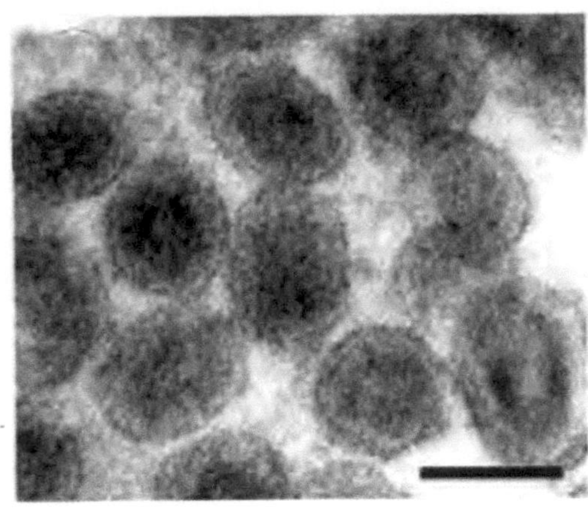

FIGURE 18.1 Virion structure (left) and electron micrograph (right) Fauquet, C.M., et al. (Ed.), 2005.

VIRION STRUCTURE

Artiriviruses are 50–60 nm in diameter with a helical or filamentous nucleocapsid ~39 nm in diameter as recently described by cryo-EM studies of PRRSV. Arteriviruses lack the notable spike proteins of the coronaviruses; in fact the envelope surface is very smooth. There are seven proteins associated with the envelope (GP2, GP3, GP4, M, E, 5a, and GP5). M, E, and 5a are nonglycosylated. GP2, GP3, and GP4 associate for form a heterotrimer. GP5 is a large glycoprotein with three membrane-spanning domains. All of the envelope proteins appear to be required for formation of infectious virions.

REPLICATION CYCLE

The overall replication cycle of the arteriviruses is very similar to that of the coronaviruses. Entry occurs via endocytosis. The infecting genome is translated to produce the polyprotein 1a and polyprotein 1ab products; these are cleaved by viral proteases to generate the nonstructural proteins (nsps). Replication is cytoplasmic and takes place in association with double membrane vesicles. Virions form by budding into membrane vesicles (endoplasmic reticulum/Golgi) and are released by exocytosis. Arterivirus genomes are considerably smaller than coronavirus genomes, but the overlapping ORFs encoding the nsps account for almost 3/4 of the total coding capacity (Box 18.1).

Structural proteins are expressed from a 3′ coterminal set of subgenomic (sg) mRNAs. The model for mRNA synthesis is the same as previously described for coronaviruses (Chapter 17: Family *Coronaviridae*). All sg mRNAs have the same leader sequence, produced by a process of discontinuous transcription. It is believed that discontinuous transcription occurs during synthesis of sg minus strands and these serve as templates for mRNAs.

Arteriviruses nucleocapsids assemble in the cytoplasm and bud into the ER or Golgi compartments. Viruses accumulate in vesicles and are released by exocytosis at the plasma membrane. All arteriviruses replicate in macrophages in vitro and in vivo. They often establish persistent infections in their natural hosts. To date they have been isolated only from vertebrate hosts and there are no known human pathogens among the arteriviruses (Box 18.2).

DISEASES CAUSED BY ARTERIVIRUSES

Equine Viral Arteritis

EAV is a common infection of horses worldwide. In the United States, over 70% of horses are seropositive. Many infections are asymptomatic or cause mild upper respiratory tract disease; however, some infections are more severe. An infrequent outcome of infection is abortion in pregnant mares. Some infected stallions become persistent shedders of virus but can be cured of the infection by castration. This indicates a hormonal link (testosterone) to viral persistence although the molecular events associated with persistence have not been determined. During acute infection virus is transmitted via aerosols and the primary sites of replication are in epithelial cells of the respiratory tract. The virus also infects macrophages and some lymphocytes

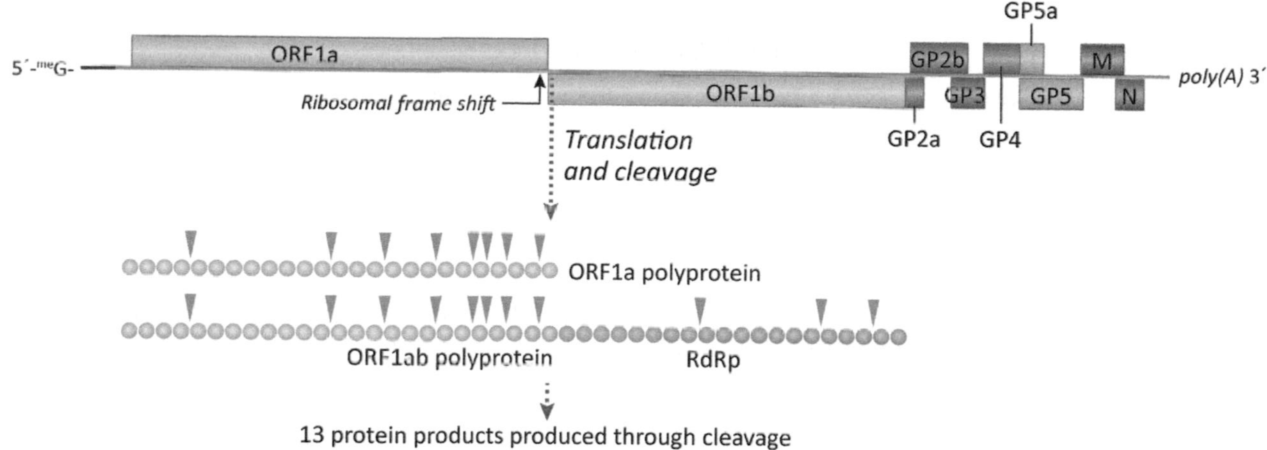

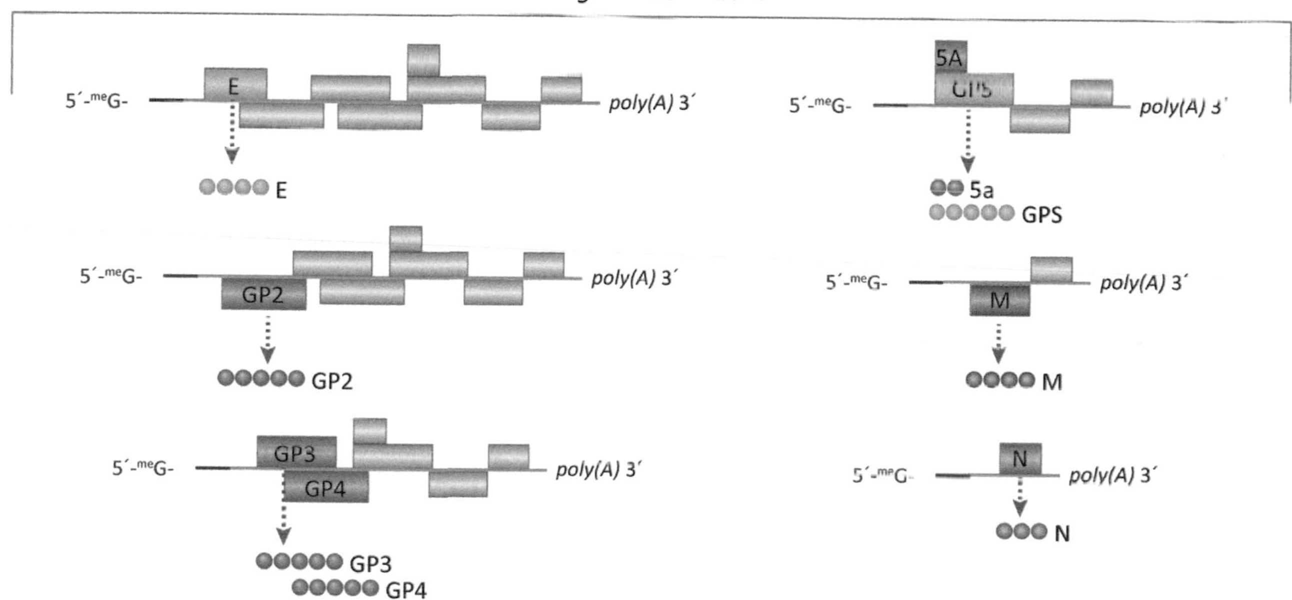

FIGURE 18.2 Genome organization and gene expression strategy of an arterivirus.

thereby moving to regional lymph nodes. Infected cells then disseminate EAV throughout the horse. Virus is present in feces, urine, vaginal secretions, and semen.

Porcine Reproductive and Respiratory Syndrome

PRRSV is an economically important virus that made its appearance in the United States in 1987. The virus is found in both farm-raised and wild pigs. Once introduced into a naïve herd the virus most often spreads by direct contact but can also be spread by the aerosol route. The virus can be introduced into a herd via infected animals, semen, or contaminated fomites. Depending on the virulence of the infecting strain, health of the herd may be unaffected or animals may experience mild to severe disease.

PRRS has two distinct clinical presentations: reproductive failure and postweaning respiratory disease. Reproductive disease causes increased numbers of stillborn piglets, mummified fetuses, premature births, and weak piglets. Reproductive problems arise because PRRS can cross the placenta. If the virus crosses the placenta in the third trimester of gestation piglets may be viremic when born, and may transmit virus for 2–3 months. The respiratory form of the disease affects piglets, causing pneumonia and respiratory distress. Pneumonia develops because the virus

BOX 18.1

CHARACTERISTICS

Genomes are positive-strand RNA, 12.7–13.7 kb. Genomes have a 5′ cap and a 3′ poly(A) tail. Proteins required for RNA synthesis and polyprotein processing are encoded in two large overlapping ORFs, ORF1a and ORF1b that are expressed from genome length RNA. ORF1a and ORF1b encode long polyproteins that are cleaved by viral proteases to generate a set of nonstructural proteins (nsps). ORF1b is expressed via ribosomal frameshifting. Nsps include an endonuclease (NendoU) common to all members of the order *Nidovirales*.

A leader sequence at the 5′ end of the genome is found on all subgenomic mRNAs, supporting a model of discontinuous transcription similar to the coronaviruses. Structural and accessory proteins are encoded from a nested set of 3′ coterminal mRNAs. Genome recombination by a copy choice mechanism is not uncommon.

Virions are enveloped, 50–60 nm diameter with a helical nucleocapsid. Structural proteins include an RNA-binding nucleoprotein (N) and seven envelope-associated proteins (M, E, 5a, GP2, GP3, GP4, and GP5). GP5 is the major envelope glycoprotein; it spans the envelope three times. None of the envelope-associated proteins have a long ectodomain, thus the surface of the virion is smooth in appearance.

BOX 18.2

TAXONOMY

Order *Nidovirales*
Family *Arteriviridae*
Genus *Arterivirus* (to date contains 13 species isolated from horses, pigs, rats, and monkeys).

infects alveolar macrophage and the infection is cytopathic. PRRSV may also predispose pigs to secondary infections due to widespread destruction of macrophages. The emergence of PRRSV has resulted in the need for increased biosecurity on pig farms. To avoid introducing the virus into a herd, farmers use strict quarantine measures when introducing new animals. They must also purchase breeding stock and semen that is known to be virus-free. Sanitation of transport vehicles and strict protocols of fomite and personnel movement between farms are also critical components of an effective program.

Simian Hemorrhagic Fever Virus and Related Viruses

SHVF was first identified in 1964 in association with an outbreak of hemorrhagic fever that affected several species of captive Asian macaques. The disease was clinically similar to human hemorrhagic fevers and the mortality was quite high, approaching 100%. Macrophages are the primary target for SHFV and cytolytic infection is probably related to viral pathogenesis. It was long suspected that the natural hosts for SHFV were African monkeys. Recent studies have revealed that in fact many species of wild African monkeys are persistently infected with arteriviruses related to SHFV. However there is a great deal of genetic diversity with viruses from different monkey species sharing only about 50% nucleotide sequence identity. In some areas, up to 40% of monkeys are persistently infected and virus titers in the blood are quite high. Thus it appears that many monkey arteriviruses are quite well adapted to their natural hosts but have the potential to cause severe disease if transmitted to another species.

In this chapter we learned that:

- Arteriviruses are smaller than coronaviruses but their genome organization and overall replication strategy is very similar to the coronaviruses. They replicate in the cytoplasm and use a discontinuous mode of transcription to generate subgenomic RNAs.
- Arteriviruses lack a distinctive spike protein. The N protein associates with genomic RNA to form a helical/filamentous nucleocapsid. Seven proteins are associated with the viral envelope (three unglycosyalted, four glycosylated).

- To date no arterivirus has been isolated from humans. EAV and PRRSV are economically important animal pathogens. SHFV and related monkey arteriviruses seem to cause little disease in their "natural hosts" but can cause fatal hemorrhagic fever on cross-species transmission.
- EAV can cause a long-term persistent infection in stallions. However castrating stallions allows the infection to be cleared indicating a link to the hormone testosterone. The molecular mechanism by which testosterone facilities viral persistence is unknown.

References

Fauquet, C.M. (Ed.), 2005. Virus Taxonomy Eighth Report of the International Committee on Taxonomy of Viruses. Elsevier Academic Press, pp. 965.

CHAPTER

19

Family *Rhabdoviridae*

OUTLINE

After studying this chapter, you should be able to provide answers to the following questions:

- What are the main characteristics of the members of the family *Rhabdoviridae*?
- What is the general replication scheme of the unsegmented, negative-stranded RNA viruses?
- How does rhabdovirus N protein participate in the switch from mRNA synthesis to genome replication?
- How is rabies virus (RABV) transmitted and how does it reach the central nervous system (CNS)?
- Why is *post*-exposure rabies vaccination an important part of rabies prevention?
- What type of disease is caused by vesicular stomatitis virus (VSV) and why it of concern to food animal producers?
- What is "pseudotyping" and why the VSV glycoprotein often the envelope glycoprotein of choice?

Viruses in the family *Rhabdoviridae* have unsegmented, negative-strand RNA genomes. Enveloped virions are bullet- or rod-shaped with helical nucleocapsids (Fig. 19.1). A single glycoprotein decorates the outer surface of virions. There are well over 100 named rhabdoviruses (the family name derives from rhabdos, Greek for rod), isolated from hosts as diverse as insects, vertebrate animals, and plants.

The most notorious rhabdovirus is RABV, a member of the genus *Lyssavirus*. The name RABV is based on the Latin *rabidus* meaning raving, furious, or mad, a name very descriptive of animals with this almost invariably fatal neurologic infection. The name of genus to which RABV belongs (*Lyssavirus*) also reflects symptoms of the disease, as in Greek mythology Lyssa was the goddess of mad rage, fury, and crazed frenzy (Box 19.1).

VSV is a rhabdovirus in the genus *Vesiculovirus*. Not nearly as well known as RABV, it is nonetheless an important pathogen of hoofed-stock. Its name derives from the blister-like vesicles that form on the mouths, hooves, and teats of infected animals. VSV can be introduced into a herd by insects, but can subsequently spread by contact. VSV is economically important, in part, because it is clinically indistinguishable from foot and mouth disease (Chapter 11: Family *Picornaviridae)* thus its appearance in a herd is disruptive until confirmatory testing can be done. VSV can also infect humans, causing a generalized febrile illness. VSV is the most thoroughly studied of the rhabdoviruses and serves as the model for their replication. Taxonomy of this virus family is presented in Box 19.2.

Viruses. DOI: http://dx.doi.org/10.1016/B978-0-12-803109-4.00019-2

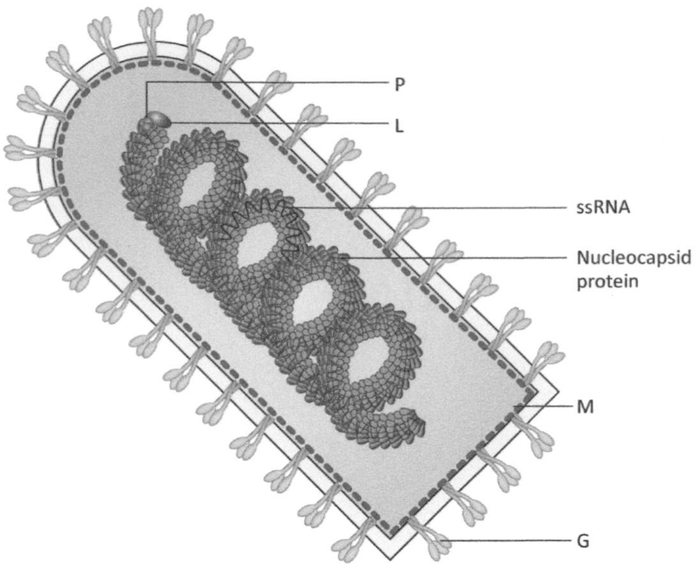

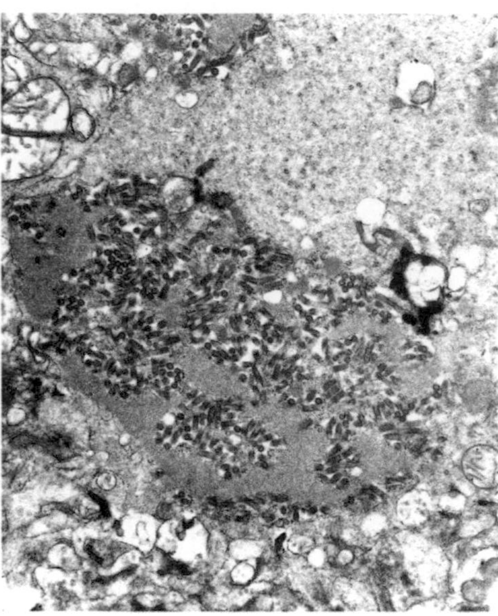

FIGURE 19.1 Virion structure (left) and electron microscopic (TEM) image (right) revealing the presence of numerous dark, bullet-shaped rabies virions within an infected tissue sample. *From CDC/F. A. Murphy.*

BOX 19.1

GENERAL CHARACTERISTICS

Rhabdovirus genomes are a single strand of negative-sense RNA, 11–15 kb. Genomes contain five open-reading frames. Short untranslated regions are found at the 3′ and 5′ ends of the genome and cis-acting regulatory sequences that instruct mRNA synthesis are present between each gene (Fig. 10.6). Transcription initiates at a single site at the 3′ end of the genome. Genome replication is cytoplasmic and requires ongoing synthesis of N and P proteins.

Enveloped virions are bullet or rod shaped (~75 nm wide) with helical nucleocapsids. The nucleocapsid consists of the genome, bound by the nucleoprotein (N), a phosphoprotein (P), and the large (L) protein (the RdRp). The matrix (M) protein is associated with the inner side of the envelope. The glycosylated envelope protein (G) assembles into homotrimers to form the protein spikes extending from the outer surface of the envelope.

BOX 19.2

TAXONOMY

Order *Mononegavirales*
Family *Rhabdoviridae*
Genera containing animal rhabdoviruses include:
Genus *Ephemerovirus* (Bovine ephemeral fever virus)
Genus *Lyssavirus* (rabies and rabies-related viruses)
Genus *Novirhabdovirus* (fish viruses)
Genus *Perhabdovirus* (fish viruses)
Genus *Sprivivirus* (fish viruses)
Genus *Tupavirus* (hosts include shrews, birds)
Genus *Vesiculovirus* (Vesicular stomatitis virus)

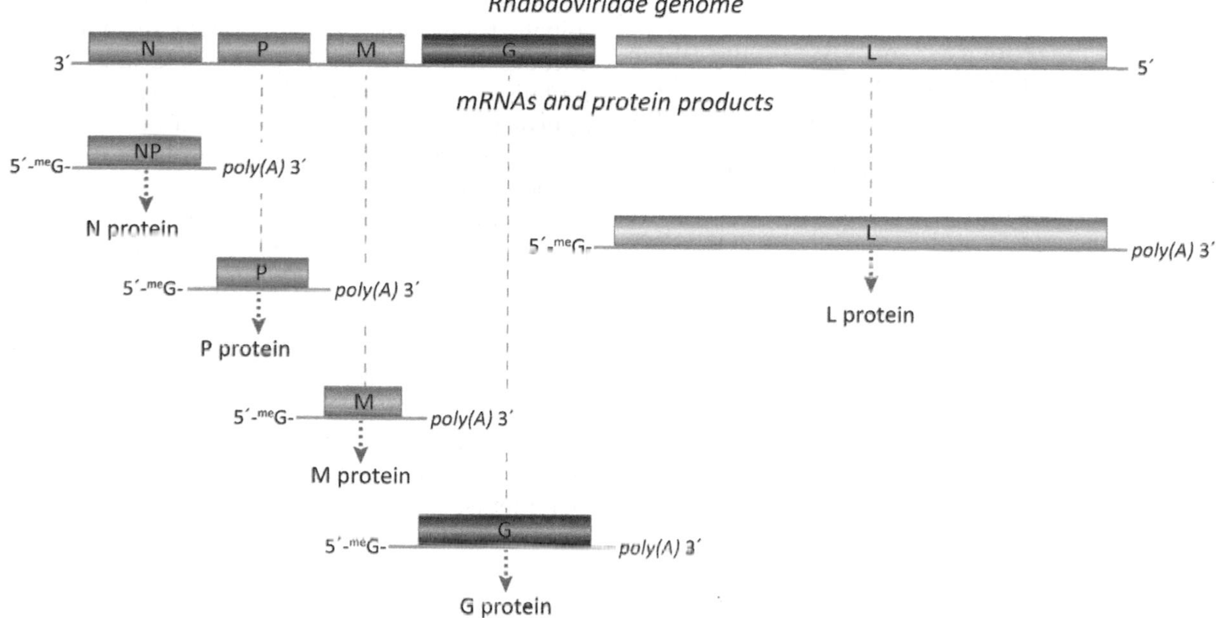

FIGURE 19.2 Rhabdovirus genome organization and gene expression strategy.

GENOME ORGANIZATION

Rhabdoviruses have unsegmented, negative-strand RNA genomes containing five open-reading frames. Rhabdovirus genomes range in size from 11 to 15 kb (Fig. 19.2). There are short (~50 nt) noncoding regions at the 3′ and 5′ ends of the genome. The genomes of rhabdoviruses (and other negative-sense RNA viruses) are often shown in diagrams with the 3′ end to the *left*, opposite the usual convention of illustrating a strand of nucleic acid. Sequences at the 3′ ends of genome serve as transcription promoters. There are important regulatory sequences between each gene (~18 nt) that include a U-rich polyadenylation signal, a 2 nt untranscribed termination signal followed by a transcription reinitiation signal (Fig. 10.6). These sequences direct the transcription complex to cap and polyadenylate individual mRNAs.

VIRUS STRUCTURE

Rhabdoviruses are enveloped particles with a rod or bullet (flat on one end) shape. Animal rhabdoviruses are 75–80 nm wide and ~180 nm long. The length of the virion is instructed by genome length. Virions contain a helical nucleocapsid. The major structural protein is the nucleocapsid protein, N, which interacts with the genome at a ratio of one N protein to nine bases of RNA. Nucleocapsids also contain several hundred copies of the phosphoprotein, P, and about 50 copies of the RdRp (L protein). A single glycoprotein, G, is associated with the envelope. G forms trimers that appear as spikes. G is uncleaved and serves as the attachment and fusion protein.

OVERVIEW OF REPLICATION

Rhabdoviruses share a common replication strategy with other families in the order *Mononegavirales*. Key features of the rhabdovirus replication cycle summarized below are based, for the most part, on studies of VSV and RABV.

Attachment and Penetration

Attachment and penetration (fusion) are mediated by the surface glycoprotein, G. G homotrimers form spikes on the surface of the virion. G is an uncleaved, membrane anchored protein. Attachment leads to endocytosis and within endosomes, low pH causes a conformational change to expose hydrophobic loops that participate in fusion. Fusion releases the ribonucleoprotein particle (RNP) into the cytosol, the site of transcription and genome replication. Due to its role in attachment and penetration G is an important target of protective immune responses. Vaccines targeting RABV G are highly effective.

Amplification

Synthesis of rhabdovirus mRNAs occurs upon release of the RNP into the cytosol. The RNP is a helical structure formed by the N protein and genomic RNA. A few copies of the transcription complex (P and L proteins) are also associated with the RNP. Rhabdovirus transcription does not require ongoing protein synthesis, and is a process that can take place in a test tube (Fig. 10.4). Transcription generates a set six RNAs, five of which are capped and polyadenylated mRNA (Fig. 19.2). The sixth RNA is the short (~47 nt), uncapped, unpolyadenylated leader (*le*) RNA. Transcription of rhabdovirus RNA does not require a primer and proceeds sequentially from the 3′ end of the infecting genome. Rhabdovirus L protein is the RdRp and also has cap synthesis and methyltransferase activities. Cis-acting RNA sequences that flank each gene provide signals for capping, polyadenylation, and transcription termination. Transcription "attenuation" is a general feature of nonsegmented, negative-strand RNA viruses. The term attenuation is used to describe the process by which the abundance of transcripts is regulated by their relative proximity to the 3′ end of the genome. N mRNA is the most abundant transcript and L mRNA is the least abundant transcript (Fig. 10.5). The process of attenuation occurs when L completes a transcript and then *dissociates* from the genome before starting transcription of the downstream gene. Studies with VSV have demonstrated that changing the locations of genes changes the relative abundance of mRNAs and reduces the overall efficiency of virus replication.

For the most part, rhabdovirus mRNAs each encode a single protein thus five mRNAs are translated to produce N, P, M, G, and L. An exception is the P gene of vesiculoviruses, which encodes multiple proteins by use of alternative start codons and is sometimes called the P/C gene. An upstream start codon initiates translation of the P protein, whereas two downstream start codons initiate translation of an alternate reading frame that encodes two small basic proteins, C and C′. However mutations that truncate C and C′ (without affecting P) appear to have no detectable effect on virus replication in cell culture or pathogenesis in mice, so the functions of these products are not obvious.

N is the nucleoprotein, the major structural protein. N oligomerizes to form stable nucleocapsid-like structures in the absence of RNA, perhaps forming a scaffold for the RNA.

P is a phosphorylated protein present at small amounts in the virion. Phosphorylated P binds both N and L and is required for the RNA synthesis activities of L. Interactions between P and N are dynamic, and P is proposed to transiently bind and dissociate as the polymerase complex moves along the RNA template. In this model the viral RNA always remains associated with N, even as it is transcribed.

M is a matrix protein associated with the inner side of the viral envelope. M serves as a bridge between the RNP and the virion envelope, interacting with both N and the cytosolic domain of G.

The single glycoprotein, G, assembles as trimer and is found in plasma membrane of infected cells, where it serves to mark the location of virion assembly and budding. G serves both in attachment and fusion, but unlike many other envelope glycoproteins, G is not cleaved. VSV G has been extensively studied as a "model" integral membrane protein (Box 19.3). G is synthesized by ER-associated ribosomes and most of the G protein is transferred to the inside of the ER during synthesis. Translocation across the ER stops with the synthesis of the membrane anchor near the C-terminus of the protein. Only the final few amino acids of G remain on the cytoplasmic side of the ER membrane. G is initially glycosylated in the ER and associates with molecular chaperones that assist in folding. After release from chaperones, G monomers associate

BOX 19.3

PSEUDOTYPING WITH VSV G

VSV infects most types of cells, suggesting that it uses a very common molecule for attachment. It has even been proposed that *nonspecific* electrostatic and hydrophobic interactions mediate attachment. Attachment is followed by endocytosis and membrane fusion at low pH. The ability of VSV to infect virtually all cell types has been harnessed as a means to get other, "fussier" viruses into a variety of cells by a process called pseudotyping. Pseudotyping is a general term describing the use of a "foreign" viral envelope glycoprotein to alter the tropism of virus. Pseudotyping is used extensively in basic virology research as well as for clinical applications. In order for a viral glycoprotein to be useful in pseudotyping, it must be readily incorporated into the envelopes of heterologous viruses, should attach to most cell types, and should be able to efficiently mediate fusion. VSV G meets all three criteria. Of course if you wished to direct a virus to a very specific cell type, VSV G would not be the glycoprotein of choice.

into trimers and are transported to the Golgi where further protein modifications take place (for example, glycosylation is modified). G then traffics to the plasma membrane of infected cells where it is found as clusters that serve as site of virus budding.

At this point in the replication cycle of a rhabdovirus, the infected cell contains an RNP and newly synthesized viral mRNAs and proteins. However, in order for virus replication to proceed, mRNA synthesis must give way to genome synthesis. Genome synthesis requires that instead of mRNAs, a *full-length* copy RNA (cRNA) must be synthesized (Fig. 10.3). The cRNA is neither capped, nor polyadenylated and serves at the template for synthesis of new genomes.

The Switch from mRNA Synthesis to Genome Replication

It was recognized early on that translation is required to trigger the switch from transcription to genome replication. Thus purified nucleocapsids can support mRNA synthesis in a test tube (Fig. 10.4), but cannot support synthesis of cRNA or genomic RNA unless additional N protein is added to the mix. Thus N protein is the critical molecule enabling the switch from transcription to genome replication. The process requires that *soluble* N encapsidate the 5′ end of a newly initiated RNA strand. Short sequences present at the 3′ ends of both genomic RNA and cRNA serve as important cis-acting signals as they are the initial binding sites of N. What is less clear, in terms of the model, is just *how* the encapsidation of RNA signals the L polymerase to ignore signals for capping, polyadenylation, and termination of transcription.

Assembly/Release

N binds genomic RNA at a ratio of one N protein to nine nucleotides to form the basic rhabdovirus RNP. P binds to N and L. RNPs traffic through the cytosol to the plasma membrane where they associate with M and G. Rhabdovirions bud through the plasma membrane. M plays a critical role in virion release as it contains a so-called *late domain*. Late domains are short amino acid motifs that interact with cellular proteins to allow release of budding virions.

DISEASES

Rabies and Rabies-Related Viruses

Rabies is a very ancient disease and human cases were clearly linked to the bite of a "rabid" dog. In the early 1800s it was formally shown that saliva from a rabid dog could transmit the disease. This provided the basis for studying the causative agent and by the mid-1800s Louis Pasteur and Emile Roux were searching for a cure. They passaged the rabies agent through rabbits and generated a vaccine from dried rabbit spinal cords. The vaccine worked to protect animals from rabies infection and in 1883 proof of effectiveness in humans was obtained when Pasteur used his vaccine to treat 9-year-old Joseph Meister. Meister had been badly bitten by a rabid dog and it was almost certain he would develop rabies. Fortunately for Joseph Meister (and Pasteur), the vaccine was successful. Although Pasteur's vaccine was not completely safe, it was used for decades as an effective postexposure treatment. Much safer and more effective vaccines are available today. In many parts of the world routine vaccination of domestic animals significantly reduces the risk human exposure. In the United States, RABV is found in bats, raccoons, skunks, and foxes so it is important that dogs and cats receive regular vaccination for their own, and their owner's protection.

RABV is most often transmitted by bites but can also be contracted after scratches or any contact with saliva. RABV enters peripheral nerves and slowly travels up axons to reach the CNS (Fig. 19.3). After reaching the CNS, RABV replicates very rapidly and brains of suspect animals are examined for the presence of virus (Figs. 19.4 and 19.5). Initial signs of rabies are nonspecific and include anorexia, lethargy, fever, and vomiting. These signs are followed by more obvious signs of CNS disease, including hyperexcitability and aggression (or sometimes extreme depression and lethargy).

After replicating in the CNS, virus moves back down the nerve fibers to locations such as the eye and salivary glands. Saliva is a source of abundant infectious virus. The neurologic changes brought about by RABV infection change animal behavior. Thus previously friendly animals become vicious, increasing the likelihood of transmission by bites. In contrast, wild animals that usually fear humans will become approachable or may even attack.

After a bite, it may take days, weeks, or months for RABV to reach the CNS and disease prevention in humans depends, in part, on postexposure vaccination. Postexposure vaccination is a critical part of postexposure treatment as it allows development of protective immune responses during the time that the RABV is "hidden" within neurons. (Postexposure treatment often also includes administration of rabies immune globulin.) In addition to routine vaccination of dogs and cats, baits containing recombinant rabies vaccines have successfully been used in the United States and Europe to reduce RABV in wild animals such as coyotes, foxes, and skunks.

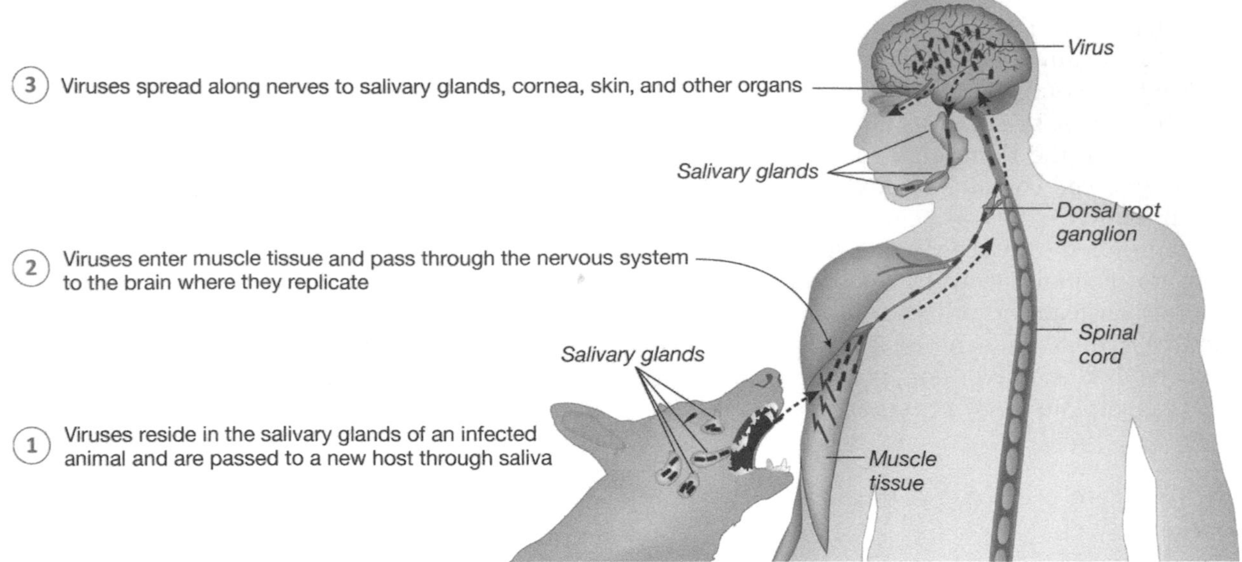

FIGURE 19.3 RABV transmission. RABV enters peripheral nerves and travels to the brain.

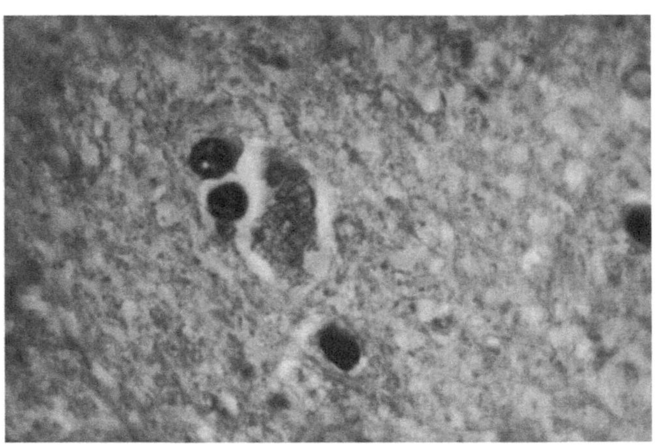

FIGURE 19.4 Hematoxylin-eosin (H&E) stain of brain tissue. Note the dark staining Negri bodies which are cellular inclusions associated with RABV replication. *From CDC/Dr. Daniel P. Perl. Public Health Image Library ID# 14559.*

The lethality of rabies infection requires that the public fully understand the risks of interacting with wildlife (including bats) as well as stray dogs and cats. Any/all animal bites should be reported in order that people and animals involved can be evaluated by health care professionals. If necessary, animals will be quarantined or tested (in the United States laws vary by state). A good rule of thumb is to be aware of any animal that is exhibiting unusual behavior. That includes pets that become aggressive as well as wild animals that appear sick and approach, rather than flee, from humans.

There are very few cases of human rabies in the United States most years due to aggressive postexposure prophylaxis when exposure is suspected. The few cases that do occur usually involved contact with bats (thus *never* handle a bat without using proper infection control measures). Bats in the home are of particular concern as they may bite sleeping individuals. Thus it is advisable to contact a physician or public health authorities if a bat is found in a room where someone has been sleeping.

Within the genus *Lyssavirus* are several species; RABV is the type species. RABV is found worldwide and is the only species found in the United States. RABV has a wide host range that includes many mammals, including bats. However, there is also genetic variation within RABV and in the United States genome sequences are used to distinguish RABVs circulating primarily among coyotes, bats, raccoons, fox, and skunks. The ability to genetically differentiate among subgroups of RABV is important for epidemiologic studies; however, the subgroups are not strictly host specific and all produce the clinical disease we know as rabies. Thus humans or other animals (including livestock) infected with bat, coyote, skunk, or dog RABV can develop clinical rabies (Box 19.4).

Steps to take upon possible exposure to RABV are as follows: Thoroughly wash any bites (15–20 minutes) to remove infected saliva from the wound. Washing is very effective and can be done even when other medical care is not immediately available. Then seek medical care and/or contact public health officials

(A)

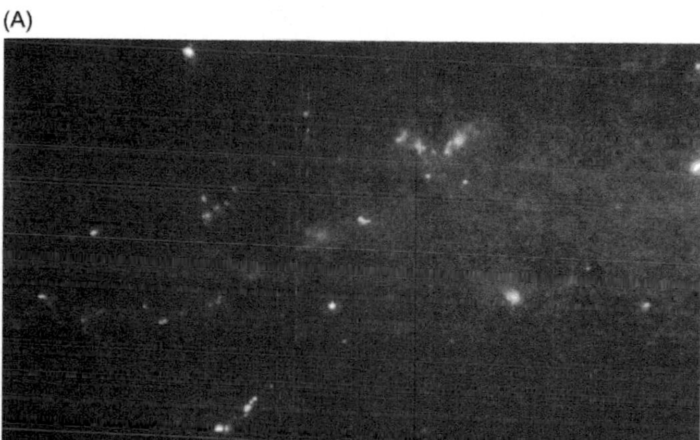

(B)

FIGURE 19.5 (A) Rabies viral antigen is present in a sample from a fox's brain. The viral antigen is revealed by staining with fluorescent antibody directed against the viral antigen. (B) Photograph of a rabid fox. The vast majority of rabies cases reported to the Centers for Disease Control and Prevention (CDC) each year occur in wild animals like raccoons, skunks, bats, and foxes. *From (A) CDC/Dr. Hicklin. Public Health Image Library Image # 12644 and (B) CDC. Public Health Image Library ID# 2628.*

to evaluate the situation. Additional postexposure treatment includes shots of immune globulin and vaccination.

As mentioned above, clinical rabies is usually fatal. The story of Jeanna Giese serves both as a cautionary tale and a rare story of survival. In 2005 Jeanna was bitten trying to "rescue" a bat. Unfortunately her family did not understand the very real danger and failed to seek immediate medical attention. Later, when Jeanna developed signs of CNS disease, rabies was suspected. She was directed to the Children's Hospital of Wisconsin in Wauwatosa where she was treated by Dr. Rodney Willoughby Jr., a pediatric infectious disease specialist. He induced a coma to partially halt Jeanna's brain functions with the goal of protecting the brain from damage while giving the immune system time to defeat the virus. Antiviral drugs were also administered (however it is not clear that they had any beneficial effect). Jeanna's outcome was good and she was released from the hospital after 76 days. She suffered few permanent sequelae, attended college, and became a vocal advocate for rabies prevention education. However it should be noted that the Milwaukee protocol is not a panacea, as survival rates of treated patients are less than 20% despite intensive hospital care. Therefore control of rabies in dogs and cats remains a high priority, as does education about the very real dangers of encounters with wild animals.

Vesicular Stomatitis Virus

VSV (family *Rhabdoviridae*, genus *Vesiculovirus*) affects hoofed livestock in North and South America. Infections are generally mild but outbreaks can have significant economic impact. VSV infection causes formation of blister-like lesions of the mouth, tongue, teats, and hooves. VSV is one of several "vesicular" diseases of livestock that cannot be distinguished from foot and mouth disease virus (FMDV, a picornavirus) without specific testing. Thus affected animals are quarantined and veterinary health officials put on alert until FMDV can be ruled out (see Chapter 11, Family *Picornaviridae*, for a discussion of FMDV). Thus a VSV outbreak creates considerable angst among livestock producers.

VSV outbreaks are sporadic, and are rare in the United States. Hosts of the virus include horses, cattle, swine, deer, and humans. VSV has been isolated from biting insects and this is how the virus is probably introduced into a herd. However, once an animal has been infected, transmission can continue as a result of contact. Human infections are generally associated with exposure to infected animals and in humans disease is a general febrile illness.

Chandipura Virus

Chandipura virus (CHPV) is endemic to India and has been isolated from a variety of insects and animals, including humans. CHPV was first identified in 1965 after isolation from the blood of two patients. CHPV is associated with neurologic disease and those most at risk for developing encephalitis are children under 15 years of age. While there is some disagreement about exact disease burden of CHPV in India, outbreaks continue to be reported as diagnostic methods improve. Case fatality rates range from 55% to 77%. There is

BOX 19.4

WORLDWIDE ELIMINATION OF HUMAN RABIES

Rabies is a disease that is feared worldwide. However the state of rabies today is one of contrasts. While clinical rabies is rare in the United States and Europe, the World Health Organization (WHO) estimates the number of human deaths as high as 55,000 per year in Asia and Africa combined. In the United States, very rare cases of clinical rabies are most often associated with bats. In contrast the vast majority of cases worldwide originate with dog bites and 40% of those bitten by suspect rabid dogs are children less than 15 years of age. It is difficult to image tens of thousands of cases of this dreaded and dreadful disease occurring annually, but the vast majority occur "out of site" in poor, rural populations. Cases are sporadic with no transmission among humans, thus no fear of the epidemic spread that so often is the catalyst for national or international attention. In the poor rural areas most affected by rabies, dogs may be largely feral and unvaccinated and postexposure treatment of human bite victims prohibitively expensive or completely unavailable. The zoonotic nature of RABV requires coordination and cooperation between human and animal health sectors to bring it under control. Vaccines for animals must be readily available and affordable enough to use routinely and consistently. Safe vaccines and protective immunoglobulin must also available to bite victims. To address these needs, the WHO and the World Organization for Animal Health (OIE) announced plans in 2015 to work together to eliminate human rabies deaths by the year 2030. We have the medical and scientific expertise to drastically reduce cases of human rabies, if we acknowledge that is threat is ongoing and real in many parts of the world.

some evidence the virulence of CHPV is increasing. CHPV neuropathogenesis is being actively studied using a suckling mouse model.

Other Rhabdoviruses associated with Disease

Other rhabdoviruses associated with disease include bovine ephemeral fever virus (BEFV) that affects cattle and buffalo in tropical and subtropical regions of the world. BEFV is transmitted by insects. Rhabdoviruses also cause fish diseases. These include spring viremia of carp virus, infectious hematopoietic necrosis virus, and viral hemorrhagic septicemia virus among others. These viruses cause significant mortality and morbidity in both wild and cultured fish.

In this chapter we learned that:

- Members of the family *Rhabdoviridae* are rod-shaped enveloped viruses with an unsegmented negative-sense RNA genome.
- As a family, rhabdoviruses have a wide host range, from plants to insects to animals. Some are transmitted by insects and they cause a variety of diseases.
- Their overall replication strategy is similar to other viruses in the order *Mononegavirales*: The L polymerase is packaged in the virion and begins transcribing mRNAs upon penetration of the RNP into the host cell. Synthesis of N protein triggers a switch from mRNA to genome synthesis as N encapsidates new transcripts.
- RABV is a dangerous pathogen of humans and animals and is found throughout most of the world. It is transmitted via infectious saliva (often as a result of a bite). In order to infect the recipient, RABV must gain access to neurons. Virus replicates very little until it completes its travels up neurons to reach the CNS where is replicates very efficiently in nerve bodies.
- The goal of postexposure RABV vaccination is to stimulate protective immune responses during the time it takes for the virus to reach the CNS.
- VSV causes a vesicular disease of livestock. VSV has a very broad cell tropism, thus its G protein is often used to pseudotype foreign viruses.

CHAPTER

20

Families *Paramyxoviridae* and *Pneumoviridae*

OUTLINE

After studying this chapter, you should be able to answer the following questions:

- What are the main characteristics of the members of the families *Paramyxoviridae* and *Pneumoviridae*?
- How do the two families differ? Why was the new family (*Pneumoviridae*) recently established by the International Committee on the Taxonomy of Viruses?
- What is the general replication scheme of the unsegmented, negative-stranded RNA viruses?
- How does paramyxovirus N protein participate in the switch from mRNA synthesis to genome replication?
- Why do paramyxovirus-infected cells fuse with uninfected neighbors?
- Name two human diseases caused by a paramyxovirus.

The family *Paramyxoviridae* is a large group of enveloped, unsegmented, negative-strand RNA viruses (Fig. 20.1). The family is one of six within the order *Mononegavirales*. Paramyxoviruses replicate in the cytoplasm of infected cells. Hosts include mammals, birds, reptiles, and fish. Paramyxoviruses are notable pathogens of humans and other animals. Some paramyxoviruses cause flu-like illnesses and together with the influenza viruses were initially referred to as myxoviruses (myxo meaning mucus or slime). When it was determined that distinct types of viruses caused flu-like illnesses they were divided into two families: *Paramyxoviridae* and the *Orthomyxoviridae*. (*Orthomyxoviridae* is the family that contains the "true" influenza viruses, Chapter 23: Family *Orthomyxoviridae*.) As we see later in this chapter, paramyxoviruses cause a variety of diseases in addition to respiratory disease.

In 2016 the International Committee on the Taxonomy of Viruses created a new family, *Pneumoviridae*. The pneumoviruses are not "newly discovered" viruses but were previously included as a subfamily within the family *Parmyxoviridae*. Viruses in these families share enough characteristics that for the purposes of this text, they will be considered in a single chapter (Boxes 20.1 and 20.2).

GENOME STRUCTURE/ORGANIZATION

Paramyxovirus and pneumovirus genomes range in size from 13 to 15 kb and encode 6–10 proteins (Fig. 20.2). Both families encode a fusion protein (F) that mediates fusion at neutral pH. Thus these viruses enter cells by fusion at the plasma membrane. The 3′ untranslated region (3′ UTR or 3′ leader) is short, ~50

Viruses. DOI: http://dx.doi.org/10.1016/B978-0-12-803109-4.00020-9

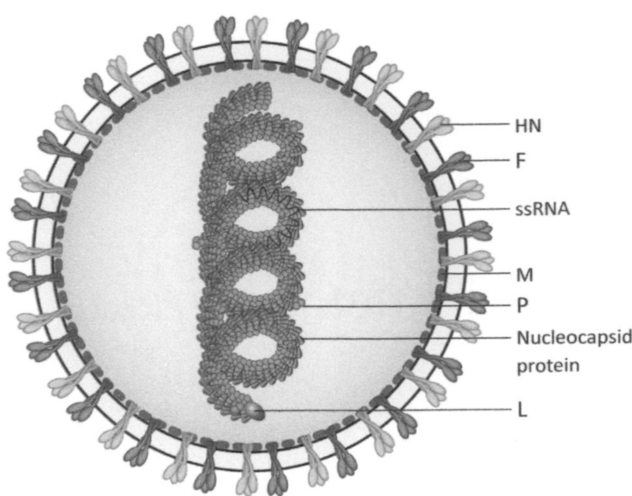

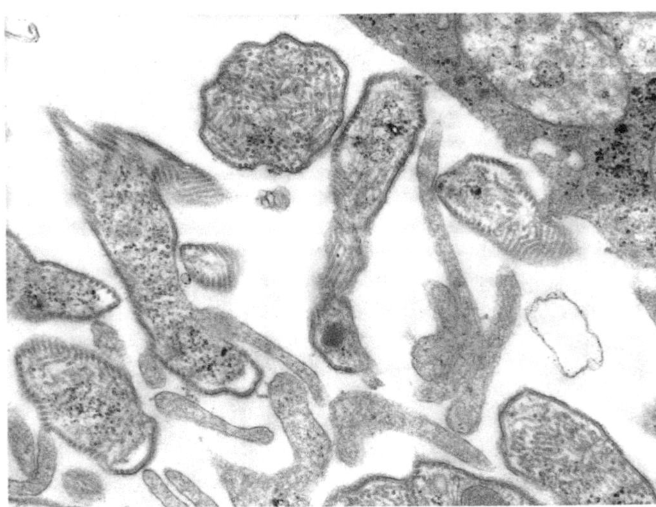

FIGURE 20.1 Virion structure (left). Transmission electron micrograph of mumps virus (right). *From CDC/Dr. F. A. Murphy. CDC Public Health Image Library Image #8758.*

BOX 20.1

GENERAL CHARACTERISTICS

Paramyxoviruses and pneumoviruses have monopartite (unsegmented), negative-strand RNA genomes (~15–20 kb) encoding 6–10 proteins from seven to nine ORFs. Genome replication and transcription are cytoplasmic. Transcription (L protein is the RdRp) initiates at a single site at the 3′ end of the genome. As a completed transcript is released, the next transcript is initiated. This process is controlled by conserved sequences flanking each ORF. mRNAs are capped and polyadenylated by the L protein.

Virions are enveloped with helical nucleocapsids. They are spherical to pleomorphic and ~150–350 nm in diameter. Paramyxoviruses and pneumoviruses encode two surface glycoproteins. These are visualized as spikes extending 8–12 nm from the envelope. One serves as the attachment protein while the second is the fusion (F) protein.

BOX 20.2

TAXONOMY

Order *Mononegavirales*

Family *Paramyxoviridae*

Subfamily *Paramyxovirinae*:

Seven genera

Genus *Morbillivirus* (includes measles, rinderpest, and canine distemper viruses)

Genus *Henipavirus* (Hendra and Nipah viruses)

Genus *Respirovirus* (includes Sendai virus and human parainfluenza)

Genus *Rubulavirus* (includes mumps and some human parainfluenza viruses)

Genus *Avulavirus* (includes avian parainfluenza viruses, Newcastle disease virus)

Genus *Ferlavirus* (reptile viruses)

Family *Pneumovirinae*: Two genera

Genus *Pneumovirus* (includes Human respiratory syncytial virus and bovine respiratory syncytial virus)

Genus *Metapneumovirus (includes Human metapneumoviruses)*

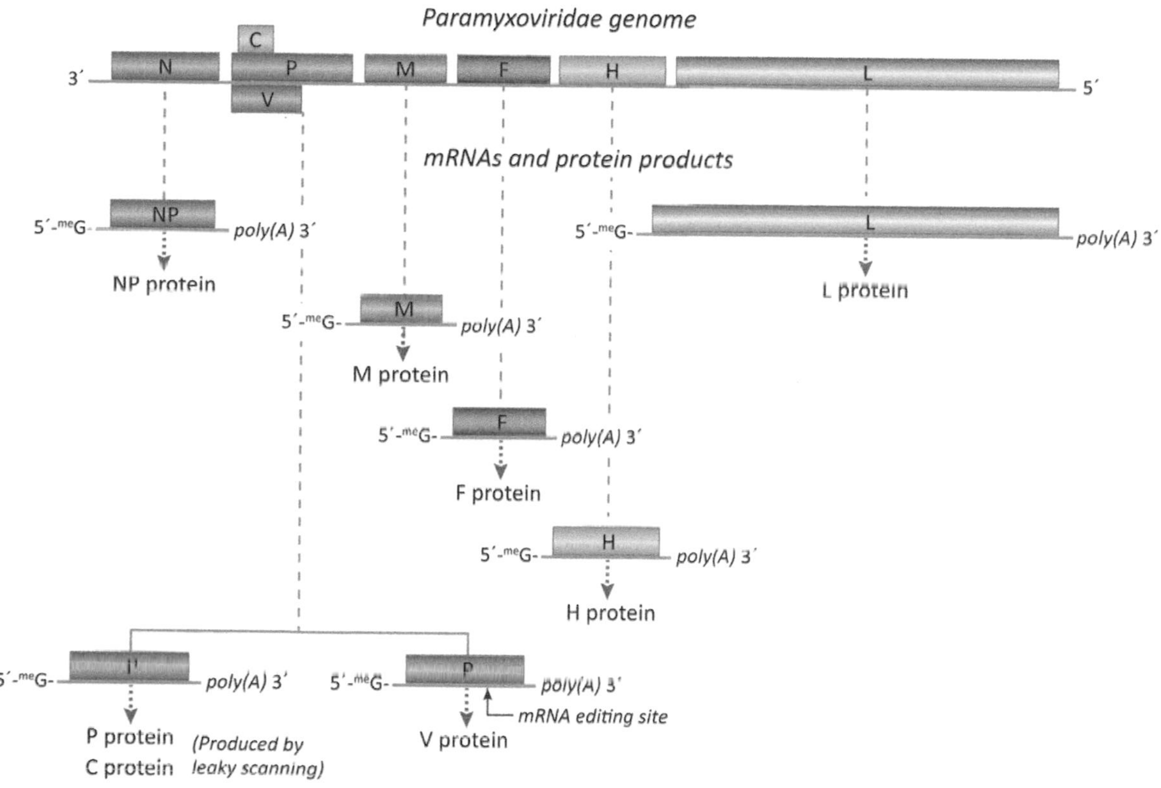

FIGURE 20.2 Genome organization and gene expression, paramyxovirus.

nt and the 5′ UTR (or 5′ trailer) is 50–161 nt. Flanking each gene is cis-acting sequences that direct the RdRp to initiate and terminate transcripts (Fig. 10.5).

VIRION STRUCTURE

Virions are enveloped with helical nucleocapsids. They are spherical to pleomorphic (sometime filamentous) and ~150–350 nm in diameter. All paramyxoviruses encode two surface glycoproteins that extend 8–12 nm from the envelope.

The major structural proteins encoded by both paramyxoviruses and pneumoviruses include the nucleocapsid protein (N) (an RNA-binding protein and the major protein of the ribonucleoprotein particle) and the matrix (M) protein that lines the inner layer of the envelope, forming a bridge between the nucleocapsid and the cytoplasmic tails of the two surface glycoproteins. A phosphoprotein (P) and the large (L) protein (RdRp) are also associated with the nucleocapsid (Fig. 20.1).

Paramyxoviruses and pneumoviruses encode two surface glycoproteins: an *attachment protein* (variously called H, HN, or G) and a separate *fusion* protein (always called F). The attachment proteins of some paramyxoviruses bind to and agglutinate red blood cells (thus they are called "hemagglutinin" and are designated H); some also have neuraminidase activity thus are called HN (hemagglutinin-neuraminidase). However the attachment proteins of some paramyxoviruses are simply designated G, to denote they are glycosylated, surface proteins. All paramyxoviruses also encode an F protein that mediates fusion at the plasma membrane, at neutral pH. The presence of F in the plasma membrane causes neighboring cells to fuse, resulting in the formation of large multinucleate cells called syncytia. Syncytia form in both cultured cells and in infected tissues.

Members of the family *Pneumoviridae* also encode a small hydrophobic (SH) protein of 64–65 amino acids (Fig. 20.3). A small amount of SH is incorporated into the viral envelope but the majority is found associated with the Golgi and at the PM during infection. SH has a single hydrophobic domain and a very small ectodomain. SH is found in glycosylated and nonglycosylated forms. It assembles as pentamers to form a cation-selective ion channel. SH is an accessory protein that is not absolutely needed for replication but viruses lacking SH are somewhat attenuated.

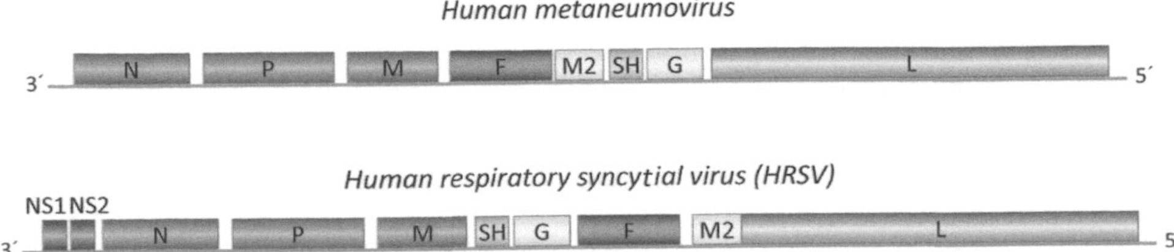

FIGURE 20.3 Genomes of human respiratory syncytial virus (HRSV) and human metapneumovirus (family *Pneumoviridae*). Both viruses encode SH and M2 proteins. HRSV encodes two additional NS proteins at the 3′ end of the genome. Note the absence of edited P gene products.

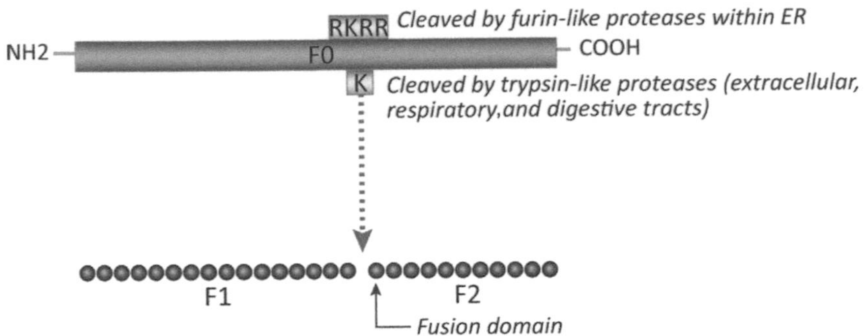

FIGURE 20.4 Cleavage of F protein.

OVERVIEW OF REPLICATION

Attachment and Penetration

The replication cycle of paramyxoviruses and pneumoviruses begins with interaction of the viral attachment protein (H, HN, or G) to cell receptors. Paramyxoviruses with H and HN attachment proteins bind to sialic acid residues. A general model for paramyxovirus/pneumovirus entry is that attachment leads to important conformational changes in the F glycoprotein. The evidence for this is that fusion does not occur unless the virion has interacted with a receptor. After attachment the hydrophobic fusion domain of F interacts with the cell plasma membrane to begin the fusion process. A key feature of the paramyxovirus replication cycle is that in order to be infectious, the F protein must be proteolytically cleaved to release the fusion domain (Fig. 20.4). Paramyxoviruses/pneumoviruses with uncleaved F proteins are not infectious. M is released from the nucleocapsid during entry.

Transcription

Transcription of paramyxovirus/pneumovirus genomes follows the same general process described for rhabdoviruses. Transcription begins when the nucleocapsid is released into the cytoplasm. The bulk of the nucleocapsid consists of the N, tightly bound to RNA at a ratio of 1 N protein to 6 nt of RNA. The complex of N protein and RNA is quite stable and serves as the substrate for transcription. The active viral transcription complex of proteins, also part of the nucleocapsid, consists of P (homo-tetramers) and L protein. There are about 50 copies of L in the virion.

The set of mRNAs produced in the infected cell can also be synthesized from nucleocapsids in a test tube (Fig. 10.4). These include a short leader (*le*) RNA, neither capped nor polyadenylated, that is transcribed from the very 3′ end of the genome followed by 8–10 capped and polyadenylated mRNAs. Most mRNAs encode a single protein, with the exception of the P mRNA from which *several* proteins can be produced. The abundance of each transcript is dependent on its position on the genome. Thus the *le* RNA is the most abundant, followed by the N mRNA. The process that regulates the relative abundance of mRNAs is sometimes called attenuation, reflecting the fact that genes positioned further along the genome (closer to the 5′ end) are produced at lower quantities (thus their transcription is attenuated) (see Fig. 10.5).

Attenuation of transcription can be explained by the fact that there is a single promoter on the genomic RNA, present at the very 3′ end, and this is the only site where transcription can initiate. Following the short *le* RNA is a cis-acting sequence that directs cap synthesis for the N mRNA. At the 5′ end of the N gene, U-rich regulatory sequences that direct polyadenylation and termination. Following the termination signal for the N mRNA is a cis-regulatory sequence that directs initiation and capping of the P mRNA.

BOX 20.3

RNA EDITING OR PSEUDOTEMPLATED TRANSCRIPTION

Most eukaryotic nuclear mRNAs are modified by capping, splicing, and polyadenylation. A rare modification of transcripts, addition of nontemplated nucleotides, is a process often called RNA editing. The term nontemplated refers to nucleotides present in an mRNA that are not present in the genome sequence. Many paramyxoviruses edit their P mRNAs to produce multiple mRNAs from a single gene. Editing was first proposed for paramyxoviruses in the 1980s when cloned P mRNAs were found to have sequences that differed from the viral genome by insertion of 1 to 6G residues. Further studies showed that the differences were consistent for a given paramyxovirus and were only seen for the P mRNA. It is now clear that insertions occur at very specific sites and that the ratios of edited mRNAs to nonedited mRNAs are constant for a given paramyxovirus P gene. Thus insertions are driven and controlled by the *specific sequence of the viral RNA*. Because the process depends on cis-acting genome sequences, the process is also referred to as "pseudotemplated" transcription, a more specific term than "editing." Additional examples of pseudotemplated transcription have been found among the members of the order *Mononegavirales*: Ebola virus (Chapter 21: Family *Filoviridae*) edits its glycoprotein gene.

After release of the N mRNA, successful synthesis of P mRNA requires that the transcription complex remain closely associated with the genome. If the transcription complex dissociates it must relocate back to the very 3′ end of the genome, where it can interact with promoter sequences.

TABLE 20.1 Products of Paramyxovirus P Genes

Genus	Identified mRNA insertions and products[a]	ORFs expressed by use of alternative start codons[a]
Rubulavirus	+1G or +4G (W), +2G (P) note that the unedited product is V	
Avulavirus	+1G (V), +2G (I)	
Respirovirus	+1G (V), +2G (W or D)	C, C′, Y1, Y2
Henipavirus	+1G (V), +2G (W)	C
Morbillivirus	+1G (V), +2G (W)	C

[a]*Vary by species.*

Protein Synthesis

N mRNA, the most abundant paramyxovirus mRNA, is cytosolic and N is synthesized as a soluble protein (sometimes called N0 to differentiate it from molecules of N tightly associated with RNA). N ranges in size from ~53 to 57 kDa. In the virion N is found in the nucleocapsid, bound to genomic RNA. Each N protein binds six nucleotides. Expression of N protein (in the absence of other viral proteins) shows that N can self-associated to form nucleocapsid-like structures. However, N is not simply a scaffolding protein, it also has roles in RNA replication. Early after infection, newly synthesized N is soluble, and in this form can interact with P.

P is synthesized in the cytosol, is ~400–600 amino acids, and is heavily phosphorylated. P is fairly abundant, present at ~300 copies in the virion. As noted above, P is a required part of the transcription/replication complex; RNA polymerization requires association of multimers of P with L. Several important domains of P have been mapped. The carboxyl terminal region has domains required for multimerization, interaction with L, and binding to N0. The importance of the carboxyl terminal region is demonstrated by the fact that, in some experimental systems, it can substitute for full-length P to support mRNA synthesis. The amino-terminal half of P is required for genome (and cRNA) synthesis, likely through interactions with soluble N0.

In addition to encoding the P protein, the P mRNA encodes additional proteins. Some of these are produced by *cotranscriptional mRNA editing*. The process is also called pseudotemplated transcription. It occurs when G residues are added to the P mRNA at a very specific location (at approximately nt 400 in the mRNA). The addition of G residues (usually one or two) shifts the reading frame. Thus edited mRNAs still code for the amino-terminal half of the P protein but lack the carboxyl terminal domains so critical for transcription (Box 20.3).

What are the functions of these additional proteins? Most paramyxoviruses encode a cysteine rich *V protein* of approximately 25–30 kDa (Table 20.1). V is produced by addition of a single G at the P mRNA editing site. Thus most of the V protein is identical to P, but editing produces a frame shift that replaces the carboxyl terminal domains of P with a short cysteine rich sequence. One role of V is a negative regulator of RNA synthesis (perhaps by binding to N0 and sequestering it from productive interactions with RNA). V also appears to be a suppressor of innate cell responses. Measles virus V interacts with several cellular proteins involved in innate immunity to viruses, including MDA5 and LGP2 (see Chapter 6: Immunity and Resistance to Viruses).

The *W protein* is expressed from P mRNAs edited by insertion of two G residues. This results in a stop codon shortly after the editing site, thus W is essentially a truncated P protein. W may also regulate viral RNA synthesis.

The *D protein* of bovine and human parainfluenza viruses is produced by addition of two Gs at the editing site. This new open-reading frame (ORF) extends for 131 amino acid residues beyond the editing site.

The editing scheme is a bit different for members of the genus *Rubulavirus*. Here the unedited mRNA encodes the cysteine rich V protein while it is P that is produced from an edited (+1G) mRNA. Rubulaviruses also produce W by insertion of one or four Gs during mRNA synthesis.

Some paramyxoviruses also encode *additional proteins* (C and C′) from P mRNA, by use of alternative start sites to access alternative reading frames. Sendai virus (genus *Respirovirus*), an extreme example, produces a nested set of four carboxyl coterminal proteins (C′, C, Y1, and Y2) that range in size from 175 to 215 amino acids. Translation of each of the C′, C, Y1, and Y2 ORFs is initiated at a different site, although translation is terminated at the same downstream stop codon. Morbilliviruses and henipaviruses express only one C protein and rubulaviruses do not express any C proteins. C proteins are basic polypeptides involved in regulation of RNA synthesis, counteracting host antiviral responses, and facilitating release of virions from infected cells.

There are two different mechanisms used to direct ribosomes to alternative start sites. Leaky scanning, whereby a ribosome fails to assemble on a start (AUG) codon (thus continues to scan the mRNA) is common (Fig. 3.14). The sequences surrounding an AUG codon are important for defining a start site and if the AUG is not flanked by appropriate sequences the AUG will be bypassed by the ribosome. Sendai virus C and C′ are produced by leaky scanning. To produce C′ ribosomes must initiate translation at an ACG triplet at nt 81 of the P mRNA. This is not a common event, so most ribosomes bypass this site and initiate translation of P protein at an AUG at nt 104 of the mRNA. However, some ribosomes also bypass the P initiation site and initiate the C protein by using an AUG codon at nt 114.

A different mechanism to direct ribosomes to alternative start codons is called ribosome shunting or stop-start (Fig. 3.14). The mechanism differs from leaky scanning in that the ribosome completely bypasses some regions of the mRNA, effectively "jumping" over them to a downstream region. It requires two cis-acting sequences: an upstream donor and a downstream acceptor. The ribosome is shunted from the donor site directly to the acceptor site, effectively bypassing the intervening sequences. The acceptor site determines the translation initiation site, sometimes a non-AUG start codon. In the case of Sendai virus, translation of the Y1 and Y2 proteins occurs through a ribosome shunting.

What is the evidence for ribosome shunting? The process was initially proposed based on experiments that modified potential start codons in the Sendai virus P gene. In the leaky scanning scenario, changing the context of an upstream start codon should affect the amount of protein produced from downstream initiation codons. For example, if the Sendai virus C′ start codon was changed to an AUG, in a good context for ribosome initiation, we predict synthesis of more C′ protein and less of the P and C proteins. However in the case of Y1 and Y2 proteins, modification of some upstream start codons had no effect on the levels of Y1 and Y2 produced, suggesting that ribosomes were not using a scanning mechanism to reach these translation start sites. Eventually, specific donor and acceptor sites were identified and it was confirmed that mutation of the intervening sequences has no effect on the levels of the downstream protein products.

The functions of the various C and Y proteins have not been well defined. In the case of measles virus, C protein is not required for virus replication in some cultured cells, but is required for replication in nonhuman primates. Evidence points to a role in antagonizing the interferon response.

The matrix (M) protein is a major structural protein. M proteins range is size from ~38.5 to 41.5 kDa. M proteins are highly basic (have a lot of positively charged amino acids) and are somewhat hydrophobic. M is associated with the inner face of membranes and can form two-dimensional arrays (sheets and tubes) under some conditions. M also associates with nucleocapsids by interactions with negatively charged surfaces. Thus M acts as a typical matrix protein, forming a bridge between envelope proteins in a membrane and the nucleocapsid. M proteins of several paramyxoviruses have so-called late domains, short amino acid motifs that interact with host cell sorting complexes to facilitate release of budded virions.

Translation of the two viral glycoproteins takes place on rough ER. Both are glycosylated on their transit through the ER and Golgi to the plasma membrane.

The F protein is synthesized as a precursor (F_0) that must be cleaved into F_1 and F_2 fragments (Fig. 20.4). F_1 and F_2 remain associated via a disulfide bond. As noted previously, cleavage of F_0 is essential for virion infectivity. Some paramyxovirus F proteins are cleaved within the ER (by furin-like or cathepsin-like proteases); however, others are cleaved by extracellular, trypsin-like proteases (found in the respiratory and digestive tracts) to achieve particle maturation. The cellular location of F_0 cleavage plays a major role in paramyxovirus pathogenesis, as cleavage by extracellular proteases necessarily restricts productive virus replication to sites such at the lungs and gastrointestinal tract. As might be expected, paramyxoviruses that cause system-wide infections (measles virus, canine distemper virus) have F proteins that are cleaved in the ER, before they are even trafficked to the plasma membrane.

Translation of L mRNA is cytosolic and produces a protein of ~2200 aa (259 kDa). L is the catalytic portion of the transcriptase/replicase complex but it does not synthesize viral RNA in the absence of P. L is required for cap synthesis, polyadenylation of mRNAs and mRNA editing. Among the paramyxovirus L proteins, the most conserved regions are in the central portion of the protein while the amino and carboxyl termini more variable.

Pneumoviruses encode additional nonstructural (NS) proteins. Genes for two NS proteins are found at the very 3′ end of the genome (upstream of the N gene). NS1 and NS2 mRNAs are abundant in infected cells and encode proteins of 139 and 124 amino acids, respectively. Viruses with deletions of NS1 and NS2 can replicate in cultured cells, but not as well as wild-type viruses. NS1/NS2 deletion mutants are very attenuated in animals. NS1 and NS2 are multifunctional proteins with important roles in virulence. For example, they antagonize antiviral responses and suppress expression of Type I interferons.

Pneumoviruses also encodes a small M2 gene. This gene has two partially overlapping reading frames to produce M2-1 and M2-2 proteins (194 and 90 amino acids, respectively). M2-1 protein is an essential part of the transcription machinery (without it the RdRp does not transcribe beyond the NS1 and NS2 genes). M2-2 deletion mutants have increased levels of mRNA synthesis and decreased levels of genome replication. The M2-2 protein is not essential for RSV replication but the mutated virus grows slowly in cultured cells (Box 20.4).

Genome Replication

Synthesis of N and P triggers a switch from transcription to genome synthesis (as was described previously for the rhabdoviruses). Genome synthesis requires synthesis of an end-to-end copy of the genome; this copy or cRNA is neither capped nor polyadenylated. Whereas mRNAs are not bound by the N protein, both the cRNA and genomic RNA bind to N protein during their synthesis. In fact their synthesis requires that free N protein be available to encapsidate the RNA as it is synthesized. Each paramyxovirus N protein binds to exactly 6 nt of RNA. During efforts to generate efficient reverse genetic systems for paramyxoviruses, it was found that efficient virus replication requires that the genome length be an exact multiple of 6. This called the paramyxovirus "rule of six."

Assembly, Release, and Maturation

Nucleocapsid assembly occurs in the cytoplasm. The nucleocapsid consists of genomic RNA, N, P, and L proteins. The nucleocapsid travels to the plasma membrane and interacts with M. Virions bud from the PM. Virions containing uncleaved F_0 are noninfectious unless F_0 is cleaved by extracellular, trypsin-like proteases. Recall that F cleavage is a necessary maturation step as it releases the hydrophobic fusion peptide at the N-terminus of F_2, a process critical for fusion to an uninfected cell.

The cellular location of F cleavage is determined by the amino acid sequence at the cleavage site. Measles and canine distemper viruses have F proteins that are cleaved in the ER/Golgi by furin-like or cathepsin-like proteases. Thus these viruses have a wide tissue tropism in the infected animal. In contrast, human parainfluenza virus F_0 is cleaved by extracellular proteases found in the respiratory tract.

DISEASES

Measles

Measles virus (a morbillivirus) is extremely contagious by the airborne route. It is estimated that 90% of susceptible persons coming in contact with an infected person will be infected. Measles most often is a disease of children and most deaths are among children less than 5 years old. Disease symptoms usually develop 10–12 days after exposure, and persist for 7–10 days. Virus can be transmitted before symptoms appear. Measles starts with fever, runny nose, cough, red eyes, and sore throat, followed by a red, flat rash that usually starts on the face and spreads over the body. Small white spots (known as Koplik's spots) may form inside the mouth. Complications occur in about 30% of infected persons and these include diarrhea, corneal ulceration and scarring, encephalitis, deafness, and

BOX 20.4

CONGRATULATIONS, YOU ARE A FAMILY!

In 2016 the International Committee on the Taxonomy of Viruses (ICTV) created a new virus family, *Pneumoviridae* (and simultaneously dropped the subfamily *Pneumovirinae* from the family *Paramyxoviridae*). The new family contains two genera, *Metapneumovirus* and *Orthoneumovirus*. Of course a change in taxonomy does not change the characteristics of the viruses (such as human and bovine RSV) within the family.

The taxonomic change was made highlight the differences between the paramyxoviruses and the pneumoviruses. These include differences in the structures of their ribonucloprotein complexes and the fact that pneumovirus genomes contain a gene, called M2, that encodes two unique proteins involved in the regulation of virus RNA synthesis. The change in status was suggested despite the similarities between the two families, including the conservation of the fusion (F) protein.

pneumonia. A leading cause of deaths in children is the development of secondary infections that result from a transient but profound suppression of the immune system. A very rare complication of measles virus infection is subacute sclerosing panencephalitis (SSPE) a progressive neurological disorder of children and young adults. SSPE results from persistent infection of the CNS by *defective* measles virus that replicates very slowly. Cases of SSPE have essentially disappeared in countries that practice widespread immunization.

A measles virus vaccine was introduced in United States in 1963. It is a safe and effective vaccine. Numbers of measles cases in the United States were dramatically reduced to only thousands of cases per year by the 1980s. Worldwide, measles was estimated to have caused 2.6 million deaths per year during the 1980s. However, by 1990 estimated deaths were just over 600,000 and by 2013 estimated yearly deaths were less than 100,000.

Why do we continue to see *any* deaths from measles virus? Despite a safe and effective vaccine, measles virus infects an estimated 20 million (unvaccinated) people per year. The risk of death is usually very low (0.2%), but may be up to 10% among malnourished children. Unfortunately, inadequate food and inadequate medical infrastructure often go hand in hand. Death rates due to measles virus infection are significantly higher (30%) among immunocompromised persons. Measles virus is thought to cause more *vaccine-preventable* deaths than any other virus.

Mumps

Mumps virus (genus *Rubulavirus*) is highly infectious and spreads rapidly among people living in close quarters. It is transmitted by respiratory droplets, direct contact, or contaminated objects. The incubation period is usually 14–18 days and patients shed virus a few days before the onset of symptoms that often include fever, muscle pain, headache, and lethargy. Initial symptoms are followed by painful swelling of the salivary glands (the parotid gland most often). The disease is usually self-limiting; however, infection of men can result in painful testicular swellings, which can cause decreased fertility and rarely sterility. Other complications include meningitis (in up to 10% of cases), ovarian inflammation (5% of adolescent and adult females), and acute pancreatic inflammation (4%). A very rare complication (0.005%) is profound hearing loss. The mumps vaccine was licensed in the United States in 1967 and since then cases have dropped significantly. Between 2001 and 2008 the average cases per year was only 265. An exception was 2006 when over 6000 cases occurred among college students.

Respiratory Tract Diseases

Parainfluenza viruses most often cause upper respiratory tract diseases, but may cause pneumonia. Virus is restricted to respiratory epithelial cells (where proteases that cleave the F protein are expressed). Natural immunity to parainfluenza viruses is not solid and reinfection can and does occur.

Canine Distemper

Canine distemper virus (genus *Morbillivirus*) causes an often fatal multisystem viral disease that affects the respiratory, gastrointestinal, and central nervous systems of young dogs. Canine distemper occurs worldwide, and once was the leading cause of death in unvaccinated puppies. Development of a vaccine (in 1960) led to a drastic reduction in infections of dogs; however, the virus infects many other carnivores including raccoons, skunks, and foxes. In 1994 CDV wiped out a third of Africa's Serengeti lion population after an outbreak in feral dogs. CDV is now emerging as a pathogen that threatens wild tigers worldwide. Big cats in captivity have also fallen prey to CDV. While there are licensed (live attenuated) vaccines available for dogs and ferrets, they are *not* suitable for other mammals, such as big cats, in which they cause fatal disease.

Henipaviruses

Nipah virus is zoonotic and causes severe disease in swine and humans. Symptoms range from asymptomatic to acute respiratory syndrome with fatal encephalitis. The natural hosts of the virus are fruit bats. Nipah virus was first identified during an outbreak that took place in Malaysia in 1998. Other outbreaks have occurred in Singapore, Bangladesh, and India. Hendra virus was first identified in an outbreak that occurred in the suburb of Hendra, Australia in 1994. The outbreak involved 21 horses and 2 human cases. Thirteen horses (case fatality of approximately 75%) and a trainer died. The natural host of Hendra virus is fruit bats.

Pneumoviruses

Respiratory syncytial virus (RSV) is a respiratory virus that can cause repeated infections. Healthy adults usually experience mild, cold-like symptoms. However, RSV can be serious for infants where it is a common cause of bronchiolitis (inflammation of the small airways in the lung) and pneumonia. In the United States most children will be infected with RSV by 2 years of age. Hospitalization is required each winter for 0.5%–2% of infections of children less than 6 months old. RSV is also responsible for more hospitalizations in 1 to 4 year olds than influenza viruses. Immunity to RSV is not solid and reinfection can occur later in life. Symptoms are generally mild, but adults with compromised immune systems and those 65 and older are at increased risk of severe disease.

Human metapneumovirus is a respiratory pathogen that causes a range of illnesses (from asymptomatic infection to severe bronchiolitis). First identified in 2001, the virus was identified in human respiratory samples. Seroprevalence studies in the Netherlands revealed that 25% of tested children, aged 6–12 months, had detectable antibodies to human metapneumovirus. By age 5 years, 100% were seropositive. Reports from all over the world now suggest that this newly discovered virus is ubiquitous. Although this viral pathogen was first described in children, subsequent reports have highlighted the importance of human metapneumovirus as a cause of respiratory illness in adults.

In this chapter we have learned that:

- Families *Parmyxoviridae* and *Pneumoviridae* are members of the order *Mononegavirales*.
- Members of both families are enveloped viruses with a nonsegmented, negative-sense RNA genome. Replication is cytoplasmic.
- RdRp is packaged in the virion and can synthesize mRNAs in an in vitro reaction.
- During an infection, newly synthesized N triggers the switch between synthesis of mRNAs and genome-length RNAs.
- Paramyxoviruses and pneumoviruses encode two envelope glycoproteins, one serves as the attachment protein and the second (always called F) is the fusion protein. F cleavage is required for virion infectivity and the site of cleavage (intracellular versus extracellular) can greatly impact virulence. F protein is found in the plasma membrane of infected cells and can mediate fusion with neighboring uninfected cells to produce multinucleate cells called syncytia.
- Until 2016 the pneumoviruses were considered a subfamily of the *Paramyxoviridae*. They are now considered a separate family.
- Measles virus, mumps virus, and RSV are major human pathogen. There are protective vaccines available for measles and mumps however a safe and effective vaccine for RSV remains elusive.

CHAPTER

21

Family *Filoviridae*

OUTLINE

After reading this chapter, you should be able to answer the following questions:

- What are the general characteristics of members of the family *Filoviridae*?
- What is viral hemorrhagic fever (VHF)?
- What are the natural hosts for filoviruses?
- How are filoviruses transmitted?
- What special equipment and precautions are taken when working with filoviruses such as Ebola and Marburg viruses?

The family *Filoviridae* includes the genera *Ebolavirus* (EBOV), *Marburgvirus* (MARV), and *Cuevavirus* (Box 21.1). EBOV and MARV cause severe, often lethal, hemorrhagic fever in humans. During the years 2013–16, an outbreak in West African was not contained and affected several countries. In time, an international response supplied medical personnel, equipment, medicines, and facilities to control the epidemic: But not before over 28,000 infections, resulting in over 11,325 deaths decimated the health care systems of affected communities.

Filoviruses are one of five families in the order *Mononegavirales*. They are enveloped viruses with unsegmented, negative-strand RNA genomes. Filoviruses are notable or their bizarre filamentous shapes with virions that are 800–1200 nm long with a diameter of 80 nm (Figs. 21.1 and 21.2).

GENOME ORGANIZATION

The overall organization of filovirus genomes is similar to that of other nonsegmented negative-strand RNA viruses (Box 21.2). An RNA-binding nucleocapsid protein (NP) is encoded at the 3′ end of the genome and the L protein, the RdRp, is encoded at the 5′ end of the genome (Fig. 21.3). However

BOX 21.1

TAXONOMY

Family *Filoviridae*
Genus *Ebolavirus*
Species *Bundibugyo ebolavirus*
Species *Reston ebolavirus*
Species *Sudan ebolavirus*
Species *Tai Forest ebolavirus*
Genome Replication
Species *Zaire ebolavirus*
Genus *Marburgvirus*
Species *Marburg marburgvirus*
Genus *Cuevavirus*

Viruses. DOI: http://dx.doi.org/10.1016/B978-0-12-803109-4.00021-0

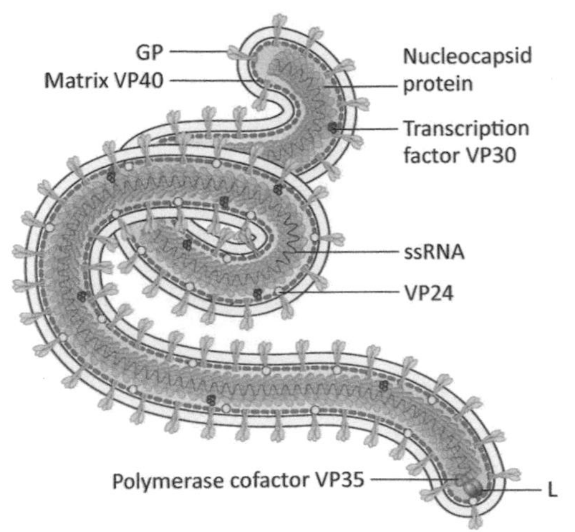

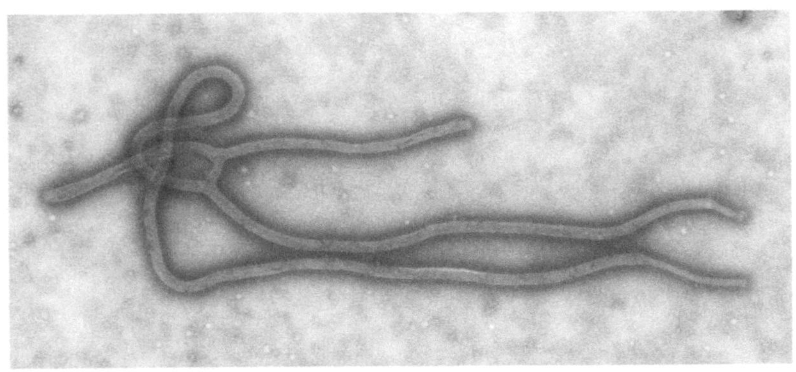

FIGURE 21.1 Virus structure (left). Transmission electron micrograph (right) of Ebola virus. *From CDC/Cynthia Goldsmith. Public Health Image Library ID#1832.*

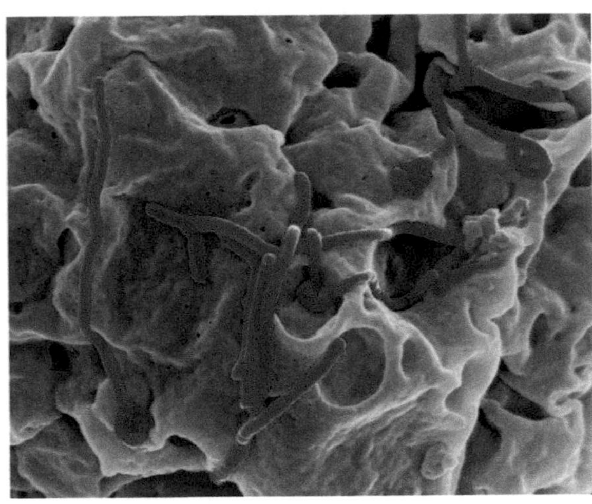

FIGURE 21.2 Digitally colorized scanning electron microscopic image depicting a number of filamentous Ebola virus particles (red) that had budded from the surface of a VERO cell (brown). Content provider: National Institute for Allergy and Infectious Diseases *Public Health Image Library ID#17777.*

BOX 21.2

GENERAL CHARACTERISTICS

Members of the family *Filoviridae* are long, filamentous-enveloped viruses with unsegmented negative-strand RNA genomes of about 19 kb. Filovirus genomes are organized in a manner similar to other members of the order *Mononegavirales* with seven genes encoding seven to eight proteins. Transcription is also similar, with the production of a set of nonoverlapping capped and polyadenylated mRNAs. L protein contains the catalytic site for RNA synthesis but is only active as a complex of NP and VP35. An active transcriptase complex synthesizes mRNAs upon release of the nucleocapsid into the cytoplasm. Genome replication and transcription are cytoplasmic

Virions are characteristically long (800–1000 nm) with a diameter of 80 nm. They are flexible and assume characteristic shapes (Fig. 21.1). Spikes on the surface of the virion consist of a glycoprotein (GP) heterodimer that is produced by cleavage of a precursor. The amino terminal portion of the precursor (GP1) contains the attachment domain. The carboxyl terminal half of the precursor (GP2) contains the fusion domain. Fusion takes place in the environment of an acidified endosome.

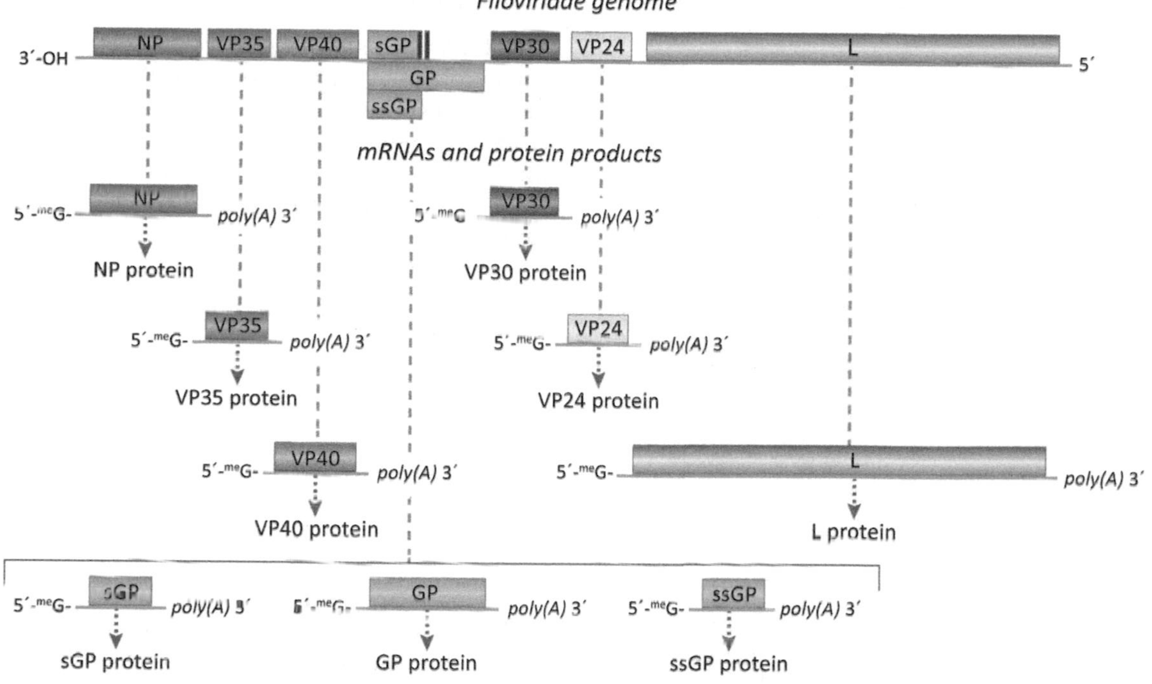

FIGURE 21.3 Genome organization and gene expression strategy of a filovirus.

filovirus genomes are longer (~19 kb) than those of other viruses in the order *Mononegavirales* and there are some additional differences. A set of capped and polyadenylated filovirus mRNAs is transcribed and cis-acting signals regulate capping, polyadenylation, and transcript termination. However, the regulatory (stop and start) signals of some genes overlap, suggesting that the polymerase must somehow backtrack on the genome after completing the upstream transcript. Filoviruses also have longer noncoding regions at the 3′ and 5′ ends of the genome than are found on rhabdoviruses and paramyxoviruses.

VIRION STRUCTURE

Filoviruses are have enveloped virions that range in size is from 800 to 1200 nm long. They have a very consistent diameter of ~80 nm. Their filamentous shapes are quite distinct from those of other enveloped viruses of animals. Two nucleocapsid proteins (NP and VP30) bind to the genome, along with the proteins (L and VP35) that comprise the active RdRp, to form the nucleocapsid. A single glycoprotein (GP) protrudes from the virion surface. The major matrix protein (VP40) is associated with the inner leaflet of the envelope.

OVERVIEW OF REPLICATION

As many filoviruses have the potential to cause severe disease, and there are no licensed vaccines, their study is restricted to high security, high containment laboratories, a very expensive undertaking (Fig. 21.4). Thus, much of the work to decipher molecular aspects of their replication is done in "surrogate" systems that do not produce infectious virions.

Attachment/Penetration

A single glycoprotein, GP, contains both attachment and fusion domains. Infectious virions contain cleaved GP. GP1 is the amino terminal region and contains the attachment domain. GP2 is the carboxyl terminal region and contains a fusion domain. Ebolavirus GP mediates attachment to a wide variety of cells, including fibroblasts, hepatocytes, adrenal cortical cells, epithelial cells, macrophages, dendritic cells, and endothelial cells. Molecules identified as receptors include carbohydrate-binding proteins (C-type lectins, for example, L-SIGN, DC-SIGN, and hMGL among others). One recently identified receptor is T-cell immunoglobulin and mucin domain-1 protein (TIM-1) found the surface of epithelial cells. Studies to identify filovirus receptors

FIGURE 21.4 Centers for Disease Control microbiologists in the process of suiting up in order to access the interior of the organization's Biosafety Level-4 (BSL-4) laboratory. The scientist on the left is attaching his supportive air hose, which provides a supply of filtered, breathable air. *From CDC/Dr. Scott Smith. CDC Public Health Image Library ID# 10723.*

often use pseudotyped viruses, for example, vesicular stomatitis virus (a rhabdovirus) expressing filovirus GP.

After attachment virions enter cells by endocytosis. Some are taken up by macropinocytosis, but virions may also enter cells by clatherin or caveolae-mediated endocytosis (Fig. 3.6). Regardless of the endocytic process used for uptake, acidification of the early endosome is required for membrane fusion. Within the acidified endosome, GP1 is first cleaved to a smaller form by the low pH-dependent cellular proteases, cathepsin L and cathepsin B. This seems to trigger an additional "attachment" event whereby processed GP1 interacts with the Niemann Pick C1 (NPC1) protein, an endosomal protein. Cellular expression of NPC1 is required for productive infection and the interaction is believed to trigger a conformational change that allows insertion of the GP2 fusion domain into the endosomal membrane. Fusion releases the nucleocapsid into the cytosol.

Transcription

mRNAs are synthesized in the cytosol by the viral transcriptase/replicase complex that consists of VP35 and L. mRNAs are capped and polyadenylated and most are monocistronic (encode a single protein). Similar to the rhabdoviruses and paramyxoviruses, mRNAs synthesis is controlled by cis-acting sequences flanking the open-reading frames. The biomechanics of transcriptase stopping and restarting is unclear as some genes contain overlapping transcription termination and initiation sites.

Translation

The phosphorylated nucleoprotein, NP, is the major structural protein that binds genome length RNAs. Ebolavirus NP oligomerizes into helical filaments, encapsidates the viral RNA genome, and serves as a scaffold for assembly of additional viral proteins to the viral nucleocapsid. As we have seen before for families in the order *Mononegavirales* soluble NP is thought to trigger the switch from transcription to replication.

Following the NP gene is VP35, a 35 kDa phosphoprotein. VP35 is part of the core ribonucleoprotein particle. VP35 binds to NP and regulates NP assembly. VP35 also interacts with L protein and serves as part of the polymerase complex. (Thus VP35 plays a role similar to the P proteins of rhabdoviruses and paramyxoviruses.) VP35 is also an interferon antagonist.

VP40 is the major matrix protein. VP40 is membrane associated and essential for virion assembly. VP40 is the most abundant protein in the virion, has an affinity for membranes, and is associated with the inner leaflet of the virion envelope. VP40 participates in budding, initiating, and driving the envelopment of the nucleocapsid at the plasma membrane.

The next gene on the filovirus genome encodes the major glycoprotein precursor, GP. GP is the membrane-anchored envelope protein that mediates both attachment and fusion. A very surprising finding for EBOVs is that synthesis of membrane-anchored GP requires mRNA editing or pseudotemplated transcription (a process described in Chapter 20: Families *Paramyxoviridae* and *Pneumoviridae*). Only about 20% of the Ebolavirus G mRNA is edited.

This is not the case for MARVs where GP is produced from a nonedited mRNA. In either case the precursor (GP0) is heavily glycosylated in the ER/Golgi by addition of both N-linked and O-linked glycans. GP0 is cleaved by furin, or a furin-like endoprotease to generate a cysteine-linked heterodimer consisting of GP1 (the amino terminal product) and GP2 (the carboxyl terminal product).

If only 20% of EBOV GP mRNA is edited, what products are produced from unedited mRNA and what function do they serve? Unedited GP mRNA encodes a soluble protein (sGP) that shares the amino terminal 300 amino acids with membrane-anchored GP. However sGP lacks the membrane anchor and is secreted from EBOV-infected cells. A lot of sGP is produced during EBOV infection and it may play a role in immune evasion, as most antibodies produced during a natural infection are targeted specifically against sGP. Antibodies targeted at sGP do not appear to be either virus neutralizing or protective. It must be noted, however, that in stark contrast to EBOV, MARV GP mRNA is not edited and no soluble forms of have been detected. As some MARVs are highly virulent, sGP cannot be the only factor contributing to filovirus virulence.

Next on the filovirus genome is viral protein 30 (VP30), a 30 kDa phosphoprotein that binds genomic RNA. VP30 is a minor nucleoprotein. It interacts with L, forming a bridge from L to the nucleocapsid. It is followed on the genome by viral protein 24 (VP24) a 24 kDa phosphoprotein. VP24 is a minor matrix-associated protein. It is a hydrophobic protein that is found at the plasma membrane. The exact role of VP24 in filovirus replication is not yet known.

The L protein, the RdRp, is encoded by the final third of the genome.

Genome Replication

The general model of filovirus genome replication is consistent with other families in the order *Mononegavirales*. There appear to be some differences in the protein requirements for EBOV and MARV genome synthesis. In the case of EBOV, NP, VP35, VP30, and L are required to synthesize minigenomes (in a recombinant model of replication) while MARV genome replication requires only NP, VP35, and L.

Assembly

Filovirus virions assemble at the plasma membrane. GP is trafficked to the PM to serve as budding sites. Matrix proteins (VP40 and VP24) are key players in assembly and budding. In fact in some systems, expression of VP40 alone results in production of virus-like particles (VLPs). However VLP production is more efficient if GP and NP are added to the system. VP40 contains a late domain that promotes budding and release of budded vesicles, but it is not clear that it is absolutely required for virus production in cultured cells.

DISEASE

MARV was the first filovirus linked to human infection and disease. It was the cause of an outbreak that began in Germany with three cases of viral hemorrhagic fever among laboratory workers processing organs from African green monkeys. By the time it was contained, the outbreak totaled 31 cases with 7 fatalities. Marburg hemorrhagic fever was next documented in Africa in 1975 and there were a handful of additional reported cases the 1980s. Larger outbreaks were recorded in 1999 (Democratic Republic of Congo, 154 cases) and 2005 (Angola, 252 cases, 227 deaths). The extremely high case fatality rate in 2005 was probably due to a large number of pediatric cases where infection may have been hospital acquired (parenteral infection).

Ebola hemorrhagic fever (EHF) was discovered with outbreaks in Democratic Republic of Congo (then called Zaire) and Sudan in 1976. The genetically distinct filoviruses associated with the outbreaks are Zaire ebolavirus (ZEBOV) and Sudan ebolavirus (SEBOV), respectively. In both outbreaks (318 cases in Democratic Republic of Congo, 284 cases in Sudan) high numbers of transmissions were caused by reuse of nonsterile needles and syringes at the treating hospitals. Sporadic outbreaks of EHF occurred during the next decades but most were limited in scale. According to the World Health Organization, the most recent outbreak (2013–15) totaled 28,657 cases with 11,325 deaths (Box 21.3). As bad as these numbers are, they do not reflect the *overall impact* of the epidemic that closed borders to travel and trade, closed schools, and decimated health care infrastructure.

Entry Into Host

EBOV infection begins with entry into the host through mucosal surfaces, breaks, or abrasions in the skin, or by parenteral introduction (needlestick). Most cases occur by direct contact with infected patients or cadavers, as large amounts of virus are present in blood and body fluids (perhaps as much as 10^7–10^8 infectious particles per mL of serum). So those most at risk of contracting the disease are heath care workers or family members without access to appropriate infection control equipment. While virus is present in respiratory secretions, transmission still requires close contact with a symptomatic patient (or cadaver). There is no evidence that EBOV is transmitted by the "aerogenic" route in the manner of the highly transmissible measles virus. In fact, in the United States and Europe, transmissions have only been

BOX 21.3

THE WEST AFRICAN EBOLA EPIDEMIC

Ebola disease is an African disease. It was first described in 1976 when the disease broke out in South Sudan and the Democratic Republic of the Congo. These outbreaks, while highly lethal, were relatively small and were controlled quickly. The source of Ebola virus is unknown but it is generally believed that it is acquired from wild animals. Two of the leading candidates have been monkeys and bats. At the present time, bats are the most likely candidates since one Ebola outbreak occurred in miners in Angola and the mine was populated by bats. Other visitors to the same caves also contracted Ebola virus.

It was therefore a surprise when a case of Ebola was described, far away, in Guinea in West Africa in December 2013. The index case was a 1-year-old boy. The disease eventually killed his mother, sister, and grandmother. Other villagers spread the infection. They lived in the vicinity of a large bat colony. However none of the bats from the village ever tested positive for the virus. It took a long time for the disease to be recognized as Ebola (there are many causes of infectious hemorrhagic fever in Africa) and was thus permitted to spread for several months unchecked. It was first reported as Ebola in March 2014 when almost 60 deaths had occurred. The disease spread into the capital of Guinea, Conakry, and by the end of May the death toll had climbed to 180.

The disease spread into two neighboring countries, Sierra Leone and Liberia in April. By July there were 442 cases in Sierra Leone. As the epidemic grew, cases were reported in another neighboring country, Mali, and more distant Nigeria. Eventually small numbers of cases were eventually transmitted to Senegal, Spain, the United Kingdom, and the United States.

Ebola is a hemorrhagic disease. Affected individuals bleed from all their body orifices and this blood contains large amounts of virus. It is also abundant in vomit and diarrhea. As a result, individuals who come into contact with sick patients are at high risk. One especially important mode of spread is from dead bodies. In much of Africa, custom requires that the dead be washed before burial. As a result, the infection is readily transmitted to family members who prepare the body. It has proven very difficult to prevent these burial practices and attempts to do so have met with resistance and hostility.

In order to manage and nurse Ebola cases, health care workers must wear full protective clothing to ensure that they do not come into contact with infected body fluids. This type of protective equipment is not readily available in much of Africa. Breaking the chain of transmission involves total isolation of the patient, a difficult feat when health care infrastructure is poor. Likewise many villages have essentially no access to health care thus disease can spread in remote areas, without the authorities being aware of it. Monthly deaths peaked in Guinea and Liberia in August 2014 and in Sierra Leone in December. Subsequently prevalence declined steadily to very low levels by late 2015. While the World Health Organization declared the epidemic over in January 2016, sporadic cases continue to occur.

Like many viral diseases, there are no established successful treatments for Ebola disease. During the recent epidemic some patients received blood transfusions from survivors in an attempt to passively immunize them by providing neutralizing antibodies. Subsequent analysis revealed no benefits of such treatments. Because Ebola hemorrhagic fever was, prior to this epidemic, a very uncommon sporadic disease, little effort had been made to develop a vaccine. As a result of the epidemic, this has changed. Many different experimental vaccines are under development including DNA vaccines, oral vaccines, recombinant viral vectors, and virus like particles (VLPs). Unfortunately, due to the unpredictable nature of Ebola outbreaks, such a vaccine may be very difficult to test reliably in the field. Who should receive the vaccine—health care workers? Perhaps vaccine should be given to individuals in regions surrounding an outbreak in an attempt to prevent its spread. One feature of concern is the suspicion of Western medicine in parts of Africa. Local populations often suspect that the appearance of medical teams is the *cause* rather than a consequence of the disease.

However it should also be noted that not all filoviruses cause disease in humans. Reston Ebolavirus (RBOV) was identified in an outbreak at an animal facility in Reston, Virginia (not far from Washington, DC) in 1989. The outbreak began among monkeys (macaques) imported from the Philippines. All of the macaques died from viral hemorrhagic fever and a filovirus was identified by electron microscopy examination of tissues from affected monkeys. The Centers for Disease Control and Prevention (CDC) was alerted and immediately dispatched an infection control team to the scene. The facility was closed and all remaining monkeys were euthanized. The prime concern for the infection control team was that personnel in contact with the monkeys might become ill. In the end, no people became ill although two animal handlers had antibodies to REBOV, a finding highly suggestive of infection. Apparently REBOV was deadly for monkeys, but caused asymptomatic infections in humans. More recently, REBOV has been isolated from pigs in the Philippines, but again the virus was not linked to cases of human disease. This is a good example of the species specificity, particularly as regards pathogenesis, displayed by many viruses.

documented in a health care setting, never by causal contact. However one surprise revealed in recent epidemic is that filoviruses can be shed for pronged periods by recovered patients, and transmission by semen is possible.

Pathogenesis

Viral hemorrhagic fever, an outcome of many human filovirus infections, is a fever and bleeding disorder. Leakage of blood from the vascular system and damage to clotting mechanisms may result in flushing of the skin, petechial rash, conjunctival bleeding, and hemorrhage of visceral organs. Edema results in swelling and is accompanied by low blood pressure and shock. Virus replication results in considerable direct organ damage as evidenced by necrotic lesions in the liver, spleen, kidneys, lungs, testes, and ovaries. Other symptoms of filovirus infection include malaise, muscle pain, headache, vomiting, and diarrhea (symptoms common to many viral infections).

Mechanisms contributing to symptoms include liver damage, disseminated intravascular coagulation (DIC), and bone marrow dysfunction. In DIC, small blood clots form in blood vessels throughout the body, removing platelets, and reducing overall clotting ability. Damage to macrophages results in release of cytokines, chemokines, and other inflammatory mediators.

EBOV and MARV Antagonize Innate Immune Responses

It is clear that runaway virus replication is associated with severe disease. So how are some filoviruses able to so completely subvert immune responses? The overall answer is that EBOV and MARV interfere with many proteins that are involved in development of a protective response. For example, there is severe suppression of interferon stimulated genes. But other genes are affected as well, including many involved in immune regulation, coagulation, and apoptosis.

The Type I IFN response is particularly important for establishing an antiviral state in host cells, to control virus replication during the period before development of adaptive or specific immune responses (Chapter 6: Immunity and Resistance to Viruses). The Type I IFN response is also important for activating an adaptive response. Filovirus proteins that disrupt innate immune responses include VP35, VP40, and VP24. For example, VP35 suppresses production of IFN-α by binding and sequestering viral RNA. During infection with RNA viruses, it is the job of the cellular protein RIG-I to detect viral RNA by binding double-stranded RNA present during genome replication. But VP35, because it binds to viral dsRNA, blocks the activities of RIG-I. Other mechanisms by which filoviruses antagonize innate immune responses include:

- EBOV VP35 inhibits the phosphorylation of IRF-3/7 by the TBK-1/IKKε kinases.

FIGURE 21.5 This image taken in Uganda (2012) shows these responders from the Ugandan Red Cross in the process of donning their personal protective equipment before responding to a reported Ebola death in a village. *From CDC. Public Health Image Library ID#:19916.*

- MARV VP40 inhibits STAT1/2 phosphorylation by inhibiting the JAK family kinases (Fig. 6.4).
- EBOV VP24 inhibits nuclear translocation of activated STAT1 by karyopherin-α.

The end result of these protein interactions is to strongly inhibit the expression of Type 1 IFNs.

The virulence of filoviruses in humans is highly variable and multifactorial. Strain or species of virus is of primary importance (for example, ZEBOV is highly virulent while Reston ebolavirus (REBOV) is on the very low end of the virulence spectrum for humans). Dose and route of infection also contribute to virulence and disease outcome. Higher fatality rates seem to be associated with parenteral infection, and a high-infecting dose is more likely to overwhelm immune responses. Of course host genetics, nutrition, overall patient health, and availability of supportive care all play important roles as well (Fig. 21.5).

In this chapter we have learned that:

- Filoviruses are filamentous-enveloped viruses with unsegmented negative-strand RNA genomes. Their overall replication strategy is similar to other viruses in the order *Mononegavirales*.
- Humans are not the natural hosts for filoviruses and many human infections result in development of severe disease. Viral hemorrhagic fever is caused by a number of factors including damage to vascular endothelium, disregulation of clotting factors, and release of excessive amounts of proinflammatory molecules.
- Multifocal organ damage is common, and is the direct result of filovirus replication.
- Filovirus virulence is associated with the ability to evade innate immune responses.
- The range of natural hosts of filoviruses has not been determined, but there is good evidence that it includes bats.
- Index cases of human filovirus infection are often associated with exposure to infected nonhuman primates (through hunting/butchering monkeys, for example) or bats. However most human cases result from exposure to body fluids of symptomatic patients.

CHAPTER

22

Family *Bornaviridae*

OUTLINE

After reading this chapter, you should be able to answer the following questions:

- What are the defining characteristics of members of the family *Bornaviridae*?
- In what ways are bornaviruses different from other viruses in the order *Mononegavirales*?
- How would you describe the interactions of bornaviruses with infected cells?
- How were avian bornaviruses discovered?
- What diseases are caused by bornaviruses?

The bornaviruses are enveloped, monopartite, negative-strand RNA viruses. The family *Bornaviridae* is in the order *Mononegavirales* and shares many basic characteristics with other viruses in the order. However bornaviruses have several unique traits. They replicate in the nucleus, use cellular splicing machinery to produce some mRNAs, and are highly cell-associated. They cause persistent infections with little or no cell pathology and are often highly associated with neurons in an infected host. Overall, bornaviruses are quite stealthy in their life styles and this is reflected in the history of the agent (Fig. 22.1 and Box 22.1).

Borna disease was first described in horses in the 18th century in central Germany. Large outbreaks among cavalry horses in the 19th century, near the city of Borna, were the basis of the name. In the early decades of the 20th century the infectious nature of the disease was confirmed using animal passage experiments. It would be another ~50 years before persistently infected cell cultures were established for the virus. However the study of the agent remained difficult due to lack of any obvious cell pathology. While persistently infected cultures provided infectious material, they did not provide sufficient cell-free virus for convincing biochemical or structural studies. (In addition, the rarity of the disease and the limited geographic area of outbreaks did not move this agent to the top of most researcher's "must study" list.) Thus it was not until 1994 that the agent was cloned and sequenced. Genome sequence and organization revealed the similarities of Borna disease virus (BoDV-1) to other unsegmented, negative-strand RNA viruses (Fig. 22.2).

GENOME ORGANIZATION

Bornaviruses have monopartite, negative-strand RNA genomes of ~9 kb. They encode six proteins (N, P, X, M, G, and L) and the gene order is reminiscent of other members of the order *Mononegavirales*. However the transcription strategy is unique in that splicing is used to generate some mRNAs. mRNA splicing is in line with the nuclear location of virus transcription.

Viruses. DOI: http://dx.doi.org/10.1016/B978-0-12-803109-4.00022-2

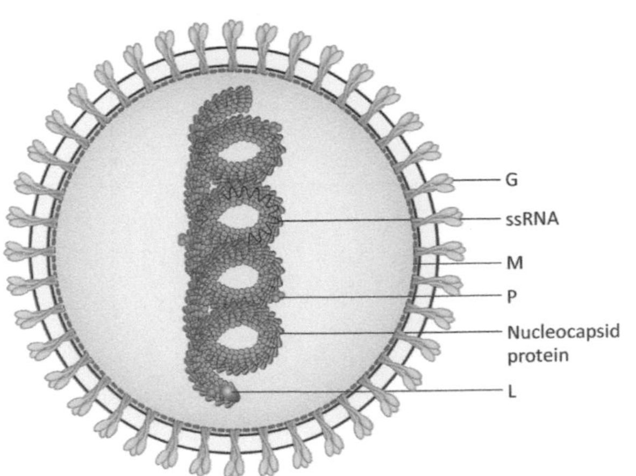

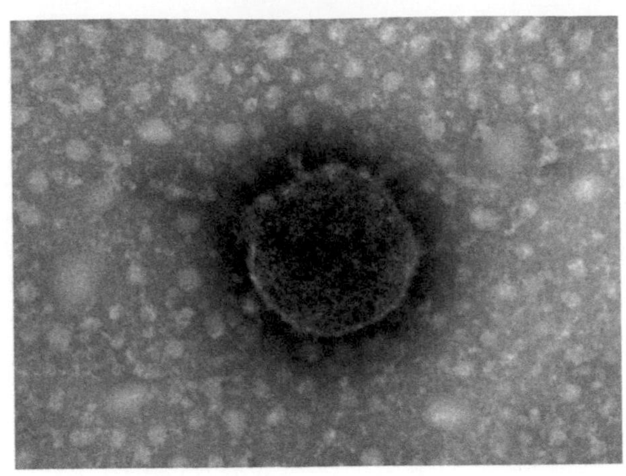

FIGURE 22.1 Virion structure (left). Virus-like particle (83 nm in diameter) from the eye fluid of an Eclectus parrot with confirmed PDD and ABV infection (right). The image was recorded with an FEI Morgagni 268 transmission electron microscope at a magnification setting of ×180,000. *Courtesy: Ross Payne.*

BOX 22.1

GENERAL CHARACTERISTICS

Bornaviruses have monopartite, negative-strand RNA genomes of ~9 kb. They encode six proteins (N, P, X, M, G, and L) and the gene order is similar to other members of the order *Mononegavirales*. Bornaviruses are unique among the nonsegmented negative-sense RNA viruses in that they replicate in the nucleus and splicing is used to generate some mRNAs.

Virions are enveloped. Based on their relationship to other monopartite negative-sense RNA viruses, the nucleocapsid is assumed to be helical, however, bornaviruses are in large part highly cell-associated thus it has not been possible to purify large numbers of virions for detailed structural analyses.

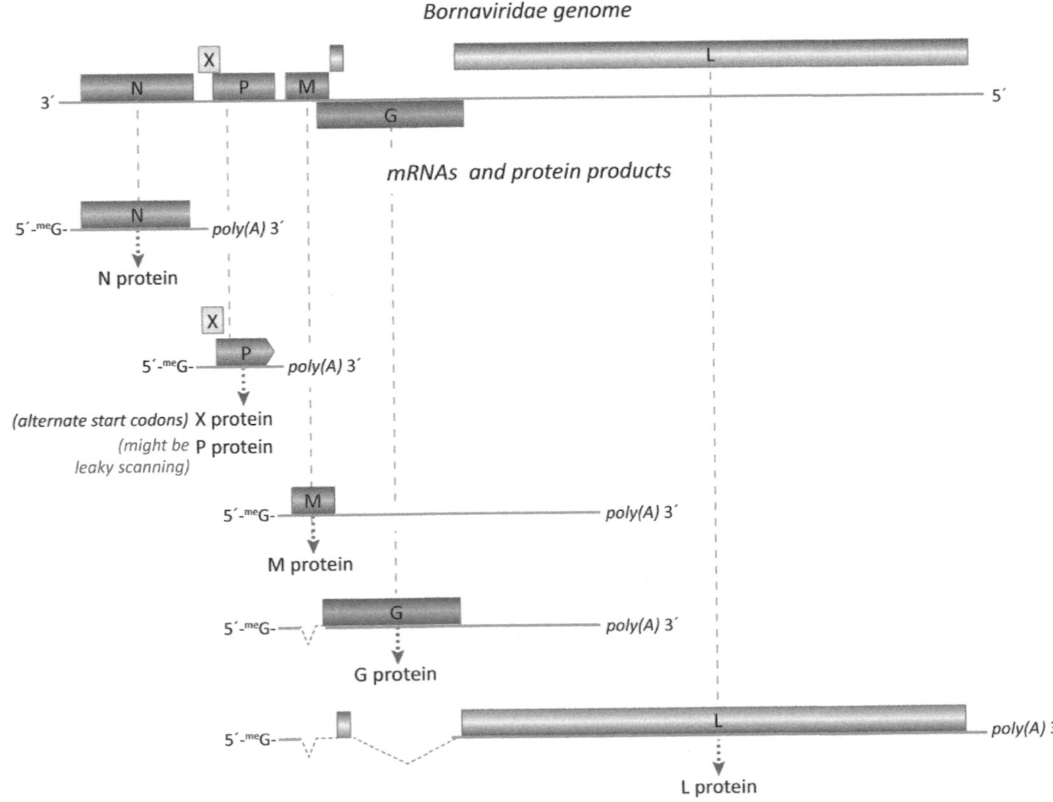

FIGURE 22.2 Genome organization.

As with other members of the order *Mononegavirales*, it is likely that there is a single promoter, near the 3′ end of the genome that drives initiation of transcription. Four transcription termination sites and 3 reinitiation sites have been mapped in the genome.

VIRION STRUCTURE

In contrast to the precise, detailed structures that are available for many viruses, details of bornavirus ultrastructure remain somewhat fuzzy due to the low amounts of complete virions produced during an infection. Bornavirions are enveloped, ~100–130 nm in diameter and are assumed to have a helical nucleocapsid (Fig. 22.1).

OVERVIEW OF REPLICATION

Attachment and Penetration

The bornavirus glycoprotein (GP) is the likely candidate for receptor attachment and subsequent fusion. Candidate host receptors have not been determined but is clear that they must be expressed on neurons. GP is also the best candidate for mediating fusion, through a hydrophobic domain present in the carboxyl terminal cleavage product, GP-C. Although the molecular details are not yet known, it is clear that the nucleocapsid (consisting of N, P, and L proteins and RNA) traffics to the nucleus.

Transcription

N, P, and L are required for transcription. It is likely that L caps and polyadenylates viral mRNAs, while cellular splicing machinery also processes some transcripts. The most abundant transcript is the mRNA encoding the N protein. mRNAs are transported back to the cytoplasm where they are translated.

Protein Synthesis

Bornaviral proteins are N, P, M, G and L (found in the virion), and X. N is the nucleoprotein and is found in two versions, p40 and p38, produced by use of alternative start codons. P is a phosphoprotein that serves as a cofactor for L. M is the matrix protein. G is the envelope glycoprotein and L is the catalytic subunit of the transcription/replication complex. In line with the nuclear location of RNA synthesis, N, P, and L proteins all have nuclear localization signals. The X protein is presumed to be nonstructural, with a role in RNA replication.

The bornavirus glycoprotein, GP, is translated on rough ER and posttranslationally modified by N-glycosylation to yield a 93–94-kDa primary product. A portion of G is cleaved to produce amino terminal (GP-N) and carboxyl terminal (GP-C) products. The virion may contain both cleaved and uncleaved glycoprotein products.

Genome Replication

Genome replication is a nuclear event and requires N, P, and L. Cloning of genomes from persistently infected cells revealed truncations at ends of the genome strand, removing promoters for genome replication and possibly transcription. It is speculated that this process leads to downregulation of replication/transcription and might play an important role in the persistent nature of bornavirus infections. Cleavage of the ends of full-length RNAs yields a 5′ monophosphate end, rather than the expected triphosphate, and this is a mechanism by which viral RNA can persist without activating innate immune responses. Recall antiviral IFN responses are strongly activated through retinoic acid-inducible gene-I recognition of 5′ triphosphates on RNA (see Chapter 6: Immunity and Resistance to Viruses).

Assembly, Release, and Maturation

Bornavirus nucleocapsids contain N and RNA, associated with P and L proteins. M is also part of the nucleocapsid. While the details are lacking, the process of nucleocapsid formation is presumed to be similar to other members of the order *Mononegavirales*. Nucleocapsid formation occurs in the nucleus and the RNP must be exported from the nucleus for the assembly of complete virions. In addition to transmission by extracellular virions, it is likely that nucleocapsids are also the transmissible particle from cell to cell (Fig. 22.3).

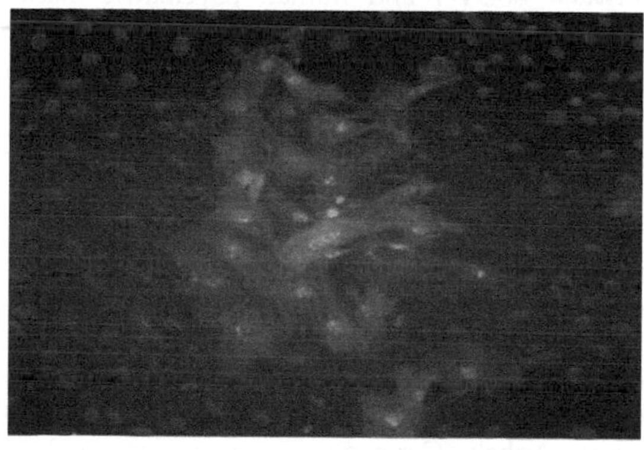

FIGURE 22.3 Fluorescent antibody labeling of duck embryo fibroblasts persistently infected an avian bornavirus. A focus of infected cells (green) sits amidst uninfected cells. The uninfected cells are visualized by DAPI staining (blue) of their nuclei.

This is most clearly seen in cultured cells where BoDV nucleocapsids associate tightly with chromatin to segregate into daughter cells during cell division.

DISEASES

Borna Disease

Prototypical Borna disease is a rare neurologic disorder of horses and sheep with outbreaks most often seen in central Germany. While the virus can be experimentally transmitted between horses and sheep, this is clearly not the natural route of infection. The natural hosts for Borna disease virus 1 (BoDV-1) are mammals (Box 22.2). The epidemiology of natural disease suggests transmission via infected urine or feces, though an olfactory route. Virus gains access to neurons and spreads by axonal transmission. BoDV-1 infection of neurons is noncytopathic, but persistent infection of an animal results in immunopathologic damage. Mammalian bornaviruses have also been recovered from other mammals with neurologic disease. In cats, infection with BoDV-1 is associated with staggering disease, a fatal neurologic condition (also see Box 22.3).

BOX 22.2

TAXONOMY

Order *Mononegavirales*
Family *Bornaviridae*
Genus *Bornavirus*

Species *Mammalian 1 bornavirus*
Species *Psittaciform 1 bornavirus*
Species *Passeriform 1 bornavirus*
Species *Waterbird 1 bornavirus*
Species *Passeriform 2 bornavirus*
Species *Psittacine 2 bornavirus*
Species *Elapid 1 bornavirus*

BOX 22.3

HUMAN INFECTION WITH BORNAVIRUSES?

Beginning in 1985 and extending over a period of about 25 years, investigators published studies linking a number of different human psychiatric syndromes to BoDV-1 infection. Evidence of infection was provided through serologic studies and use of polymerase chain reaction. In fact it was early reports of an association between BoDV-1 and human psychiatric disease that encouraged interest in the exact nature of the Borna disease agent. Unfortunately, even as the molecular and biochemical nature of the agent became clearer, the association between BoDV-1 and human psychiatric disease remained murky, as many negative studies were published. While an association between BoDV-1 and human psychiatric disease cannot be completely ruled out, there is no evidence that it is a common event.

However in 2015, a *novel* bornavirus was clearly linked to fatal, acute encephalitis in humans. The three victims were men (63, 62, and 72 years age) from the state of Saxony-Anhalt, Germany. The cases occurred between 2011 and 2013. The clinical course for all three men was that of a progressive encephalitis that ended in death. The men were screened for a number of infectious agents associated with CNS disease, but results were negative. An intriguing epidemiologic finding was that all three victims were breeders of the variegated squirrel (*Sciurus variegatoides*, a native of central America) and had exchanged squirrels on multiple occasions. Some of these squirrels had also succumbed to neurologic disease but again, initial screenings for suspect infectious agents were negative.

Eventually brain samples from the squirrels were examined by unbiased genome sequencing and sequences of a novel bornavirus were obtained from clinically affected animals. On revisiting stored human samples, a virus identical to the variegated squirrel virus was found. Variegated squirrel bornavirus is clearly distinct from BoDV-1 and was unknown before these studies. While the affected squirrels are native to Central America, it is not yet clear where, or how, the captive breeding stock was infected, as other mammalian bornaviruses are clearly endemic to Germany.

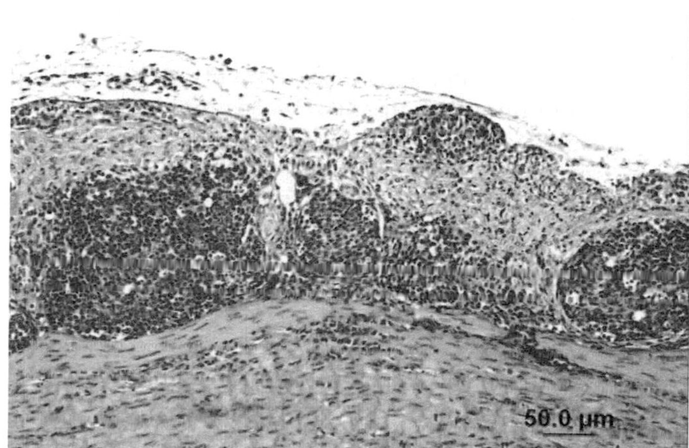

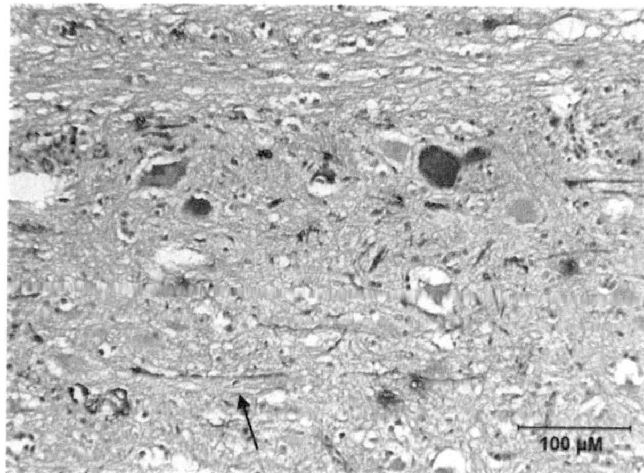

FIGURE 22.4 Microscopic lesions from a PDD bird, showing lymphocytic infiltration (dark purple cells) (left) and presence of bornavirus antigen (discrete areas of dark brown staining) in neurons (right).

The best-characterized animal model of BoDV-1 infection is the rat model, using specifically adapted strains of virus. In this model the outcome of infection depends on the genetic background, age of infection and immune status of the rat. Infection of adult, immunocompetent rats results in the development of encephalitis similar to that seen in natural infections. Early studies quickly demonstrated that disease is immune-mediated, with cytotoxic T-cells being the major players (through killing of infected cells). This is consistent with the pathology of natural infection, where microscopic examination of the brain reveals massive infiltration of lymphocytes.

In contrast, experimental infection of neonatal rats (which are not yet immunocompetent) leads to persistent infection, in the absence of acute disease. However infection is not without consequence as persistently infected rats are significantly smaller than mock-infected littermates and exhibit various behavioral, emotional, and cognitive impairments.

Birds and Bornaviruses

In the early 1970s veterinarians began reporting cases of a mysterious and deadly disease of captive parrots. The birds were eating, but were not able to digest their food and were wasting away. Some had severe neurological problems including incoordination, seizures, and blindness. The most obvious finding was *extreme* enlargement of a region of the birds' digestive tract called the proventriculus (akin to the esophagus in mammals). One of the common names for the disease is proventricular dilatation disease (PDD). The disease appeared to be caused by an infectious agent, but for over 30 years the cause of the disease remained a mystery despite numerous efforts to identify a culprit. Therefore no diagnostic tests were available. As many large parrots are highly endangered in their natural habitats, a deadly disease among those in breeding programs was devastating. In 2008 the mystery was solved using molecular techniques. Virologists using genomic-sequencing methodologies detected evidence of a bornavirus in the brains of affected birds (Box 22.2). More importantly, the viral sequences were not found in healthy birds.

Now it was obvious why the PDD "agent" had eluded detection for so long: It was a bornavirus that caused persistent, noncytopathic, highly cell-associated infections in cell cultures. Disease was primarily immune-mediated, with no evidence of virus particles in lesions. Once the initial identification of a bornavirus was made, immunological reagents developed for BoDV-1 proved useful for detecting the virus in cultured cells (Fig. 22.3) and for demonstrating that PDD birds have central nervous system (CNS) lesions containing abundant bornaviral proteins (Fig. 22.4). Most cases of PDD are seen in captive parrots but infection and (rarely) disease among wild waterfowl in the United States and Canada are also well documented. It should be noted that a high percentage of infected birds are completely asymptomatic. It is not yet known to what extent bornavirus affects wild bird populations.

In this chapter we learned that:

- Members of the family *Bornaviridae* have unsegmented, negative-sense RNA genomes that encode six proteins (N, P, X, M, G, and L). Virions are enveloped.

- Transcription of bornaviral mRNA and genomes is nuclear.
- Infections are noncytopathic and persistent, both in cultured cells and animals.
- Neonatal rats become persistently infected and immunotolerant after experimental infection. They display developmental, emotional, and cognitive impairments.
- In natural infections disease involves the CNS and is immune-mediated. BoDV-1 is highly neurotropic while avian bornaviruses seems to have a wider cell tropism.

CHAPTER

23

Family *Orthomyxoviridae*

OUTLINE

After reading this chapter, you should be able to discuss the following:

- What are the general characteristics of the orthomyxoviruses?
- What are the functions of HA?
- What is the function of neuraminidase (NA)?
- Describe genome variation and host range of influenza A viruses.
- Describe the naming scheme used for influenza A, B, and C viruses.
- What is antigenic drift? What are the consequences of influenza virus antigenic drift?
- What is antigenic shift? What are the consequences of influenza virus antigenic shift?
- Describe synthesis of influenza virus mRNAs. How does mRNA synthesis differ from genome synthesis?
- What is the function of the influenza A virus M2 protein?
- What classes of drugs are used to treat influenza virus infection? (Which viral proteins do they target?)

The family *Orthomyxoviridae* contains pathogens of humans and animals. Virions are spherical to filamentous, about 100 nm in diameter (Fig. 23.1). The family contains five genera, the best-described being *Influenzavirus A*. Influenza A virus is quite common and infects a wide variety of mammals and birds. Influenza A virus is often restricted to the respiratory or gastrointestinal tracts but the highly pathogenic avian influenza viruses (HPAIV) cause severe systemic infections of domestic poultry. Influenza A virus is genetically diverse, as a consequence of accumulating and tolerating point mutations, wide host range and the ability to reassort genome segments. In the early 20th century (1918–19) a particularly lethal strain of influenza A virus caused an estimated 40 million human deaths worldwide. Although a pandemic of

Viruses. DOI: http://dx.doi.org/10.1016/B978-0-12-803109-4.00023-4

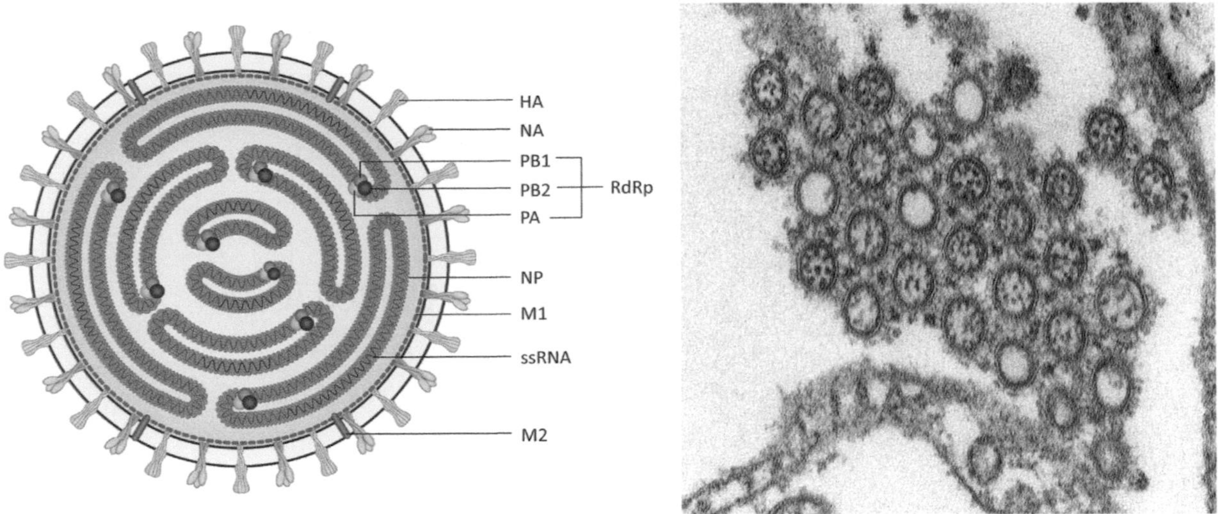

FIGURE 23.1 Virion structure (left). Transmission electron microscopic image of influenza virions (right). *From CDC/Cynthia Goldsmith. CDC Public Health Image Library Image ID#11746.*

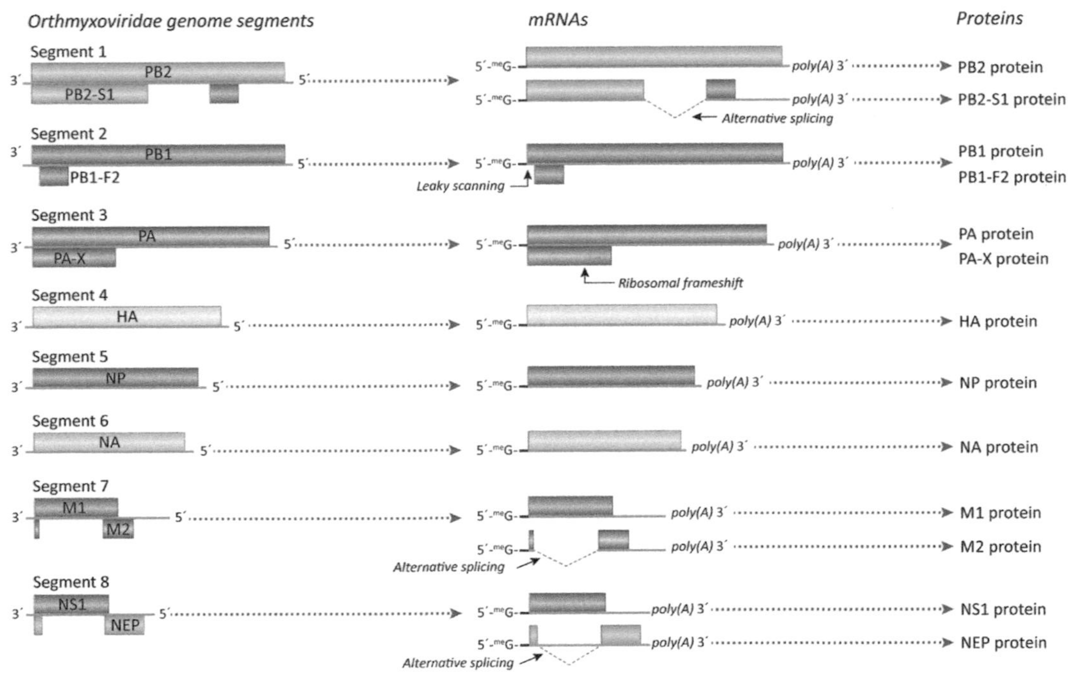

FIGURE 23.2 Example of orthomyxovirus genome organization and gene expression strategy.

that magnitude has not occurred since, epidemics and pandemics occur with regularity, assuring that influenza A viruses are carefully monitored and studied.

GENOME STRUCTURE/ORGANIZATION

Members of the family *Orthomyxoviridae* have segmented, single-stranded negative-sense RNA genomes (Fig. 23.2). Influenza viruses A and B as well as isavirus have eight genome segments, influenza virus C has seven segments and thogotovirus has six segments. Genome segments form helical nucleocapsids that contain N protein and an active polymerase (RNA-dependent RNA polymerase, RdRp) complex. Most genome segments encode a single protein; however, there is some variation (i.e., some influenza A viruses encode up to 11 proteins). The three largest genome segments (~2300 nt each) encode proteins that form the replicase/transcriptase complex. The smallest segment (~900 nt) usually encodes one or two nonstructural proteins. The nucleoprotein (NP), matrix (M1) protein, and glycoproteins are encoded by individual genome segments.

VIRION STRUCTURE

Members of the family *Orthomyxoviridae* are enveloped viruses containing 6–8 helical nucleocapsids. Virions are spherical to filamentous, about 100 nm in diameter with glycoprotein spikes that extend 10–14 nm from the outer surface of the envelope. Structural proteins include NP (forms the scaffold for RNA), M1 (forms a bridges between nucleocapsids and envelope), an integral membrane protein (M2) and one or two envelope glycoproteins that appear as spikes extending from the surface of the virion. A molecule of polymerase complex (assembled from proteins PA, PB1, and PB2) is associated with each genome segment. In transmission electron micrographs, the individual structures formed by each genome segment can be clearly identified (Box 23.1).

Members of the genera *Influenzavirus A* and *Influenzavirus B* (Box 23.2) encode two glycoproteins. Hemagglutinin (HA) forms homotrimer spikes extending from the virion surface. HA serves as both the attachment protein and the fusion protein and is a major target of neutralizing antibodies. As suggested by the name, HA binds sialic acid residues on glycoproteins and agglutinates red blood cells. NA spikes are homotetramers. NA is a *receptor-destroying* enzyme that cleaves sialic acid residues from cellular and viral glycoproteins. NA plays a key role in virion release from the infected cell. In contrast, members of the genus *Influenzavirus C* encode a single major envelope glycoprotein called hemagglutinin-esterase-fusion (HEF). HEF combines the functions of HA and NA. Thus influenza C viruses have only seven gene segments. The glycoprotein spikes of the isaviruses (Box 23.2) divide up their essential functions somewhat differently. Isaviruses encode two glycoproteins, hemagglutinin acetylesterase and F. The hemagglutinin acetylesterase binds sialic acid residues and hydrolyzes the acetyl group (thus has receptor binding and receptor-destroying activities). The second glycoprotein is the fusion (F) protein and as its name suggests, is essential for membrane fusion. Thogotoviruses are unique among orthomyxoviruses as they are arboviruses (transmitted by ticks). They encode a single envelope glycoprotein, G that serves in attachment and fusion. Surprisingly thogotovirus G shares 30% amino acid sequence identity with an envelope glycoprotein of baculoviruses, DNA viruses that infect a variety of insects.

CLASSIFICATION AND NOMENCLATURE OF ORTHOMYXOVIRUSES

There are currently five recognized genera in the family *Orthomyxoviridae* (Box 23.2). Three of these, *Influenzavirus A*, *Influenzavirus B*, and *Influenzavirus C*, infect humans. Influenza A virus is a major human pathogen and infects other mammals and birds. In mammals disease is usually respiratory. Influenza B virus is primarily a human respiratory virus, but has also been isolated from seals. Influenza B viruses do not infect birds and they usually cause milder human infections than influenza A viruses. Influenza C virus also infects humans. In fact, most human adults have antibodies to influenza C virus suggesting that it is a widespread, common, and probably very mild upper respiratory tract virus.

The influenza A viruses are significant human and animal pathogens. They exhibit extreme genetic

BOX 23.1

GENERAL CHARACTERISTICS

Influenza virus genomes are segmented, single-stranded, negative-sense RNA. Genomes contain 7 or 8 segments that range in size from 2.3 to 0.9 kb. Segmented genomes lead to high rates of reassortment. Transcription and genome replication are nuclear. mRNAs are primed by cap-snatching and are polyadenylated. Some viral mRNAs are spliced. Most RNA segments encode a single protein.

Virions are enveloped with helical nucleocapsids. The envelope displays two transmembrane glycoproteins and also contains an integral membrane ion channel protein. Virion shape is spherical to filamentous. The RNA-binding NP protein is the most abundant in the virion. A polymerase complex (PB1, PB2, and PA) is associated with each RNA segment.

BOX 23.2

FAMILY ORTHOMYXOVIRIDAE

Genus: *Influenzavirus A* (Species *Influenza A virus*)
Genus: *Influenzavirus B* (Species *Influenza B virus*)
Genus: *Influenzavirus C* (Species *Influenza C virus*)
Genus: *Thogotovirus*

Species: *Dhori virus*
Species: *Thogotovirus*

Thogotoviruses are arboviruses transmitted by ticks. A wide range of mammals and birds are infected. THOV causes disease in livestock. THOV, DHOV, and Bourbon virus can infect humans, and have occasionally been associated with human disease.

Genus: *Isavirus*
Species: *Infectious salmon anemia virus*

The etiological agent of infectious salmon anemia is the infectious salmon anemia virus.

Genus: *Quaranjavirus*

Species: *Johnston Atoll virus*
Species: *Quaranfil virus*

(Isolated from large disease outbreaks in birds).

BOX 23.3

KEEPING A CLOSE WATCH

Given the history of repeated epidemics and pandemics, influenza virus surveillance is a global health imperative. The WHO sponsors a Global Influenza Program (GIP) to collect and analyze influenza virus surveillance data from around the world. This allows the WHO to monitor global trends in influenza virus transmission and to support the selection of strains for vaccine production. In the United States, the Epidemiology and Prevention Branch in the Influenza Division at the CDC coordinates a robust influenza virus surveillance program. The CDC produces weekly reports (FluView and FluView Interactive), available at the CDC website (www.cdc.gov). The US surveillance system is a collaborative effort between CDC and state, local and territorial health departments, public health and clinical laboratories, and healthcare providers. The program includes documenting and reporting when and where influenza activity is occurring, tracking influenza-related illness and deaths, identifying what viruses are circulating and perhaps most critically, detecting changes in circulating influenza viruses that may foreshadow an epidemic or pandemic. There are ~350 laboratories in the United States that report (weekly) the total number of respiratory specimens tested and the number positive for influenza virus.

Cataloging and tracking influenza viruses generate a lot of information and it is imperative that each virus isolate be clearly named. In the case of human influenza viruses, names include the following critical information:

- Genus (Type A, B, or C),
- location the isolate was collected and the isolate number,
- year of isolation,
- HA and NA subtypes.

diversity and a wide host range. They tolerate (and benefit from) a high rate of point mutations and also undergo genome reassortment. A complex nomenclature system has been developed to track them (Box 23.3). Genome sequencing reveals that each genome segment has subtypes. For example, among influenza A viruses there are 16 known HA subtypes (H1–H16) and 9 known NA subtypes (N1–N9). All NA and NA subtypes have been found among birds; the most common HA and NA subtypes that circulate in humans are H1, H2, and H3 and N1 and N2, respectively. Human HA and NA subtypes can be found in various combination (i.e., H1N1, H2N1, H3N2). However, important genetic and antigenic variation also exists within a subtype, thus all H1N1 viruses (even those that share all other segments) are not created equal. So names of influenza A viruses include information recording exactly when and where a particular virus was isolated (if that information is known). In the case of human influenza viruses, names include the following information:

- Genus (*Influenzavirus*) A, B, or C,
- location the isolate was collected and the isolate number,
- year of isolation,
- HA and NA subtypes.

Thus "A/Texas/1/79 (H3N2)" denotes an influenza A virus isolated in Texas in 1979. It was the first isolate of the year and the HA and NA subtypes were H3 and N2. When an influenza virus is isolated from an animal (other than a human) the species is added to the name.

OVERVIEW OF REPLICATION

This section will focus almost exclusively on influenza A viruses as we walk through a general replication cycle (Fig. 23.3). Much detailed information is available for each viral protein so we will begin the discussion by describing their major functions.

Hemagglutinin

HA is a transmembrane glycoprotein critical for both attachment and penetration (it has membrane fusion activity). HA spikes are assembled as homotrimers in the viral envelope. HA binds to sialic acid

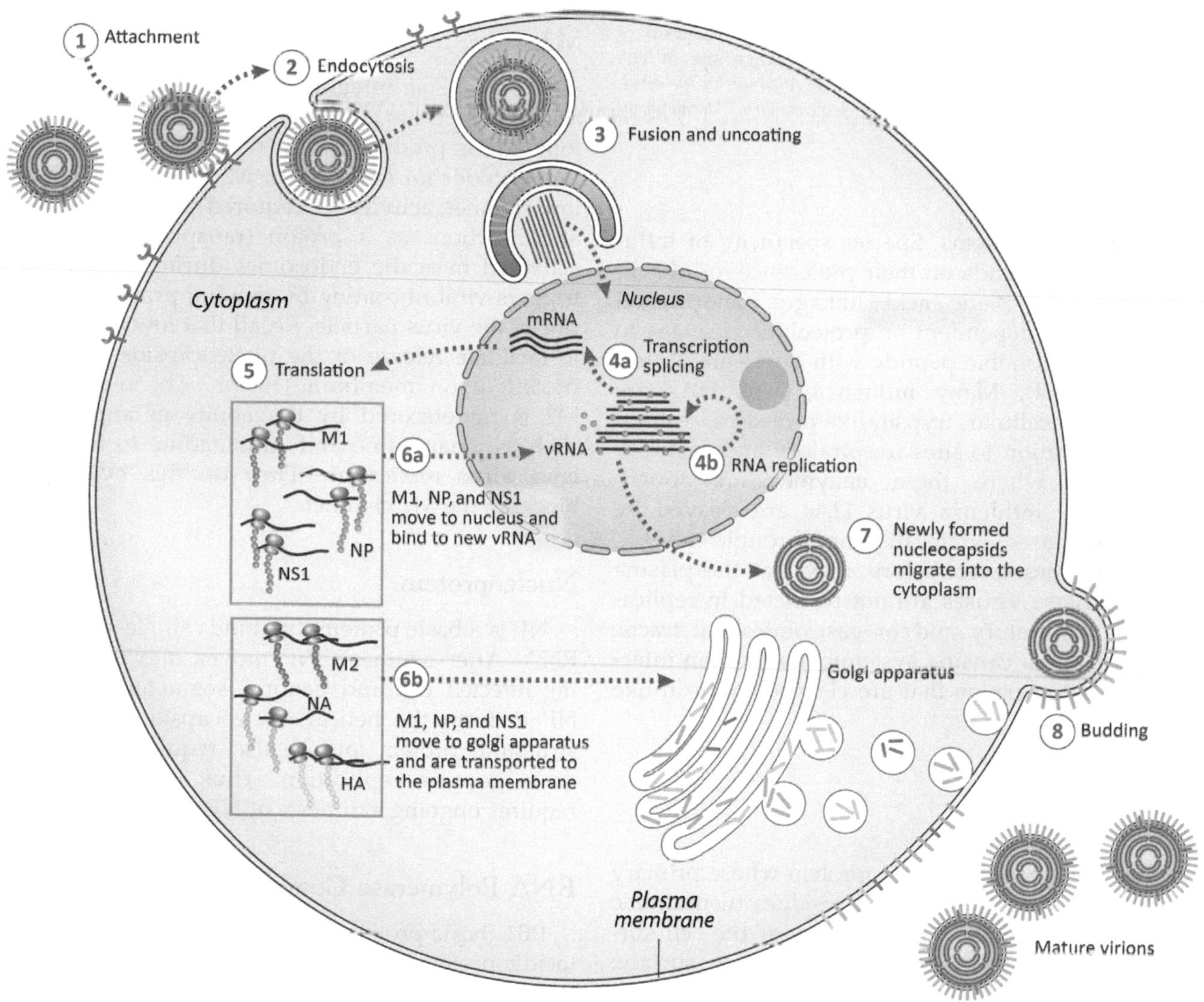

FIGURE 23.3 General replication cycle of influenza virus. After release from endosomes, the nucleocapsids move into the nucleus. Transcripts are produced by viral RdRp in the nucleus. mRNA moves into the cytoplasm. After translation some viral proteins must traffic back into the nucleus to support genome replication and nucleocapsid assembly. Nucleocapsids move back into the cytoplasm and traffic to the plasma membrane, the site of budding.

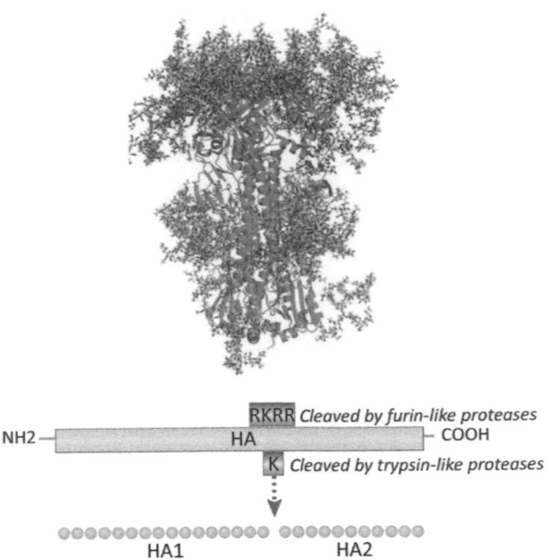

FIGURE 23.4 HA structure and cleavage. Ribbon diagram of HA molecule, showing glycosylation (top). Cleavage site of HA (bottom). *From Krammer, F. and Palese, P. (2015) Advances in the development of influenza virus vaccines. Nature Reviews Drug Development 14:167–182.*

residues on glycoproteins. Species specificity of influenza A viruses depends on their preference for attaching to particular sialic acid linkages. The fusion activity of HA is dependent on proteolytic cleavage to generate a hydrophobic peptide with a free amino terminus (Fig. 23.4). Many influenza virus HAs are cleaved by extracellular, trypsin-like proteases, restricting virus replication to sites (respiratory and gastrointestinal tract) where these enzymes are found. However some influenza virus HAs are cleaved by furin-like proteases while moving through the ER, thus are fully processed before arrival at the plasma membrane. These viruses are not restricted to replication in the respiratory and/or gastrointestinal tracts. HPAIV, capable of causing systemic, multiorgan infections, have HA proteins that are cleaved by furin-like proteases.

Neuraminidase

NA is a transmembrane glycoprotein whose primary function is cleavage of sialic acid residues (neuraminic acid) from proteins and carbohydrates at the cell surface. NA forms homotetramers on the virion surface. NA functions late in the influenza virus replication cycle, as a receptor-destroying enzyme. NA allows release of virions from the plasma membrane and limits their aggregation. Inhibiting NA activity does not prevent virion formation but does limit spread.

Matrix Protein (M1)

M1 is a major structural protein that serves as a bridge between nucleocapsids and the virion envelope. After synthesis, most M1 traffics to the plasma membrane where it binds the cytoplasmic tails of HA and NA. As virions are being assembled, nucleocapsids are transported from the nucleus to the cytoplasm where they interact with M1 (thus bringing them to budding sites). Key to M1 function is the pH dependence of its interaction with nucleocapsids. M1 binds nucleocapsids at neutral pH to facilitate virion assembly at the plasma membrane. However, early in the infection process (during entry) the low pH of the endosome disrupts M1 nucleocapsid interactions. This allows the nucleocapsids to move into the nucleus, the site of transcription and genome replication.

M2

M2 is a 97 aa integral membrane protein that forms acid activated ion channels in the influenza virus envelope. M2 is produced from the same genome segment that encodes for M1, but the M2 mRNA is spliced. M2 ion channel activity is required for virus infectivity and functions as a proton transporter. M2 becomes activated in acidic endosomes during viral entry and triggers viral uncoating by moving protons to the interior of the virus particle. Recall that low pH is required to facilitate release of the nucleocapsids from the M1 protein upon membrane fusion. The requirement of M2 is underscored by the ability of antiviral drugs such as amantadine and rimantadine to inhibit influenza virus replication. They do this by specifically blocking the M2 channel.

Nucleoprotein

NP is a basic protein that binds single-stranded viral RNA. After synthesis NP moves into the nucleus of the infected cell and genome segments wrap around NP to form the helical nucleocapsid. NP is a major structural protein and is also required for influenza virus genome replication. Thus genome replication requires ongoing synthesis of NP in the infected cell.

RNA Polymerase Complex

PB1 (basic protein 1), PB2 (basic protein 2), and PA (acidic protein) assemble to form the active polymerase (RdRp), ~250 kDa in size. RdRp associates with NP bound genome segments to form nucleocapsids packaged into virions. Thus early in infection, when nucleocapsids traffic into the nucleus, they are transcriptionally active. PB1 contains the catalytic domain

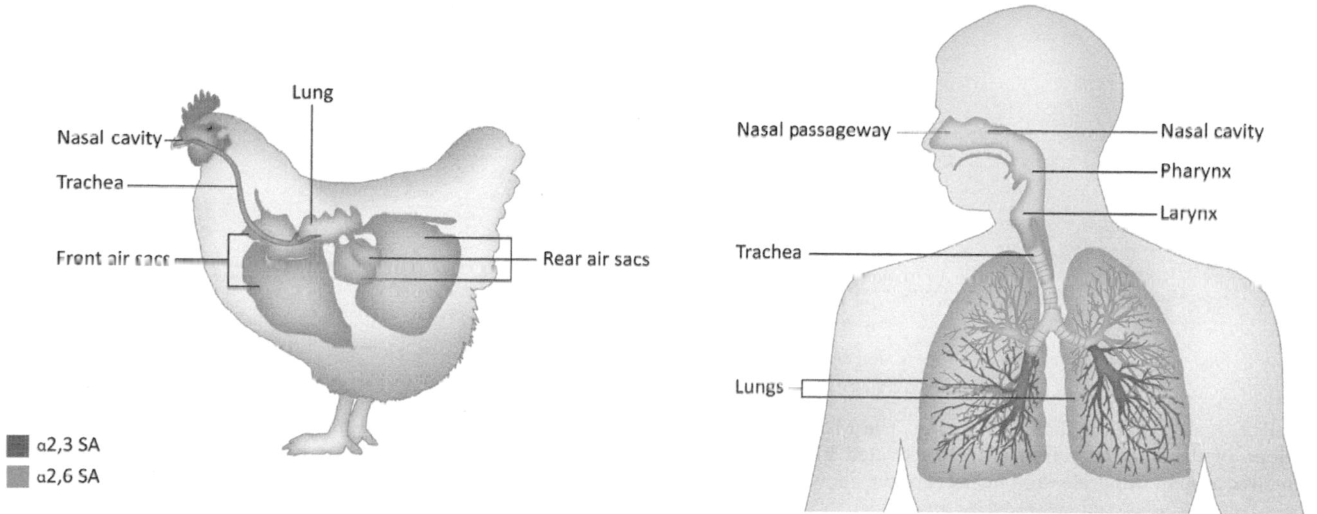

FIGURE 23.5 Avian flu transmissibility and virulence. Avian flu is poorly transmitted to humans as it does not bind A2,6 SA in the upper respiratory tract. However, if virus gets into the lung, A2,3 SA receptors support replication.

active in chain elongation, PB2 binds methyl guanosine caps of host mRNAs, and PA has endonuclease activity. After early rounds of transcription and translation, newly synthesized PB1, PB2, and PA move to the nucleus to initiate genome replication.

Nuclear Export Protein

Nuclear export protein (NEP) is nonstructural protein translated from a spliced mRNA produced from the same genome segment that codes for NS1 (discussed later). After synthesis NEP moves into the nucleus, the site of nucleocapsid assembly where it associates with nucleocapsids (and cellular proteins) to facilitate transport of RNPs from the nucleus.

Nonstructural Protein 1

Nonstructural protein 1 (NS1) is encoded from the smallest genome segment (segment 8 of influenza A virus). NS1 is abundant in infected cells but is not found in virions. NS1 functions include: (1) inhibiting release of cellular mRNA from the nucleus; (2) inhibiting cellular mRNA splicing; (3) inhibiting the interferon pathway.

Attachment/Penetration

HA spikes are critical for both attachment and fusion. HA binds sialic acid residues on cell surface glycoproteins and glycolipids. Given that sialic acid is a ubiquitous cell surface molecule it somewhat surprising that HA binding strongly influences host susceptibility. However the linkage of the terminal sialic acid to the penultimate galactose in the carbohydrate chain specifies HA binding. Human influenza viruses preferentially bind to SAα2,6Gal while avian viruses preferentially bind to SAα2,3Gal. As we might guess, SAα2,6Gal is common in the upper respiratory tract of humans and SAα2,3Gal decorates carbohydrate chains in the avian respiratory and digestive tracts (Fig. 23.5). However, SAα2,3Gal is also found on cells in the lower respiratory tract of humans and this might partially explain why some HPAIV are very poorly transmissible to humans, yet cause severe disease if an infection becomes established. It also has been demonstrated that a single amino acid substitution can suffice to change HA-binding specificity.

Binding triggers endocytosis and within endosomes acidification results in rearrangement of HA to expose its fusion domain, thereby mediating fusion of virion and endosomal membranes. (Recall that HA must be cleaved in order to be active as a fusion protein. Cleavage occurs prior to virus entry.)

For many years it was thought that the fusion process alone released nucleocapsids to traffic to the nucleus. However we now understand that before fusion, the *inside* of the virion must be exposed to low pH. Acidification of the virion requires the ion channel protein, M2. M2 channel activity is itself activated by low pH in the endosome and then allows hydrogen ions into the interior of the virion (Fig. 23.6). Only after exposure to low pH are nucleocapsids released from their tight interactions with M1. Thus inactivation of the M2 channel blocks influenza virus replication and this is the mechanism of one class of antiinfluenza drugs (adamantanes).

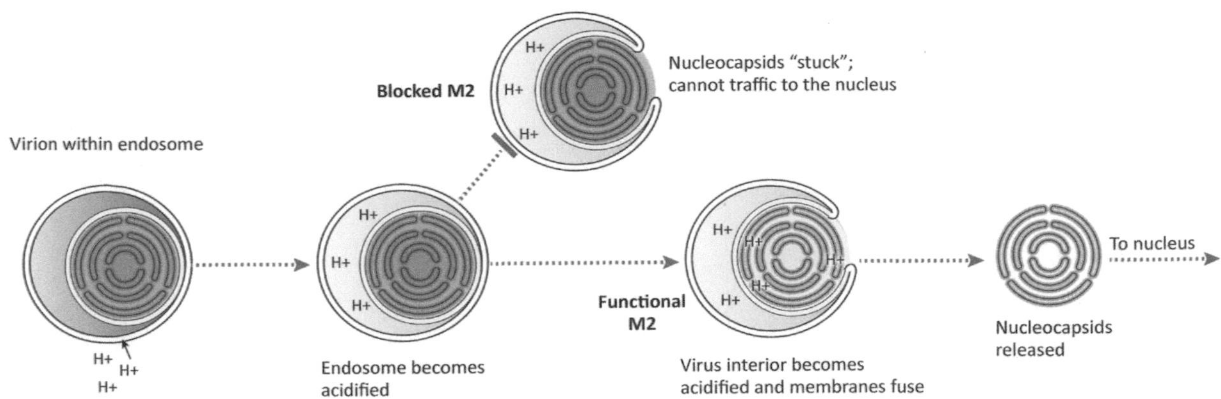

FIGURE 23.6 M2 ion channel function. The M2 protein is a membrane-associated, low pH activated ion channel. M2 is present in the envelope of the infecting virion. M2 is activated in the endosome allowing protons to move into the interior of the virion. At low pH, the matrix protein (M) "releases" nucleocapsids, allowing their transit to the nucleus.

Transcription

Nucleocapsids are actively trafficked to, and imported into the nucleus. As with other negative-strand RNA viruses (i.e., order *Mononegavirales*), transcription and genome replication are discrete processes. In the case of the orthomyxoviruses, transcription depends on an *exogenous primer* that is derived from cellular mRNAs in the nucleus. RdRp binds to a newly synthesized, capped cellular mRNA and cleaves it ~10–13 nt downstream from the cap. This capped oligonucleotide remains associated with the RdRp to serve as the primer for synthesis of an influenza virus mRNA. This process is called "cap-snatching" and the result is that the 5′ ends of influenza virus mRNAs are heterogeneous in sequence and are not exact copies of genome segments. (The need for capped cellular RNAs explains the requirement for active host RNA polymerase II to support influenza virus replication.) Prior to reaching the very 3′ end of the genome segment, the influenza virus mRNA is polyadenylated by the RdRp. Some mRNAs (notably products of the NS and M genome segments) are alternatively spliced. Viral mRNAs are exported from the nucleus.

Translation

Translation of influenza virus NP, M1, PA, PB1, and PB2 are cytosolic. The majority of the NA, PA, PB1, PB2 proteins, and some M1 protein, traffic back into the nucleus (recall that the genome of the infecting virus is in the nucleus). Translation of M2, HA and NA are on rough ER. HA and NA are glycosylated in the ER and Goli and are transported to the plasma membrane. HA is produced as a precursor (HA0) that must be cleaved to produce fusion active, disulfide-linked HA1 and HA2. HA0 is sometimes cleaved in the ER but is usually cleaved by extracellular proteases. Most M1 protein traffics to the PM where it associates with the cytoplasmic tails of HA and NA, forming sites for virion assembly.

Genome Replication

Before any new virions can be assembled, genomes must be replicated. Genome replication is a process very different from mRNA synthesis, although the same polymerase complex (PA, PB1, PB2) is used. The switch from transcription to genome replication is triggered by the presence of newly synthesized, unbound (to RNA) NP, PB1, PB2, and PA that have moved into the nucleus after translation. Replication initiates at the very 3′ end of each genome segment to produce an exact end-to-end copy (Fig. 23.7). The copy RNA strands are neither capped, nor polyadenylated but they are bound to NP. Each cRNA serves as the template for synthesis of many genomes.

Assembly, Budding, and Maturation

As genomes are synthesized, they associate with NP and the polymerase complex to form nucleocapsids. Also critical to the process is M1. Nuclear M1 may facilitate assembly of nucleocapsids and is required for their export from the nucleus. Nuclear export also requires the viral NEP and host cell proteins. NEP interacts with the cellular nuclear export machinery and with M1. Nucleocapsids traffic to the PM where they interact with M1 at that location. A major question in influenza virus biology is how seven or eight distinct genome segments are assembled into each virion. There is evidence of selective packaging that is regulated by RNA sequences unique to each genome segment. However there is also evidence of random

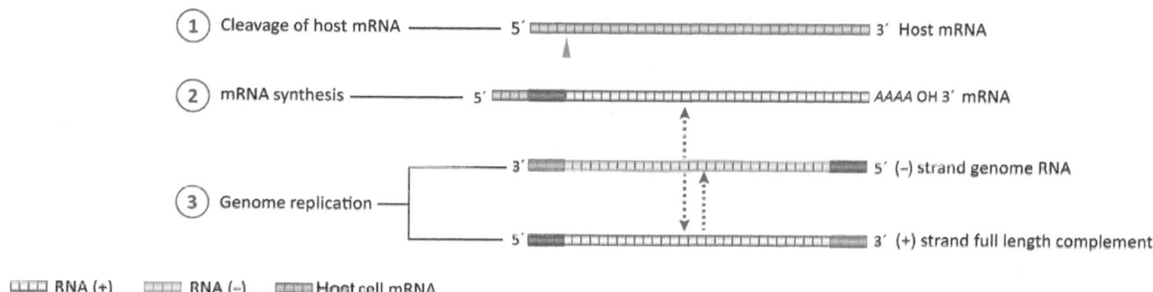

FIGURE 23.7 Comparison of processes to produce genomes and mRNAs. Genome replication produces an exact end-to-end copy of each genome segment. mRNAs are primed with a fragment of cellular mRNA. Transcripts terminate and are polyadenylated at sites upstream of the 5′ end of each genome segment.

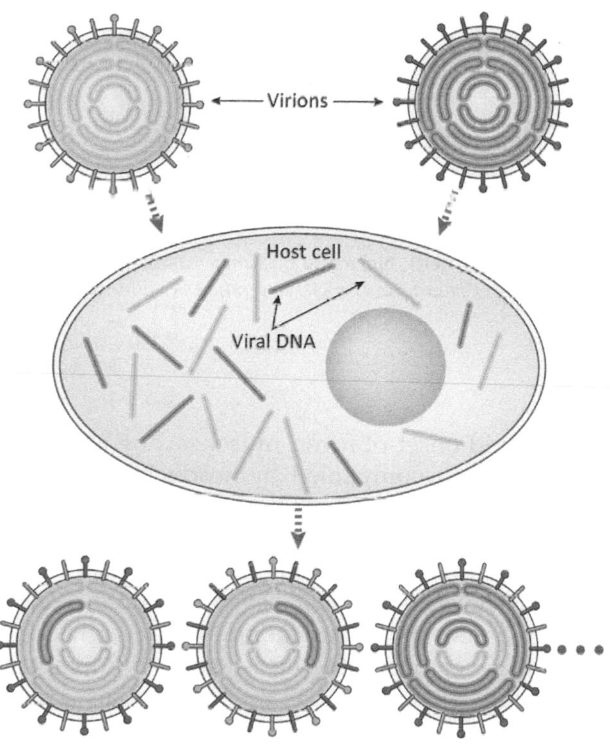

FIGURE 23.8 Reassortment of genome segments can occur in cells infected with more than one type of influenza virus. This is a powerful mechanism for generating genome diversity and leads to emergence of new strains.

packaging, suggesting that the selection mechanism may be "sloppy." It is clear that if a cell is infected with two different influenza viruses, individual virions can contain segments from both parents (a process called reassortment, Fig. 23.8).

Virions bud from the PM and release requires NA. Recall that NA cleaves sialic acids from viral and host glycoproteins at the PM to release virions from the cell surface. In the absence of NA activity, large virion aggregates are seen at the cell surface (Fig. 23.9). These aggregates are less transmissible to new hosts. An important class of antiviral drugs inactivates NA.

For most influenza viruses, production of infectious virions requires an extracellular maturation event, cleavage of HA0 by extracellular trypsin-like proteases. These enzymes are found in the respiratory and digestive tracts, thus limiting the productive sites of virus replication. When growing influenza viruses in cultured cells, the virions must usually be exposed to trypsin to generate infectious particles. However, for some influenza viruses (e.g., HPAIV), HA0 is cleaved during its journey through the ER, by furin-like proteases. The amino acid sequence at the cleavage site determines the type of protease required. HPAIV have broad tissue tropism and can cause multiorgan, systemic infection in susceptible birds.

ANTIGENIC DRIFT AND ANTIGENIC SHIFT

Genetic variation enables RNA viruses to exploit variable and changing environmental conditions that include (among other things) genetics, immune status, and life style of potential hosts (Fig. 10.6). Influenza A viruses exploit numerous hosts, but key among them are aquatic birds. All known HA and NA subtypes have been found in aquatic birds, while a limited number are found in mammals.

Two types of mutations are key to the success of influenza viruses. The first are simple nucleotide substitutions (point mutations) that arise during genome replication. In the context of human influenza viruses, the point mutations of most consequence are those that allow escape from host immune responses. These are most often seen for HA and NA, as these surface glycoproteins are key targets of the adaptive immune response. The term *antigenic drift* is often used to describe point mutations in HA and NA (the term drift implying a slow, continuous process). Antigen drift in influenza A viruses is

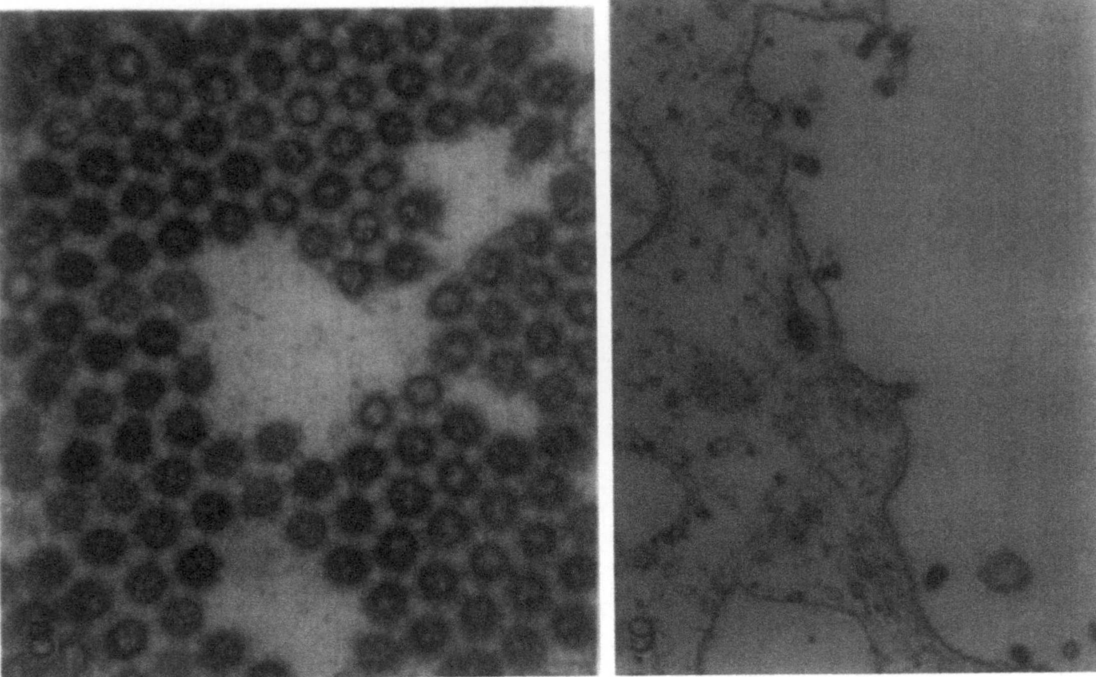

FIGURE 23.9 A packed array of influenza virions when neuraminidase is inactivated (left). Normally budding virions upon addition of exogenous neuraminidase (right). *From Palese, P., Tobita, K., Ueda, M., Compans, R. (1974) Characterization of Temperature Sensitive Influenza Virus Mutants Defective in Neuraminidase Virology 61:397–410*

the process that mandates ever changing vaccines, because a protective immune response against last year's flu virus does not provide complete protection against next year's flu virus. How can the HA and NA proteins tolerate point mutations given their important functions in the virus replication cycle? HA and NA both have unstructured regions (loops) of amino acids that are good targets for neutralizing antibodies. In contrast functional domains of HA (attachment domain, cleavage domain, fusion domain) and NA (enzyme active site) are less accessible targets of the immune response.

Genome reassortment (*antigenic shift*) provides for larger, more sudden changes in influenza viruses. Given that there are eight genome segments, infection of a cell with two different influenza viruses could yield up to 254 possible reassorted viruses, plus the two infecting types. New NA and HA combinations can generate a virus for which large segments of the human population have little preexisting immunity. Thus the emergence of "shifted" viruses well-adapted to humans cause of worldwide epidemics (pandemics).

DISEASE

In humans, influenza (the flu) is a mild to severe respiratory tract disease whose common symptoms include rapid onset of fever, muscle aches, and fatigue. These may be accompanied by headache, runny nose, sore throat, or (occasionally) gastrointestinal tract symptoms. An indication of influenza virus infection (as opposed to a head cold) is the rapid onset of aches, pain, and fever. The most severe symptoms usually last only a few days, but fatigue and cough can last longer. In severe cases the virus causes a primary viral pneumonia. Secondary bacterial infections are also common.

Transmission occurs via respiratory droplets. Often these contaminate our hands and are deposited on other objects. Thus a common pathway of infection is touching contaminated objects and then your nose or eyes. Frequent and thorough hand washing (soap and water) limits the spread of influenza virus and should be practiced by those who are sick and those wishing to remain healthy.

Mammals other than humans catch influenza. For example, influenza A viruses infect pigs, dogs, and horses. Pigs are an important host as they can be infected by both mammalian and avian influenza (AI) viruses, providing a source of novel reassortants. Influenza virus infection can cause considerable economic loss, thus pork producers use biosecurity measures to keep their animals safe from infected humans and birds.

Avian Influenza

Migratory aquatic birds are the natural hosts for influenza A viruses (mammalian influenza viruses likely originated in birds). When AI is in the news, stories are most often describing outbreaks among domestic poultry, particularly chickens and turkeys. In commercial flocks, mortality can be high and outbreak control is often accomplished by killing millions of birds.

AI is defined as low pathogenic (LP) or highly pathogenic (HP) for domestic poultry (Fig. 23.10). Low pathogenic avian influenza viruses (LPAIV) replicate primarily on mucosal surfaces of the respiratory and/or gastrointestinal tracts and disease is generally mild, with low mortality. In contrast, HPAIV replicates both on mucosal surfaces and in internal organs. These systemic infections cause severe disease with mortality exceeding 75%. Until relatively recently outbreaks of HPAIV were sporadic, regional events. However that changed with the emergence of the so-called Asian lineage of H5N1 HPAIV. This virus emerged in southern China in 1996 and eventually spread across Asia to Europe and Africa. The first outbreak to gain international attention occurred in 1997 in the live bird markets of Hong Kong. In addition to avian cases, there were 18 documented human cases, with 6 deaths. The human cases resulted from direct contact with infected birds. The Hong Kong outbreak was controlled by the closure of live bird markets and slaughter of ~1.5 million birds. However the virus continued to circulate in southern China and in 2002 another outbreak in Hong Kong involved wild birds and birds in a zoological collection. Thus the highly pathogenic H5N1 virus was clearly not confined to domestic poultry.

By late 2003 HPAIV H5N1 had spread to South Korea. By 2004 genome sequencing revealed development of multiple sublineages of its HA protein. In the winter of 2006, H5N1 was associated with deaths of waterfowl in Europe, confirming fears that the virus would spread via migratory birds. Currently, H5N1 is known to be endemic in at least four countries (Vietnam, Egypt, Indonesia, and China). Sporadic human cases of H5N1 infection continue to be reported, and as of 2016 the World Health Organization (WHO) has tallied over 850 cases of human disease, with 450 deaths. All cases continue to be linked to direct contact with infected poultry, as the virus remains poorly transmissible among mammals. The H5N1 outbreak has lead to increased surveillance for AIV among migratory birds and in domestic flocks. In the case of migratory birds, HPAIV is most frequently found associated with dead birds and is rarely isolated from healthy migrating birds. Thus there is still a lack of understanding about the basic ecology of HPAIV in wild aquatic birds.

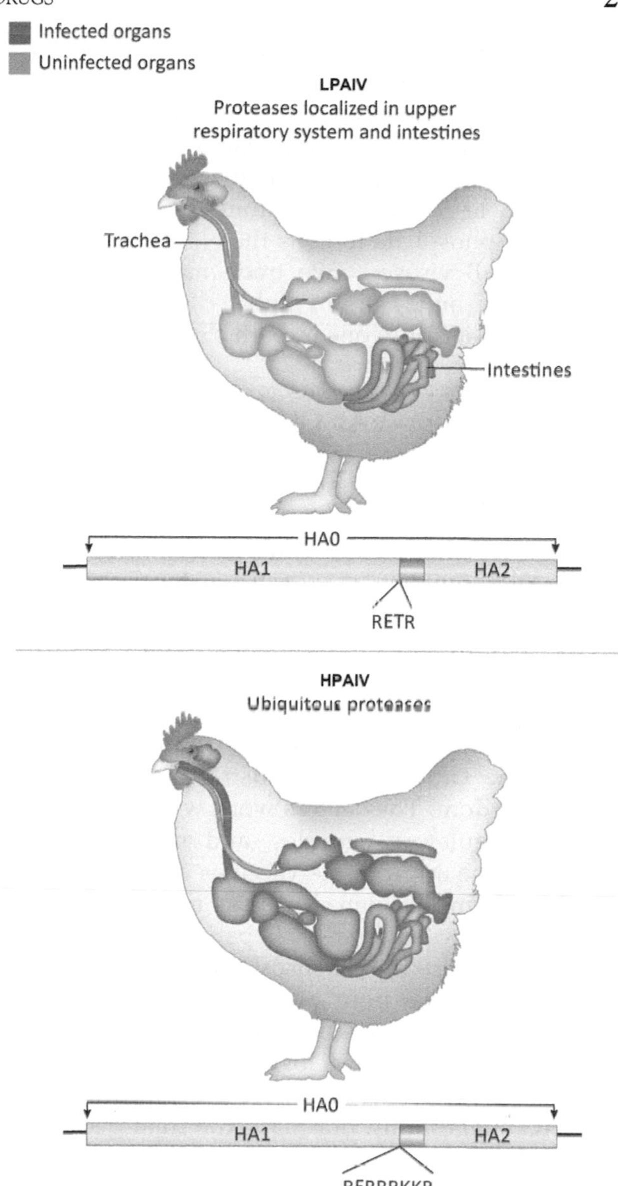

FIGURE 23.10 Tissue tropism of LPAIV and HPAIV. LPAIV are restricted to the respiratory and gastrointestinal tracts where extracellular trypsin-like proteases can cleave and activate HA. Some HPAIV have a much broader tissue tropism and can replicate in many organs as their HA protein is cleaved by intracellular furin-like proteases.

ANTIVIRAL DRUGS

There are two classes of antiviral drugs effective in limiting influenza virus replication. One class targets the M2 ion channel protein. Examples are amantidine (trade names Symadine and Symmetrel) and rimantidine (trade name Flumadine). These drugs are effective only against influenza A viruses. The second class of drugs inhibits NA. Recall that in the absence of NA, influenza viruses form large aggregates that remain attached to cells. The aggregates are less infectious and less easily transmitted to new cells. Oseltamivir phosphate inhibits the NA activity of influenza A and B viruses. Two trade names are Relenza (taken orally) and Tamiflu (inhaled). Some influenza viruses are resistant to one or both classes of drugs.

VACCINES

In the United States both killed and live attenuated influenza virus vaccines are manufactured on a yearly cycle. The vaccines include three or four different virus strains. The ability to manufacture the "correct" vaccines depends on global tracking of human influenza viruses, with the hope of predicting which strains will arrive with the next flu season. The WHO coordinates a Global Influenza Surveillance Network (GISN) that includes more that 100 National Influenza Centers based in the United States, Australia, Japan, and the United Kingdom. The centers work with academic laboratories, regulatory agencies, and manufacturers to ensure that the appropriate vaccines are available each year. Circulating influenza viruses are tracked by the collection of swabs from sick patients. Collaborating centers analyze the viruses and compare their sequences to those from previous years. Surveillance data are reviewed and experts recommend the most appropriate viruses for inclusion in vaccines. However this is always an educated guess! Countries consider the WHO suggestions, but make the final decisions on which strains to include in vaccines. Once chosen, the selected viruses must be prepared for manufacture. This usually requires generating reassorted viruses that can be grown to high titers. Historically eggs have been used for virus propagation but alternatives such as cell cultures have been established.

In this chapter we learned the following:

- Orthomyxoviruses are enveloped viruses with segmented negative-sense RNA genomes. They replicate in the nucleus of infected cells, using newly transcribed capped host cell mRNAs as a source of primers for viral mRNA synthesis.
- Influenza A viruses are major human pathogens. They have two envelope glycoproteins, HA and NA. HA is the attachment and fusion protein. NA is required for virion release as it destroys the sialic acid residues to which HA binds.
- Aquatic birds are the major reservoirs for all influenza A viruses. Mammals are infected with smaller subsets of influenza A viruses. Genome reassortment (antigenic shift) leads to generation of new viruses, with altered host range and virulence determinants.
- The names of influenza A, B, and C viruses include the viral species, host name (if other than human), the date and location of isolation, and HA and NA types.
- HA and NA of influenza viruses are major targets of the immune response. However a steady accumulation of point mutations generates new isolates that can escape (at least in part) from preexisting immune responses.
- To keep up with antigenic drift and shift, human vaccines are reformulated yearly.
- Two classes of drugs are effective at controlling influenza virus infection. One class targets the M2 protein (an ion channel) while the second class targets NA.

References

Krammer, F., Palese, P., 2015. Advances in the development of influenza virus vaccines. Nat. Rev. Drug Dev. 14, 167–182.

Palese, P., Tobita, K., Ueda, M., Compans, R., 1974. Characterization of temperature sensitive influenza virus mutants defective in neuraminidase. Virology. 61, 397–410.

CHAPTER

24

Family *Bunyaviridae*

OUTLINE

After studying this chapter, you should be able to:

- Describe the general characteristics of members of the family *Bunyaviridae*.
- Explain the "ambisense" coding strategy.
- Explain the modes of transmission of orthobunyaviruses, phleboviruses, and hantaviruses.
- List some human and animal diseases causes by bunyaviruses.

Bunyaviruses are enveloped, negative-strand RNA viruses (Fig. 24.1). The family Bunyaviridae includes over 350 named viruses isolated from plants, insects, and vertebrates. Bunyaviruses can cause severe disease and substantial economic loss through infections of domestic livestock and plants. Most bunyaviruses are transmitted by insect vectors; notable exceptions are the hantaviruses, which have rodent hosts and are occasionally transmitted to humans via rodent feces and urine.

GENOME ORGANIZATION

Bunyavirus genomes are three segments of negative-sense RNA (Box 24.1). The RNA segments are designated small (S), medium (M), and large (L). Total genome size ranges from 11 to 23 kb. The 5′ and 3′ ends of each RNA segment are complementary and base pair to form panhandle structures causing the RNA segments to circularize. There is some genomic diversity as RNA segments vary in size as well as in coding strategy. Members of the genus *Orthobunyavirus* (Box 24.2) use a negative-sense RNA coding strategy in which all mRNAs are transcribed from genome-sense RNA, as shown in Fig. 24.2. In contrast, members of the genera *Tospovirus* and *Phlebovirus* use an "ambisense" coding strategy. Ambisense refers to the fact that mRNAs are transcribed from both genome strands and their complements (cRNAs) as shown in Fig. 24.3. The L genome segment always contains one open reading frame, encoding the L protein by a negative-sense strategy. The S genome segment always encodes the N protein using negative-sense coding strategy. However some bunyaviruses also encode a nonstructural protein from the S segment (called NSs). The M genome segment of all bunyaviruses encodes the glycoprotein precursor. In some cases additional proteins are encoded by M and these are expressed as part of the precursor. Some bunyaviruses package cRNAs within virions; it believed that upon penetration into the cell cytosol, these serve as templates for mRNA synthesis by virion-associated RNA-dependent RNA polymerase (RdRp). An examination of a

Viruses. DOI: http://dx.doi.org/10.1016/B978-0-12-803109-4.00024-6

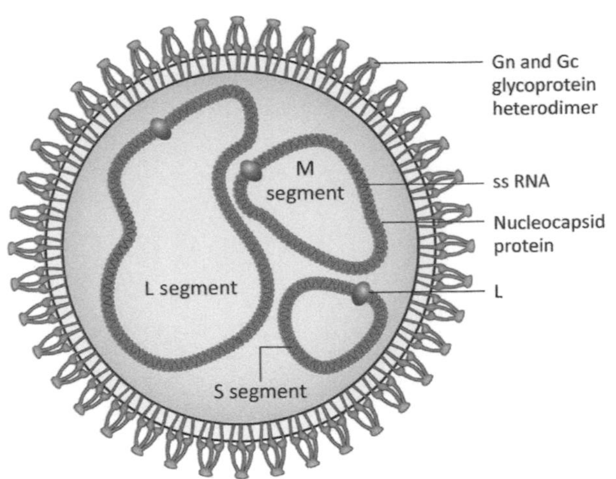

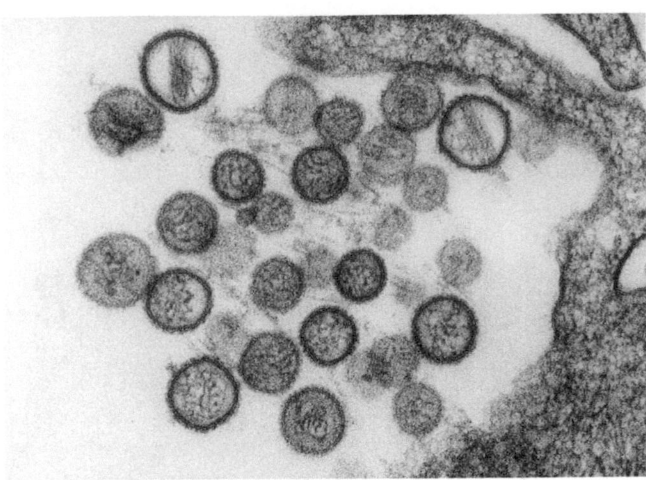

FIGURE 24.1 Virion morphology (left). Transmission electron microscopic image of a hantavirus known as the (Sin Nombre virus) (right). *From CDC/Brian W.J. Mahy, PhD; Luanne H. Elliott, M.S. CDC Public Health Image Library, Image ID #14340.*

BOX 24.1

CHARACTERISTICS

Negative or ambisense single-stranded RNA genomes contain three segments designated large (L), medium (M), and small (S). Total genome sizes range from ~11 to 20 kb. RNA segments have complementary nucleotides at their 3′ and 5′ ends. Base-pairing of 5′–3′ terminal nucleotides is predicted to form stable panhandle structures to form noncovalently closed circular RNAs.

Virions are enveloped, 75–115 nm in diameter with helical nucleocapsids.

BOX 24.2

TAXONOMY

Family *Bunyaviridae*

Genus *Hantavirus* (24 species) includes Hantaan virus (HTNV), named after the Hantaan river in Korea.

Genus *Nairovirus* (7 species). Most are tick-borne viruses.

Genus *Orthobunyavirus* (48 species). This genus contains more than 170 named viruses.

Genus *Phlebovirus* (10 species). Rift Valley fever virus (RVFV) is a medically and agriculturally important virus in Africa.

Genus *Tospovirus* (11 species). Plant viruses.

phlebovirus genome (Fig. 24.3) illustrates the ambisense coding strategy: The NSs protein can only be expressed from an mRNA transcribed from a cRNA.

VIRION STRUCTURE

Bunyavirus virions are enveloped; the envelopes are studded with glycoprotein heterodimers consisting of Gn and Gc (so-called because they are derived from the N-terminal and C-terminal halves, respectively, of the precursor polyprotein). Both Gn and Gc are transmembrane proteins. Gn and Gc associate as heterodimers that further assemble into higher order structures within the viral envelope. The structure of Rift Valley Fever virus (RVFV, genus *Phlebovirus*) has been solved and Gn/Gc are arranged in an icosahedral $T = 12$ lattice within the envelope. The genomic RNA is found

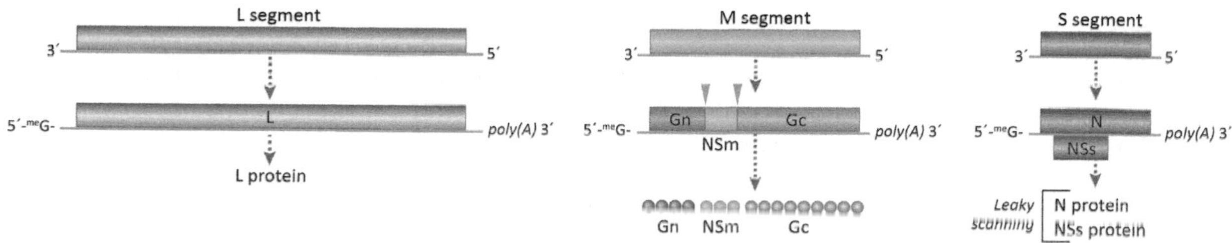

FIGURE 24.2 Genome and protein expression strategy of an orthobunyavirus.

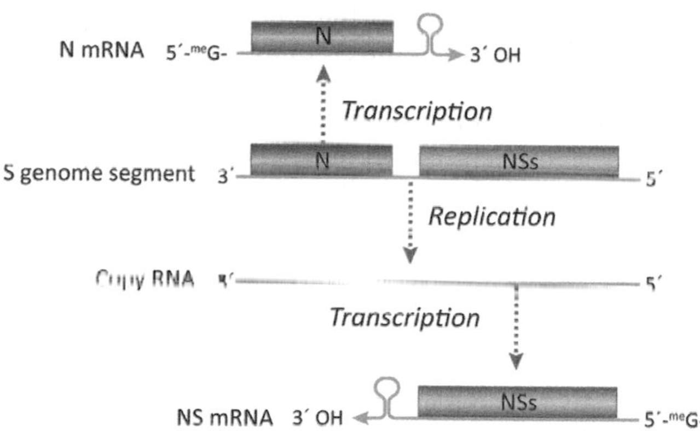

FIGURE 24.3 Organization and ambisense coding strategy of the phlebovirus S genome segment.

in association with N protein and L, forming helical nucleocapsids within the envelope.

GENERAL REPLICATION STRATEGY

Attachment

Gn and/or Gc are the viral attachment proteins. Gc appears to be the primary attachment protein for orthobunyaviruses in mammalian and mosquito cells.

Penetration and Uncoating

Many bunyaviruses enter cells by endocytosis and acidification of endocytic vesicles is followed by membrane fusion releasing nucleocapsids into the cytoplasm. However some family members fuse at the plasma membrane. Studies of different bunyaviruses suggest that there may not be a single "family-wide" model of attachment and/or penetration.

Amplification

Bunyaviruses replicate in the cytoplasm and within the family, the orthobunyaviruses, hantaviruses, and nairoviruses uses a negative-strand RNA virus coding strategy. That is, the genomic RNAs serve as templates for mRNA synthesis. (Recall that for negative-strand RNA viruses the first molecules synthesized are mRNAs.) Synthesis of bunyavirus mRNA is primed using capped oligonucleotides (10–20 nt long) cleaved from host mRNAs present in the cytosol. Thus bunyaviruses prime mRNA synthesis by a cap-snatching mechanism, similar to influenza viruses. Transcription and genome replication appear to occur in association with membranes. In contrast to other negative-strand RNA viruses, bunyaviruses require *ongoing protein synthesis* in order to produce a full complement of viral mRNAs. One model to explain this has ribosomes assembling on newly synthesized viral mRNAs to inhibit premature termination of transcription.

Bunyavirus N proteins are always encoded by the S genome segment. N proteins range in size from ~19 kDa for orthobunyaviruses to ~54 kDa for hantaviruses and nairoviruses. N proteins oligomerize and bind to both genome RNA and cRNA. N also interacts with L, Gn, and Gc proteins to facilitate virion assembly and morphogenesis. The S segments also encode a nonstructural protein called NSs (nonstructural protein from the S segment). NSs proteins range in size from 10 kDa to more than 50 kDa. There is little amino acid sequence homology among NSs proteins (even within a single genus) and they have different functions. Activities reported for NSs proteins include regulation of RdRp catalytic activity and antagonism of interferon signaling. The flexibility of NSs sequence and function may be one of the factors allowing the bunyaviruses to exploit a wide host range.

The bunyavirus M genome segment encodes the glycoprotein precursor. The precursor polyprotein is cleaved into Gn (amino-terminal) and Gc (carboxy-terminal). In the case of the orthobunyaviruses, Gn is 32–35 kDa and Gc is 110–120 kDa. Both are integral membrane glycoproteins. Gn is predicted to have the receptor-binding domain while Gc is predicted to contain a fusion domain. For some orthobunyaviruses, cleavage of the polyprotein encoded by the M segment also yields a small nonstructural protein called NSm. Orthobunyavirus NSm is located between Gn and Gc in the polyprotein. In contrast, RVFV (genus *Phlebovirus*) NSm protein is cleaved from the amino-terminus of the polyprotein and seems to function as a virulence factor that suppresses apoptosis.

The bunyavirus L genome segment encodes the L protein, the RdRp. L is multifunctional with polymerase, helicase, and endonuclease domains. Other functional motifs identified in some bunyavirus L proteins include topoisomerase, gyrase, and protease domains. The sizes of bunyavirus L proteins vary widely. Hantaviruses, orthobunyavirus, and phlebovirus L segments are ~6500 nucleotides (encoding an ~237 kDa protein). In comparison nairoviruses L segments are ~12,000 nt (encoding an ~459 kDa protein).

In contrast to mRNA synthesis, genome synthesis requires production of uncapped full-length cRNAs from each genome segment. Similar to other negative-strand RNA viruses, there must be a molecular mechanism (yet to be defined) that triggers the switch from transcription to genome replication. As with other negative-strand RNA viruses, N protein is likely involved.

Assembly and Release

Signals for the encapsidation of the vRNAs and cRNAs are present in the noncoding 3′ and 5′ regions. Bunyavirions usually assemble and mature by budding into smooth membranes of the Golgi followed by fusion of virus-containing vesicles to the PM to release virions by exocytosis. However some bunyaviruses, in some cell types, have been observed to bud directly from the plasma membrane.

As expected, bunyavirus genomes can reassort if a cell is infected with two different viruses. Theoretically a cell infected with two bunyaviruses can produce eight types of progeny: six reassortants and two parental genome types. However only closely relate viruses can reassort and certain genome segment combinations may be nonviable.

DISEASES

Orthobunyaviruses are transmitted by insects. In vertebrate cells, orthobunyavirus infection is usually lytic. In contrast, infection of arthropod cells results in the establishment of a persistent infection.

Pathogens of humans and domestic animals include:

- La Crosse virus (LACV) (pediatric encephalitis, humans),
- Oropouche virus (severe febrile illness, humans),
- Ngari virus (hemorrhagic fever, humans),
- Schmallenberg virus (mild febrile illness of adult livestock, but causes congenital malformations and stillbirths),
- Cache Valley virus (CVV) (spontaneous abortion and congenital malformations in ruminants).

La Crosse Virus

According to the Centers for Disease Control and Prevention (CDC), there are ~80–100 reported cases of encephalitis caused by LACV each year (the actual number of cases may be higher). LACV encephalitis occurs most often in children under the age of 16. The vast majority of LACV infections do not cause severe disease. (What kind of study would reveal this?)

LACV is transmitted to humans by *Aedes triseriatus*, the eastern treehole mosquito. Infected female mosquitos transmit LACV vertically to offspring, via their eggs. Studies have shown that infected females are more efficient at mating than noninfected females. *A. triseriatus* is a mosquito that usually feeds on small mammals such as chipmunks and squirrels. These same small mammals are the natural hosts for LACV. Humans, although occasionally infected with LACV, are likely dead end hosts that do not play a role in maintenance of the virus in the environment. Prevention of LACV infection centers on avoiding mosquito bites. Although *A. triseriatus* usually lay eggs in treeholes, they also use artificial containers.

Oropouche Virus

Oropouche virus is a common orthobunyavirus in Brazil and other South and Central American countries. It can cause severe acute febrile disease and over 30 epidemics were reported in the period from 1960 to 2009. Oropouche virus is considered a public health threat in tropical and subtropical areas of Central and South America.

Schmallenberg and Cache Valley Viruses

Schmallenberg and CVVs infect livestock and both orthobunyaviruses are associated with congenital malformations and abortions in ruminants. Schmallenberg virus emerged in Europe in 2011. CVV is endemic to the United States, Canada, and Mexico. Among wild ruminants, deer in the United States have a high seroprevalence for CVV. CVV is a zoonotic virus and there have been a handful of cases of human infection in the United States. Both viruses are transmitted by mosquitos and midges.

Rift Valley Fever Virus

In the genus *Phlebovirus* RVFV is a pathogen of note. RVFV is endemic to Africa. A large outbreak, with over 500 human cases occurred in Saudi Arabia in the year 2000. Many human cases were severe (hemorrhagic fever) and mortality was about 17%. The usual hosts of RVFV are ungulates (cattle and sheep) and many infected humans reported exposure to animal blood. Transmission from mosquitos to humans is also a likely route of infection.

Hantavirus Pulmonary Syndrome and Hantavirus Hemorrhagic Fever

Members of the genus *Hantavirus* infect rodents worldwide. They are transmitted from rodent to rodent, and rarely, to humans. Human exposure results from contact with rodent urine or feces (Figs 24.4–24.6). Hantavirus pulmonary syndrome (HPS) was first described in the southwestern United States in 1993. The disease presented as a flu-like illness that *rapidly* progressed to severe respiratory disease. The death rate was almost 50% among hospitalized patients (Box 24.3). Sporadic cases of HPS have been associated with at least 10 different hantaviruses, each associated with a different rodent species. The best prevention for hantavirus infection is adequate rodent control and limiting exposure to infected dust. Hantaviruses also endemic to Europe and Asia where they cause hemorrhagic fever with renal syndrome (HFRS) in humans.

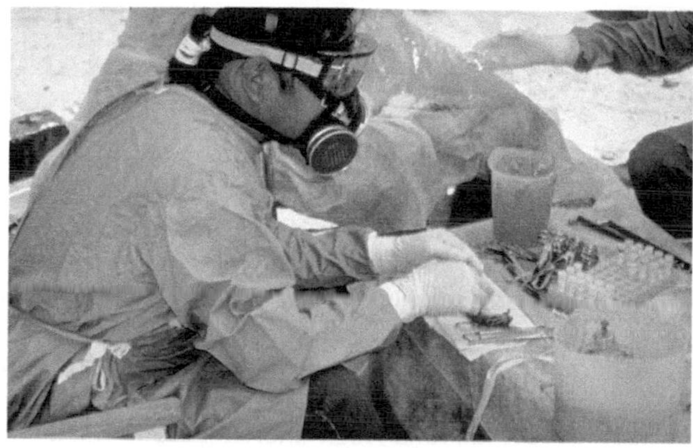

FIGURE 24.6 CDC scientist, wearing protective clothing while collecting specimens from trapped rodents. *From CDC/Cheryl Tryon, CDC Public Health Image Library, Image ID #2.*

FIGURE 24.4 Deer mouse, *Peromyscus maniculatus*, amongst deteriorating sheets of fabric and bird feathers *From CDC. Public Health Image Library, Image ID #13214.*

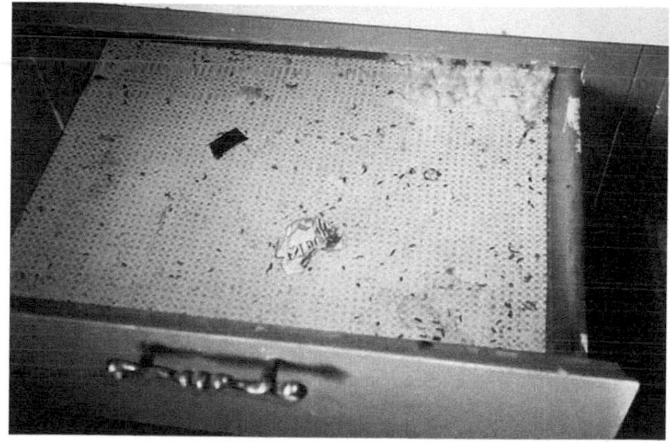

FIGURE 24.5 Dresser drawer in a home, showing evidence of rodent habitation (from home of a hantavirus patient). *From CDC Public Health Image Library, Image ID #14340.*

In this chapter we have learned that:

- Bunyaviruses are enveloped viruses with single-stranded, segmented RNA genomes (three segments).
- The overall replication strategy of bunyaviruses is similar to other negative-sense RNA viruses; however, some have ambisense genome segments whereby mRNAs are transcribed from both the genome-sense RNA and from cRNA. Some bunyaviruses may package cRNAs along with genome RNAs.
- Segmented genomes means that reassortment can lead to the emergence of "novel" viruses.
- Most bunyaviruses are transmitted by insects (often mosquitos and midges). Exceptions are the hantaviruses. Hantaviruses are not arboviruses. They are transmitted by direct contact or exposure to rodent urine and feces.

BOX 24.3

HANTAVIRUS PULMONARY SYNDROME

The 1993 Four Corners hantavirus outbreak was the first time that human hantavirus infection was documented in the United States. Four Corners refers to the quadripoint where the boundaries of Colorado, Utah, Arizona, and New Mexico meet. This is a mostly rural, arid, rugged area of the United States. While only a handful of cases occurred, they were serious and clinically notable for their very rapid onset of severe shortness of breath, and high mortality. The CDC was notified and sent a team to the region to investigate. The CDC initially identified five people with similar symptoms, all of whom had died. A total of 48 cases were identified in 1993. Most of the stricken were young Navajos living in New Mexico. Virologists at the CDC, using new molecular methods, were able to link the new disease with a previously unknown hantavirus. Hantaviruses had previously been associated with sporadic cases of human hemorrhagic fever in Asia, and were known to infect rodents. Therefore investigators trapped many species of rodents in areas where patients lived and worked. Several types of rodents, including deer mice, were infected with hantaviruses. Deer mice were probably the link to human disease, as these rodents live in rural and semirural areas, sometimes inhabiting homes, outbuildings, barns, and woodpiles.

The reason for the notable outbreak in 1993 was probably environmental. There were unusually large numbers of deer mice in 1993, due to an abundance of snow, rain, and vegetation. It was estimated that there were 10 times more deer mice in 1993 than in 1992. Once researchers had identified the viral cause of the 1993 outbreak, they tested stored samples from patients who had previously died of unexplained adult respiratory distress syndrome and found many more hantavirus positive samples. The oldest was from 1959. Thus human infections had occurred in small numbers, for years.

Today we know that many species of rodent can be infected with hantavirus. Virus is excreted in urine and feces. Human infections occur mainly by inhaling infected dust from rodent infested buildings. The rodents that carry hantaviruses (for example, the deer mouse and white-footed mouse) live in rural, as opposed to urban areas. Therefore infections, while rare, are more common in rural areas. Practices to avoid infection include airing out structures (i.e., cabins) that have been closed up for long periods of time, followed by cleaning with water containing household bleach or other disinfectant. It is very important to avoid generating airborne dust while cleaning.

In 2012, 10 confirmed cases (with 3 deaths) of hantavirus infection were identified among visitors to Yosemite National Park in California. Many of the infected had stayed in a relatively new set of "tent cabins" erected a few years earlier. Investigations revealed that manner in which the structures were built provided very attractive nesting areas for rodents, thus they were subsequently torn down to prevent future outbreaks. However, as in the 1993 Four Corners outbreak, weather probably played a role, as rodent populations were high at the time of the outbreak.

- Hantaviruses are rodent viruses that likely do little harm to their rodent hosts. However human infections can result in severe diseases including HPS (North America) or HFRS (Europe and Asia). The numbers of human cases each year are relatively small.
- Insect vectored bunyaviruses that are associated with human disease include LACV, Oropouche virus, and RVFV.

CHAPTER

25

Family *Arenaviridae*

OUTLINE

After reading this chapter, you should be able to discuss the following:

- What are the key features of all members of the family *Arenaviridae*?
- How is arenavirus genome replication similar to negative-strand RNA viruses?
- What is an ambisense genome segment?
- At what site in the cell do arenaviruses replicate?
- What are the natural hosts of arenaviruses?
- Under what conditions are people infected with arenaviruses?

The name arenavirus comes from the Latin *arenosus* (sandy). Ribosomes are packaged within arenavirus virions and these are responsible for the "sandy" appearance of the particles that gives these viruses their name. (Functions of the packaged ribosomes remain unknown). Rodents are natural hosts of many arenaviruses though some can infect humans through contact with infected rodents or inhalation of infectious urine or feces. While many human infections are asymptomatic, some arenaviruses are pathogenic for humans causing systemic infections with hemorrhagic fever and/or neurological involvement.

GENOME ORGANIZATION

Arenaviruses are segmented RNA viruses that replicate using an overall strategy common to negative-strand RNA viruses. However, the two genome segments (large (L) and small (S)) are ambisense: Each genome segment contains two open reading frames, separated by an intergenic region (IGR) that serves as a transcription termination signal (Fig. 25.1). IGRs are predicted to form hairpin structures that signal the release of the polymerase from the substrate during transcription. The L genome segment encodes the L (RNA-dependent RNA polymerase, RdRp) and Z proteins. The S genome segment encodes the RNA-binding nucleoprotein (NP) and viral glycoprotein complex (GPC). Arenavirus mRNAs are not polyadenylated but do have a hairpin structure at their 3′ end. Sequences at the 3′ end of the L and S segments are highly conserved; they are likely important cis-regulatory elements (i.e., promoter for polymerase entry). There is base complementarity between the 5′ and 3′ ends of each genome segment and they are predicted to form panhandle structures. Viral mRNAs are capped at the 5′ end followed by 4–5 nucleotides not found in the genome, a finding suggestive of a cap-snatching mechanism to prime mRNA synthesis.

VIRION STRUCTURE

Arenavirus virions are enveloped with helical nucleocapsids and are pleomorphic, ranging in size from 100 to 130 nm in diameter (Fig. 25.2). Within

Viruses. DOI: http://dx.doi.org/10.1016/B978-0-12-803109-4.00025-8

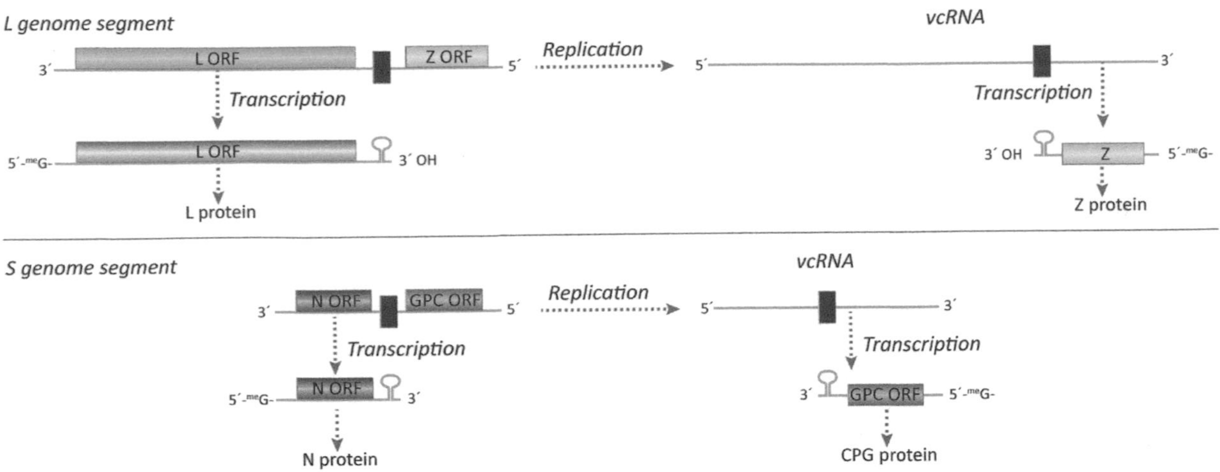

FIGURE 25.1 Genome and protein expression strategy.

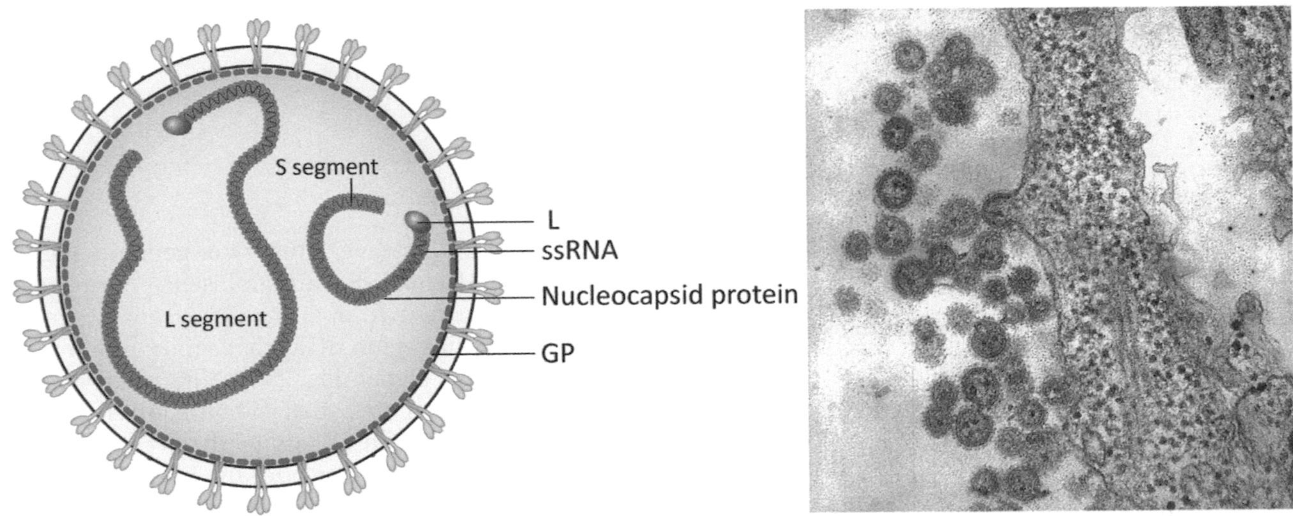

FIGURE 25.2 Virion morphology (left). Electron micrograph of the Machupo virus (right). Machupo virus is a member of the Arenavirus family, isolated in the Beni Province of Bolivia in 1963. *From CDC/Dr. Fred Murphy; Sylvia Whitfield. Public Health Library Image #1869.*

virions, the L and S genome segments are organized into circular structures. Nucleocapsids consist of genome segments bound to the NP protein. As with other negative-sense RNA viruses, the L protein (RdRp) is associated with nucleocapsids. Glycoprotein spikes extend from the virion surface. Spikes are a complex of GP1 and GP2 (products of the GP precursor protein). Z protein is a zinc-binding protein that functions as a matrix protein.

OVERVIEW OF REPLICATION

Attachment/Penetration

Arenavirus attachment to and entry into the host cell is mediated by the GPC (described later). The virion is endocytosed and membrane fusion is initiated in the acidified endosome (Box 25.1).

Transcription/Genome Replication

Arenavirus transcription and genome replication are cytoplasmic. After penetration, L protein initiates transcription to produce NP and L mRNAs. The presence of short, nontemplated sequences at the 5′ ends of arenavirus mRNAs suggests that L uses a cap-snatching mechanism to prime transcription. mRNAs terminate at the noncoding IGR found in all arenavirus genomes. As the concentration of NP in the cell increases, genome segments are replicated, to produce full-length copies (cRNA). cRNAs are now templates

BOX 25.1

CHARACTERISTICS

Virions are enveloped with helical nucleocapsids. Virions are pleomorphic, ranging in size from 100 nm to more than 130 nm in diameter. Virions have a "sandy" or granular appearance due to the presence of host ribosomes within virions. Transcription and genome replication are cytoplasmic.

Genomes: Arenaviruses have segmented RNA genomes. The large (L) segment is ~7.2 kb and the small (S) segment is ~3.5 kb. Arenavirus replication follows the pattern of negative-sense RNA viruses; however, both genome segments are ambisense. There are two open reading frames on each segment. mRNAs have 4–5 nontemplated nucleotides and a cap structure at their 5′ ends, which are likely obtained from cellular mRNAs via cap-snatching mechanisms (whose details remain to be determined). mRNAs have a hairpin at the 3′ end. The S genome segment encodes the NP, N, and the viral glycoprotein precursor (~75 kDa). The L segment encodes the L (RdRp) and Z (matrix) proteins.

for transcription of GPC mRNA (from the S segment) and Z mRNA (from the L segment).

Viral Proteins

NP is the most abundant arenavirus protein. It has two distinct structural domains, and N-terminal domain and a C-terminal domain connected by a flexible linker region. The N-terminal domain appears to bind RNA. As with other negative-sense RNA viruses, interactions between N and RNA probably trigger a switch from transcription to genome replication. The C-terminal domain of lymphocytic choriomeningitis virus (LCMV) counteracts the interferon response by preventing activation and nuclear translocation of interferon regulatory factor 3 (Fig. 6.3).

L protein is the RdRp. It has characteristic protein motifs found in the L proteins of other negative-sense RNA viruses. L has an endonuclease activity that may have a function in cap-snatching, to prime mRNA synthesis.

Z is a zinc binding protein. It shares no overall amino acid sequence homology with proteins of other negative-strand RNA viruses. However Z is myristoylated (myristic acid, a fatty acid, is linked to the protein near the N-terminus). Z is found at the plasma membrane, thus functions as the arenavirus matrix protein. Z drives budding of virions.

Glycoprotein complex. A glycoprotein precursor is encoded on the S genome segment. A signal peptidase cleaves the precursor at about amino acid 58. Unique to arenaviruses, this long leader peptide is myristoylated and is retained as part of the mature GPC. The glycoprotein precursor is further cleaved by an ER protease to produce GP1 (receptor-binding protein) and GP2 (fusion protein). The leader peptide, GP1, and GP2 form a complex that traffics to the PM.

Assembly/Release

Nucleocapsids (genome length RNA, NP, and L) associate with Z and CPG at the plasma membrane where progeny virions are released by budding. Budding is driven by the Z protein.

In cultured cells, many arenavirus infections are noncytolytic and long-term persistent infection is a common outcome. During persistent infection few virions are released, instead virus spreads when nucleocapsids partition into daughter cells during cell division.

DISEASES CAUSED BY ARENAVIRUSES

Arenaviruses are commonly found in association with rodents. However some can be transmitted to humans and cause disease. Humans become infected by through contact with infected rodents or through inhalation of infectious rodent urine and feces. In humans disease often begins with a gradual onset of fever and myalgia. More serious disease, such as viral hemorrhagic fever and/or neurologic impairment, may develop.

Lassa Fever

Lassa fever virus (LFV) is endemic to West Africa where its natural hosts are rats of the genus *Mastomys*. LFV is zoonotic and is transmitted to humans via contact with materials contaminated with rodent urine or feces. LFV can also be transmitted from person-to-person, thus controlling outbreaks must include contact tracing and good infection control measures. Sexual transmission of LFV has also been reported. About 80% of people who become infected with LFV have no symptoms but one in five infections result in

BOX 25.2

TAXONOMY

Family *Arenaviridae*

Genus *Arenavirus*

The genus contains many species (>24) that cluster into two groups based on antigenic properties and genome sequences. NW (or Tacaribe serocomplex) includes Junin virus, Machupo virus, and Tacaribe virus among others. Old World viruses include LFV and LCMV.

severe disease, where the virus affects organs such as the liver, spleen, and kidneys. Disease may include gastrointestinal involvement (abdominal pain, nausea and vomiting, diarrhea, or constipation). Other complications include bleeding, edema, pharyngitis, and conjunctivitis. Estimated numbers of cases range from 300,000 to 500,000 per year (Box 25.2).

Lymphocytic Choriomeningitis

LCMV infections in humans are rarely fatal, but serious disease (viral hemorrhagic fever or neurological complications) can occur. The natural host for LCMV is the common house mouse, *Mus musculus*. Other rodents, such as hamsters sold in the pet trade can be infected through contact with mice. LCMV infections of humans have been reported worldwide, wherever infected rodent hosts are found. As with other arenaviruses, contact with infected rodents, their urine or droppings is associated with most human infections. It is likely that most human infections with LCMV are asymptomatic or mild, seldom resulting in the severe neurologic disease for which the virus is named. Infection of pregnant women can cause fetal congenital defects. In the United States, infection through solid organ transplantation has been reported.

While LCMV is a rare disease of humans, this mouse virus has been a valuable model for dissecting diverse mechanisms of viral pathogenesis. LCMV infects mice (*Mus musculus*), a good laboratory model animal, with variable disease outcomes. Complex interactions between host and virus determine the outcome of LCMV infection. Acute disease can range from mild to lethal. Persistent infections can also develop; some with no pathology and some accompanied by immune-mediated disease. Some of the factors that determine LCMV outcome of infection include the virus strain, breed of mouse, route of inoculation, and age of mouse. Both direct effects of viral infection and the effects of immune responses are important in the outcome of infection. Many important concepts in viral immunology were first described for the LCMV system. It continues to be an important model for viral immunologists (Box 9.3).

New World Arenaviruses

Several New World (NW or Tacaribe complex arenaviruses) have been associated with human disease. These include Junin (Argentine hemorrhagic fever), Machupo (Bolivian hemorrhagic fever), and Guanarito (Venezuelan hemorrhagic fever). Other NW Arenaviruses have been isolated from sporadic cases of hemorrhagic fever. Each of the aforementioned arenaviruses is associated with one or a few closely related species of rodents. NW arenaviruses continue to be isolated, with outbreaks depending on complex interactions humans, rodents, and the environment.

In this chapter we learned that:

- Arenaviruses are segmented RNA viruses.
- Virions are enveloped and ribosomes are packaged into virions giving a "sandy" appearance.
- The overall replication strategy is similar to the negative-strand RNA viruses.
- Arenaviruses have ambisense genomes. Each genome segment is transcribed to produce an mRNA. However after genome replication, mRNAs are also transcribed from full-length cRNAs.
- Many rodents are infected with arenaviruses and most of these infections are nonpathogenic. Humans can be infected by some arenaviruses (through contact with rodents).
- LFV, LCMV, Junin virus, and Machupo virus can cause severe disease in humans.

CHAPTER

26

Family *Reoviridae*

OUTLINE

After reading this chapter, you should be able to discuss the following:

- Describe the structure of reovirus genomes.
- Describe the structure of reovirus capsids. Do all reoviruses have the same capsid type?
- What is the "viroplasm" and why is it important in the reovirus replication cycle?
- Describe the unique features of reovirus transcription and genome replication.
- Discuss the transmission strategies of orthoreoviruses, rotaviruses, and orbiviruses.
- What are some important human and animal diseases caused by reoviruses?

Members of the family *Reoviridae* infect hosts ranging from microorganisms to insects to plants to vertebrates. The family *Reoviridae* is large and diverse, containing two subfamilies and numerous genera. The best-known animal reoviruses are those in the generus *Rotavirus* (major enteric pathogens of humans and other mammals), *Orbivirus* (insect vectored animal pathogens), and *Orthoreovirus* (mild upper respiratory tract disease of mammals and poultry diseases) (Box 26.1).

The family name, *Reoviridae*, derives from the fact that the first isolates were of respiratory and enteric origin, and were not associated with known disease (thus were respiratory, enteric orphanviruses). Reoviruses have segmented, double-stranded (ds) RNA genomes. Genome segments number from 9 to 12. Genome reassortment has likely contributed to their wide host range and great success as a family. Virions are unenveloped with icosahedral symmetry and are unique in having two or three discrete capsid layers (Fig. 26.1). The innermost capsid layer encloses dsRNA segments and enzymes for RNA synthesis. Upon infection, transcriptionally active genome segments remain within cores. Newly synthesized mRNAs are released into the cytosol through pores at the fivefold axes of symmetry (Box 26.2).

GENOME ORGANIZATION

Reovirus genomes contain from 9 to 12 segments of linear, dsRNA. For the most part, each segment encodes a single protein (Fig. 26.2). For example, rotavirus and orbivirus genomes have 11 segments encoding 12 proteins. Genome segment sizes range from ~600 to 3000 bp for a total coding capacity of approximately 18,500–19,200 bp. Positive genome strands and mRNAs are capped but not polyadenylated. Negative strands have a diphosphate at their 5′ end. There are short untranslated regions at the 5′ and 3′ ends of each genome segment. Replication is cytoplasmic,

Viruses. DOI: http://dx.doi.org/10.1016/B978-0-12-803109-4.00026-X

BOX 26.1

TAXONOMY

Family *Reoviridae*
Subfamily *Sedoreovirinae*
Genera include:

Genus *Orbivirus*
Orbiviruses are transmitted by insects. The genus contains 22 named species that infect a variety of mammals. Includes Blue tongue virus (BTV) and Epizootic hemmorrhagic disease virus (EHDV). Virions are double-layered particles containing 10 genome segments.

Genus *Rotavirus*
Rotaviruses are transmitted by the fecal oral route. Virions are triple-layered particles containing 11 genome segments. Rotaviruses are noted for causing severe diarrhea in young animals. There are eight species (Rotavirus A–H). Most disease is associated with the so-called Group A rotaviruses (GARV).

Subfamily *Spinareovirinae* (the name refers to the large turrets at the fivefold axes of symmetry)
Genera include:

Genus *Coltivirus* (Colorado tick fever virus, CTFV)
Genus *Orthoreovirus* (hosts include birds, mammals, and reptiles)

Ubiquitous viruses, present in feces. Seldom cause disease but sometimes associated with mild upper respiratory tract disease or gastroenteritis.

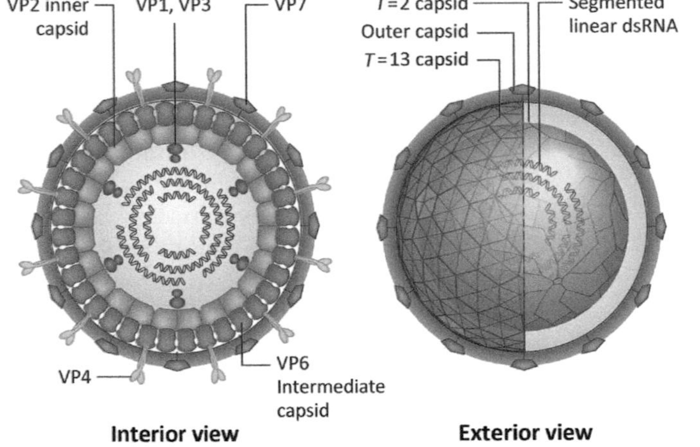

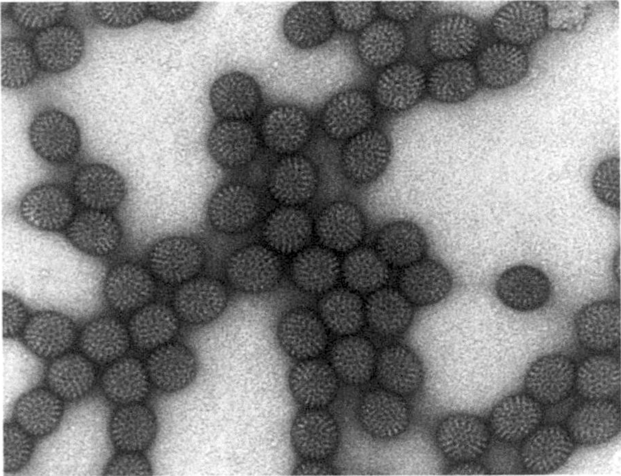

FIGURE 26.1 Virion structure of a triple-layered rotavirus (left) and electron micrograph (right). This negatively stained transmission electron microscopic (TEM) image revealed some of the ultrastructural morphology displayed by rotavirus particles. *From CDC/Dr. Erskine Palmer. CDC Public Health Image Library Image #197.*

BOX 26.2

GENERAL CHARACTERISTICS

Members of the family *Reoviridae* have segmented, dsRNA genomes. The number of genome segments ranges from 9 to 12. Reovirus replication is cytoplasmic and a unique feature of these viruses is that transcription takes place from within capsids. Virions are unenveloped, with two or three capsid layers (double or triple layer capsids). Middle and outer capsid layers are $T = 13$ icosahedral lattices. The inner capsids or cores are $T = 1$. As a group, reoviruses are very successful with a wide host range that includes fungi, plants, invertebrates, and vertebrates.

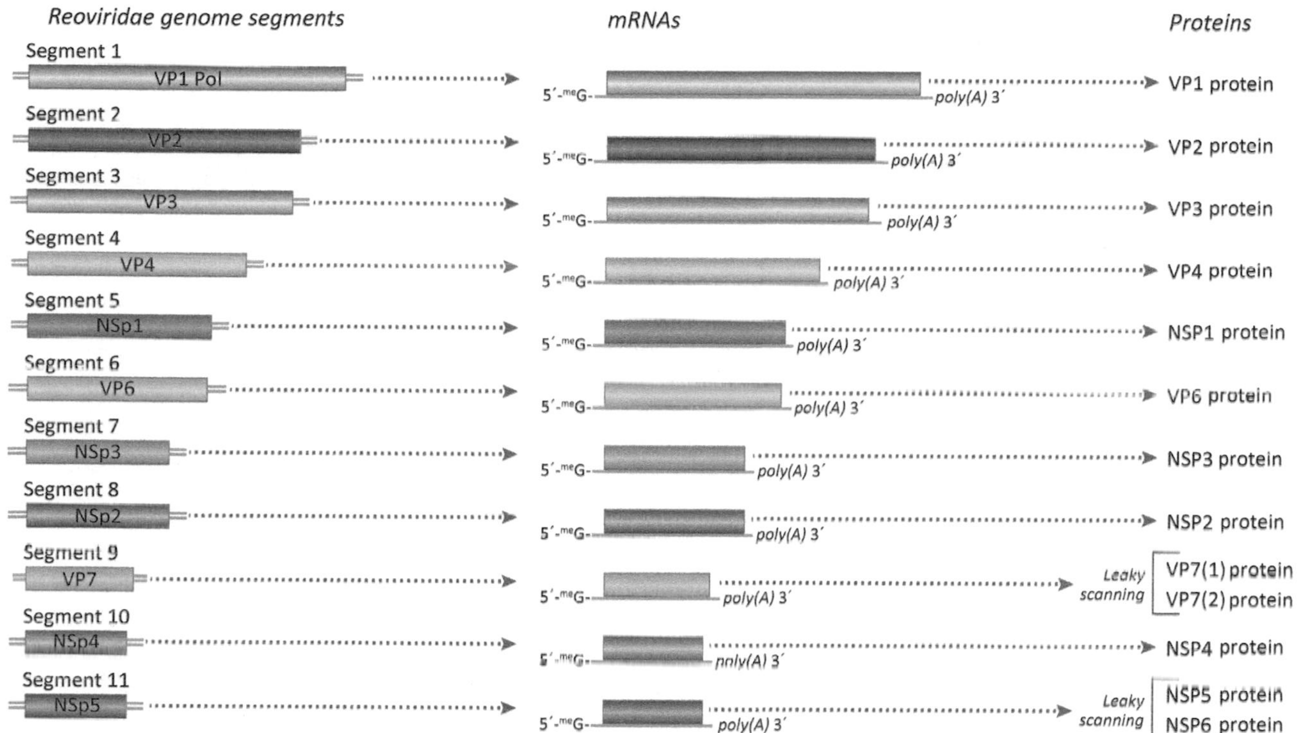

FIGURE 26.2 Depicting genome segments, mRNAs, and protein products of a rotavirus. See Table 26.2 for a comparison of the naming schemes used for different genera in the family.

TABLE 26.1 Characteristics of Selected Reoviruses

Genus	*Orthoreovirus*	*Rotavirus*	*Orbivirus*	*Coltivirus*
Subfamily	*Spinareovirinae*	*Sedoreovirinae*	*Sedoreovirinae*	*Spinareovirinae*
RNA segments	10	11	10	12
Capsid structure	2 layers	3 layers	3 layers	2 layers
Hosts	Vertebrates	Vertebrates	Vertebrates, insects	Vertebrates, insects

but associated with discrete pseudo-organelles variously called virus factories, virus bodies, or viroplasm.

VIRUS STRUCTURE

Ultrastructural studies of reoviruses reveal capsids made up of two or three discrete layers (Table 26.1). The innermost layer (core) has $T=1$ icosahedral symmetry and is assembled from 60 dimers of single type of structural protein. The core contains genome segments, each associated with enzymes needed for transcription. It is thought that each genome segment is located at a five-fold axis of symmetry. During transcription, mRNAs exit the intact cores through channels in the structure. The core is surrounded by a capsid layer with $T=13$ icosahedral symmetry. Some reoviruses (members of the genus *Rotavirus*) have a third protein layer, also with $T=13$ icosahedral symmetry. Rotaviruses have glycosylated spike proteins extending from the surface giving the particles a wheel-like appearance.

The family *Reoviridae* contains two subfamilies: Members of the subfamily *Sedoreovirinae* (genus *Rotavirus*, genus *Orbivirus*) have relatively smooth outer layers (or short spikes) while members of the subfamily *Spinareovirinae* (genus *Coltivirus*, genus *Orthoreovirus*) have large turreted projections at the fivefold axes of symmetry.

OVERVIEW OF REPLICATION

While the overall reovirus replication scheme is conserved (Fig. 26.3), it is not surprising that the details

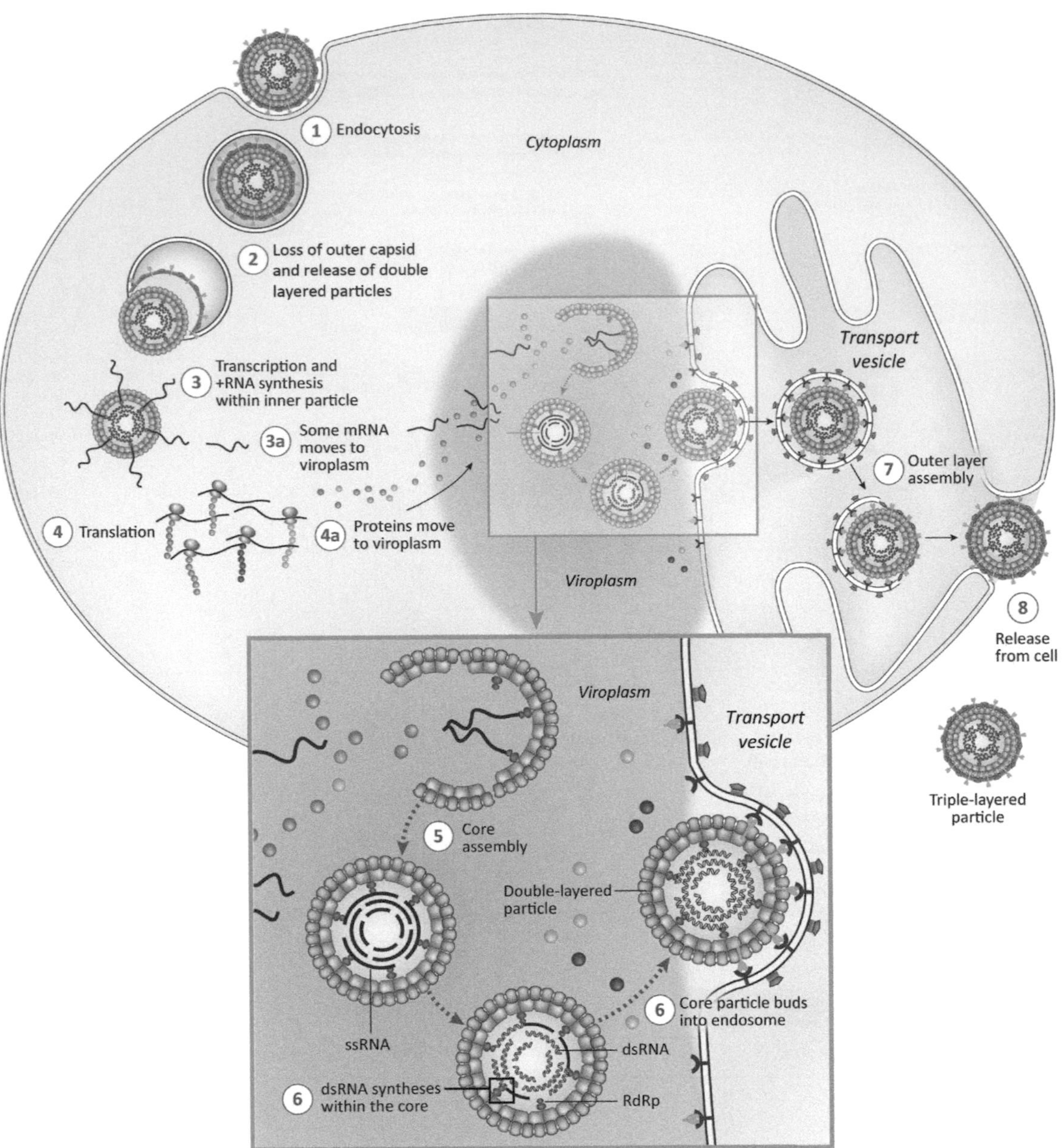

FIGURE 26.3 General overview of rotavirus replication cycle. Note that transcripts are synthesized within the core and mRNAs are released into cytoplasm through openings in the capsid. NSPs are important for development of the viroplasm, an electron dense area in the cell that contains a high concentration of viral proteins and mRNAs. Capsid assembly takes place in the viroplasm. In the model shown here, mRNAs are selected and packaged; RNA synthesis within the newly forming capsid generates the double-stranded genome segments. The outermost layer of the rotavirus triple-layered particle is obtained upon budding into ER/transport vesicles.

vary among genera. For example, we expect that rotaviruses (enteric viruses) and orbiviruses (transmitted by insects) might display some difference in replication, in light of their adaptation to very different host environments. The next sections highlight conserved features of reovirus replication but should be considered only a general introduction. Given that rotaviruses are a major cause of human infant mortality in some areas of the world, they have been studied in great detail. It should be noted that accepted schemes for naming/labeling reovirus genome segments and proteins vary by genus, as shown in Table 26.2. Reverse genetic systems have been developed for some reoviruses, as described in Box 26.3 and Fig. 26.4.

Attachment/Entry

Reovirus attachment is mediated by capsid proteins (rotavirus VP4, orthoreovirus σ1, orbivirus VP2). A variety of cell surface molecules act as receptors. Some

TABLE 26.2 Within the Family *Reoviridae*, Nomenclature for Genome Segments and Protein Products Differs by Genus

Genus *Orthoreovirus*			Genus *Orbivirus*			Genus *Rotavirus*		
Genome segment	Protein	Function	Genome segment	Protein	Function	Genome segment	Protein	Function
L1	λ3	RdRp	1	VP1	RdRp	1	VP1	RdRp
L2	λ2	Core spike	2	VP2	Outer shell spike	2	VP2	Inner shell
L3	λ1	RNA helicase	3	VP3	Inner core scaffold protein	3	VP3	Capping enzyme
M1	μ2	RNA binding	4	VP4	Capping enzymes	4	VP4	Outer shell spikes
M2	μ1	Major outer capsid protein	5	VP5	Outer shell protein (fusogenic)	5	NSP1	Interferon antagonist
M3	μNS	NSP, associates with viral mRNA	6	NS1	Nonstructural, enhances protein synthesis	6	VP6	Middle shell
M3	μNSC	NSP, unknown function	7	VP7	Core surface protein	7	NSP3	Nonstructural, required for systemic spread
S1	σ1	Attachment protein	8	NS2	Nonstructural, forms inclusion bodies	8	NSp2	Nonstructural forms inclusion bodies (viroplasm)
S1	σ1s	NSP, required for ematogenous dissemination	9	VP6	Helicase	9	VP7	Outer capsid shell
S2	σ2	Major capsid protein	9	NSP4	Nonstructural	10	NSP4	Enterotoxin, ER receptor
S3	σNS	Nonspecific RNA-binding protein	10	NS3	Nonstructural, virus trafficking and release	11	NSP5	Nonstructural forms inclusion bodies (viroplasm)
S4	σ3	Major outer capsid protein	10	NS3A	Unknown	11	NSP6	Nonstructural forms inclusion bodies (viroplasm)

BOX 26.3

REVERSE GENETICS SYSTEMS: ALL REOVIRUSES ARE NOT CREATED EQUAL

Reverse genetics systems allow the production of infectious virus from cloned nucleic acids. Such systems are instrumental in allowing researchers to determine the functions of both proteins and genome regulatory regions. However, when it comes to reverse genetics systems, all reoviruses are not equal. Systems based entirely on cloned genes have been developed for some orthoreoviruses and orbiviruses. The mammalian orthorovirus system involves transfecting cells with plasmids that encode (and express) each viral gene. The system for bluetongue virus (an orbivirus) differs only slightly. In this case, mRNAs are transcribed in vitro (in a test tube) and the single-stranded RNA products are transfected into cells. While there are important technical considerations, such as the need to produce the correct 5′ end of each mRNA, the systems are conceptually simple (as illustrated in Fig. 26.4). In contrast, similar strategies have not been successful when applied to rotaviruses. Current rotavirus reverse genetic systems require use of coinfecting helper viruses and this limits their utility. This is certainly frustrating, as the rotaviruses are widely accepted as the most important and widespread human pathogens in the family. It appears that simply expressing or introducing rotavirus mRNAs into a cell is not sufficient to produce infectious virus. The reasons remain elusive!

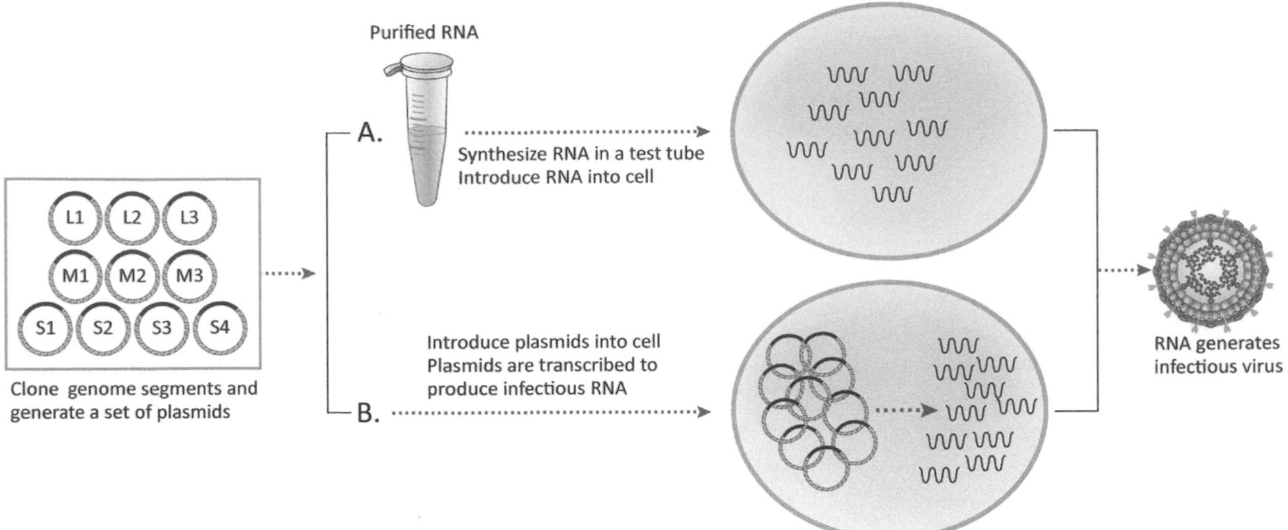

FIGURE 26.4 Reverse genetics systems have been developed for orthoreoviruses and orbiviruses. Each genome segment is cloned into a plasmid designed to produce mRNAs with defined 5′ and 3′ ends. In one strategy, the plasmids are introduced into cells and transcription of a set of viral mRNAs initiates the reovirus replication cycle. In an alternative strategy, mRNAs are synthesized in vitro and are introduced into cells. Note that the rotavirus reverse genetics system (for reasons that remain unknown) requires coinfection with helper virus.

reoviruses use protein receptors, others use carbohydrate receptors (for example, sialic acid) and some use a combination of the two. Rotavirus VP4 is glycosylated and is cleaved by extracellular proteases found in the digestive tract. Receptors for some rotaviruses are histo blood group antigens (HGBAs). HGBAs are complex carbohydrates found on epithelia of the respiratory, genitourinary, and digestive tracts. Intact rotavirus VP4 binds to HGBAs but VP4 cleavage products (designated VP5* and VP8*) interact with additional protein coreceptors.

Reoviruses encode proteins that interact with membranes to form pores or channels (rotavirus NSP1, orbivirus VP5, orthoreovirus μ1 protein). Penetration across plasma membrane or endosomal membranes is accompanied by disassembly of the outer capsid layers. The innermost core, containing genome segments, undergoes molecular rearrangements, but remains intact. In the case of rotaviruses, disassembly of outer layers is triggered by the low calcium concentration of the cytosol. Orthoreoviruses uncoat with the help of endosomal proteases such as cathepsins B and L.

Amplification

Penetration and uncoating activate the transcription activities of the core. Viral mRNA is synthesized within the core and newly formed (capped) transcripts are released through pores found at the vertices of the inner capsid. It is postulated that each genome segment is associated with its own pore and transcription complex. mRNAs are not polyadenylated, but form a hairpin structure at the 3′ end. In the case of rotaviruses, synthesis of viral transcripts is mediated by a complex of VP1 and VP3. VP1 contains the catalytic RNA synthesis domain while VP3 binds ssRNA and has nucleotide phosphohydrolase, guanyltranserase, and methyltransferase activities (thus VP3 is the cap-synthesizing component of the transcription complex). Transcriptionally active cores can be prepared from purified virions (by removal of the outer capsid layers) and in vitro these can produce up to milligram quantities of mRNA, given an adequate source of precursors.

Upon exiting the cores, mRNAs are translated. In the case of rotaviruses, nonstructural protein 3 (NSP3) binds to eIF4F, the cytoplasmic cap-binding complex to facilitate translation. Thus NSP3 seems to serve the function of host polyA-binding protein in translation initiation. Some rotavirus mRNAs are translated in the cytosol but mRNAs for rotavirus VP4, VP7, and NSP4 are translated on rough endoplasmic reticulum (ER), and the proteins are glycosylated therein. Cytosolic proteins form dense viral organelle-like structures called virus factories or the viroplasm. Cores assemble in the viroplasm and these are capable of transcription to generate additional mRNAs.

Genome Replication

The exact mechanism used to select and package reovirus genome segments is not well understood, but clearly the process occurs in the viroplasm. One model has mRNAs associating with pentamers of the inner core. mRNA secondary structures might provide signals for selective packaging of 9–12 different segments. It is clear that each mRNA is the template for synthesis of *single* complementary (negative) strand to form a ds genome segment.

Assembly, Maturation, and Release

Given their importance as worldwide human pathogens rotaviruses have been studied in great detail. Rotaviruses have triple layer particles and their outer layer displays glycosylated proteins, synthesized in the ER. However the virion is not enveloped, so how does it assemble and acquire its outer coat of glycoproteins?

Rotavirus double layer particles (DLPs) are synthesized in the viroplasm concurrent with genome replication. It is thought that pentamers of VP2 dimers (the inner core protein) interact with VP1 and VP3 and a single genome segment. Assembly of pentamers forms the inner core. Subsequently VP6 is added to form the DLP. (This is very simplified description of the process.) Rotavirus DLPs then leave the viroplasm and bud into the ER lumen by virtue of interactions with molecules of NSP4 located in the ER membrane. In this process, NSP4 is serving as an ER "receptor." While budding into the ER, the particle transiently acquires an envelope and VP4 and VP7. However the lipid bilayer is subsequently lost (by an unknown biochemical mechanism). The high Ca^{++} concentration in the ER is necessary for assembly of VP4 and VP7 into the outmost shell. Virions are released upon cell death. Rotavirus replication in the gastrointestinal tract is quite efficient and robust, as virions are shed in concentrations of up to 10^{10} particles per gram of feces.

An interesting feature of some orthoreoviruses is that they are fusogenic and induce formation of large syncytia (fused cells). Fusion is not mediated by capsid proteins, but by proteins that are not packaged into virions. These small transmembrane proteins are called FAST proteins (fusion-associated small transmembrane proteins). As cell fusion is proapoptotic, this may be a mechanism to enhance virion release.

DISEASES

Genus *Rotavirus*

Rotaviruses cause gastrointestinal infections of humans and other mammals. While many infections are mild to asymptomatic, some rotaviruses can cause severe gastrointestinal disease, with diarrhea and vomiting. There are eight species of rotavirus, designated A through H. The group A rotaviruses (GARV) are those most often associated with severe disease in infants and young children. There are many different GARV, due to their propensity for genome reassortment. In agricultural settings it is not uncommon to find reassortants among from bovine and porcine rotaviruses. Human rotaviruses can also exchange gene segments with animal rotaviruses.

In humans, most severe disease (life-threatening diarrhea) is seen among children 3 months to 2 years of age. Rotaviral disease is also most severe in young animals. The exact nature of the age restriction is not clear. Virions are highly contagious and are very stable in the environment. Transmission is most often through direct contact although water-borne outbreaks have been documented. The primary treatment for rotavirus-induced dehydration is to replace fluids and electrolytes. This can be accomplished through delivery of intravenous fluids or by use of oral fluid replacement therapies.

The primary site of rotavirus replication is the small intestine where virions infect nondividing enterocytes found at the tips of villi. Rotavirus infection is cytopathic and loss of intestinal epithelial cells results in malabsorption of nutrients accompanied by mild diarrhea. In cases of severe diarrhea the culprit appears to be rotavirus NSP4. NSP4 is a multifunctional protein critical in several aspects of the rotavirus replication cycle. As mentioned above, some NSP4 is located in the ER membrane and directs budding of the newly assembled DLPs into the ER. However, some NSP4 is secreted from infected cells. Secreted NSP4 can be found at both apical and basolateral surfaces of polarized epithelial cells. NSP4 attaches to basolateral integrins to trigger mobilization of intracellular calcium (release of calcium stores from the ER) that in turn triggers release of chloride through calcium-activated channels. Release of chloride is followed by sodium and water, resulting in secretory diarrhea that can be quite severe. (Recall that the rotavirus outer capsid layer disassembles at low Ca^{++} concentrations. Is it possible that one role of NSP4 is to trigger release of Ca^{++} from intracellular storage compartments to stabilize newly formed virions?)

Rotavirus vaccines are live-attenuated products generated by reassortment. These vaccines protect against severe disease, but do not prevent infection. In the United States, live-attenuated products that are recommended for all infants and it is estimated that vaccination prevents millions of episodes of gastroenteritis, and thousands hospitalizations, among young children each year. Worldwide the numbers are staggering, with an estimated 2 million hospitalizations and ~450,000 childhood deaths from rotovirus-induced diarrheal disease per year, many of which could be prevented through vaccination.

Genus *Orbivirus*

There are 22 named species of orbivirus that infect animals and insects. Orbiviruses are "arboviruses" as they are transmitted by insects such as ticks, mosquitoes, sand flies, and biting midges. They have a wide host range among animals and birds. Transmission requires that orbiviruses replicate in both insect and animal hosts. In animal cells they cause cytopathic infections but in insects infections are noncytopathic.

The type species in the genus *Orbivirus* is bluetongue virus (BTV), spread by biting midges of *Culicoides* sp. BTV originated in Africa but since the 1920s the virus has moved worldwide. Infections can range from entirely asymptomatic to quite severe (edema of lips, tongue and head, fever, depression, nasal discharge, excess salivation, pain, and death). In extreme cases, edema causes cyanosis of the tongue, the "blue tongue" for which the disease is named. Among domestic animals, the disease affects sheep, goats, and cattle. Sheep are usually the most severely affected.

Before 1950, BTV was reported only from parts of Africa and Cyprus. However the virus spread rapidly across the world (in conjunction insect vectors) in a broad band between ~40°N and 35°S. In 2006, BTV made news by spreading north to make its first appearance in Western Europe (the Netherlands, then Belgium, Germany, and Luxembourg). In 2007 BTV was reported in the Czech Republic and the United Kingdom. There is no evidence of transovarial transmission, and cold winters of northern Europe kill most adult *Culicoides*, but some overwintering of adults appears to occur. Some carryover of virus may also result from uterine transmission that may allow persistent infection of newborn lambs or calves. Livestock vaccines are available; however, there are multiple serotypes of BTV.

Another orbivirus of note is Epizootic hemorrhagic disease virus (EHDV), also transmitted by *Culicoides* midges. EHD is a disease of wild ruminants, particularly white-tailed deer in North America. EHD is one of the most important diseases of deer in North America. Disease onset is sudden and in extreme cases is characterized by high fever, weakness, excessive salivation, rapid pulse and respiration rate, and hemorrhage. Within only 8–36 hours of the onset of symptoms, deer become prostrate and die.

African horse sickness virus is an orbivirus endemic to Africa. Hosts include horses, donkeys, mules, and zebras. The most serious disease occurs in horses and mules while zebras are often asymptomatic and are thought to the natural reservoir hosts. Transmission is through bites of *Culicoides* sp.

Genus *Orthoreovirus*

Orthoreoviruses are in the subfamily *Spinareovirinae*. Hosts include birds, mammals, and reptiles. These viruses are ubiquitous and seldom cause severe disease. They are so common in feces that their presence is used as an indicator of fecal contamination. Diseases associated with these orthoreoviruses include mild upper respiratory tract disease, gastroenteritis, and biliary atresia.

Genus *Coltivirus*

Members of the genus *Colitivirus* are in the subfamily *Spinareovirinae*. Their genomes contain 12 segments. In North America the vertebrate hosts of Colorado tick fever virus (CTFV) include rodents, ungulates (deer, elk, sheep), and canids (coyotes). CTFV has also been isolated from humans where infection can result in flu-like symptoms, meningitis, encephalitis, and occasionally other severe complications. (CTFV disease was initially confused with a mild form of Rocky Mountain spotted fever, which is caused by *Rickettsia rickettsii*.) CTFV is an arbovirus, transmitted via the bite of an infected tick (Rocky Mountain wood tick (*Dermacentor andersoni*)). Most human infections occur at 4000–10,000 feet above sea level and during the spring and summer months when ticks are most active.

In this chapter we have learned that

- Members of the family *Reoviridae* have segmented, dsRNA genomes. Virions are unenveloped and have double or triple-layered capsids.
- Genomes are enclosed within the innermost capsid layer (core) and transcription occurs within intact cores. mRNAs exit cores through pores at the fivefold axes of symmetry.
- The viroplasm is an electron dense region of the cell that is formed by reovirus proteins. It is the site of genome replication and core assembly.
- Human rotaviruses are a major cause of severe, debilitating diarrhea in young children. Recently developed vaccines can reduce the morbidity and mortality associated with these infections.
- Members of the genus *Orthoreovirus* are ubiquitous viruses transmitted by the fecal oral route. Their presence in water is an indicator of fecal contamination.
- Members of the genus *Orbivirus* are transmitted by insects. Animal pathogens include BTV, EHDV, and African horse sickness virus.

CHAPTER

27

Family *Birnaviridae*

OUTLINE

After reading this chapter, you should be able to answer the following questions:

- What are the major characteristics of the members of the family *Birnaviridae*?
- What does the family name signify?
- In what ways do birnavirus structure and replication differ from the other dsRNA viruses such as reoviruses?

Birnaviruses are unenveloped double-stranded (ds) RNA viruses that infect birds, fish, and insects (Fig. 27.1). To date, none have been isolated from mammals. One birnavirus, infectious bursal disease virus (IBDV), is an important pathogen of young chicks. Birnaviruses have a number of interesting molecular features. Virions contain a filamentous nucleocapsid, somewhat reminiscent of negative-strand RNA viruses. In contrast to the reoviruses (Chapter 26: Family *Reoviridae*), RNA is uncoated for transcription and genome replication. A precursor polyprotein encoding major structural proteins is cleaved by a viral protease and the 5′ ends of the birnavirus genomic RNA are covalently linked to a protein primer, features similar to some positive-strand RNA viruses. While there is great interest in studying these viruses for their economic impact, many details of their molecular biology remain elusive (Box 27.1).

GENOME STRUCTURE

Birnavirus genomes are bisegmented (two segments) of dsRNA. (Hence the family name is *Birnaviridae*.) Segments are designated A and B and range from ~2800 to 3400 kbp. The 5′ ends of each genome segment are covalently linked to viral protein 1 (VP1). VP1 is the RNA-dependent RNA polymerase (RdRp) as well as the primer for synthesis of genomic RNA. Segment A, the larger of the two, encodes an ~100 kDa precursor polyprotein (VP2-VP4-VP3). VP4 is a protease and mature VP2 and VP3 are capsid proteins. Segment A also encodes a nonstructural protein of ~17 kDa from an overlapping open reading frame. Genome segment B encodes VP1, the virion-associated RdRp. Some VP1 is free in the virion; some is linked to the 5′ ends of genomic RNA. Each genome segment contains conserved untranslated sequences at the 5′ and 3′ ends. It is likely that these regions are important for genome replication and packaging (Fig. 27.2).

VIRION STRUCTURE

Birnaviruses have icosahedral capsids with $T=13$ symmetry, formed by 780 copies of VP2 (organized into 260 trimers). Mature virions are ~60 nm in diameter. Four small peptides (7–46 amino acids) are also associated with capsids. These are cleavage products derived from the carboxyl terminus of VP2. The longest peptide is Pep46, a 46 amino acid peptide that deforms membranes. VP3 is found on the inner face of the capsid. VP3 binds RNA and functions as scaffolding protein.

Viruses. DOI: http://dx.doi.org/10.1016/B978-0-12-803109-4.00027-1

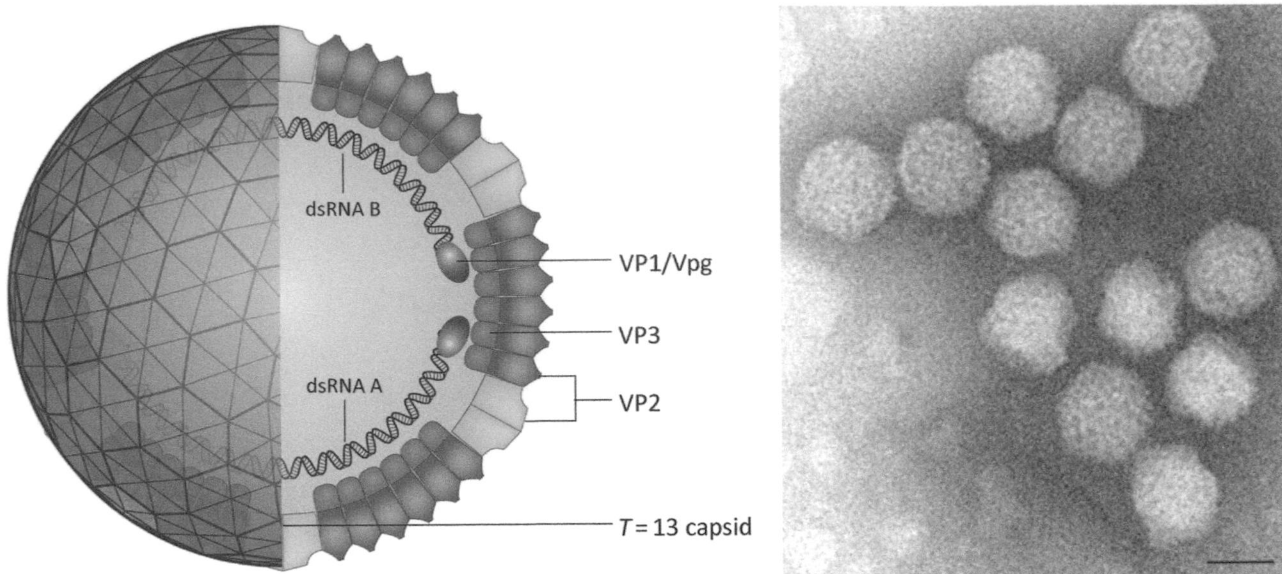

FIGURE 27.1 Virus structure (left). Cryo-electron micrograph of purified particles of Espirito Santo Virus, a birnavirus that replicates in insect cells (right). *From Vancini, R., et al., 2012. J. Virol. 86, 2390–2399.*

BOX 27.1

CHARACTERISTICS

Birnaviruses have bisegmented dsRNA genomes. Segments are ~2800–3400 kbp and are named A and B. The 5′ end of each RNA strand is covalently linked to the protein primer VP1. In addition to serving as the transcription primer, VP1 contains the RdRp catalytic site. Genome segment A encodes a precursor polyprotein encoding a protease (VP4) and major structural proteins VP2 and VP3. Segment B encodes the RdRp, the VP1 protein. Genome replication and transcription are cytoplasmic.

Virions are unenveloped icosahedral capsids ~70 nm in diameter (proposed capsid structure $T = 13$) with 260 spike proteins. VP2 is the spike protein while VP3 forms the inner capsid layer and interacts with dsRNA. Virions contain a transcriptionally active polymerase (VP1).

REPLICATION CYCLE

Attachment/Penetration

It is likely that attachment is mediated by the VP2 trimers that extend from the surface of the capsid and is followed by endocytosis. Within endosomes, low pH may trigger release of pep46 to produce pores in endosomal membranes.

Amplification

Following penetration, viral RNA is found associated with vesicular compartments in the cytoplasm. VP1, the RdRp, transcribes a single full-length mRNA from each genome segment. mRNAs are not polyadenylated but have a base-paired hairpin the 3′ end. It is not entirely clear if mRNAs have a 5′ cap, or are covalently linked to the VP1 primer. One mRNA encodes a polyprotein precursor with domains for VP2 (capsid protein), VP3 (RNA binding and scaffolding protein), and VP4 (viral protease). The second mRNA encodes VP1, the RdRp, as shown in.

VP1 contains the catalytic site for RNA polymerization (RdRp activity) and another site that functions to self-guanylylate a site near its amino terminus. The guanylylated form of VP1 serves to prime transcription, thus generating RNAs covalently attached to a protein at the 5′ end. It is possible to generate infectious virus by transfecting cells with capped, in vitro transcribed full-length mRNAs suggesting that once they are translated, the resulting protein products are competent to complete the virus life cycle.

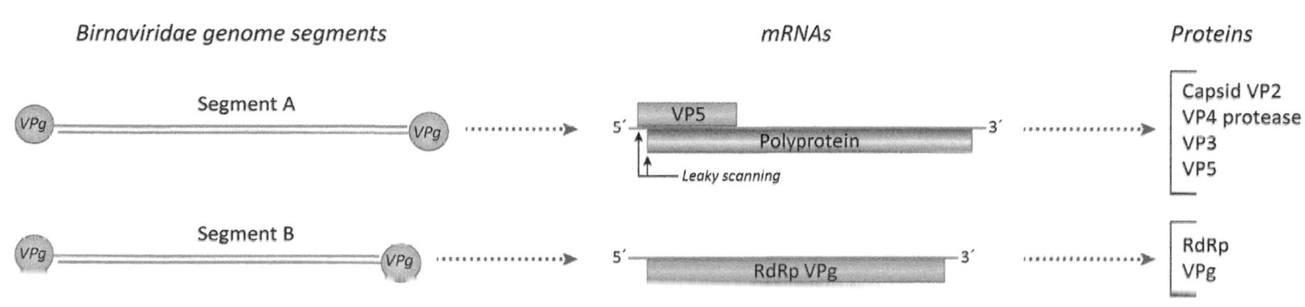

FIGURE 27.2 Genome organization and protein expression strategy.

BOX 27.2

TAXONOMY

Family *Birnaviridae*

Genus *Aquabirnavirus* (includes several important fish pathogens)

Genus *Avibirnavirus* (IBDV and other pathogens of chickens).

Assembly and Release

VP2 is the major capsid protein. Mature VP2 is the product of several cleavages that produce four peptides from its carboxyl terminal end. VP3 is a dsRNA-binding protein that serves both as a scaffold for capsid assembly and a transcriptional activator. VP3 is the major protein component of the nucleocapsid. VP3 localizes to vesicular structures bearing features of early and late endocytic compartments.

In addition to the polyprotein precursor, an addition protein (VP5) is translated from an overlapping reading frame at the 5′ end of the A segment mRNA. VP5 is a small, basic, cysteine-rich protein. VP5 is not found in cell-free virions but is present in infected cells, localized to the inner face of the plasma membrane. Among the roles ascribed VP5 are inhibitor of early apoptosis, stimulator of apoptosis, and virion release.

DISEASE

Infectious Bursal Disease (Gumboro)

IBDV causes immunosuppression in young chicks as the virus targets and destroys the bursa. The bursa is a site of hematopoiesis and is critical for correct development of the avian immune system. Removal of the bursa from newly hatched chicks severely impairs the ability of the adult birds to produce antibodies, while removal of the bursa from adult chickens has little effect on immune function. An interesting feature of IBDV is that development of disease is age related. Chicks infected a less than 3 weeks of age usually have no acute signs while those infected between the ages of 3 and 6 weeks have high mortality. However all infections of young chicks can cause immunosuppression, making birds susceptible to any number of secondary infections. Impairment of the immune system also results in poor responses to vaccines.

There are several birnaviruses that cause serious disease of natural and farmed fish. One of these is infectious pancreatic necrosis virus (IPNV). IPNV infects freshwater trout and Atlantic salmon. Highly virulent strains can cause >90% mortality in hatcheries while survivors can remain life-long carriers, serving as infectious reservoirs (Box 27.2).

In this chapter we learned that:

- Members of the family *Birnaviridae* have unenveloped icosahedral virions ($T = 13$) that contain two segments of dsRNA. Genome replication (and possibly mRNA synthesis) is initiated with a protein primer.
- In contrast to the members of the family *Reoviridae* the birnaviruses uncoat their genomes upon entry into a cell, in a manner reminiscent of negative-strand RNA viruses. Capsid proteins are synthesized as a precursor polyprotein which is cleaved by the virus-encoded protease.
- Birnaviruses cause economically important diseases of fish and poultry, but have not been isolated from mammals.

CHAPTER

28

Introduction to DNA Viruses

OUTLINE

After studying this chapter, you should be able to answer the following questions:

- What is a DNA virus?
- Which families of DNA viruses replicate in the cytoplasm?
- Which families of DNA viruses replicate in the nucleus?
- How is DNA virus replication linked to cell cycle?
- What is the general mechanism by which papillomaviruses, adenoviruses, and polyomaviruses transform cells? For these viruses, what is the relationship between cell transformation and virus replication?

DEFINITION AND BASIC PROPERTIES OF DNA VIRUSES

DNA viruses have DNA genomes that are replicated by either host or virally encoded DNA polymerases. There is considerable diversity among DNA virus genomes and the relative stability of DNA allows for genomes much larger than possible for RNA viruses. Genomes of DNA viruses that infect animals range in size from less than 2 kb of single-stranded DNA to over 375 kb of double-stranded DNA. There are even larger DNA viruses that infect eukaryotic microorganisms.

A common theme among DNA viruses is separation of gene expression into early and late phases (Fig. 28.1). Early transcription occurs before DNA synthesis to provide the protein products needed for DNA replication. These are generally called "early genes." After DNA replication the gene expression profile changes to expression of structural proteins needed to package DNA and form virions. These are generally called "late genes." Herpesviruses encode a number of "late proteins" that are packaged into virions to support transcription upon infection of a new cell (Chapter 34: Family Herpesviridae).

VIRAL DNA REPLICATION

The genomes of DNA viruses are quite variable in structure as well as in size: Single- or double-stranded linear molecules, single or double-stranded circular molecules, and even a highly base-paired single-stranded circular DNA that is often described as a linear double-stranded molecule with "closed ends" (Chapter 35: Family *Poxviridae*) (Fig. 28.2). As might be expected, genome structure and genome replication strategy are linked. Therefore DNA viruses employ variety of genome replication strategies. There is also considerable variation as to the source of DNA replication machinery: Smaller DNA viruses depend almost entirely on host replication machinery while larger DNA viruses encode much of their own (some even replicating in the cytoplasm of eukaryotic cells).

DNA viruses have adopted a variety of strategies to prime genome synthesis: Adenoviruses use a *protein* to prime synthesis of their linear dsDNA genomes. The single-stranded linear genomes of parvoviruses adopt base-paired structures at the ends. Thus the 3′ end of the parvovirus DNA genome serves as the primer to initiate DNA replication (Chapter 29: Family

Viruses. DOI: http://dx.doi.org/10.1016/B978-0-12-803109-4.00028-3

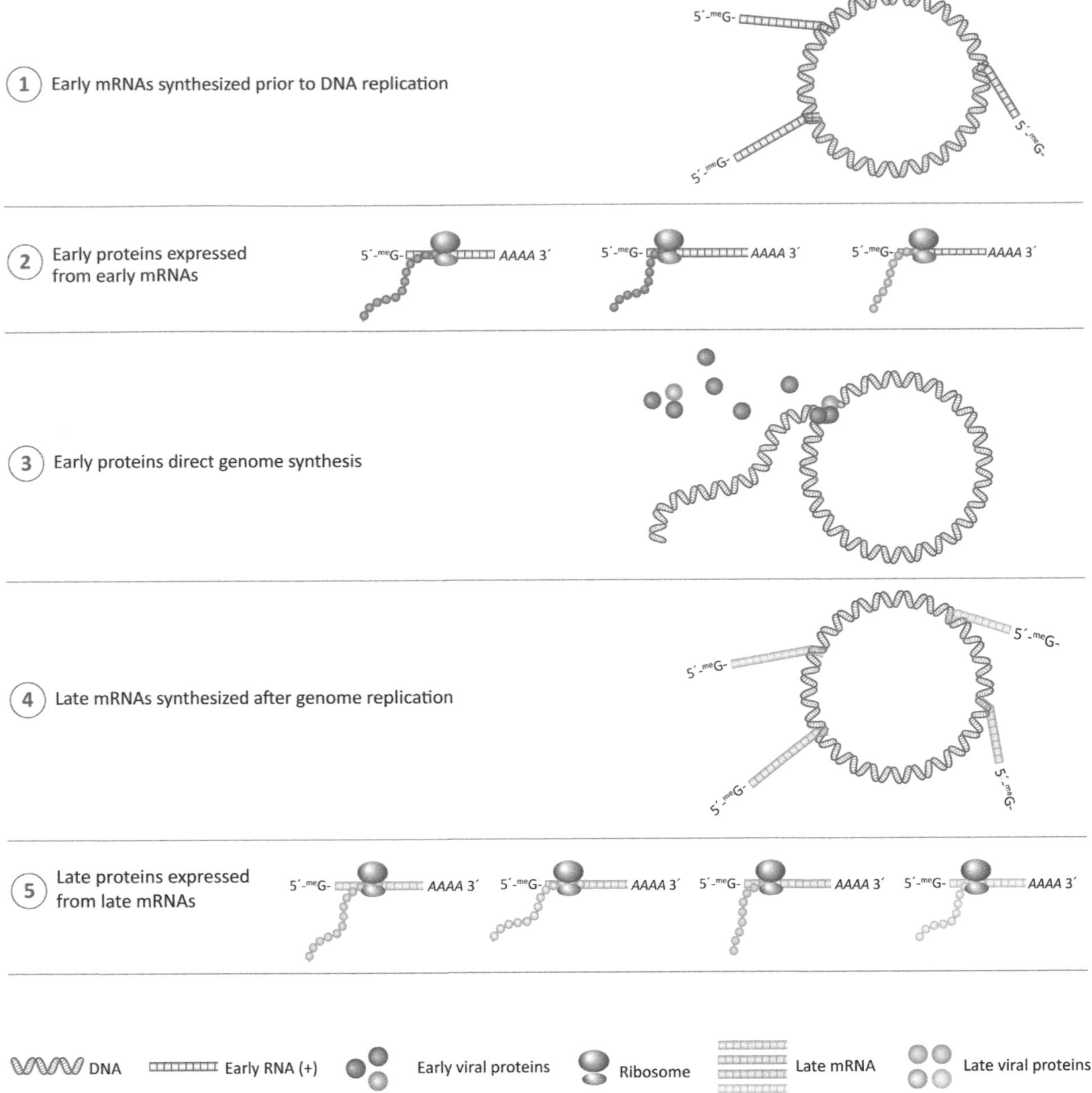

FIGURE 28.1 Upon uncoating, the genomes of DNA viruses are transcribed to produce an "early" set of mRNAs. Early mRNAs typically encode for proteins that modulate the host cell environment and/or are required for viral genome replication. After genome replication another set of mRNAs, the "late" mRNAs are expressed. Late genes encode structural proteins (and other proteins that are packaged within virions).

Parvoviridae). Other DNA viruses make use of host cell primase to synthesizes RNA primers.

DNA viruses use a variety of DNA polymerases for genome synthesis. The smaller DNA viruses use host DNA replication machinery while the larger DNA viruses encode their own DNA polymerases (Table 28.1). Herpesviruses encode a viral DNA polymerase and this has been exploited as a target for antiviral drugs. Of course DNA viruses that use host DNA synthetic machinery must replicate their genomes in the nucleus. Adenoviruses and herpesviruses encode their own DNA polymerases but also replicate in the nucleus as they do not encode a DNA-dependent *RNA polymerase*. DNA viruses that replicate in the nucleus have evolved mechanisms to direct viral DNA into and out of the nucleus. The process often uses both viral and host proteins.

Poxviruses and asfariviruses have large genomes and encode DNA and RNA polymerases. They belong to a group of DNA viruses called the "nuclear and

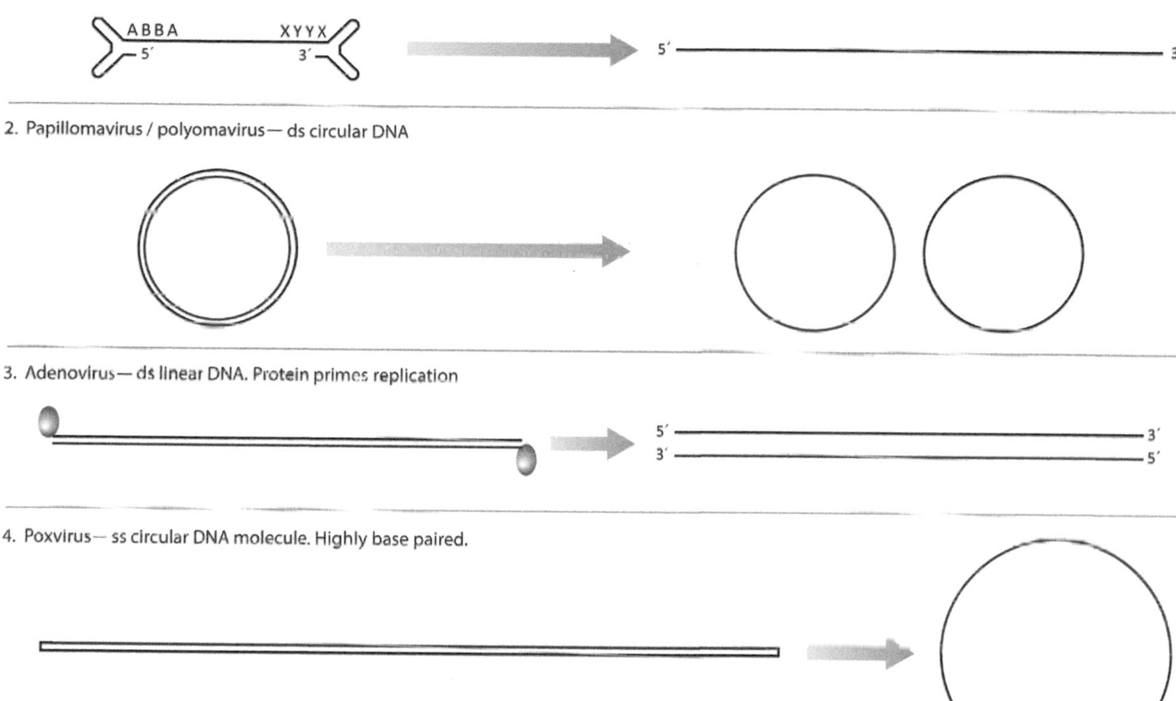

FIGURE 28.2 Examples of viral DNA genomes. Parvoviruses (top panel) have single-stranded DNA genomes that base pair to form double-stranded regions at the ends. The base-paired 3′ OH serves as the primer for DNA replication (Chapter 29: Family *Parvoviridae*). Several virus families have circular, double-stranded DNA genomes. Adenoviruses have double-stranded linear DNA genomes. Adenoviruses use a protein to prime DNA replication (Chapter 33: Family *Adenoviridae*). Poxviral genomes are often described as linear double-stranded DNA molecule. However, the ends are covalently closed, forming a single-stranded circular DNA (Chapter 35: Family *Poxviridae*). This is most obvious when the DNA is denatured.

TABLE 28.1 DNA Viruses of Animals[a]

Virus family	Genome	Genome size (kb)	Virion morphology	Source of DNA polymerase	Source of RNA polymerase	Site of replication
Parvoviridae	ss linear	4–6	Naked, icosahedral	Host	Host	Nucleus
Circoviridae	ss circular	1.8–3.8	Naked, icosahedral	Host	Host	Nucleus
Anelloviridae	ss circular	3.8	Naked, icosahedral	Host	Host	Nucleus
Polyomaviridae	ds circular	5	Naked, icosahedral	Host	Host	Nucleus
Papillomaviridae	ds circular	8	Naked, icosahedral	Host	Host	Nucleus
Adenoviridae	ds linear	35–36	Naked, icosahedral	Viral	Host	Nucleus
Herpesviridae	ds linear	120–240	Enveloped, icosahedral capsid	Viral	Host	Nucleus
Poxviridae	ds linear[b]	130–375	Enveloped, complex	Viral	Viral	Cytoplasm
Asfariviridae	ds linear	170–190	Enveloped, icosahedral capsid	Viral	Viral	Cytoplasm

[a]*Many of these families contain members that infect a wide range of organisms.*
[b]*Ends of the molecule are linked (single-stranded circle upon denaturation).*

cytoplasmic large DNA viruses." Poxviruses and asfariviruses replicate and transcribe their genomes in the cytoplasm: In addition to polymerases, they encode topoisomerases, transcription factors, and processivity factors that link DNA polymerase to the template, such as proliferating cell nuclear antigen.

The process of actually synthesizing new DNA strands is also variable. Some strategies, such as the DNA displacement method used by adenoviruses (Chapter 33: Family *Adenoviridae*), are relatively easy to describe. In other cases the process can be quite complex, relying on DNA nicking enzymes and methods for transitioning or resolving structures from double-stranded to single-stranded molecules (Chapter 29: Family *Parvoviridae*). Two of the more common strategies for replication of double-stranded circular DNA genomes are illustrated in Fig. 28.3: Theta replication and rolling circle replication. Herpesviruses package a molecule of linear double-stranded DNA that circularizes in the nucleus. Herpesvirus genome replication seems to initiate with a theta replication and then switch to a rolling circle mechanism that produces long genome concatamers.

DNA VIRUSES AND CELL CYCLE

A feature common to many DNA viruses is their dependence on host DNA replication to provide enzymes (for example, DNA polymerases) and pools of deoxynucleotide triphosphate precursors. While RNA turnover is constant in metabolically active cells, DNA synthesis is tightly regulated. Cells not mitotically active may be poor hosts for DNA viruses. In adult animals there are many cells that are terminally differentiated and seldom, if ever divide (for example, neurons). Some cells divide in response to damage, while others (for example, epithelial cells) divide regularly. In contrast, the mitotic activity of a developing fetus and growing animal is much higher than in the mature adult. There are three basic strategies used by

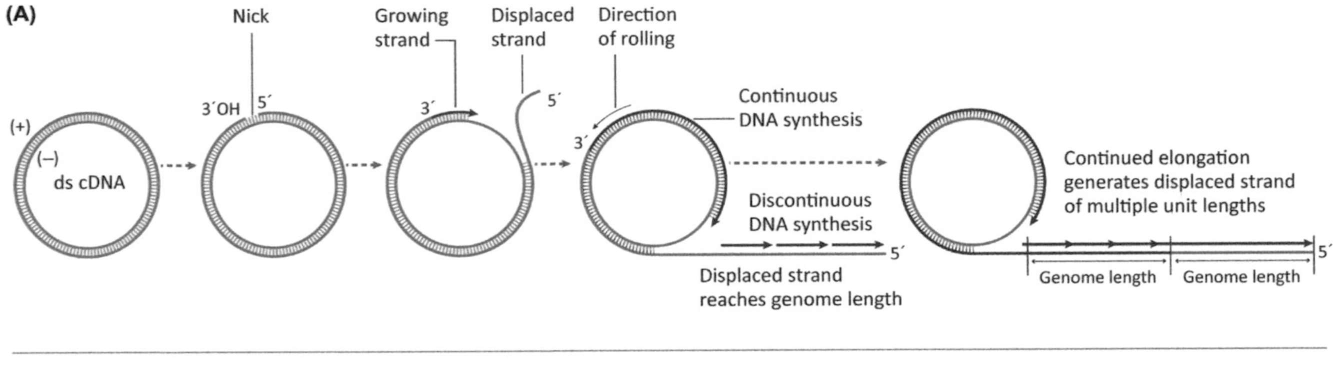

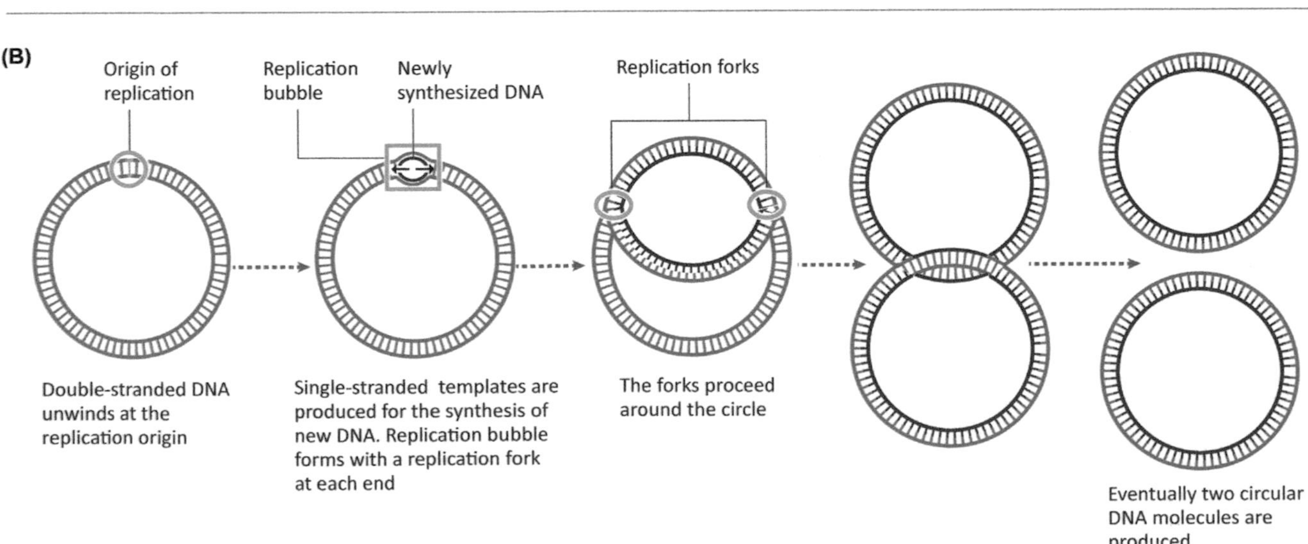

FIGURE 28.3 General models of rolling circle DNA replication (panel A) and theta replication (panel B).

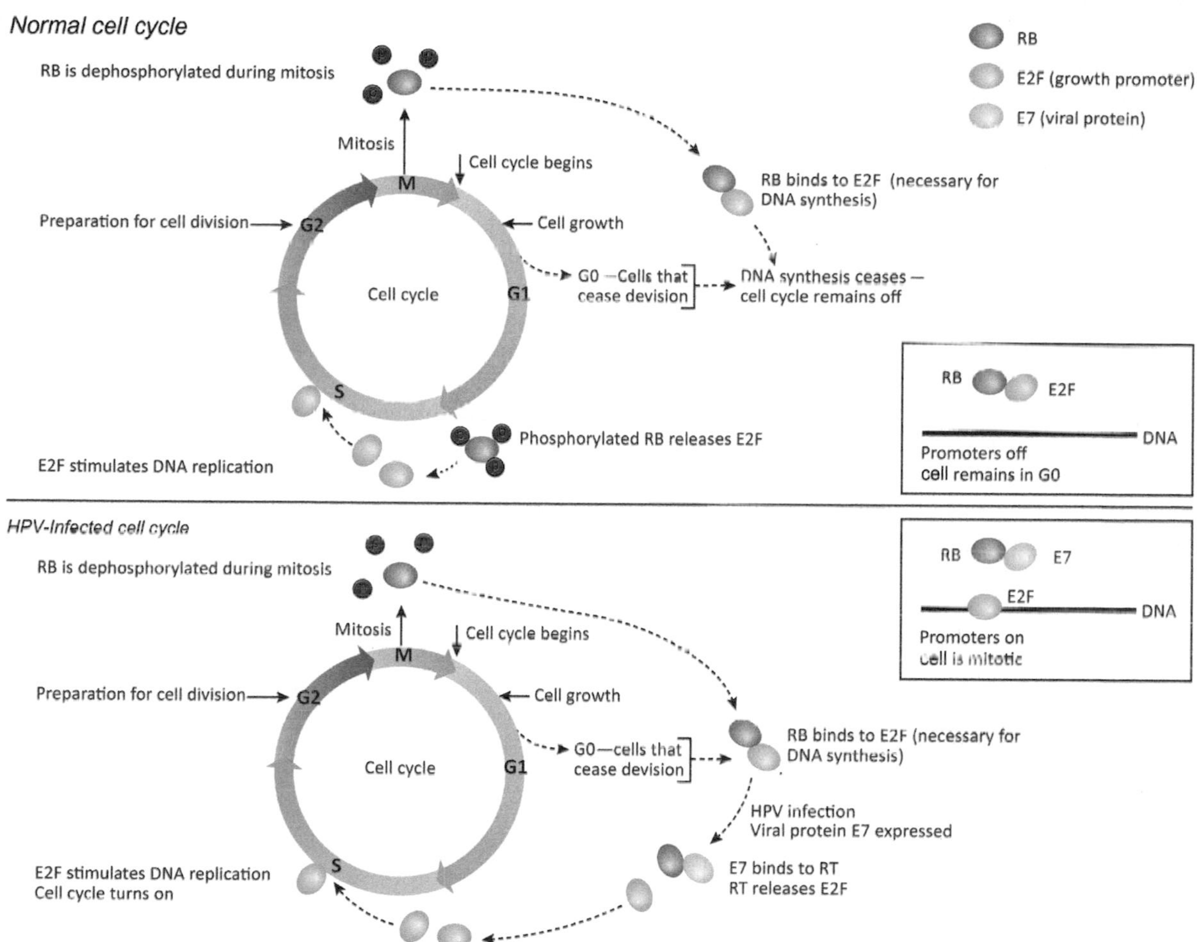

FIGURE 28.4 Comparison of normal cell cycle and HPV-infected cell cycle. Top panel: Summary of cell cycle showing contributions of pRB family proteins in regulating DNA replication. In cells that are not mitotically active (G_0), pRB is bound to the transcription factor E2F. E2F is a major regulator of genes required for DNA synthesis. In a normal cell, external signals (for example, growth factors) may initiate mitosis, leading to the phosphorylation of pRB by cyclin-dependent kinases. Phosphorylated pRB releases E2F to turn on genes required for DNA synthesis. At the end of the cell cycle, pRB is dephosphorylated and rebinds to E2F. Bottom panel: Some DNA viruses (papillomaviruses, polyomaviruses, and adenoviruses) encode proteins that disrupt normal RB function. In this example papillomavirus E7 protein binds to RB causing the release of E2F, thereby inducing continued DNA replication in the infected cell.

TABLE 28.2 Three families of DNA viruses use similar mechanisms to affect cell cycle.

Virus family	Proteins that interact with pRB	Proteins that interact with p53
Polyomaviridae	Large T	Large T
Papillomaviridae	E7	E6
Adenoviridae	E1A	E1B

DNA viruses to ensure that they can replicate their genomes:

- Infect only mitotically active cells.
- Induce nondividing cells to enter the cell cycle by use of viral proteins.
- Encode most or all enzymes required for dNTP synthesis and DNA replication.

The smallest DNA viruses (circoviruses and parvoviruses) productively infect only mitotically active cells. As we will see in coming chapters, this impacts cell tropism and pathogenesis. The larger DNA viruses (polyomaviruses, papillomavirus, adenoviruses, and herpesviruses) encode proteins that induce resting cells to become mitotically active. Even the larger poxviruses, that encode much of the enzymatic machinery required for DNA synthesis, tend to stimulate cell proliferation.

ONCOGENESIS

Several families of DNA viruses are associated with cell transformation and oncogenesis. The polyomaviruses, papillomaviruses, and adenoviruses transform cells using a similar strategy. All encode proteins that alter cell cycle and inhibit apoptosis (Table 28.2). The common cellular targets are retinoblastoma protein (pRB) and p53. pRB and related proteins tightly control cell cycle by sequestering the transcription factor E2F. E2F activates transcription of a set of host genes that drive DNA replication and mitosis. By inactivating pRB, polyomaviruses, papillomaviruses, and adenoviruses induce host cell DNA synthesis (Fig. 28.4). Eucaryotic cells also have mechanisms to sense DNA damage, and either pause the cell cycle or activate apoptotic pathways. A key regulator of these processes is the protein p53. In order to complete their replication cycles, polyomaviruses, papillomaviruses, and adenoviruses encode proteins that inactivate p53. However, in a productive infection, neither polyomaviruses, papillomaviruses nor adenoviruses transform cells, as the end result is virion production and cell death. It is only when infection is restricted or nonproductive that cell transformation is a possible outcome. This is why adenoviruses and polyomaviruses do *not* produce tumors in their normal hosts; they replicate with a resulting lytic infection. However in nonpermissive cells their viral oncogenes may be expressed and initiate the process of transformation. Some papillomaviruses *do* contribute to tumor formation in natural hosts. The best-studied are the human papillomaviruses (HPVs) associated with cervical cancers. Papillomaviruses may be oncogenic (occasionally) in their natural hosts because of their overall replication strategy: Virions are only produced in differentiated epithelial cells (Chapter 32: Family *Papillomaviridae*). However there is considerable viral DNA replication, without virion production, in nondifferentiated epithelial cells. This provides an environment in which transformation may occur. When one examines the tumors initiated by papillomavirus infection, it is usual to find that they are monoclonal (derived from a single precursor cell) and contain part of a papillomavirus genome integrated into host DNA. Thus, while virally induced cancers are of great importance to the host, they are actually a "dead end" for the virus involved.

In this chapter we learned that:

- DNA viruses have DNA genomes that are replicated by a DNA-dependent DNA polymerase. Some encode a viral polymerase but other use host enzymes.
- *Poxviridae* and *Asfariviridae* are two families of large DNA viruses that replicate in the cytoplasm.
- Most families of DNA viruses replicate in the nucleus. These include *Circoviridae*, *Parvoviridae*, *Anelloviridae*, *Polyomaviridae*, *Papillomaviridae*, *Adenoviridae*, and *Herpesviridae.*
- Replication of several DNA viruses is tightly linked to the cell cycle. In order to replicate viral genomes, the cell must provide DNA polymerase and dNTPs. Some small DNA viruses infect only mitotically active cells; other viruses induce cells to become mitotically active.
- Papillomaviruses, adenoviruses, and polyomaviruses encode proteins that promote cell division (by inactivating pRB family members). They also encode proteins that inactivate p53, to inhibit apoptosis induced by DNA damage.

C H A P T E R

29

Family *Parvoviridae*

OUTLINE

After reading this chapter, you should be able to answer the following questions:

- What are the major characteristics of the members of the family *Parvoviridae*?
- What cell environment does a parvovirus require for productive infection and how does this impact pathogenesis?
- What common childhood disease is caused by a parvovirus?
- What life-threatening disease of cats is caused by a parvovirus?

Parvoviruses are small, unenveloped, linear single-stranded (ss) DNA viruses whose genome size is ~5000 nt. Parvoviruses have T = 1 icosahedral symmetry with capsids assembled from two related capsid proteins, VP1 and VP2 (Fig. 29.1). The family *Parvoviridae* contains two subfamilies, *Parvovirinae* and *Densovirinae* (invertebrate parvoviruses). There are eight named genera in the *Parvovirinae* subfamily including *Erythroparvovirus* (human B19 parvovirus), *Dependovirus* (nonautonomous adeno-associated viruses (AAVs)) and *Protoparvovirus* (includes feline panleukopenia virus (FPV) and canine parvovirus) (Box 29.1). Given their small genome size, parvo viruses require much cellular assistance to replicate their genomes. For example, they *require* dividing cells to provide adequate pools of dNTPs; but parvoviruses do not encode proteins that activate the cell cycle therefore many replicate only in actively dividing cells. The dependoviruses rely upon helper viruses (larger DNA viruses such as adenoviruses or herpesviruses) to induce cell division. While the small size of parvovirus genomes is somewhat limiting, parvoviruses virions are extremely stable in the environment. They resist both high and low pH and are stable for an hour at 56°C (132.8°F).

GENOME ORGANIZATION

Parvoviruses genomes are a molecule of linear ssDNA of ~5000 nt. At each end of the genome are noncoding regions that form stable hairpins. These hairpin structures are essential for genome replication as they serve as primers for DNA replication. Two large open reading frames (ORFs) encode a nonstructural protein (NS or REP) and a structural protein (VP or C). Additional small ORFs may also be present. The number and positions of transcription promoters vary among genera as well. Use of variable transcription start sites, alternative splicing and alternative polyadenylation sites produce a diverse set of mRNAs from these relatively simple genomes (Box 29.2).

Viruses. DOI: http://dx.doi.org/10.1016/B978-0-12-803109-4.00029-5

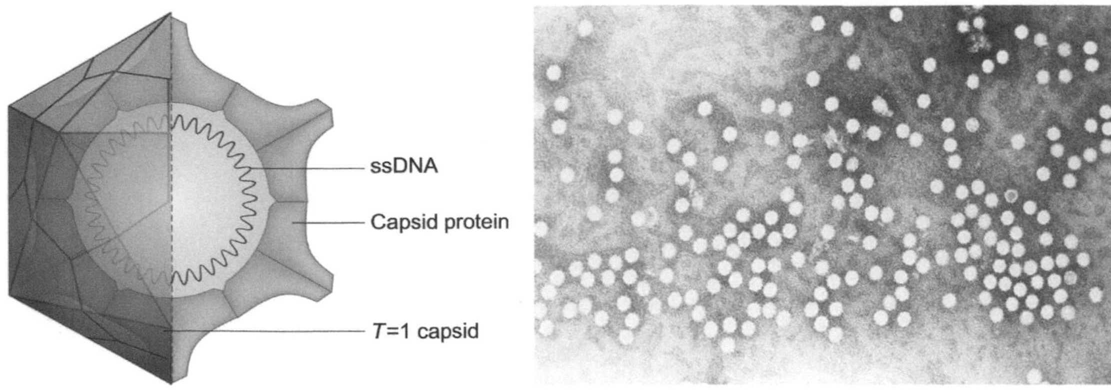

FIGURE 29.1 Parvovirus structure (left) and electron micrograph (right). *From CDC/ R. Regnery; E. L. Palmer CDC Public Health Image Library Image 5618.*

BOX 29.1

TAXONOMY

Family: *Parvoviridae*
Subfamily: *Parvovirinae*
Genus: *Amdoparvovirus* (Aleutian mink disease virus)
Genus: *Aveparvovirus* (Chicken parvovirus)
Genus: *Bocaparvovirus* (Bovine parvovirus)
Genus: *Copiparvovirus* (Ungulate copiparvovirus)
Genus: *Dependovirus* (Commonly called AAVs they require coinfection with an adenovirus or (less commonly) a herpesvirus. The dependoviruses are most closely related to autonomous avian parvoviruses.)
Genus: *Erythroparvovirus* (The type species is primate erythroparvovirus 1, usually referred to as human parvovirus B19)
Genus: *Protoparvovirus* (Includes Minute virus of mice (MVM), Feline panleukopenia virus (FPV), canine parvovirus (CPV), and mink enteritis virus)
Genus: *Tetraparvovirus* (primate tetraparvovirus 1)
Subfamily: *Densovirinae* (insect viruses).

BOX 29.2

GENERAL CHARACTERISTICS

Parvovirus genomes are linear, ssDNA, 4–6 kb. Most have two reading frames. (Some have additional short ORFs.) Transcription and genome replication are nuclear. Parvoviruses replicate only in dividing cells or in the presence of a helper virus.

Virions are unenveloped with $T=1$ icosahedral symmetry with spike-like protrusions, ~25 nm diameter. They assemble from 60 copies of two or three alternative forms of VP. Virions are very stable in the environment.

VIRION STRUCTURE

Parvoviruses are small, unenveloped viruses with $T=1$ icosahedral symmetry. Virions are about 25 nm in diameter with spike-like protrusions. Capsids are assembled from 60 copies of between two and four forms of VP. Most autonomous parvoviruses and the AAVs assemble from three capsid proteins that share a common carboxyl-terminus. They are produced from alternative splicing of the VP or capsid gene. The capsid of human parvovirus B19 is assembled from two capsid proteins. For most or perhaps all parvoviruses, VP1 (the largest of the capsid proteins) is a minor component of the capsid. However VP1 is critical for infectivity as it contains an essential enzymatic domain with calcium-dependent phospholipase A2 (PLA_2) activity.

OVERVIEW OF REPLICATION

The so-called autonomous parvoviruses replicate only in actively dividing cells while members of the genus *Dependovirus* have broader cell tropism but require coinfection with a helper virus. This section will focus primarily on replication of the autonomous (helper-independent) parvoviruses.

Attachment/Penetration

Parvoviruses use a variety of protein and/or carbohydrate receptors. For example, human parvovirus B19 (B19V) binds to a glycolipid receptor (called P antigen or globoside) present on primary erythroid progenitor cells. (Erythroid progenitor cells are committed stem cells that give rise to red blood cells.) Minute virus of mice (MVM) binds sialic acid receptors, thus hemagglutinates mouse erythrocytes. There are 12 serotypes of AAVs that are targeted to different cells by virtue of different receptor usage. Receptors for AAVs include both carbohydrate and protein receptors. The variety of receptors used by AAV has facilitated their use as gene delivery vehicles.

Parvoviruses enter cells by endocytosis and the low pH of environment of the endosome leads to critical structural changes that expose both a PLA_2 domain and a nuclear localization signal present at the amino terminus of VP1. The PLA_2 domain is a calcium-dependent phospholipase (hydrolyzes phospholipids into fatty acids) that is required to liberate the virion from the endosome. After a virion leaves the endosome, it associates with microtubules and is actively transported to the nucleus. VP1 then appears to interact with cellular proteins for translocation across the nuclear pore.

The 3′ end (20–30 nt) of parvoviral DNA appears to be exposed on the outside of essentially intact capsids. It has been suggested that the exposed DNA could serve as the initiation site for DNA polymerase, such that the viral DNA is actively removed from the capsid without its disassembly.

Genome Replication of Autonomous Parvoviruses

The genomes of autonomous parvoviruses are replicated by host cell DNA polymerases. NS1 is the key viral player in DNA replication. The current model for autonomous parvovirus DNA replication is that cellular DNA polymerase begins synthesis at the base paired 3′ end of the genome, forming a linear duplex (Fig. 29.2).

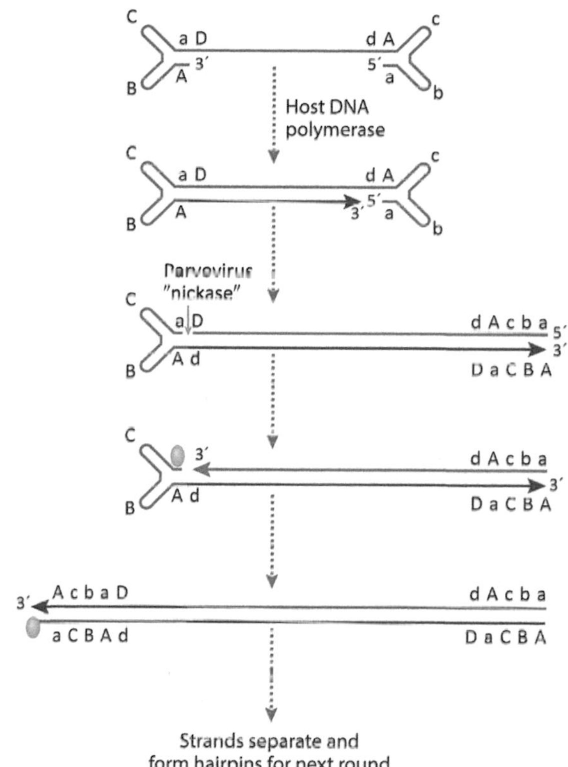

FIGURE 29.2 Parvovirus genome replication strategy.

After formation, the linear duplex is nicked by the NS1 protein. In the process of nicking, NS1 is covalently linked to the 5′ end of the viral DNA. DNA synthesis continues at the new 3′ end liberated by nicking. NS1 is the major parvovirus regulatory protein and is a helicase, adenosine triphosphatase (ATPase), and site-specific DNA nicking enzyme (nickase). Parvovirus DNA replication takes place in discrete subnuclear structures that contain cellular DNA polymerases alpha and delta.

Transcription and Translation

Transcripts and proteins appear early in the course of infection. The transcriptional schemes of the autonomous parvoviruses are highly variable. Many use two promoters and one or two polyadenylation signals. Human parvovirus B19V uses a single promoter. There are also one or two splice donor sites and multiple splice acceptors sites (Fig. 29.3). These regulatory signals serve to generate a variety of mRNAs.

All parvoviruses encode an NS protein (called REP in the case of the dependoviruses). NS protein is important for both DNA replication and transcription

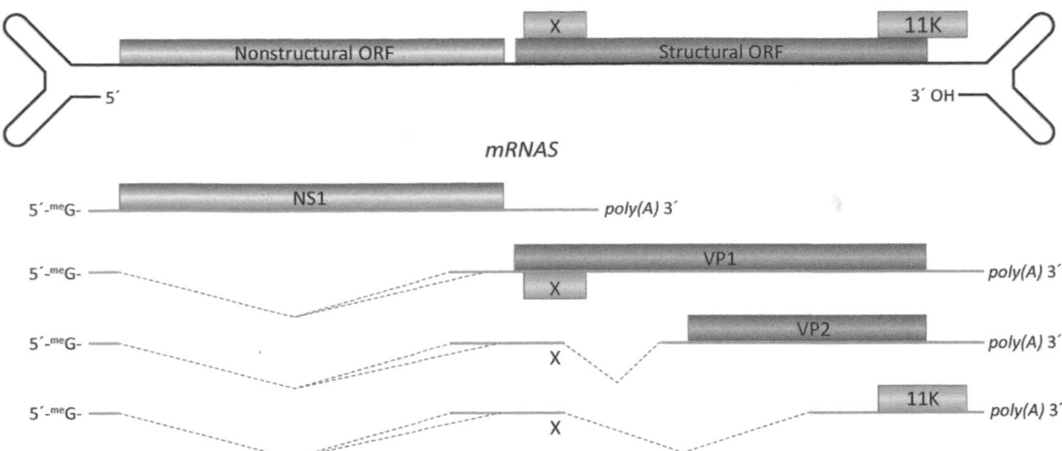

FIGURE 29.3 Parvovirus genome organization.

regulation. NS is an ~77 kDa phosphoprotein that has endonuclease (nicking) and helicase activities. It is localized to the nucleus of infected cells where its nicking and helicase activities are critical for genome replication. Some parvoviruses produce spliced versions of NS. In the process of nicking the parvovirus DNA strand, NS is covalently linked to the 5′ end of the DNA strand and in this context it facilitates DNA packaging. B19V NS also affects host cell transcription and induces apoptosis in some cell types.

Spliced mRNAs are used to express parvovirus structural proteins (VP1 and VP2). VP1 and VP2 share a common carboxyl-terminus; VP1 of B19V is ~84,000 kDa while VP2 is ~58,000 kDa. The majority of the B19V capsid is constructed from VP2. B19V VP1 has an additional 227 amino acids at the amino terminus and this domain contains the phospholipase activity essential for virus entry into cells. (In the case of dependoviruses, the VP ORF is often called CAP, for capsid.)

Assembly

Parvovirus capsids assemble in the nucleus. DNA incorporation occurs by insertion of viral ssDNA into the preassembled capsid and requires the helicase activity of NS1 to translocate ssDNA into the capsid. Capsids may be retained within the nucleus, translocated into the cytoplasm, or actively transported out of the cell.

DISEASES

Human B19 Parvovirus

Human B19 parvovirus (B19V) infects school age children and is the cause of erythema infectiosum or fifth disease (it was fifth on a list of six childhood rash-forming illnesses). It is also called slapped-cheek disease due to bright red rash on face that is a common symptom. The disease begins with general symptoms of a stuffy or runny nose, headache, and a low-grade fever. After these initial symptoms resolve a rash appears, typically on the face and then spreads to the trunk, arms, and legs.

We mentioned earlier that B19V infects erythroid progenitor cells and is found replicating in the bone marrow. Why then is the typical course of acute infection in children a mild febrile illness followed by a rash? Based on limited clinical studies, it appears that in a healthy child the virus initially replicates in the upper respiratory tract before spreading to other sites. In the bone marrow, erythroid progenitor cells are infected and produce an abundance of virus, resulting in viremia (and the clinically apparent "mild febrile illness"). By this time the immune system is responding and immune complexes (antibody and virus) are deposited in the skin, leading to the characteristic rash. Immune complex deposition could also explain symptoms of joint pain and arthritis associated with B19V infections in adult women.

It is likely that many people are infected with B19V without developing any clinical symptoms at all. However the disease course can be very different in patients with underlying conditions that affect the life span of their red blood cells. In these patients the transient loss of erythroid progenitor cells can cause severe, life-threatening anemia. B19V is also implicated in a variety of rare, but severe blood disorders in immunocompromised individuals. Finally, there are suggested links between B19V infection and spontaneous abortion.

Metagenomics studies have led to the identification of additional human parvoviruses. In 2005 a new human parvovirus, related to bovine and canine parvoviruses (genus *Bocaparvovirus*), was identified and named human bocavirus (HBoV). HBoV is found worldwide, present in both respiratory tract samples and stool samples. The most frequently described clinical presentation of HBoV1 infection is upper respiratory tract disease.

Feline Panleukopenia Virus

FPV infects all felids (domestic and wild cats) and in some members of related families (e.g., raccoon, mink). FPV is found worldwide and the virus is endemic in unvaccinated domestic cat populations. FPV is associated with the diseases feline panleukopenia, feline infectious enteritis, and feline distemper. These diseases are most common in recently weaned kittens, at risk of infection as levels of maternal antibody decline. Disease occurs when virus replicates in mitotically active tissues including intestine, bone marrow, thymus, lymph nodes, and spleen. Numbers of circulating white blood cells may decrease by 90% resulting in panleukopenia. FPV is transmitted by respiratory and/or fecal-oral routes. Virus is shed in feces, saliva, urine, vomit, and blood. Virus can be shed for long periods (many weeks) and the virus is very stable in the environment.

Canine Parvovirus

Canine parvovirus-2 (CPV-2) is closely related to FPV. While FPV does not infect dogs, CPV-2 emerged as a dog pathogen in the 1970s and quickly spread worldwide. CPV-2 was initially associated with very high morbidity and mortality due to heart failure in utero or shortly after birth. However the clinical picture changed as more dogs were exposed to the virus and acquired immunity. Now CPV-2 infection of puppies most often leads to enteric disease.

Aleutian Mink Disease

Aleutian mink disease is also caused by a parvovirus genetically similar to FPV and CPV-2. (Aleutian refers to the gray coat color of the farmed minks at the center of the first reported outbreak. However the virus infects minks of all colors, as well as ferrets.) The disease is worthy of mention because its pathology is quite different from the other "carnivore" parvoviruses. The infection is chronic, not acute, and ferrets can be long-term healthy carriers. When symptoms do appear, they include poor reproduction, weight loss, oral and gastrointestinal bleeding, and renal failure. The novel feature of Aleutian mink disease is that its pathogenesis is immune-mediated. Chronic infection is asymptomatic until high levels of virus and antibodies combine to form immune complexes, deposited in many organs.

In this chapter we learned that:

- Parvoviruses are small, unenveloped viruses with linear ssDNA genomes.
- Some parvoviruses are autonomously replicating viruses that infect mitotically active cells. Others rely on coinfection with a larger DNA virus (adenovirus or herpesvirus).
- In some cases the pathogenesis of parvoviral infections is related to direct killing of mitotically active cells while in other cases disease is immune-mediated.
- The most common human parvoviral infection is B19V. It causes erythema infectiosum or fifth disease in children. The infection is usually well controlled by the immune system, but can cause chronic infection in immunosuppressed individuals. In rare cases, severe blood disorders have been associated with B19V infection.
- FPV infects wild and domesticated cats worldwide. Disease is usually associated with recently weaned kittens. The virus can cause gastrointestinal disease (enteritis) or a global reduction in white blood cells (panleukopenia).

CHAPTER

30

Other Small DNA Viruses

OUTLINE

After reading this chapter, you should be able to answer the following questions:

- What are the general characteristics of circoviruses?
- What are the general characteristics of anelloviruses?
- How do the capsid structures of circoviruses and anelloviruses differ?
- How does the genome organization of circoviruses differ from genome organization of anelloviruses?
- Why are these viruses restricted to replication in mitotically active cells?
- What is apoptin and why is it of interest to cancer researchers?

FAMILY *CIRCOVIRIDAE*

Circoviruses are small, unenveloped viruses ($T = 1$) that are extremely stable in the environment. Their capsids assemble from 60 copies of the ~28 kDa capsid (C) protein (Fig. 30.1). Virions have a rather smooth surface appearance and a size of ~17–20 nm. Circovirus genomes are covalently closed, circular, single-stranded DNA of ~1700–1800 nt. They have two major open reading frames (ORFs) transcribed in opposite directions from a single promoter site (Fig. 30.2). One ORF encodes the capsid (C) protein, the other encodes the nonstructural replicase-associated (Rep) protein. Circoviruses replicate only in mitotically active (dividing) cells as they use host cell polymerases for genome replication and do not encode proteins that can stimulate cell division (Box 30.1).

The only virally encoded proteins involved in genome replication are Rep and Rep′. Rep is 314 amino acids whereas Rep′, synthesized from a spliced mRNA, is 178 amino acids. The amino terminal ends of Rep and Rep′ are the same and contain conserved sequences found in proteins that initiate rolling circle replication.

Detailed models for circovirus replication are lacking but it appears that they may favor carbohydrate receptors. It is assumed that entry is mediated by endocytosis and the virion makes its way to the nucleus. The single-stranded DNA must next be converted to a double-stranded molecule to support transcription by a host RNA polymerase and finally replication by host DNA synthesis machinery. Details of DNA packaging, assembly and release are sketchy, but will certainly be forthcoming.

DISEASES

Beak and Feather Disease

Beak and feather disease virus (BFDV) infects psittaciform birds (parrots). The virus seems to have originated in Australia where it continues to threaten wild parrot populations (Fig. 30.3). The virus is also found worldwide in captive parrots. The virus infects mitotically active cells of the immune system as well as the cells that produce feathers and beak. Affected birds

Viruses. DOI: http://dx.doi.org/10.1016/B978-0-12-803109-4.00030-1

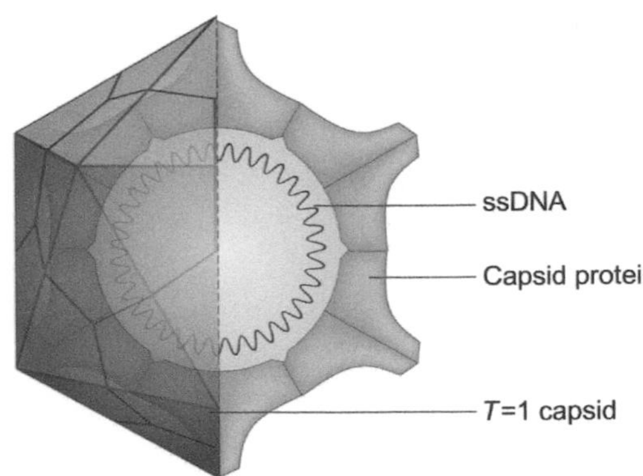

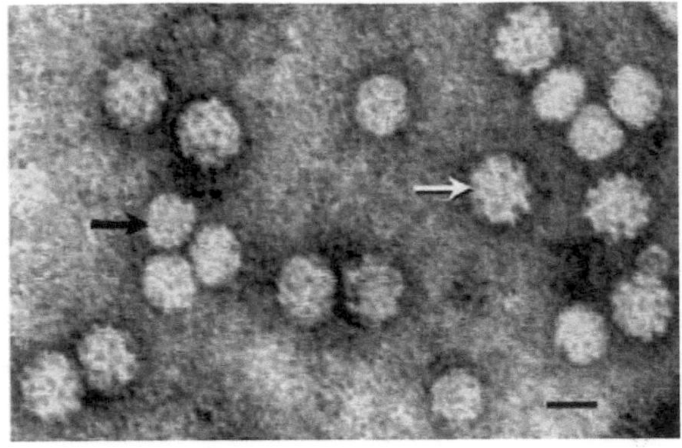

FIGURE 30.1 Circovirus virion structure (left) and negative contrast electron microscopy (right) of particles of an isolate of Chicken anemia virus (CAV) (*black arrow*) and Beak and feather disease virus (BFDV) (*white arrow*), stained with uranyl acetate. The bar represents 20 nm. *Anon, 2005. The single stranded DNA viruses. In: Fauquet, C.M., Mayo, M.A., Maniloff, J., Desselberger, U., Ball, L.A. (Eds.), Virus Taxonomy. Academic Press, San Diego, p. 327.*

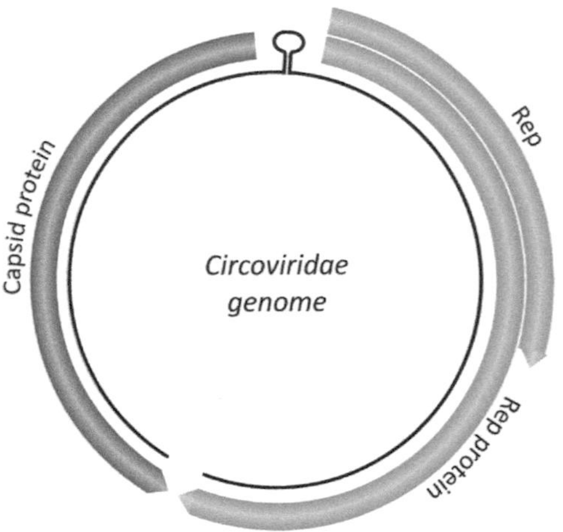

FIGURE 30.2 Circovirus genome organization.

gradually lose their feathers, develop beak abnormalities and become immunocompromised, often dying of infections by a variety of microorganisms. As might be expected for a virus that depends upon mitotically active cells for replication, young birds are much more susceptible to infection than are adults birds. There is no treatment and the disease is usually fatal.

Postweaning Multisystemic Wasting Syndrome

Two porcine circoviruses have been isolated from pigs. Porcine circovirus 1 (PCV1) is ubiquitous, and nonpathogenic in swine. Porcine circovirus 2 (PCV2) emerged as a major swine pathogen about 20 years ago and in now endemic worldwide. PCV2 is associated with postweaning multisystemic wasting syndrome (PMWS), a disease characterized by jaundice, diarrhea,

BOX 30.1

GENERAL DESCRIPTION AND TAXONOMY OF CIRCOVIRUSES

Genomes are single-stranded circular DNA of ~1700–3800 nt. The genome contains a single promoter (with binding sites for known cellular transcription factors). Two major transcripts are synthesized in opposite directions. One mRNA encodes the 28 kDa capsid (C) protein that assembles to form the $T=1$ capsid. The second ORF encodes Rep and Rep′ (a product of mRNA splicing). The Rep proteins are likely the key viral players in initiating rolling circle replication to produce more genomes. Genome replication and transcription take place in the nucleus.

Family: *Circoviridae*

Genus: *Circovirus*

Species include *Porcine circovirus type 1 and 2 and beak and feather disease virus (BFDV)*. ICTV Taxonomy (2015) lists 22 species detected in animals and birds. Some have been identified as a result of sequencing studies to characterize fecal or environmental microbiota and a few are bird pathogens.

Genus: *Cyclovirus*

Several species associated with human fecal samples. Also identified from insects, birds, and other mammals.

respiratory disease, failure to thrive, and sudden death. The origins of PCV2 are not clear (it did not descend directly from PCV1) but it may have been a long-time ubiquitous agent in pigs. Its exact role in disease development is still somewhat murky, as PCV2 is necessary, but not sufficient, on its own, to cause disease. In experimental systems, cofactors such as infections with porcine parvovirus or immune stimulation greatly exacerbate disease. What is clear is that use of PCV2 vaccines reduces outbreaks of PMWS.

FAMILY ANELLOVIRIDAE

In 1997 a novel, partial viral clone was isolated from a patient (initials TT) with posttransfusion hepatitis of unknown etiology. This was the first report of TT virus (TTV) but follow-up epidemiological surveys revealed that TTV-like viruses were ubiquitous among humans, worldwide. Sequencing studies also revealed that TTV is genetically quite diverse. Now known at torque teno virus, TTV is a member of the family *Anelloviridae* (Box 30.2). As of 2015, the family contains 12 genera containing viruses isolated from a variety of animals. One genus, *Cyclovirus*, contains chicken anemia virus (CAV), a well-recognized avian pathogen that was once thought to be a close relative of the circoviruses (see Fig. 30.1).

FIGURE 30.3 Parrots with beak and feather disease.

Anelloviruses are small, unenveloped viruses with $T=1$ icosahedral symmetry (~25 nm diameter). The best studied is CAV. Capsids have a spiky appearance as the result of being assembled from trumpet-shaped capsomers (12 pentamers of VP1 capsid protein, Fig. 30.4). Anelloviruses genomes are circular single-stranded DNA, range from 2 to 4 kb and encode three or four proteins (Fig. 30.5). A single untranslated region of the genome contains a promoter that directs mRNA synthesis in one direction. VP1 is the 52 kDa capsid protein (60 copies assemble to generate the capsid). VP2 is a nonstructural protein with phosphatase activity. VP2 catalyzes the removal of phosphate from both phosphoserine/phosphothreonine and phosphotyrosine substrates. VP3 is a nonstructural protein that induces apoptosis (thus is also called apoptin).

Anelloviruses infect only mitotically active cells and they have gained notice and recognition as causing lytic infections of cancer cells, but not normal cells. This characteristic is linked to the activities of VP3/apoptin. Apoptin is a so-called *oncolytic protein* due to its ability to lyse cancer cells while leaving normal cells unaffected. Apoptin is required for anellovirus replication and virion formation. The exact

BOX 30.2

GENERAL CHARACTERISTIC AND TAXONOMY OF ANELLOVIRUSES

Anelloviruses are small, unenveloped viruses with $T=1$ icosahedral symmetry (~25 nm diameter). Capsids are assembled from 60 copies of the capsid protein VP1. Anellovirus genomes are a single strand of circular DNA ranging in size from ~2 to 4 kb. The best studied anellovirus is Chicken anemia virus (CAV). In addition to VP1, CAV encodes two nonstructural proteins, VP2, a phosphatase, and VP3, a nonstructural protein that induces apoptosis in transformed cells.

Family: *Anelloviridae*
Genus: *Alphatorquevirus*
Genus: *Betatorquevirus*
Genus: *Deltatorquevirus*
Genus: *Epsilontorquevirus*
Genus: *Etatorquevirus*
Genus: *Gammatorquevirus*
Genus: *Gyrovirus* (one species, *Chicken anemia virus*)
Genus: *Iotatorquevirus*
Genus: *Kappatorquevirus*
Genus: *Lambdatorquevirus*
Genus: *Thetatorquevirus*
Genus: *Zetatorquevirus*

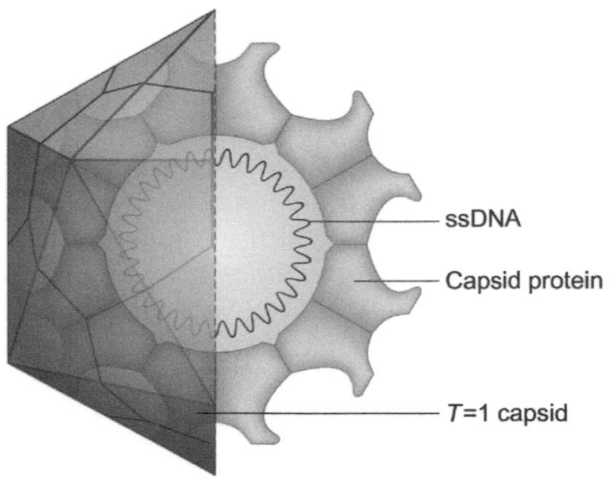

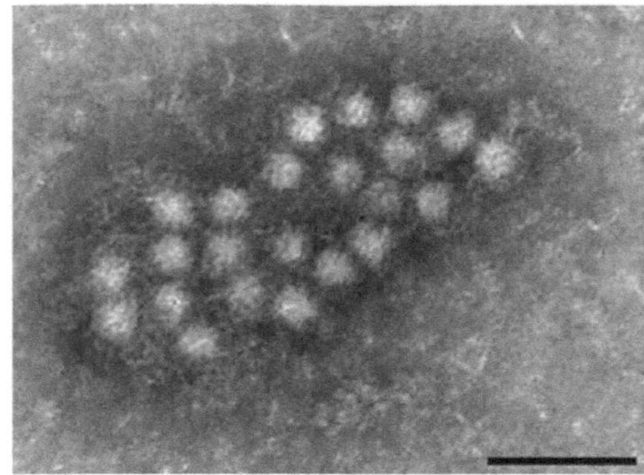

FIGURE 30.4 Anellovirus virion structure (left) and negative contrast electron microscopy (right) of particles of an isolate of Torque teno virus, stained with uranyl acetate (From Itoh et al., 2000). The bar represents 100 nm.

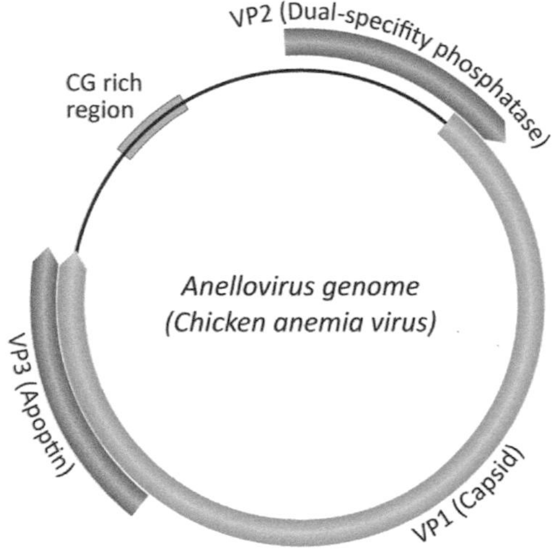

FIGURE 30.5 Anellovirus genome organization and protein expression strategy.

mechanism of action of apoptin is not yet understood, but recent experiments show that in transformed cells it is phosphorylated by the β isozyme of protein kinase C. As a result of phosphorylation, apoptin becomes localized to the nucleus, triggering apoptosis and cell death.

In this chapter we learned that:

- Circoviruses and anelloviruses are unenveloped single-stranded DNA virus with circular genomes.
- Circoviruses and anelloviruses replicate only in mitotically active cells.
- The two families can be differentiated based on genome organization and capsid structure.
- Among the circoviruses, Beak and feather disease virus (BFDV) and Porcine circovirus 2 (PCV2) are significant pathogens.
- Anelloviruses are ubiquitous and most infections are not associated with disease.
- Apoptin, encoded by some anelloviruses, is a nonstructural protein that causes apoptosis in tumor cells, but not in normal cells. It is being studies for its oncolytic properties.

References

Anon, 2005. The single stranded DNA viruses. In: Fauquet, C.M., Mayo, M.A., Maniloff, J., Desselberger, U., Ball, L.A. (Eds.), Virus Taxonomy. Academic Press, San Diego, p. 327.

Itoh, Y., Takahashi, M., Fukuda, M., Shibayama, T., Ishikawa, T., Tsuda, T., et al., 2000. Visualization of TT virus particles recovered from the sera and feces of infected humans. Biochem. Biophys. Res. Commun. 279, 718–724. Available from: http://dx.doi.org/10.1006/bbrc.2000.4013.

CHAPTER

31

Family *Polyomaviridae*

OUTLINE

After reading this chapter, you should be able to discuss the following topics:

- What characteristics of polyomavirus genomes made them a good model for studying processes of host-cell DNA replication and transcription?
- What characteristics of polyomaviruses made them a good model for studying cell cycle control and transformation?
- Discuss some of the functions of polyomavirus large T and small T proteins.

Polyomaviruses are unenveloped DNA viruses with $T=7$ icosahedral symmetry and a diameter of ~45 nm (Fig. 31.1). The name polyomavirus is derived from the Greek words poly (many) and oma (tumor) as the first polyomavirus (murine polyomavirus) was isolated from laboratory mice with multiple tumors. Simian virus 40 (SV40, now called *Macaca mulatta polyomavirus 1*) was discovered as a contaminant of cell cultures being used to manufacture human vaccines (from about 1955 to 1963). A few polyomaviruses are associated with disease, but more often polyomaviruses have been studied as productive and insightful models for understanding the processes of DNA replication, cell cycle control, and cell transformation.

GENOME STRUCTURE

Polyomaviruses have small (~5200 bp) double stranded circular DNA genomes. The polyomavirus genome is associated with host-cell derived histone proteins (H2A, H2B, H3, and H4) thus assumes a nucleosome structure similar to that found in cellular chromatin. Polyomavirus genomes are often called "minichromosomes" due to their shared structure with eukaryotic chromosomes (Box 31.1).

The genome organization of all polyomaviruses is similar, although the exact number of encoded proteins can vary from 5 to 9 (Fig. 31.2). All polyomavirus genomes contain a noncoding control region (NCCR) that serves as the origin of replication and a bidirectional promoter controlling early and late transcription. Early transcription produces the mRNAs for the large T and small T proteins (also called T antigens; their names derives from "tumor antigen," originally identified as unique proteins found in tumor cells). The late transcript encodes the capsid proteins, VP1, VP2, and VP3.

VIRION STRUCTURE

Polyomaviruses have unenveloped virions with $T=7$ icosahedral symmetry and a diameter of

Viruses. DOI: http://dx.doi.org/10.1016/B978-0-12-803109-4.00031-3

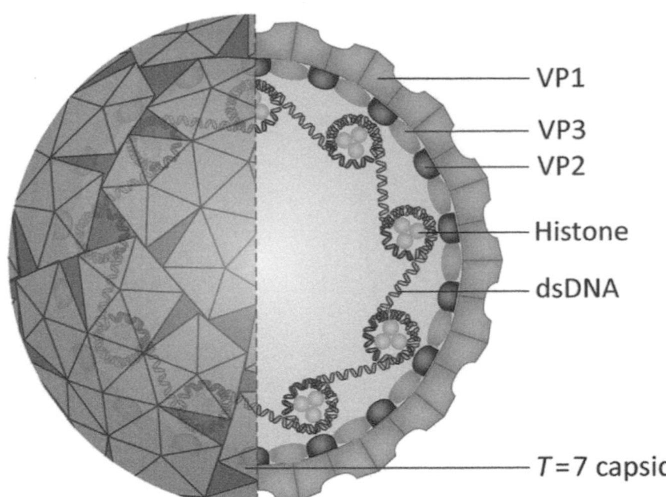

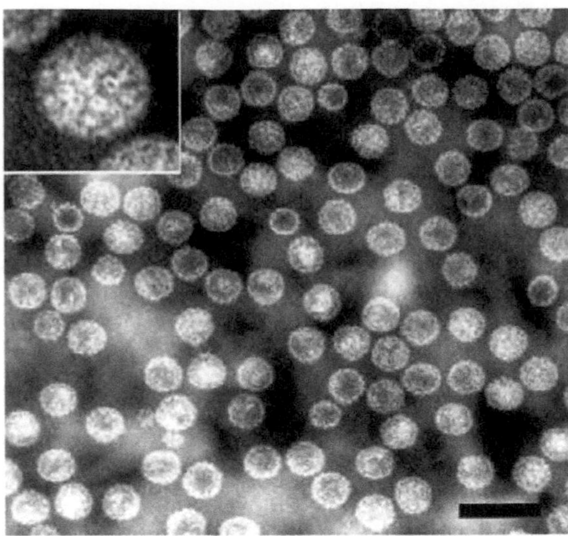

FIGURE 31.1 Virion structure (left). Transmission electron microscopic (TEM) image of polyomavirus virions (right). *Source: Content provider CDC/ Dr. Erskine Palmer. CDC Public Health Image Library Image ID#5629.*

BOX 31.1

GENERAL CHARACTERISTICS

Polyomavirus genomes are circular, double stranded DNA molecules of ~5000 bp that encodes 5–9 proteins. Genomes take the form of minichromosomes with the same nucleosome structure of cellular chromatin. There is a single noncoding control region (NCCR) that serves as the replication origin and a bidirectional promoter. Polyomaviruses encode T (tumor) antigens (proteins) that interact with host-cell proteins to promote DNA synthesis as polyomaviruses rely on DNA cellular polymerases for their replication.

Virions are unenveloped icosahedral structures ($T = 7$), ~45–50-nm in diameter. VP1 is the major capsid protein. Capsids contain 360 molecules of VP1 arranged in 72 pentamers.

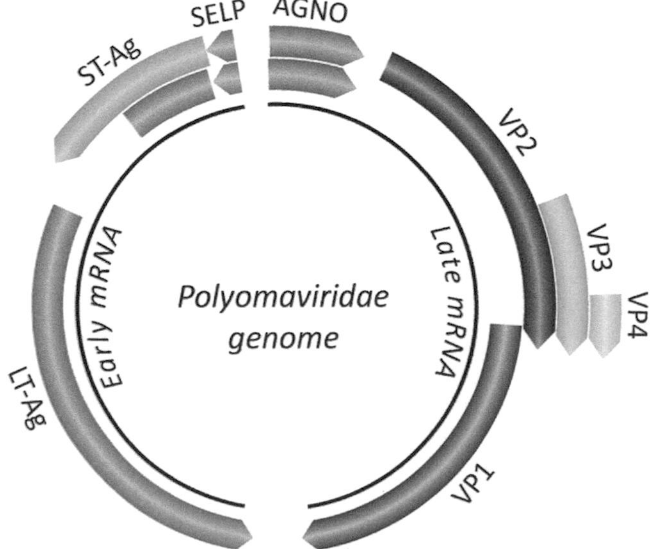

FIGURE 31.2 Genome structure of SV40, a betapolyomavirus.

~45 nm. Capsids contain three structural proteins, VP1, VP2, and VP3. The bulk of the capsid is constructed from 360 molecules of VP1 (arranged at 72 pentamers). Virions are relatively resistant to heat or formalin inactivation.,

REPLICATION

Polyomaviruses encode proteins (T antigens) that stimulate cell to divide, so as to provide the appropriate environment for the polyomavirus genome replication. During a productive infection, virions are formed and infection is often cytopathic. However, under circumstances where cells are not fully competent to support productive replication polyomavirus infection can transform cells (in this case virions are usually not produced). In this section we will review the productive replication cycle of a polyomavirus. The virus–cell interactions that lead to transformation will be discussed later in this chapter.

Attachment and Penetration

Receptors have been identified for several polyomaviruses and they include proteins and carbohydrates. Receptor interactions are mediated by VP1. Following attachment virions are endocytosed, however, the specific endocytic pathway varies among polyomaviruses. The endosome is only the first step in the voyage of the polyomavirus to its ultimate destination, the nucleus. Virions are delivered by endosomes to the endoplasmic reticulum (ER). In the case of SV40, uncoating begins in the ER and involves breakage and reforming of disulfide bonds that link capsid proteins. Following structural rearrangements of capsid proteins to expose hydrophobic surfaces on VP1, virions are transported across the ER membrane. Alternatively is has been proposed that VP2, a myristoylated protein, interacts with the ER lipid bilayer to support movement of virions out of this compartment. SV40 VP4 is a viroporin that may play a similar role.

Within the virion the SV40 genome is associated with four core histones (H2A, H2B, H3, and H4). These four core histones form a bead around which the DNA is wound. However, once the viral genome is delivered into the nucleus, histone H1 joins the process. H1 stabilizes the DNA wrapped around the core histones and binds to the linker DNA in regions between nucleosomes. Within the nucleus, the SV40 genome contains 24 nucleosomes while the NCCR remains free of nucleosomes.

Early Transcription

Early gene expression begins when host transcription factors and RNA polymerase II interact with the early promoter on the infecting genome. For example, the SV40 early promoter contains GC-rich sequences that are bound by transcription factor SP1. Thus the early promoter contributes to host range, as productive infection is restricted to cells expressing appropriate DNA-binding proteins. Also within the NCCR are AT-rich sequences (TATA boxes) that direct RNA polymerase II to the proper transcription initiation site. Early transcripts are alternatively spliced to produce large T and small T proteins. Large T is a multifunctional protein. Among the activities of large T are the following:

- Binds to the viral origin of replication.
- Helicase domain unwinds the viral dsDNA.
- Recruits cellular proteins (including replication protein A (RPA), DNA polymerase/primase, and topoisomerases I and II).
- Induces formation of viral replication centers or foci within the nucleus.
- Induces cells to exit from the quiescent state and enter into S phase (by binding pRB, p107, and p130).
- Counters the cell's apoptotic responses (by binding p53).
- Represses early transcription by binding to the replication origin thus preventing binding of cellular transcription factors to the region.

Small T of SV40 and human polyomaviruses overlaps the amino terminus of large T. However, small T has a unique C terminus that binds to cellular phosphatase 2A (PP2A). Binding of small T inhibits, and or changes, the substrate specificity of PP2A. Expressed on its own, small T is not active in transformation but it can contribute to the transforming properties of large T.

In summary large T and small T have a complex set of activities that coordinate cell cycle and polyomavirus replication. When introduced into cells, large T and small T are transforming, however, we should remember that cell transformation is not the usual outcome of a *productive* polyomavirus infection.

Genome Replication

The process of polyomavirus genome replication begins with binding of two molecules of large T to the origin of replication in the NCCR. Each molecule of large T recruits an additional five molecules of large T to form two hexamers that surround the origin of replication. These initiate melting and unwinding of the DNA. Large T also recruits the cellular DNA synthesis machinery and genome replication proceeds in a bidirectional manner.

Late Transcription

Late gene transcription begins with the onset of DNA replication. Large T has a role in this process as well. Large T stimulates late transcription via interactions with the basal transcription machinery (e.g., TATA-binding protein). Late mRNAs encode the polyomavirus capsid proteins VP1, VP2, and VP3. Some polyomaviruses encode an additional nonstructural protein called agnoprotein. SV40 also encodes a VP4 protein (a viroporin) from the late transcript.

VP1 is expressed from a spliced mRNA. The protein is ~350–370 aa and is the major capsid protein. VP1 functions in attachment and entry. VP1 associates with VP2 and VP3 (which are found inside the mature capsid). VP2 is myristoylated at its amino terminus. VP3 overlaps the carboxyl terminus of VP2.

Another important product of late transcription is a microRNA (miRNA) whose sequence is complementary to the polyomavirus early transcript. Thus it functions to downregulate early gene transcription.

Assembly and Release

Assembly begins with the translocation of VP1, VP2, and VP3 into the nucleus and the formation of VP1 capsomeres (each capsomere contains five molecules of VP1). VP1 alone is sufficient to drive capsid assembly but VP2 and VP3 are needed for production of infectious virions. In some cases polyomavirus infections are cytopathic and virion release can accompany cell death. However, infections can be persistent and there is evidence for regulated virion release (perhaps by exocytosis).

TRANSFORMATION

Murine polyomavirus (MPyV) was the first polyomavirus identified, found to be the cause of tumors in laboratory mice. It was subsequently found that SV40 (a primate polyomavirus) could induce tumors if injected into newborn hamsters. Thus began the study of polyomaviruses for the purpose of understanding their robust transforming potential .

Key to understanding the transforming activities of polyomaviruses is to recall that these small DNA viruses induce cells to enter S phase to facilitate viral genome replication. In a productive infection the balance favors virion production and cell death rather than transformation. What is the process by which some cells are transformed? Transformation is associated foremost with restricted polyomavirus replication. The general model for transformation is that T antigens are expressed but that viral DNA replication is blocked (perhaps because of species-specific differences among DNA replication factors). Thus few or no virions are formed. Rarely the polyomavirus genome becomes integrated into the host-cell chromosome in a manner that supports *continued expression of T antigens.* Now the cell is maintained in a mitotically active state. The activities of large T that support transformation are its ability to bind Rb family members and to inactivate the tumor suppressor p53 (refer to Chapter 28: Introduction to DNA viruses). The vast majority of natural human polyomavirus infections do not result in transformation. However, as will be discussed below, a human polyomavirus (Merkel cell polyomavirus) has recently been implicated in tumor formation.

HUMAN POLYOMAVIRUSES AND DISEASE

Human polyomaviruses appear to be ubiquitous and infections are generally benign. There is strong evidence to support that polyomaviruses cause persistent infections in healthy individuals, without any evidence of disease. Serosurveys indicate that most of the world's population is infected by early to mid-childhood.

The first identified human polyomaviruses were JC polyomavirus (JCPyV) and BK polyomavirus (BKPyV) recovered from respiratory and lymphoid tissues respectively, of diseased patients. Other human polyomaviruses include Merkel cell carcinoma polyomavirus (MCPyV) and human polyomaviruses 6, 7, and 8 all isolated from normal skin (Box 31.2).

Polyomavirus involvement in human disease is usually associated with immunocompromised patients. JCPyV can cause lytic infections in the brain while BKPyV can cause lytic infections of the kidney, bladder, and ureter. BKPyV is often associated with kidney failure after organ transplantation. Quite simply immunocompromised patients are unable to control replication of these viruses and lytic infection leads to tissue damage. Unfortunately one outcome of increased use of immunosuppressive regimens is increased incidence of polyomavirus-associated progressive multifocal leukoencephalopathy (PML). PML is a rare and usually fatal brain infection.

BOX 31.2

TAXONOMY

Family *Polyomaviridae*

Genus *Alphapolyomavirus* [Type species *Mus musculus polyomavirus 1*, also includes Merkel cell polyomavirus (MCPyV)]

Genus *Betapolyomavirus* [Type species *Macaca mulatta polyomavirus 1* (formerly Simian Virus 40)] also includes BK polyomavirus (BKPyV) and JC polyomavirus (JCPyC)

Genus *Deltapolyomavirus* (Type species *Human polyomavirus 6*)

Genus *Gammapolyomavirus* (Type species *Aves polyomavirus 1*)

MCPyV is so named because of its association with a rare skin cancer, Merkel cell carcinoma. Risk for developing Merkel cell carcinoma is increased in immunocompromised patients. Integrated MCPyV DNA is present in tumor tissues. Integration is clonal, meaning that all tumor cells have arisen from a single progenitor in which the integration event occurred. The integrated DNA encodes mutated versions of large T.

In this chapter we have learned that:

- Polyomaviruses are small double stranded DNA viruses whose genomes associate with histones. Thus polyomavirus genomes are often called "minichromosomes" and historically they provided a robust model for probing eukaryotic DNA replication and transcription.
- Polyomaviruses encode T antigens (proteins) that interact with cellular proteins to alter cell cycle control. Polyomaviruses drive cells into S phase in order to facilitate virus genome replication.
- In a productive polyomavirus infection, cells are often killed. In nonpermissive or poorly permissive cells, transformation is a rare outcome. Transformed cells usually have integrated copies of genes for polyomavirus T antigens.
- T antigens are multifunctional. Some activities include: Orchestrating DNA replication, controlling early versus late transcription, binding and inactivating pRb family members, binding and inactivating p53.

CHAPTER

32

Family *Papillomaviridae*

OUTLINE

After reading this chapter, you should be able to answer the following questions:

- What are the major characteristics of viruses in the family *Papillomaviridae*?
- Discuss host and tissue tropism of papillomaviruses (PVs). Why are we unable to grow these viruses in continuous cell cultures?
- What are the major functions of PV E6 and E7 proteins?
- Discuss the relationship between human PVs (HPVs)and cervical cancer. Why do some HPVs pose a higher risk of cancer development?

PVs are unenveloped icosahedral viruses with circular double-stranded (ds) DNA genomes. PVs have a strict tropism for epithelial cells and induce benign hyperplasia of skin to form common warts. PVs have been isolated from mammals, birds, and reptiles. In general, PVs are quite host specific and cross-species transmission is very rare. A notable feature of PVs is their requirement for differentiating epithelial (skin) cells in order to complete their replication cycle (Box 32.1). Thus PVs cannot be grown in transformed cells in culture and systems for replicating PVs are limited. The study of human PVs accelerated in the 1970s when advances in molecular biology provided methods for characterizing and mapping PV genomes and transcripts from lesions. Over 100 types of human PVs (HPVs) have been identified. Warts rarely progress to malignant cancers although cervical and other anogenital cancers are highly associated with a small subset of human PVs.

GENOME STRUCTURE

PV genomes are ds circular (~8000 bp) DNA molecules. Viral DNA associates with cellular histones to form a chromatin-like complex (Fig. 32.1). PV genomes have a single untranscribed regulatory region (URR) (also called the long control region) of about 1000 bp that contains the DNA replication origin (ori) and promoters. Transcription proceeds in one direction (one strand of DNA serves as the coding strand, Fig. 32.2). mRNAs are transcribed by host cell RNA polymerase II and genome replication is catalyzed by host DNA polymerases. PV early proteins interact with the URR to direct and regulate these processes. The nucleus is the site of viral genome replication and virion assembly.

Viruses. DOI: http://dx.doi.org/10.1016/B978-0-12-803109-4.00032-5

BOX 32.1

GENERAL CHARACTERISTICS

Genomes are ds, circular DNA, ~8 kbp. Transcription and genome replication are nuclear. mRNAs are transcribed by RNA polymerase II and are extensively spliced. PV genome replication is directed by early proteins (E1 and E2) and catalyzed by cellular enzymes.

Early genes are expressed prior to genome replication (and expression continues throughout the replication cycle). Late genes are expressed after DNA replication. Late gene expression requires epithelial cell differentiation. Products of the late genes (L1 and L2) are capsid proteins.

Virions are unenveloped icosahedral ($T = 7$) capsids containing two capsid proteins. L1 is the major capsid protein and when expressed alone, assembles to form VLPs. The production of infectious virions is tightly linked to epithelial cell differentiation, thus PVs do not replicated in transformed cells in culture.

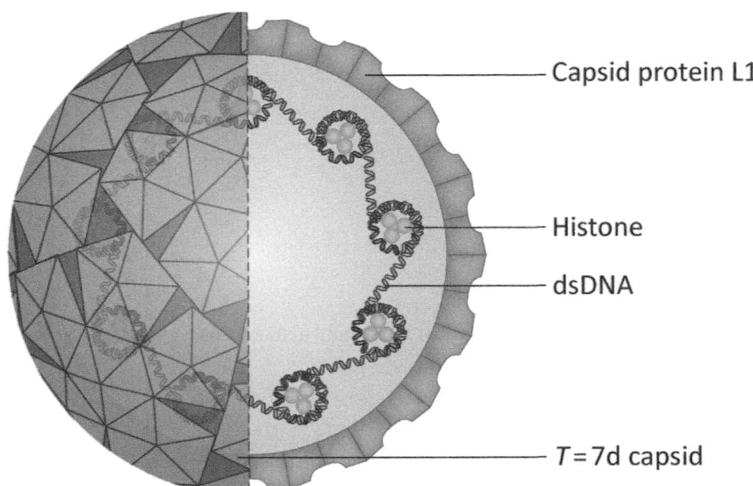

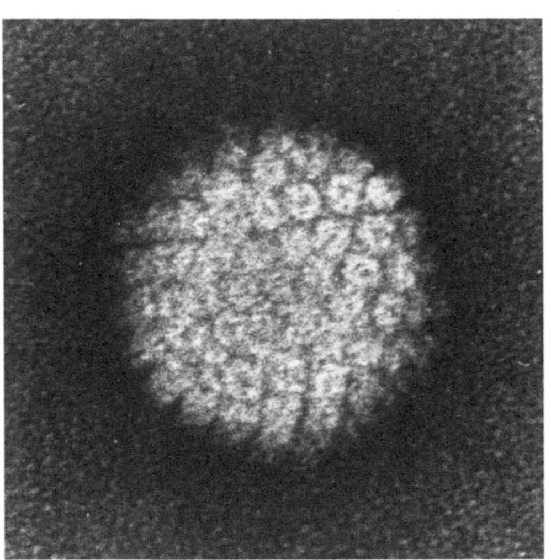

FIGURE 32.1 Virus structure (left). Negative stain electron micrograph of a human papillomavirus (right). *From NIH-Visuals Online# AV-8610-3067; Laboratory of Tumor Virus Biology.*

VIRION STRUCTURE

Virions are unenveloped and have $T = 7$ icosahedral symmetry (Fig. 32.1). Virions are ~60 nm diameter. Two structural proteins, L1 and L2, assemble to form capsids. Capsomers contains five molecules of L1 and one molecule of L2, arranged with L2 facing the lumen and associated with viral DNA. When expressed alone, L1 can assemble to form virus-like particles (VLPs). L2 is required for the formation of infectious virions.

REPLICATION CYCLE

Key to the PV replication cycle is that productive infection is dependent on the differentiation stage of the epithelial cell. PV begins the replication cycle by infecting mitotically active basal cells where PV DNA is initially replicated. When the cell divides, daughter cells inherit copies of the PV genome. The process continues, to generate a population of infected cells. However, capsid protein synthesis and production of virions must wait for cell differentiation (cells must be mitotically active to be infected, but then must undergo differentiation). Thus the PV life cycle is intimately linked to skin development (Fig. 32.3).

Attachment/Penetration

PVs infect the dividing basal cells of the skin. PVs access these cells through physical breaks in the skin. Initial attachment of HPVs is thought to involve heparan sulfate proteoglycans exposed on the surface of the basement membrane. It is likely that additional interactions are required before virions can access the basal cells themselves.

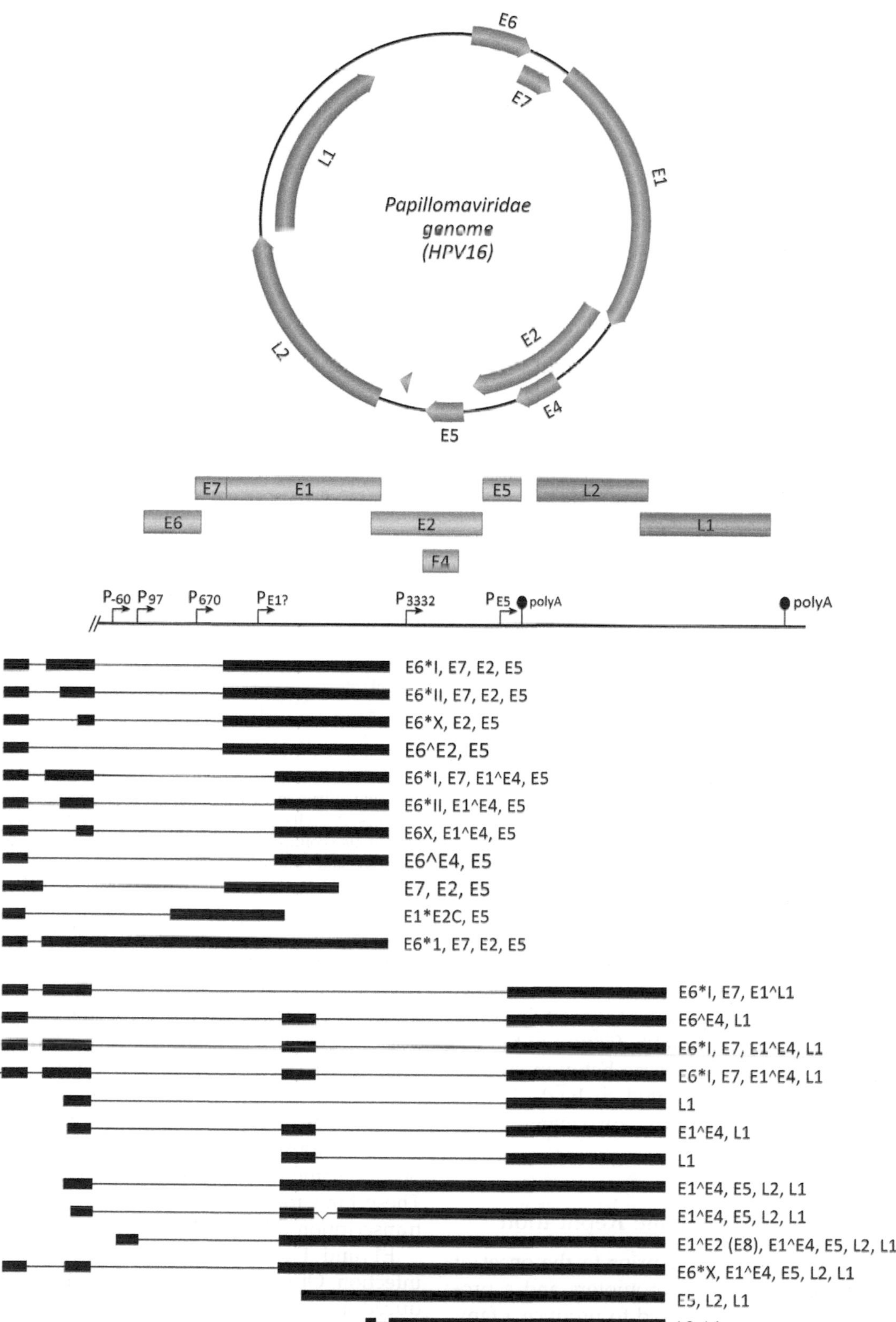

FIGURE 32.2 Genome organization and gene expression strategy of human papillomavirus 16 showing open reading frames and alternatively spliced transcripts. Note that alternative splicing generates a large number of mRNAs.

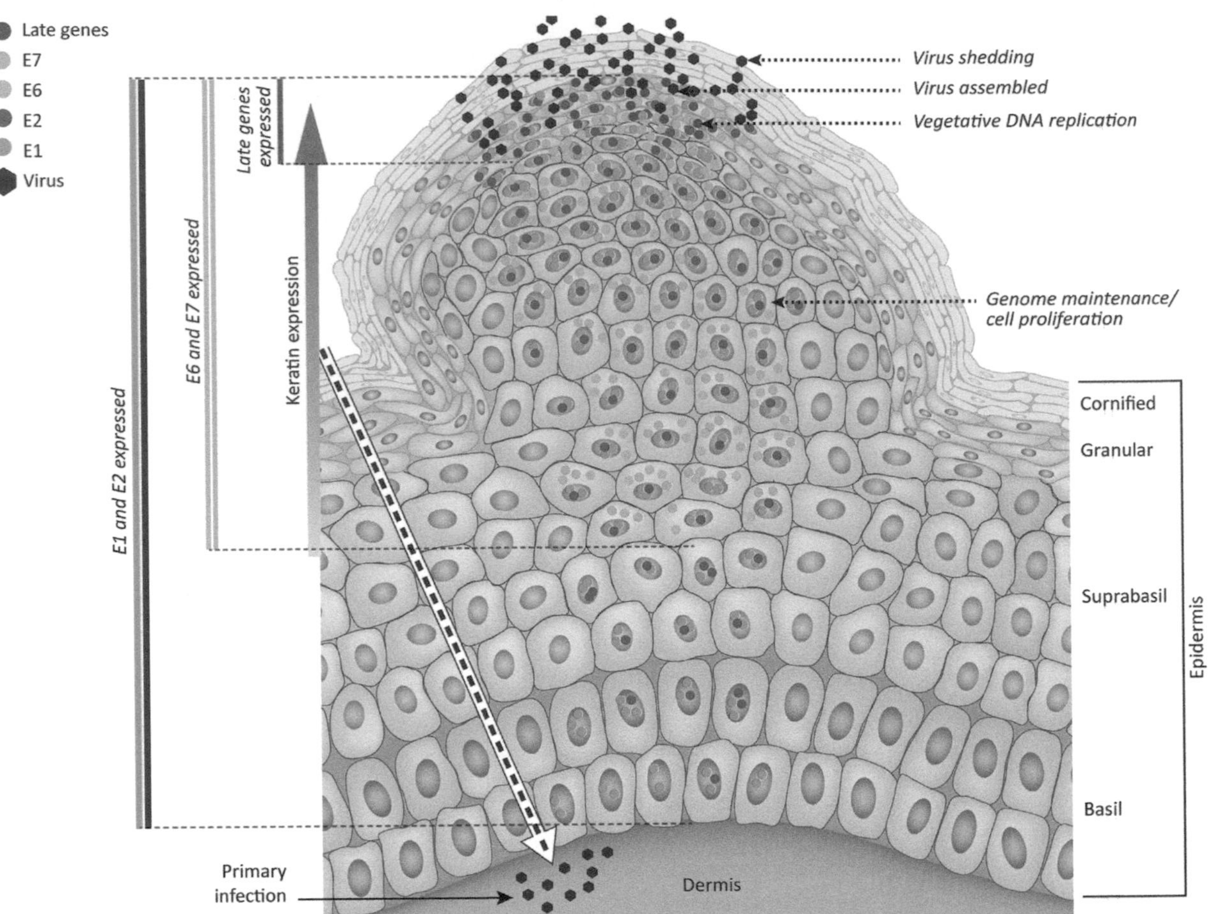

FIGURE 32.3 PV replication is dependent upon epithelial cell differentiation. Virus infects mitotically active basal cells in the skin and genome replication begins (replicative phase) to produce 50–100 genome copies. As cells divide, genomes in each daughter cell replicate to reestablish a level of 50–100 genome copies per cell. Infected cell proliferate, but also continue on a pathway to differentiate. Expression of cellular keratin genes triggers additional PV genome replication (so-called vegetative state) and expression of late transcripts.

Binding is followed by endocytosis. Details of endocytosis and escape from endosomes are very sketchy due to limits of productive replication systems. It appears that L2-genome complexes escape endosomes and traffics to the nucleus. A notable feature of the PV replication cycle is the length of time between surface binding and transcription, estimated to be 1–3 days (compared to minutes to hours for many other viruses).

Early Transcription and Genome Replication

Early transcription is initiated prior to the onset of viral DNA replication. Multiple promoters and a program of alternative splicing are used to produce a family of early (E) proteins necessary for initial rounds of PV genome replication. The first viral protein products produced are E1 and E2.

PV E1 is a phosphoprotein that binds the DNA replication ori. E1 has ATPase and helicase activities. E1 unwinds the DNA at the ori and interacts with the p180 subunit of cellular primase to recruit the cellular DNA replication machinery to the ori. Binding of E1 to the ori is enhanced by the E2 protein.

E2 is actually a small set of proteins that regulate viral transcription and DNA replication. E2 has sequence-specific DNA-binding and dimerization domains as well as a transactivating domain that enhances transcription. All versions of E2 bind the PV ori, but some E2 proteins lack transactivating domains. These bind to the ori, but *inhibit*, rather than activate, transcription (Fig. 32.4).

E1 and E2 are produced at the earliest stages of infection. Other early proteins (E5, E6, and E7) are produced a bit later, as the basal cells begin to differentiate (Fig. 32.3). E5, E6, and E7 cooperate to stimulate replication and proliferation (hyperplasia) of the infected tissue. Later, when infected cells begin to differentiate, these early proteins continue to support viral genome synthesis. PV E6 inactivates p53 to block

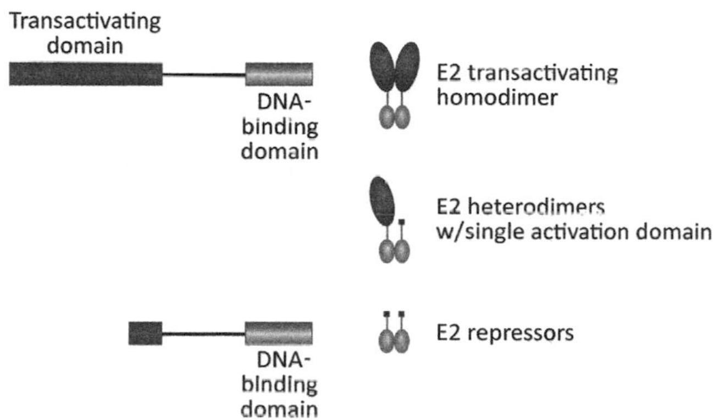

FIGURE 32.4 PV E2 is both an activator and inhibitor of transcription. Full-length E2 (top) has an amino terminal transactivating domain followed by a linker region and a carboxyl terminal DNA-binding domain. Full-length homodimers (with transactivating domain) are transcriptional activators. Dimers of the DNA-binding domain are repressors of transcription (bottom).

apoptosis while E7 interacts with pRb family proteins to prevent sequestration of the cellular transcription factor E2F (Fig. 28.3). E6 and E7 proteins are conserved among all PVs, E5 is not always present.

The initial phase of PV DNA replication occurs within dividing basal skin cells. This is often referred to as the replicative phase, during which infecting PV genome is amplified to about 50–100 copies. These viral "plasmids" associate with chromatin and segregate into daughter cells during cell division. In the so-called "maintenance phase" each copy of the PV genome is replicated approximately once per cell division to maintain a stable number of PV genomes.

Late Transcription and Vegetative DNA Replication

As skin cells move closer to the surface, they undergo major changes in gene expression. Key among these is the expression of various keratin genes. Productive PV replication is tied to this process, which triggers an increase in viral DNA replication (the so-called vegetative stage of DNA replication) and the expression capsid proteins (L1 and L2) needed to package that DNA. Virions are shed from the uppermost layers of the skin, along with dead skin cells.

In the vast majority of PV infections the process of tumor formation does not proceed past benign hyperplasia. Normally the immune system brings the infection under control and warts are resolved. Conversely, certain types of immune impairment favor growth of large papillomas. However, it is also clear that PVs can be present in small amounts in the absence of a visible wart. Virions are quite stable and can be transmitted by procedures such as ear-tagging or tatooing livestock.

PV also infects mucosal surfaces such as the mouth, gastrointestinal, and urogenital tracts. The types of PVs that infect mucosal surfaces are distinct from those that infect other body parts. Lesions that form on mucosal surfaces are called condylomas.

DISEASE

Human Papillomaviruses

There are at least 100 known types of HPV (Box 32.2). They cause warts, epidermal, and epithelial lesions at different body sites. Most lesions are benign and rarely progress to a malignant state. In healthy individuals virus replication is controlled by T-cells thus warts are self-limiting and will eventually resolve.

However, a few HPVs are highly associated with carcinomas of cervix, anus, larynx, and lung (Box 32.3). These include HPVs 16, 18, 31, 33, 35, 39, 45, 51, 52, 56, 58, 59, and 68. Types 16 and 18 account for about 70% of cervical cancers and these were the targets of the first cervical cancer vaccines. These tumors may take years or decades to develop. The lesions do not produce virions but it is common to find a fragment of the HPV genome, containing genes for E6 and E7, integrated into the chromosome. Recall that all PVs have E6 and E7 proteins that support DNA synthesis during productive HPV infection. Why are only a few types of HPV associated with development of malignant cells?

The high-risk HPVs encode E6 and E7 proteins that are more robust and varied in their ability to inactivate p53 and pRB family member proteins to promote cell growth and division. They do this by a variety of regulatory mechanisms. The E6 and E7 genes of the high-risk HPVs are also expressed differently than those of the low-risk HPVs. A single promoter drives expression of high-risk HPV E6 and E7, while the low-risk HPVs use two independent promoters. The HVP16 E6 protein has also been specifically shown to activate telomerase in keratinocytes (recall that maintenance of telomers is important in the process of cellular immortalization and transformation).

Other Cancers Caused by Human Papillomavirus

At the cellular level, the oral mucosa is structurally similar to the vagina and cervix and most oral cancers are squamous cell carcinomas, the same type found in cervical cancers. Those facts prompted surveys of head and

BOX 32.2

TAXONOMY

Family *Papillomaviridae*

The family contains over 45 genera, named using letters of the Greek alphabet (*alpha, beta, gammapapillomaviruses*). Species names are based on the genus name and individual species are numbered. For example, within the genus *Alphapapillomavirus* (includes HPVs associated with genital and mucosal cancers) there are 14 species, numbered 1–14. There is no correspondence between the new species number designation and the historical "type" number associated. Thus HPV16, the type found most frequently in cervical cancer, is a member of *Alphapapillomavirus* species 9.

BOX 32.3

HPV (CANCER) VACCINES

The history of cervical cancer as a sexually transmitted disease began in 1842 with publication of mortality statistics of women dying of cancer in the city of Verona, Italy. Italian physician, Domenico Rigoni-Stern, noted that "cancer of the uterus" was more common among married women and widows than among virgins and nuns. This was the first suggestion of a sexually transmitted cancer agent. The first links between cervical cancers and HPVs came in the middle of the 20th century with the recognition that genital warts were a sexually transmitted disease. Cytologists studying cervical cancers and precancers noted the similarities between cells associated with malignancies and those associated with cervical warts. By the early 1980s, multiple studies showed a direct link between HPV infection and cervical cancer. Over the next decade, it became increasingly clear that virtually *all* cervical cancers, and many other anogenital cancers, were the result of prior HPV infection. As molecular techniques facilitated HPV typing, it also became clear that a few "high-risk" HPV types were associated with cervical cancer.

The identification of specific viruses as the cause of cervical cancers set the stage for vaccine development. A huge hurdle was the inability to grow large amounts of virus. However, the same technologies that facilitated studies of PVs (DNA cloning and sequencing technologies) also provided the tools for vaccine development. By the 1990s, methods were being developed to produce recombinant HPV capsid proteins and coax them to assemble into VLPs that were antigenically similar to virions but contained no HPV genetic material. In 2006 the US Food and Drug Administration licensed a Merck & Co (Kenilworth, NJ, USA) vaccine that marketed specifically for the prevention of cervical cancer. The first HPV vaccines were developed specifically for women but newer vaccines, marketed for men and women, protect against HPVs that are the major cause of genital warts in men. Because of the high prevalence and transmissibility of genital HPVs, the US Centers for Disease Control and Prevention currently recommends vaccination of boy and girls, before they become sexually active.

neck cancers for the presence of HPV genes. In one study, 25% of 253 patients diagnosed with head and neck cancers, the tissue taken from tumors was HPV positive (predominantly HPV16). Smoking and alcohol consumption can promote oral cancers and these activities may enhance to rate of transformation by high-risk HPVs.

Epidermodysplasia Verruciformis

Epidermodysplasia verruciformis (EV) or "tree man illness" is an *extremely* rare genetic disorder that causes an abnormal susceptibility to HPVs. The inability to resolve HPV infections results in development of numerous scaly lesions, particularly on the hands and feet. The HPVs associated with these lesions normally cause asymptomatic infections in healthy individuals. By about 20 years of age, the lesions progress to disfiguring and incapacitating nonmelanoma skin cancers. EV is a primary immunodeficiency and the affected genes have been identified. These genes seem to regulate zinc distribution but how this affects HPV infection is not understood.

ANIMAL PAPILLOMAVIRUSES

Shope papillomavirus (now called cottontail rabbit papillomavirus) was discovered by Richard Shope in the 1930s. He purified virus particles from papillomas removed from wild-caught rabbits and successfully transmitted warts and tumors to laboratory rabbits. This was the first virus demonstrated to cause cancer.

Bovine PVs (BoPVs) are common, and they were used early models for PV replication due to the relatively large amounts of virus present in warts. BoPVs can be spread by modern ranching techniques such as use of ear tags. The virus is difficult to inactivate and equipment must be carefully cleaned avoid transmission of BoPV. Warts are also relatively common among young dogs but these usually resolve without treatment. Many animals have PV-associated warts; however, none are transmissible to humans.

In this chapter we have learned that

- PVs are undeveloped viruses with ds circular DNA genomes. Their replication cycle is tightly linked to the differentiation program of epithelial cells. Most papillomavirus infections result in self-limiting, benign warts that are formed by epithelial cell hyperplasia.
- PVs have narrow host and cell tropism.
- A few HPVs are associated with cancers and some of these can be prevented by vaccination.
- Papillomavirus E6 and E7 are growth promoting/transforming proteins. E6 interferes with normal functions of p53 and E7 interferes with pRB. These interactions inhibit apoptosis and support DNA replication. The so-called "high-risk" HPVs encode E6 and E7 proteins that have stronger anti-p53 and anti-pRB activities, respectively.

CHAPTER

33

Family *Adenoviridae*

OUTLINE

After reading this chapter, you should be able to answer the following questions:

- What are the general characteristics of adenoviruses (AdVs)?
- Describe AdV genome structure and mechanism of replication.
- How do AdVs stimulate host cell DNA replication?
- Explain why human AdVs can cause tumors in rodent models, but do not cause tumors in humans.
- What are the diseases most often associated with human AdV infection?
- Explain how oncolytic AdVs are modified to kill cancer cells while not replicating in normal cells.

AdVs are naked icosahedral viruses with prominent fiber proteins extending from their capsids (Fig. 33.1). AdV genomes are linear double-stranded (ds) DNA molecules ranging in size from 24 to 40 kbp. Human AdVs were first isolated from cell cultures of adenoid tissue (hence the name) and most often cause mild, upper respiratory tract infections. AdVs produce large amounts of progeny and they efficiently infect cells regardless of cell cycle status (Box 33.1). Studies to physically map AdV genomes and transcripts using electron microscopy provided the first evidence of the process of RNA splicing. AdVs were valuable models of cell transformation, as human AdVs transform rodent cells (Box 33.2). AdVs also grow to high titer, making them good models for a variety of biochemical and molecular studies. Currently, AdVs are being exploited as efficient gene delivery vehicles and so-called oncolytic AdVs are being developed and tested as tumor-cell killing agents.

GENOME ORGANIZATION

AdV genomes are a molecule of linear ds DNA. The 5′ end of each strand is covalently linked to a protein called the terminal protein (TP) as a result of protein-primed DNA replication. AdV genomes range from 24 to 40 kbp. The most conserved region of AdV genomes is the central portion, which encodes the structural proteins and viral enzymes (i.e., AdV DNA polymerase (Pol)). Regions outside the central core code for regulatory proteins (early proteins) and they are more variable in both sequence and size. The genomes of all AdVs have inverted terminal repeat sequences ranging in size from 36 to over 200 bp. Base pairing of these

Viruses. DOI: http://dx.doi.org/10.1016/B978-0-12-803109-4.00033-7

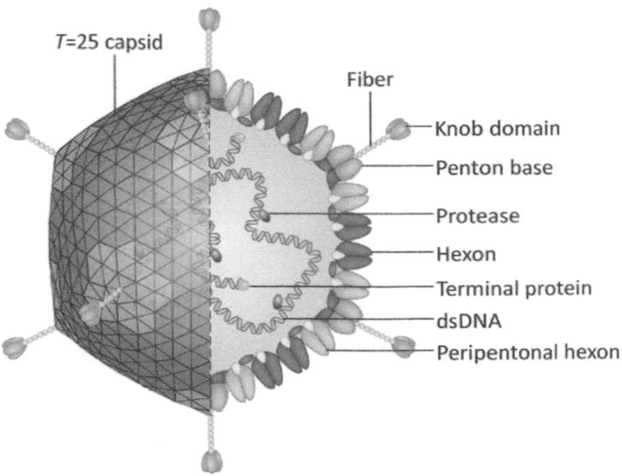

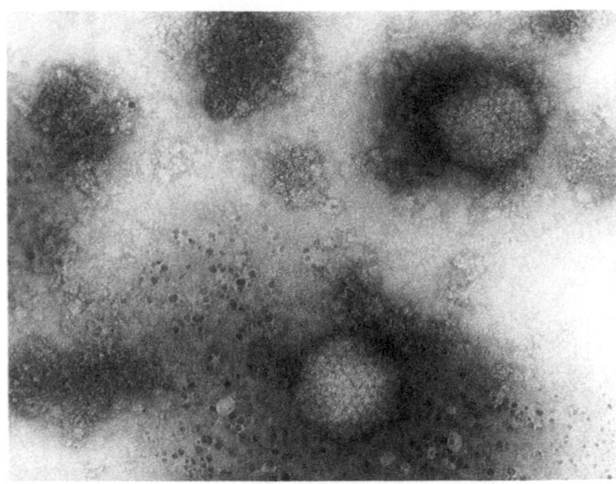

FIGURE 33.1 Virus structure (left). TEM image reveals some of the ultrastructural morphology exhibited by AdV-2 virions. Visible at this high magnification are the capsomeress, which in this case are hexagonally shaped, also called hexons (right). *From CDC/Dr. John Hierholze. CDC Public Health Image Library Image ID#14768.*

BOX 33.1

GENERAL CHARACTERISTICS

AdVs are unenveloped viruses with linear ds DNA genomes. Genomes range from 24 to 40 kbp and the 5′ end of each genome strand is covalently linked to a 55 kDa viral protein (terminal protein, TP) that served as the primer for DNA replication. The best conserved region of AdV genomes is the central portion, which encodes the structural proteins and viral enzymes. Ends of the genomes encode regulatory proteins (early proteins) and exhibit greater variation in both sequence and size. Transcription and genome replication and assembly are nuclear. Transcripts are capped and polyadenylated and undergo extensive splicing. Most AdV infections are cytopathic.

Virions are naked with icosahedral capsids, 70–90 nm in diameter. Long fiber proteins extend from each of the 12 fivefold axes of symmetry.

BOX 33.2

TAXONOMY

Family: *Adenoviridae* (AdVs have been isolated from mammals, birds, and fish.)

Genus: *Mastadenovirus* (infects only mammals, contains most human AdVs)

Genus: *Aviadenovirus* (birds)

Genus: *Siadenovirus* (reptile and birds)

Genus: *Ichtadenovirus* (fish)

regions is critical for DNA replication. AdV transcription and genome replication are nuclear. Transcripts are capped and polyadenylated and undergo extensive splicing (Fig. 33.2).

VIRION STRUCTURE

AdVs are naked icosahedral viruses (~95 nm diameter) with long, distinct fiber proteins protruding from the 12 pentamers of the $T=25$ icosahedral capsid. Fibers are trimers of the fiber protein. The fiber shaft is formed from a repeated structural motif and terminates with a distinct knob region at the tip (Fig. 33.1). Capsids contain at least nine polypeptides and are constructed from distinct pentamers and hexamers. Pentamers are assembled from two different proteins, the penton base protein and the fiber protein. The major capsid protein is the hexon protein. The hexon protein has two structural domains and 720 copies of

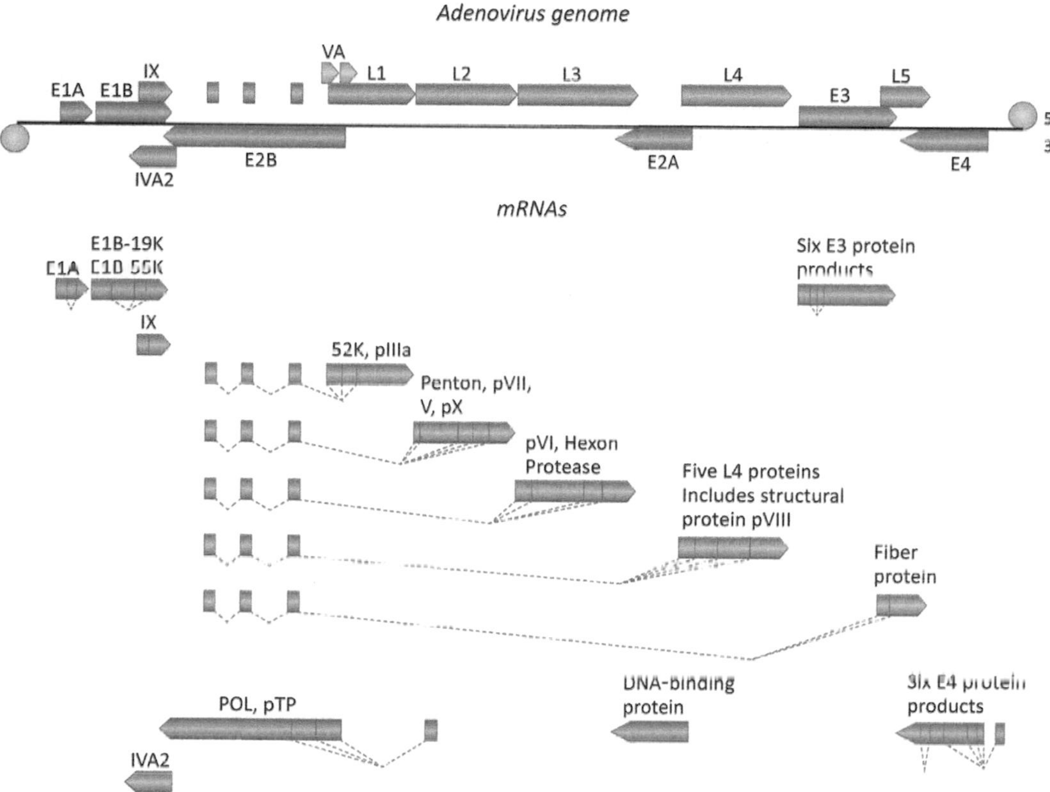

FIGURE 33.2 AdV genome organization and gene expression pattern. Note that transcripts for the major structural proteins all share a spliced, tripartite leader sequence. Alternative splicing and use of different termination signals generate the full set of mRNAs.

hexon protein assemble into 240 trimers that further form 120 hexamers.

ADENOVIRUS REPLICATION CYCLE

Attachment/Penetration

AdV entry is a two-step process. The knob region of the long fiber protein makes the initial interaction to tether virions to the cell surface. This is followed by interactions of the penton base protein with cell-surface integrins, resulting in endocytosis of the virion. The process of AdV uncoating has been studied extensively. The process begins very early, with detachment of fibers, such that a fiberless virion is endocytosed. Acidification of the endosome continues the step-wise uncoating process. The final steps of uncoating occur at the nuclear pore, at which time AdV DNA is imported into the nucleus.

Early Transcription

After entering the nucleus, a set of early (E) genes are transcribed by host RNA Pol II. AdV transcripts are capped, polyadenylated, and extensively spliced. The activities of the early proteins set the stage for the later processes in AdV replication. The major functions of the early protein products are described below. (These activities summarized here are based on studies of human Ad2 and human Ad5.)

E1A

A key function of E1A is to bind and inactivate cellular pRb family members, thus stimulating cellular DNA synthesis (Fig. 28.4). AdV E1A inhibits phosphorylation of Rb and the dephosphorylated protein cannot bind/sequester transcription factor E2F. Thus E2F is free to activate genes that support DNA synthesis. E1A also interacts with transcriptional modulators (i.e., histone acetyltransferases, histone deacetylases, and other chromatin remodeling factors). These interactions affect transcription of cellular genes to (among other things) inhibit the interferon response and downregulate genes for MHC Class 1 proteins. E1A protein also activates transcription of the other early transcription units.

E1B

The E1B transcription unit produces two mRNAs that encode proteins called E1B-19K and E1B-55K. The major role of these proteins is to block induction of apoptosis by the tumor suppressor (TS) p53. This is essential to allow time for virus to complete its replication cycle. E1B-19K and E1B-55K have discrete functions. E1B-19K inactivates two proapoptotic proteins (BAK and BAX) while E1B-55K directly inhibits p53 function by binding to its activation domain.

E2

The E2 transcription unit encodes three proteins required for AdV DNA synthesis: The precursor to the terminal protein (pTP), which primes DNA synthesis, AdV DNA Pol, and AdV ss DNA-binding protein. The E2 transcription unit is highly expressed, in part because its promoter is activated by the cellular transcription factor E2F. Recall that interaction of AdV E1A with pRB releases E2F, allowing it to activate promoters for cellular genes involved in DNA synthesis.

E3

Products of the E3 transcription unit function to antagonize host antiviral defenses. Among the characterized proteins are:

E3-gp19K protein is a glycoprotein that resides in the endoplasmic reticulum where it binds to a subunit of the major histocompatibility complex class I (MHC-I). This interaction prevents trafficking of MHC-I to the cell surface. E3-gp19K also binds to transporter associated with processing protein (TAP) and this interaction inhibits loading of peptides into MHC-1. Both of these activities prevent killing of AdV-infected cells by cytotoxic T-lymphocytes. (Recall that receptors on cytotoxic T cell must bind antigen-presenting MHC-I to mediate killing of target cells.)

E3-14.7K protein inhibits tumor necrosis factor (TNF)-induced cytolysis of AdV-infected cells in culture by binding to cellular IKKγ to modulate the activity of NF-kB.

E3-10.4K and E3-14.5K form a complex that internalizes receptors from the cell surface including the epidermal growth factor receptor, FAS, TRAIL receptor 1, and TNF receptor 1. Based on these activities, the proteins have been renamed receptor internalization and degradation (RID)α and RIDβ.

Viral DNA Replication

AdV DNA replication is notable because the ends of the linear ds DNA genome are covalently linked to a molecule of TP that served as the primer for viral DNA synthesis. Synthesis of AdV DNA can be reconstituted in in vitro reactions (in a test tube) and the

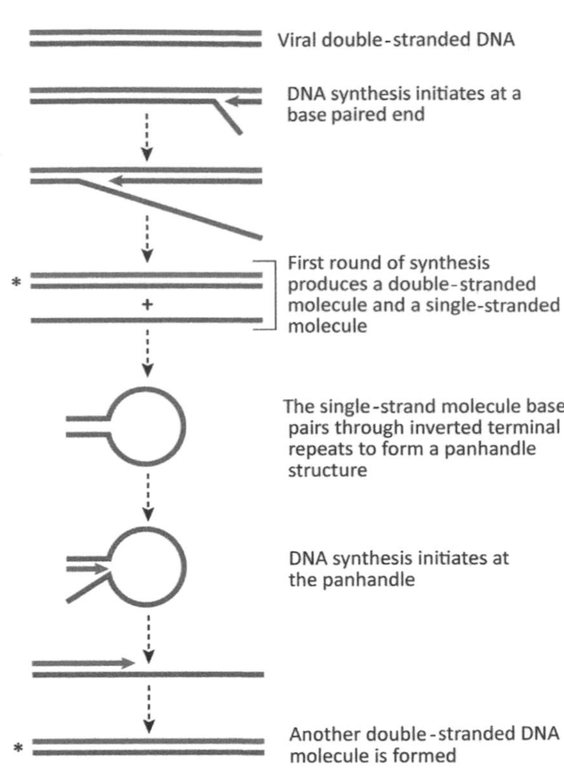

FIGURE 33.3 Replication of the linear ds DNA genome takes place by a strand displacement mechanism. A protein primer (not shown in this figure) primes DNA synthesis. Initiation of DNA synthesis begins at a base-paired end of the ds DNA molecule. A single strand is displaced during synthesis. Sequences at the end of the displaced strand are complementary and base pair to form an initiation site for DNA synthesis.

process requires three virally encoded proteins: pTP, AdV DNA Pol, and AdV DNA-binding protein. The addition of cellular proteins stimulates the process.

Initiation of DNA synthesis is asymmetric, beginning at one end of the viral genome. As one parental strand is copied, it base pairs with the newly synthesized strand. The other parental strand is displaced and circularizes as the result of base pairing of its terminal inverted repeat regions to form a "panhandle" that serves now serves as a replication origin (Fig. 33.3).

Late Transcription

The so-called major late promoter drives expression of the late (L) genes, most of which are capsid proteins. There are five distinct families of L transcripts (L1–L5) all of which contain the same spliced, tripartite nontranscribed leader sequence at their 5′ end. L1 transcripts are produced by polyadenylation at the L1 polyadenylation site. Longer transcripts (L2–L5) are generated through use of downstream polyadenylation signals.

The E4 transcription unit is also expressed after DNA replication. Functions of E4 proteins include regulation of transcription, RNA splicing, and translation. E4 products also modulate DNA replication and apoptosis.

VA RNAs

In addition to mRNAs, AdV produces short (~150–200 nt) noncoding RNAs called VA RNAs. These abundant RNAs are transcribed by cellular RNA Pol III from promoters similar to those found in tRNA genes. VA RNAs fold into very stable structures and accumulate to high levels in cells. The predominant type is VA RNAI that accumulates to up to 10^8 molecules per infected cell. An important function of VA RNAI is binding and inhibition of the ds RNA-activated protein kinase (PKR) (see Chapter 6: Immunity and Resistance to Viruses). Thus in the presence of VA RNAI, PKR is inhibited from doing its job as an interferon-stimulated inhibitor of viral protein synthesis. VA RNAs also interfere with other host cell processes, for example, they compete with pre-micro (mi)RNAs for export from the nucleus and processing by Dicer (see Box 33.3).

Assembly and Release

AdV assembly takes place the nucleus. Genome packaging requires a cis-acting domain at the left end of the AdV genome. This region contains a series of repeated sequences, termed A repeats due to their AT-rich character. A protein from the E3 transcription unit (E3 11.6-kDa or the AdV death protein) kills cells as it accumulates during the late stage of infection and promotes cell lysis.

ADENOVIRUSES AND HUMAN DISEASE

Many AdV infections are probably asymptomatic and persistent, and virus shedding can last for months. AdVs are spread by the fecal oral route. When associated with disease, human AdVs usually cause only mild illnesses. They are most often associated with acute respiratory, gastrointestinal, urinary, and eye infections. It is thought that about 10% of human colds are caused by AdVs. However disease can sometimes be severe, with respiratory symptoms including pneumonia, croup, and bronchitis. Infants and people with weakened immune systems are at higher risk for severe complications.

AdVs are a common cause of acute respiratory illness in military recruits, probably due to a crowded and stressful environment. The problem is significant enough that the US military has sponsored development of AdV vaccines.

ADENOVIRUSES AND TRANSFORMATION

Productive infections by AdVs are cytopathic; however, infection of nonpermissive cells can result in transformation. The transformation model is infection of rodent cells with human AdVs. Transformed cells always contain the E1A and E1B genes, but may not have any other viral proteins. If introduced into cells

BOX 33.3

COMPETITION AMONG SMALL RNAS IN ADV-INFECTED CELLS

miRNAs are small (~22 nt) cellular RNAs that inhibit translation of mRNAs to which they base pair. miRNAs are actually synthesized as long precursors that are cleaved by the cellular endoribonuclease Drosha to produce ~65 nt pre-miRNAs that are exported from the nucleus. In the cytoplasm, pre-miRNAs are further cleaved by another cellular ribonuclease, Dicer, to generate ~22 nt miRNAs that are incorporated into multiprotein RNA silencing complexes (RISC). RISC use miRNAs as a point of interaction with cellular mRNAs (via imperfect base pairing) and initiate cleavage and degradation (thus "silencing") of the bound mRNA.

Because AdV VA RNAs are synthesized at high levels in the nucleus of infected cells, they compete directly with cellular pre-miRNAs for export to the cytoplasm. Once in the cytoplasm VA RNAs also bind to Dicer, once again competitively inhibiting processing of cellular pre-miRNAs. Once incorporated into RISC, the ~22 nt VA RNA fragments do *not* target viral mRNAs (because they are transcribed from noncoding regions of the AdV genome). They may, however, bind and target a subset of cellular mRNAs. Whether or not VA RNAs directly silence cellular mRNAs, they certainly interfere with normal cellular control of translation.

together, E1A and E1B are sufficient for transformation. The activities of E1A and E1B that result in transformation were described above. The possibility that human AdV could cause human cancers has been extensively investigated and there is no credible evidence that this occurs.

ONCOLYTIC ADENOVIRUSES

Based on their unique properties, AdVs have been modified to enable them to kill cancer (tumor) cells while sparing nondividing cells. Recall that the AdV E1A and E1B proteins are produced by wild-type AdVs. These proteins inactivate the host cell TSs Rb and p53 (Table 28.2) thereby stimulating host DNA replication. Without E1A and E1B, AdV cannot replicate in normal cells! But the vast majority of tumor cells lack antioncogenes such as Rb and p53. Thus in constitutively dividing tumor cells AdVs lacking E1A and E1B do cause cytopathic infection. In practice, oncolytic AdVs are proving to be relatively safe and somewhat effective when injected into solid tumors. (But recall that the immune system will respond to these viruses; will eventually generate an antiviral response, limiting their repeated use.)

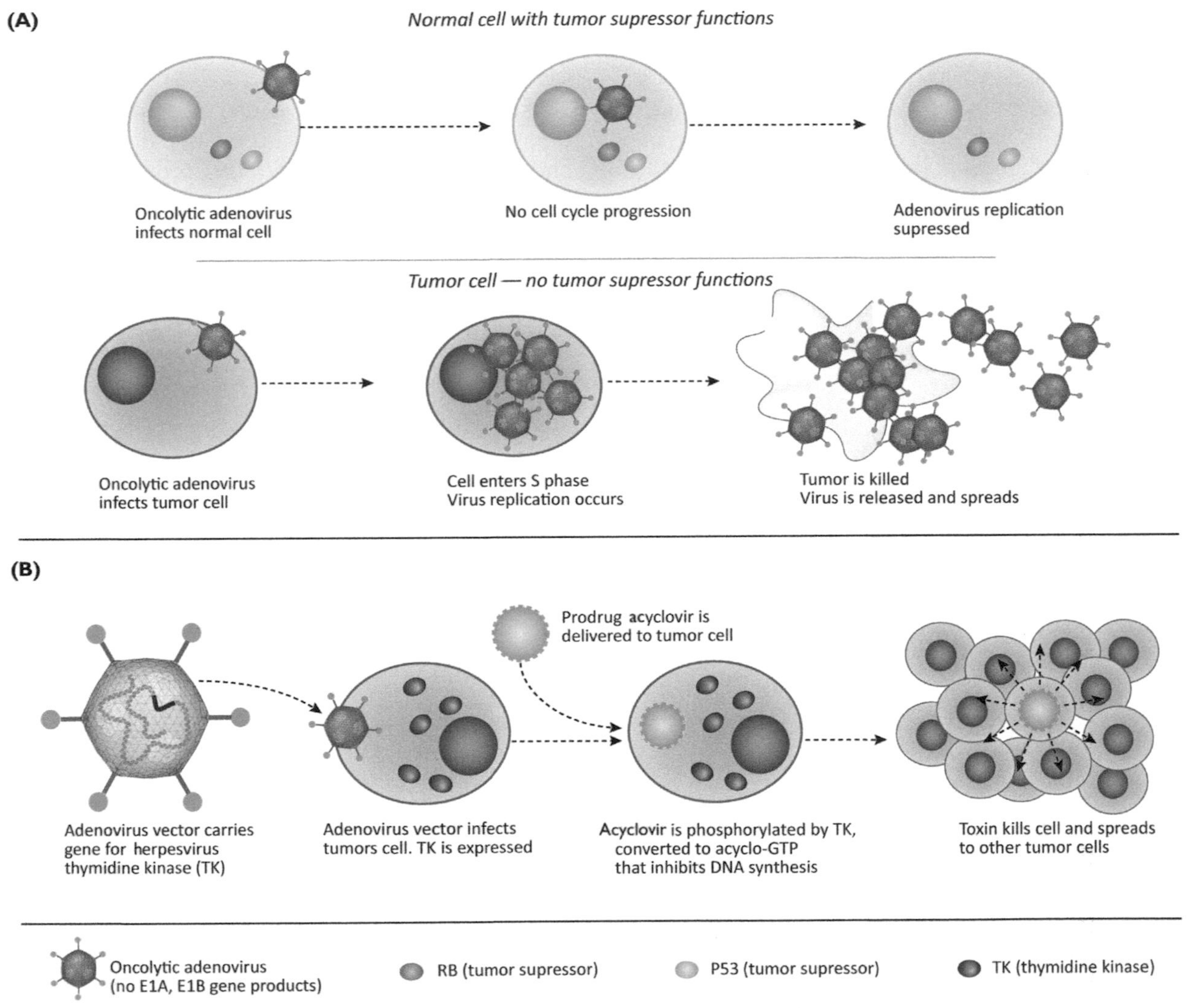

FIGURE 33.4 (A) Oncolytic AdVs lack E1A and E1B proteins therefore cannot replicate in normal, nondividing cells. However cancer cells are mitotically active and no longer express TS such as the p53 protein therefore the oncolytic AdV can and does replicate, killing the tumor cells. (B) In this scenario the AdV carries a foreign gene (for example, herpesvirus thymidine kinase) into the cancer cells. The herpesviral TK (Fig. 34.6) activates the prodrug (acyclovir in this example) to produce an active drug that inhibits tumor-cell DNA replication. The activated drug can enter nearby, uninfected cells.

ADENOVIRAL VECTORS

AdVs have proven to be useful vectors for gene delivery, with applications in basic research, gene replacement therapy, and cancer therapy. Human AdVs package about 40 kbp of DNA and the newest generation of vectors can contain up to 37 kbp of foreign DNA. To make a vector, the foreign DNA is delivered to so called "packaging" cells that contain the genes for AdV DNA Pol, TP, and capsid proteins. Within packaging cells, the foreign DNA is replicated and packaged into virions. The virions are now ready to deliver the foreign DNA to target cells. The DNA delivered by AdV vectors does not replicate in the recipient cell and it remains extrachromosomal (it is not integrated into the host genome). The foreign genes can be expressed, but as recipient cells divide, the recombinant gene is lost. This can be a disadvantage; however, it provides a measure of safety as well.

In this chapter we learned that:

- AdVs are unenveloped, icosahedral ds DNA viruses. AdV genomes are linear and covalently linked at their 5′ ends to the viral protein (TP) that served as a primer for replication.
- AdV DNA replication is asymmetric. One parental strand is copied while the other is displaced. The displaced strand forms ds hairpin structures that then serve as replication origins.
- AdV stimulates host cell DNA replication by early gene products that inactivate pRB family members and p53.
- Human AdV cannot productively infect rodent cells. Nonproductive infection occasionally results in integration of DNA encoding E1A and E1B, leading to transformation. Productive infection of human cells is cytolytic.
- Human AdV infections are often asymptomatic but can cause mild respiratory, urinary tract, and eye infections.
- Genetically modified oncolytic AdVs lack E1A and E1B thus they cannot replicate in normal cells (they cannot stimulate DNA replication or counter the apoptotic activities of p53). As cancer cells are mitotically active, and most do not express p53, they can support the lytic growth of the genetically modified oncolytic AdVs.

CHAPTER

34

Family *Herpesviridae*

OUTLINE

After reading this chapter, you should be able to discuss the following:

- What characteristics are shared by all herpesviruses?
- What are the basic features of herpesvirus virions?
- What is cellular latency? How does latent infection compare with productive infection?
- What types of cells are productively infected by human herpesviruses (HHV)-1 and 2? What types of cells are latently infected by these viruses?
- What diseases are caused by HHV-1 and 2?
- What are some of the diseases caused by HHV-4 (Epstein–Barr virus or EBV) and how are they related to the cell tropism of this virus?
- What is shingles and what is the purpose of the shingles vaccine? How is the shingles virus related to the chickenpox virus?
- How does acyclovir work and why is it specific for herpesvirus-infected cells?

Members of the family *Herpesviridae* infect a wide range of vertebrates including mammals, birds, and reptiles. The family name is derived from the Greek herpein "to creep," referring to the latent, recurring infections typical of this group of viruses. After initial productive infection, these large enveloped double-stranded (ds) DNA viruses (Fig. 34.1) probably persist for the life of the organism. Infections are often completely silent, a process called latency (Box 34.1). For example, the chickenpox virus that infects a child may remain latent for decades, but may reactivate in old age to infect a new generation!

The family *Herpesviridae* contains three subfamilies. The subfamily *Alphaherpesvirinae* contains the human herpesviruses (HHV)-1 and 2, more commonly known as herpes simplex viruses. These viruses productively infect epithelial cells, producing blister-like lesions and establish latent infections in neurons.

The members of the subfamily *Betaherpesvirinae* tend to replicate slowly; a productive infection cycle can take as long as a week in cultured cells. Persistently infected cultures are easy to establish and infected cells may become enlarged. Betaherpesviruses establish latency in leukocytes; human betaherpesviruses include HHV-5 (cytomegalovirus, CMV, named for the production of cytomegalia or enlarged cells), HHV-6A, HHV-6B, and HHV-7.

The members of the subfamily *Gammaherpesvirinae* infect mammals. Gammaherpesviruses productively infect epithelial and/or lymphoid cells. These viruses

Viruses. DOI: http://dx.doi.org/10.1016/B978-0-12-803109-4.00034-9

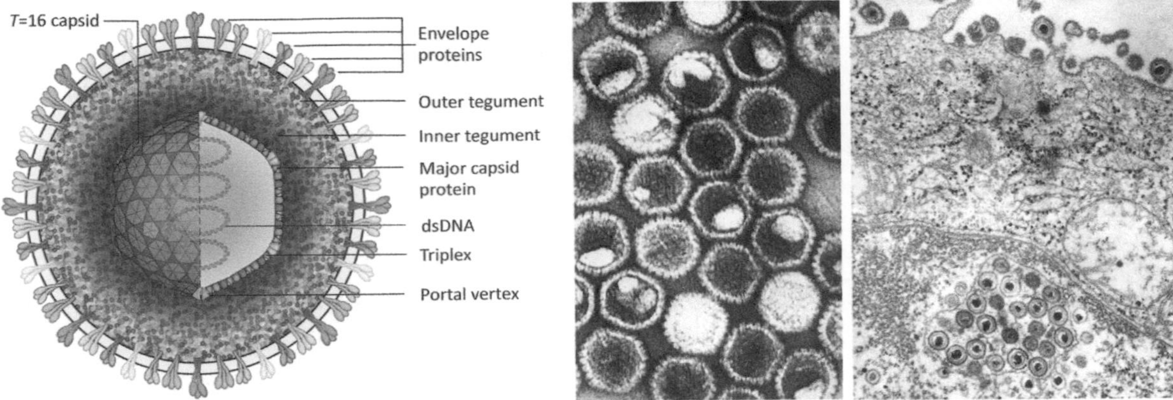

FIGURE 34.1 Herpesvirus structure (left). Negatively stained transmission electron microscopic image (right) showing numerous herpes simplex (HHV-1) virions, located both inside the nucleus, and extracellularly. *From CDC/Dr. Fred Murphy; Sylvia Whitfield. CDC Public Health Image Library Image ID#10260.*

BOX 34.1

GENERAL CHARACTERISTICS

Herpesvirus genomes found within infectious virions are linear ds DNA ranging in size from ~150 to 225 kbp. However, soon after entry into the nucleus, the DNA becomes a covalently closed circle. Genomes encode over 100 genes. Most mRNAs are unspliced but they are capped and polyadenylated. There are clear patterns of early versus late gene transcription. DNA replication occurs after early transcription. Structural proteins are synthesized from late transcripts. Transcription and genome replication are nuclear. Capsids assemble in the nucleus.

Virions are enveloped with an icosahedral capsid. Surrounding the capsid is a dense layer of viral proteins called the "tegument." There are several glycoproteins associated with the envelope.

Herpesviruses establish of life-long latent infections.

are usually specific for replication in either T cells or B-cells; latency is established in lymphoid cells that may proliferate in response. Human gammaherpesviruses include HHV-4 (EBV) and HHV-8 (Kaposi's sarcoma-associated herpesvirus) (Box 34.2).

GENOME ORGANIZATION

Herpesvirus virions contain a molecule of linear ds DNA. Herpesvirus genomes range in size from ~124 to 259 kbp and they encode anywhere from 70 to 200 proteins. Some types of herpesviruses have genomes with extensive repeat regions. Upon delivery to the nucleus, viral DNA circularizes prior to transcription and replication. During latency herpesvirus genomes remain episomal for the most part.

Transcription and genome replication are nuclear. Transcripts are synthesized by RNA polymerase II (RNAP II) and are capped and polyadenylated. Most viral mRNAs are not spliced and gene overlaps are common (Fig. 34.2). Genome replication requires a multisubunit herpesvirus DNA polymerase. Some herpesviruses also encode enzymes required to generate adequate pools of dNTPs. For example, HHV-1 encodes thymidine kinase (TK), dUTPase, and ribonucleotide reductase.

VIRUS STRUCTURE

Virions range in size from 120 to 260 nm in diameter. Virions are complex and assembled from a large number of proteins. Estimated numbers of structural proteins range from 4 to 7 capsid proteins, 9 to 20 tegument proteins, and 4 to 19 envelope proteins.

Herpesvirus capsids are ~125 nm in diameter. Capsids are $T=16$ icosahedral structures assembled from 150 hexons, 11 pentons, and single portal complex. The specialized portal complex takes the place of the 12th pentamer. Hexons and pentons are assembled from the major capsid protein (MCP). The basic

BOX 34.2

TAXONOMY

Order: *Herpesvirales*

The order *Herpesvirales* contains three virus families. This chapter focuses on the family *Herpesviridae*, whose members infect mammals, birds, and reptiles.

Family Herpesviridae

Subfamily: *Alphaherpesvirinae*

Genus: *Iltovirus* (avian herpesviruses)
Genus: *Mardivirus* (avian herpesviruses)
Genus: *Scutavirus* (turtle herpesvirus)
Genus: *Simplexvirus* (mammalian herpesviruses including human herpesviruses (HHV) 1 and 2, the herpes simplex viruses.)
Genus: *Varicellovirus* (mammalian herpesviruses including HHV3, Varicella zoster virus)

Subfamily: *Betaherpesvirinae*

Genus: *Cytomegalovirus* [includes HHV-5 (cytomegalovirus, CMV)]
Genus: *Muromegalovirus* (murid betaherpesviruses)
Genus: *Proboscivirus* (elephant betaherpesvirus)
Genus: *Roseolovirus* (human betaherpesviruses 6 and 7)

Subfamily: *Gammaherpesvirinae*

Genus: *Lymphocryptovirus* [includes human gammaherpesvirus 4 (Epstein-Barr virus, EBV)]
Genus: *Macavirus* (mammalian gammaherpesviruses)
Genus: *Percavirus* (equid and mustelid gammaherpesviruses)
Genus: *Rhadinovirus* (includes human gammaherpesvirus 8)

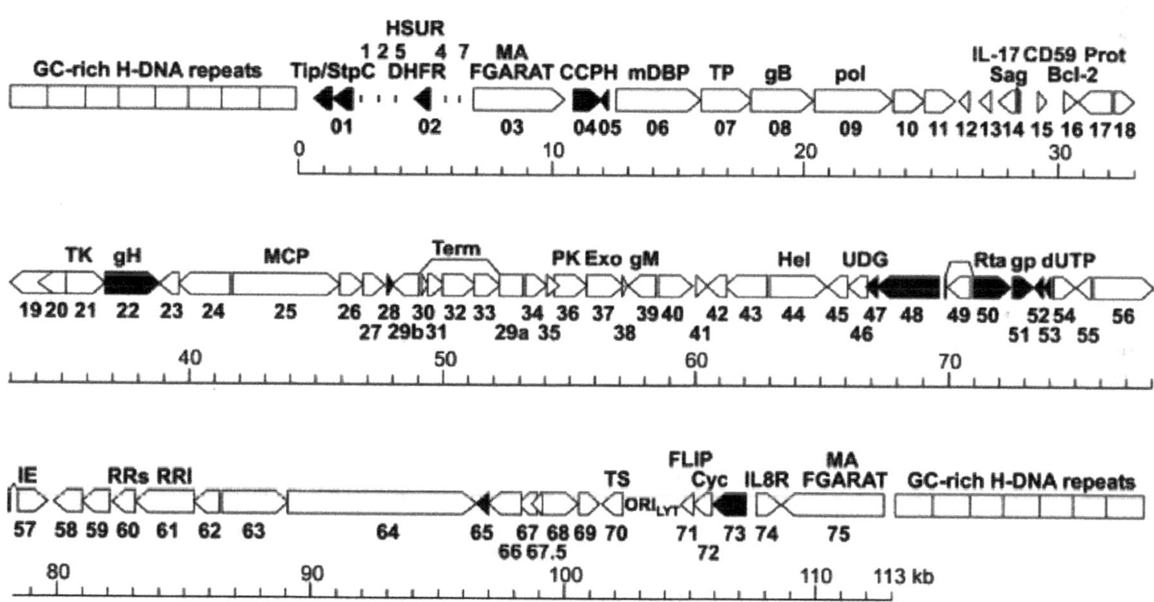

FIGURE 34.2 An example of herpesvirus genome organization showing densely packed transcripts. (From Ensser et al., 2003)

structure of the herpesvirus capsid protein is distinct from those of most animal viruses. Herpesviral MCP is structurally related to the capsid proteins of some DNA phages. The portal complex is assembled from 12 copies of the capsid portal protein. This is the route by which the genome enters and exits the capsid. A so-called "portal capping protein" (PCP) is associated with mature DNA-containing capsids. Within the capsid, DNA is densely packed, aided by the anion spermine.

Virions are enveloped and contain several different envelope glycoproteins. Herpesvirus envelopes are densely packed with glycoproteins; the copy number of individual glycoproteins can exceed 1000 per virion.

Between the capsid and envelope is a dense protein layer called the tegument. Tegument proteins are key players in the early phases of the herpesvirus replication cycle. Functions of tegument proteins include inhibition of cell defenses, shut off host protein synthesis and stimulation of early viral gene expression.

REPLICATION CYCLE

As might be expected of large, complex viruses that switch between productive and latent forms of infection, there is much variability in the details of herpesvirus replication. Add to that the importance of HHV as pathogens, and the literature becomes extensive! The following sections address only the most basic aspects of herpesvirus replication. It should also be noted that herpesvirus protein nomenclature is variable, complex, and confusing at times. In the following sections proteins will be described only by their most obvious functions (for example, "DNA polymerase").

Attachment/Penetration

Attachment and penetration are probably achieved by multiple viral glycoproteins. Penetration into cells requires fusion; in some cases at the plasma membrane, but in other cases following endocytosis. Released capsids and some tegument proteins move to nuclear pores with the help of the cell cytoskeleton. Tegument proteins may be active during the transport process, altering host cell signaling pathways to favor virus replication.

DNA moves out of the capsid and through the nuclear pore in an energy requiring process. It is likely that some tegument proteins enter the nucleus as well. In the nucleus the genome is circularized.

Transcription/DNA Replication

Once in the nucleus, the DNA may be transcribed and enter a pathway to productive infection. Alternatively the cell environment may favor the establishment of latency. In the case of productive infection, the first genes transcribed are those that support DNA replication. Viral DNA replication requires many viral proteins including: Viral DNA polymerase, single-stranded DNA-binding protein, Ori (replication origin)-binding protein, ds DNA-binding protein, primase/helicase, TK, ribonucleotide reductase, dUTPase, uracil DNA glycosylase among others. Enzymes involved in nucleotide biosynthesis are among those that are often dispensable for replication in cultured cells. During a productive infection, there is an ordered schedule of transcription similar to that seen for other DNA viruses. In general, transcription of structural proteins, including the tegument proteins, begins during or after DNA replication.

DNA replication is through a rolling circle mechanism that results in formation of head to tail concatemers (Chapter 28: Introduction to DNA Viruses). These are cleaved into unit length genomes during packaging.

Assembly/Release

Capsid assembly takes place in the nucleus, the site of genome replication. Capsid assembly is complex, and occurs with the help of scaffold proteins. Nascent capsids are filled with viral DNA (through the portal complex) in a process that requires energy. DNA is packaged in a "headful" mechanism whereby concatemers are cleaved at conserved sequences that define the genome ends. The DNA is tightly packed, producing a rigid capsid and the capsid is "sealed" by a PCP.

DNA-containing capsids leave the nucleus and traffic through the cytoplasm. Some studies suggest that capsids bud from the nucleus and then lose the nuclear envelope. Other studies suggest that naked capsids leave the nucleus. In either case, capsids associate with tegument proteins as the capsid travels through the cytosol. Virions then obtain a final envelope by budding into a cytoplasmic organelle (for example, recycling endosomes or Golgi). Virions are then transported, within a vesicle, to the plasma membrane and are released by exocytosis. Productive infection is usually accompanied by cell death.

Latency

Herpesviruses are noted for their ability to establish cellular latency. Cellular latency is a process during which genomes are maintained in cells, with limited expression of viral proteins and no virion production. Cellular latency is regulated at the level of the individual cells. Thus, one or a few cells may break from latency, while other cells remain latently infected. Use of the term latency implies that virus has the *ability to reactivate* (enter a productive replication cycle) in response to certain stimuli. In this way latency is differentiated from *abortive* infection (during which virus is never produced). The precise molecular mechanisms that lead to establishment of, and reactivation from, the latent state are not fully understood and can differ from one herpesvirus to another. Finally, it should be noted that cellular latency is distinct from clinical latency. The latter term refers to the overall state of the infected organism. Thus clinical latency is the absence of disease, not the absence of virus.

Cellular latency permits herpesviruses to remain in the infected host for long periods while avoiding detection by the immune system. Members of the subfamily *Alphaherpesvirinae* (for example, HHV-1, 2, and 3) replicate productively in epithelial cells but establish latent infections in neurons. The process begins when virus released from epithelial cells and enters the axonal terminus of a nearby neuron. Fusion releases the capsid which uses cytoskeletal components to actively move (by retrograde axonal transport) to the nucleus in the cell body of neuron. DNA is delivered to the nucleus but transcription is limited to a set of noncoding RNAs called *latency-associated transcripts*. If circumstances permit, the HSV genome may become transcriptionally active (reactivated). Infectious virus is formed in the cell body and moves by anterograde axonal transport back to the site of the original infection.

Beta- and gammaherpesviruses establish latency in leukocytes. HHV-4 (EBV) latency has been studied extensively. HHV-4 productively infects epithelial cells and establishes latency in B-lymphocytes. HHV-4 latency is quite complex, with variable numbers of proteins and small RNAs expressed. In cultured cells, EBV infection can immortalize B-cells to produce lymphoblastoid cell lines (LCL). Most LCLs are latently infected and they express a small number of HHV-4 proteins and noncoding RNAs. The proteins expressed during latency include the Epstein–Barr nuclear antigens (EBNAs) and the latent membrane proteins (LMPs). Small RNAs include EBV-encoded small RNA (EBERs) and micro (mi) RNAs.

EBNA-1 acts as a bridge to couple the HHV-4 episome to the cellular chromosome. This enables replication and stable maintenance of the genome during cell division. EBNA-2 is a transactivating protein that works through association with cellular DNA-binding proteins to modulate expression of cellular and EBV genes. The EBNA-3 genes are required for transformation of primary B-cells. The LMPs are membrane receptors. LMP1 mimics active tumor necrosis factor receptor, TNFR. LMP2A mimics an activated B-cell receptor. Acting in concert these receptors activate signaling pathways necessary for survival and growth of LCL.

EBERs are small noncoding RNAs. They are not expressed during lytic infection but are abundant in latently infected cells. Both EBER1 and EBER2 are transcribed by RNAP III; EBER1 is also transcribed by RNAP II thus is quite abundant in infected cells. EBER RNAs assume stable secondary structures and localize to the nucleus where they are found in complexes with cellular proteins La and L22. La protein acts as an RNAP III transcription factor in the nucleus and as a translation factor in the cytoplasm. L22 is a component of the 60S ribosomal subunit. One function of EBERs is to block interferon-induced PKR phosphorylation and (recall that phosphorylated PKR inhibits translation initiation factor eIF-2α). EBV also encodes a group of miRNAs that likely have roles in establishing or maintaining EBV latency, targeting both viral and cellular mRNAs.

The pattern of EBV latent gene expression in LCLs is called "latency 3" to distinguish it from other patterns of gene expression that are observed in transformed cells or in B-cells of normal persons. In normal B-cells (in the immunocompetent patient) few if any, EBV genes are expressed and this state is called latency 0. In Burkitt lymphoma (BL) tumors only the EBNA-1 protein is expressed (latency 1). In other EBV-related tumors EBNA-1, LMP1 and LMP2 are expressed (latency 2). In summary, EBV latency is multifaceted and complex! The common factor is that virions are not produced and episomal DNA is maintained in dividing cells.

DISEASES CAUSED BY HUMAN HERPESVIRUSES

Most herpesviruses are very host specific. They are well adapted to their hosts and their lifestyle is to remain within an infected host for life. Thus many herpesvirus infections are asymptomatic. Table 34.1 lists the human herpesviruses and their associated diseases. Occasionally herpesviruses cause severe systemic and/or central nervous system infections. Neonates and immunocompromised individuals are the most at risk for severe infection. There are also clear associations between some herpesviruses and cancer.

HHV-1 and 2 Disease

The herpex simplex viruses cause blister-like lesions. The viruses are found worldwide and transmission occurs by contact. Primary infection is usually mild and localized (or asymptomatic). While virus is obviously present in lesions, it can also be present (and is transmissible) by asymptomatic persons.

HHV-1 and 2 are serologically related, sharing ~80% sequence identity. The two viruses usually infect different parts of the body, with HHV-2 more frequently infecting genital mucosa. However the incidence of genital HHV-1 is rising; about 50% of primary genital infections involve HHV-1. HHV-1 and 2 establish latency in neurons. Reactivation is often associated with compromise of the immune system, for example, by the extreme stress of final exams!

Herpes simplex virus infections of neonates can be very serious and life threatening. Infants can be infected in utero, during birth, or shortly after birth. Infections of neonates are frequently fatal due to

TABLE 34.1 Human Herpesviruses

ICTV[a] name	Vernacular name	Subfamily	Associated diseases[b]
HHV[c]-1	Herpes simplex 1	*Alphaherpesvirinae*	Cold sores
HHV-2	Herpes simplex 2	*Alphaherpesvirinae*	Genital herpes
HHV-3	VZV/chickenpox/shingles virus	*Alphaherpesvirinae*	Chickenpox (acute)/shingles (upon reactivation from latency)
HHV-4	EBV	*Gammaherpesvirinae*	Infectious mononucleosis; various B-cell cancers
HHV-5	CMV	*Betaherpesvirinae*	Infectious mononucleosis
HHV-6A	Roseolovirus	*Betaherpesvirinae*	Unknown
HHV-6B	Roseolovirus	*Betaherpesvirinae*	Roseola infantum (rose rash of infants, ES, or the sixth disease)
HHV-7	Roseolovirus	*Betaherpesvirinae*	Roseola infantum (rose rash of infants, ES, or the sixth disease)
HHV-8	Kaposi-associated herpesvirus	*Gammaherpesvirinae*	Usually asymptomatic; associated with KS in select populations

[a]*International Committee on the Taxonomy of Viruses.*
[b]*Many infections are asymptomatic.*
[c]*Human herpesvirus.*

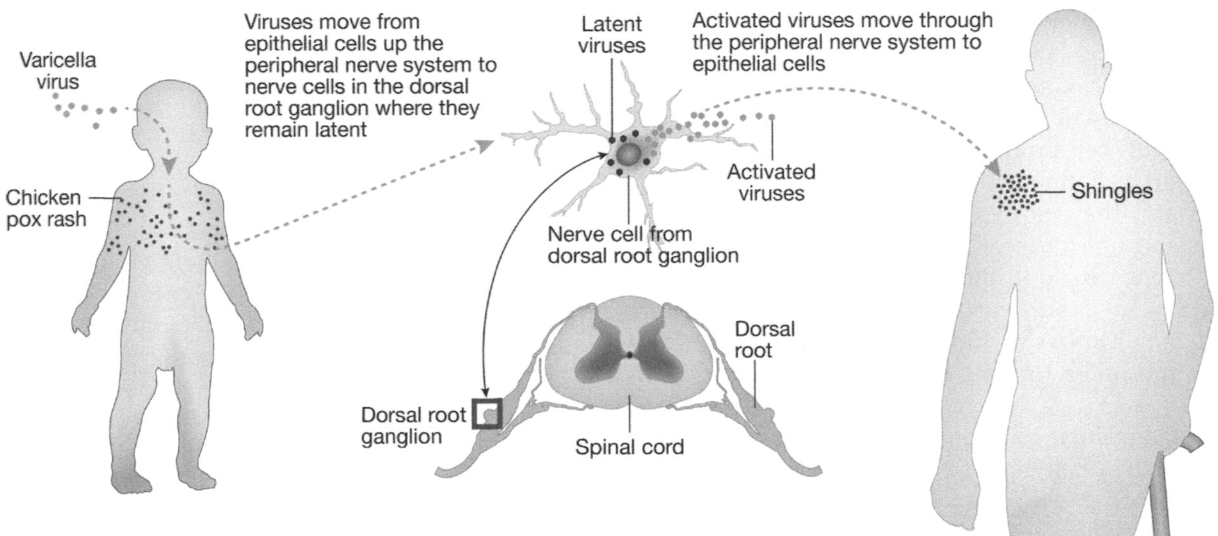

FIGURE 34.3 Disease cycle of chickenpox/shingles.

disseminated infection and encephalitis. The incidence of such infections is estimated to be 1 in every 2000–5000 live births. To date there has been little success developing protective vaccines for either HHV-1 or 2.

HHV-3 (Varicella Zoster Virus)

HHV-3 causes chickenpox, a common childhood rash. HHV-3 is spread by the respiratory route with replication beginning in the upper respiratory tract. Virus then spreads throughout the body, ending up in the skin. The virus replicates in skin cells causing a rash and itchy skin lesions due to lytic infection of the epidermis. Skin lesions contain large amounts infectious virus. Virus is shed from the skin and is passed to a new host via inhalation. In children, chickenpox is usually a mild, self-limiting infection. However, infection of adults can be more serious, resulting in pneumonia, male sterility, and/or liver failure.

HHV-3 establishes latency in dorsal root ganglia of the spinal cord (Fig. 34.3). During latency, no viral proteins are produced. Latency can last for decades. Reactivation most frequently occurs in response to stress or immunosuppression. Older adults are at risk for reactivation as their immunity to HHV-3 gradually wanes. Reactivation of HHV-3 results in a painful rash called shingles that consists of large blister-like lesions that contain infectious virus. The rash may last for weeks, and debilitating nerve pain can continue for months.

Varicella zoster virus (VZV) is the only human herpesvirus for which a vaccine is available and widely recommended. The VZV vaccine contains live attenuated virus (strain Oka, developed in Japan in 1974) and was approved for use in the United States in 1995. Current recommendations are for universal vaccination of children. Childhood vaccination is meant to prevent the more serious complication of chickenpox that can result when naïve adults are infected. There are also indications that the vaccine strain is less prone to reactivation than wild-type VZV. It is hoped, but remains to be seen, if vaccination in childhood protects against shingles decades later. The Oka vaccine is also licensed for use in elderly adults, to provide protection against shingles.

HHV-4 (Epstein–Barr Virus)

HHV-4 infects epithelial cells and B-lymphocytes. Only infection of epithelial cells is productive. HHV-4 establishes latency in B-cells. When HHV-4 breaks out of latency, virions seed the oral mucosa. Studies of bone marrow transplantation patients have confirmed the "life-cycle" of HHV-4. Prior to a bone marrow transplant, the recipient's bone marrow and immune cells are destroyed. After a transplant, the HHV-4 from the *donor* is found in the transplanted B-cells and in epithelial cells of the recipient. This suggests that the only HHV-4 reservoir was in the recipient's B-cells. In the event that bone marrow from an uninfected individual is transplanted into an HHV-4 positive recipient, the recipient will now be negative for HHV-4. These findings provide good evidence that over the long term, HHV-4 is maintained in B-cells and only periodically and transiently infects epithelial cells.

During periods of productive replication, HHV-4 is produced (usually with no symptoms or lesions) and can be transmitted via saliva. HHV-4 is a very common virus and most humans are infected by age 3. Once infected, always infected, and more than 90% of the human population is infected with HHV-4. Most often there is no disease. Infection of older individuals (teenagers or older) can be symptomatic and the disease is infectious mononucleosis. Symptoms include fatigue, fever, rash, swelling of lymph nodes, and enlargement of spleen and liver. The disease is characterized by the presence of large numbers of mononuclear cells (T-lymphocytes) in the blood. These T cells are responding to HHV-4 infection. This strong T-cell response (a cell-mediated response to the infection) is an immunopathologic event.

The ability of HHV-4 to stimulate B-cell proliferation is a potentially (but not common) oncogenic event. (HHV-4 is routinely used to produce cultures of transformed B-cells however.)

HHV-4 and Burkitt Lymphoma

BL is a childhood lymphoma (a tumor of lymphocytes) common throughout equatorial Africa. Early studies of BL-derived cell lines showed that they produced herpesviruses (HHV-4). Children with BL also have very high levels of antibody to HHV-4, and high antibody titers to HHV-4 are associated with higher risk of developing BL. However, HHV-4 infection is only *one* step in the development of BL. BL tumor cells also have a chromosomal abnormality. In fact most BL tumors have a very specific chromosomal translocation. Another risk factor for development of BL is coinfection with the malarial parasite *Plasmodium falciparum*.

The currently proposed scenario for the development of BL, summarized in Fig. 34.4, is as follows:

A child is infected with HHV-4, establishing an infection of B-cells. The immune system initially functions normally to suppress B-cell proliferation. However, if the child is infected with *P. falciparum*, the immune system can be suppressed, allowing for uncontrolled proliferation of HHV-4 infected B-cells. These proliferating cells are at increased risk of mutation. A chromosomal translocation that activates an oncogene is the final event: B-cells with mutated chromosomes proliferate even more, leading to the

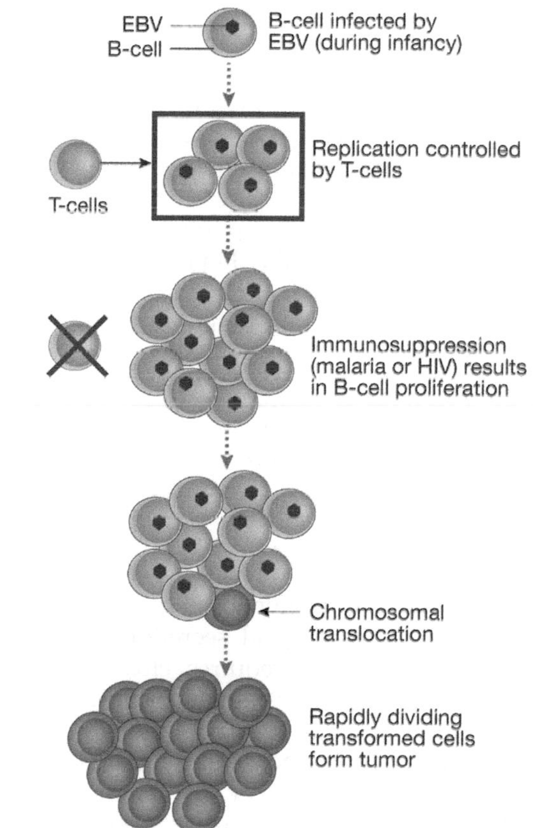

FIGURE 34.4 Model of development of BL.

development of the B-cell tumor. BL tumors are monoclonal, that is, all tumor cells originated from a single HHV-4 infected cell that underwent a chromosomal translocation.

HHV-4 and Nasopharyngeal Carcinoma

Nasopharyngeal carcinoma (NPC) is a malignancy of epithelial cells of the nasopharynx. NPC is rare in North America and Europe but the incidence in Asia (southern China) is as high as 1 case per 1000 persons per year. There is also a high incidence in Mediterranean, African, and Eskimo populations. Given that over 90% of humans are infected with HHV-4, how can it be associated with NPC among only certain populations? First we must consider the evidence that HHV-4 is involved in tumor development. The link is based partly on seroepidemiologic evidence. Persons with NPC have *very high levels* of antibody to HHV-4. Healthy individuals with high levels of antibody to HHV-4 are also *more likely* than the general population to develop NPC. The NPC tumor cells are HHV-4 infected and the tumors are clonal, suggesting a rare mutation event driving cell proliferation. At this time there is no simple model to explain the development of NPC. Disease seems to result from a combination of environmental, genetic, and virologic factors.

HHV-4 is also suspected of causing some cases of B-cell lymphoproliferative disease (B-cell lymphoma) in immune deficient humans. There is also an association between HHV-4 and Hodgkin disease, the most common malignant lymphoma in the Western hemisphere. Peak age of onset is 25–30 and then beyond age 45. Certain types of Hodgkin lymphomas (but not all) are strongly associated with EBV. Finally, HHV-4 in AIDS patients causes oral hairy leukoplakia, wart-like lesions of the lateral borders of the tongue are epithelial foci of EBV replication.

HHV-5 (Cytomegalovirus)

HHV-5 infects epithelial cells in many different tissues. Infection stimulates host cell DNA, RNA, and protein synthesis; infected cells become enlarged (hence the name cytomegalo or big cells). HHV-5 is transmitted by oropharyngeal secretions, breast milk, and blood. Transmission requires close contact. The virus is common, with 40%–100% infection rates by adulthood. Infection is usually asymptomatic, but may cause mononucleosis. Neonates of seronegative mothers are at highest risk of developing disease as they lack protective material IgG. Severe neurologic complications have occurred in immunocompromised patients. CMV is a common cause of AIDS retinitis.

HHV-6 and 7

HHV-6 and 7 are betaherpesviruses (genus *Roseolovirus*) associated with a febrile, rash illness called roseola infantum, exanthem subitum (ES), or sixth disease. HHV-6 and 7 are *ubiquitous viruses*, with horizontal infection commonly occurring during infancy. More than 85% of adults have antibody to both viruses. (HHV-6 contains two subtypes, A and B, which are actually separate species. HHV-6B is most commonly associated with ES.)

HHV-7 infects CD4+ T-lymphocytes (believed to be the site of latent infection) and salivary gland epithelial cells (productive infection). HHV-6B infects lymphoid and endothelial cells. Latent infection seems to be in monocytes, macrophage, and CD34-positive progenitor cells.

The HHV6 genome is quite interesting in that it has an array of sequences similar to mammalian telomeres and these may be involved in site-specific integrations of HHV-6. Chromosomal integrations of HHV-6 (called ciHHV-6) can lead to germline inheritance. While most HHV-6 infections occur via horizontal routes, approximately 1% of the population inherits an HHV-6 genome from a parent.

ES is a common disease of infants worldwide. It is characterized by sudden fever, which lasts for a few days, followed by a rash that appears on the trunk and face and spreads to lower extremities as the fever subsides. However it is likely that many HHV-6 primary infections of infants are asymptomatic. In adults primary infections by HHV-6 can cause mononucleosis-like disease and hemophagocytic syndrome.

HHV-8 (Kaposi Sarcoma Herpesvirus)

Before the HIV pandemic, Kaposi sarcoma (KS), a slowly developing endothelial cell tumor, was thought to be limited to older men in the Mediterranean region. KS can occur in various sites in the body, but skin lesions are common. In the early 1980s, KS emerged as an AIDS-defining cancer. Epidemiologic data suggested the involvement of a sexually transmitted agent, other than HIV, in the development of KS, and this lead to the identification of the KS herpesvirus (HHV-8) in 1994. Initial findings were that HHV-8 was present in cells within KS lesions, but not in normal surrounding tissues.

KS lesions are complex, containing multiple cell types. A key cell is the so-called spindle cell, of endothelial cell origin. KS herpesvirus infection, through a number of viral products (including proteins and miRNAs), contributes to the transformation of normal endothelial cells. It is thought that tumorigenesis is due to genes expressed during latency. One set of HHV-8 proteins, the Kaposins, can transform Rat-1 fibroblasts.

HHV-8 is present in all KS lesions, but spindle cells loose the viral genome when placed in culture.

In addition to spindle cells, KS lesions contain T cells, B-cells, and monocytes within aberrant slit-like neovascular spaces. These spaces are lined with infected and uninfected endothelial cells and are structurally fragile. They are very susceptible to leakage and rupture, causing the hemorrhages typical of KS lesions.

Although HHV-8 was first identified in KS lesions (endothelial cell tumors), it also associated with rare B-cell lymphoproliferative diseases.

Monkey B Herpesvirus

Monkey B herpesvirus infection is caused by Macacine herpesvirus (formerly Cercopithecine herpesvirus 1). The virus is also commonly referred to as herpes B, herpesvirus simiae, and herpesvirus B. In monkeys this virus causes a disease similar to HHV-1 in humans, that is, localized blister-like lesions. However, if a person becomes infected with the monkey B herpesvirus, it causes serious neurologic damage with high fatality rates. Persons in contact with Old World monkeys are at highest risk of contracting monkey B virus. The virus is transmitted via body fluids and most research laboratories in the US laboratories use only animals that are antibody negative for B virus. (But according to the CDC: Macaques housed in primate facilities frequently become B virus positive by the time they reach adulthood!)

ANTIHERPESVIRAL DRUGS

Safe and effective treatments for HHV-1, 2, and 3 are widely available. Acyclovir was the first antiherpes drug developed and it provided the basis for development of additional antiherpesvirus drugs also in use today.

Acyclovir is a competitive inhibitor of dGTP. It is recognized by herpesvirus DNA polymerase and its incorporation prevents DNA chain elongation. An important feature of acyclovir is low toxicity for humans. This is due to the production of the *active* phosphorylated form of the drug *only* in herpesvirus-infected cells. Herpesviruses encode a viral TK that readily phosphorylates acyclovir while normal cells do not phosphorylate the prodrug (Fig. 34.5). In addition, the herpesvirus DNA polymerase has a higher affinity for phosphorylated acyclovir than do cellular DNA polymerases. Resistance to acyclovir resistance is low, but mutants can arise. Early studies of resistant mutants were helpful in delineating the mechanism of action of the drug: The finding that resistance mapped to mutations in herpesviral TK and/or DNA

Acyclovir

Herpesvirus thymidine kinase (TK)

Acyclo-GMP

Cellular (TK)

Acyclo-GTP

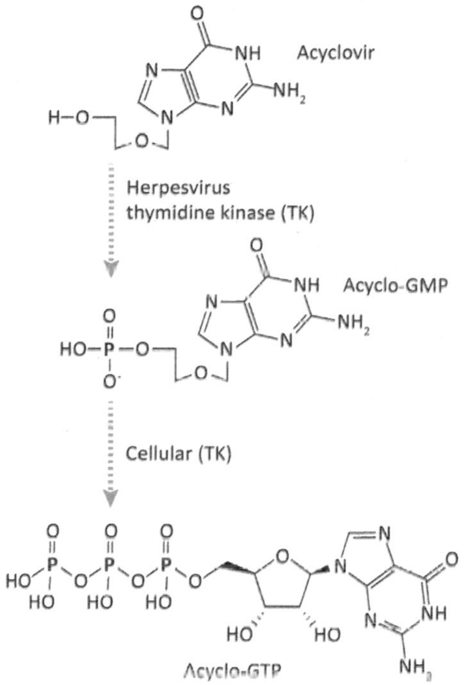

FIGURE 34.5 Mechanism of action of acyclovir. Acyclovir is an inactive "prodrug" until phosphorylated by herpesvirus TK. Monophosphorylated acyclovir is further phosphorylated by host kinases.

polymerase led to the realization that specificity for infected cells was due to the requirement for the viral TK to produce an active drug.

Ganciclovir and the prodrug valganciclovir are also competitive inhibitors of dGTP. They are often used for treating CMV infections (HHV-5), particularly in immunodeficient patients. Both drugs must first be phosphorylated to ganciclovir monophosphate by the CMV kinase. Additional antiherpesviral drugs include famciclovir and valacyclovir. Valacyclovir is a prodrug for acyclovir that can be taken orally.

In this chapter we learned that:

- Herpesviruses are large ds DNA viruses that are notable for the establishment of latent infections.
- Virions are enveloped with an icosahedral capsid. The capsid is surrounded by a protein layer called the tegument. Within the capsid, the DNA is tightly packaged. The capsid is formed by the MCP, notable for its structural relationship to the capsid proteins of certain types of DNA phages. The capsid contains a special "portal complex" through which the DNA genome is packaged and released.
- Cellular latency is the maintenance of herpesviral genomes in cells, in the absence of productive infection. Few or no viral proteins are produced in the latently infected cells. Some herpesviruses (for example, HHV-4) stimulate replication of latently infected cells. Most productive infections are cytolytic.

- HHV-1 and 2 productively infect epithelial cells but latently infect neurons.
- HHV-1 and 2 cause cold sores and genital lesions. Painful blister-like lesions are filled with infectious virus. In neonates these viruses can produce severe systemic infections. It is increasingly recognized that infectious HHV-1 and 2 can be present on skin in the absence of any obvious lesions.
- HHV-4 or EBV is one cause of infectious mononucleosis although most infections are asymptomatic. HHV-4 is a ubiquitous virus that infects most of the world's population. HHV-4 establishes latency in B-cells and induces them to proliferate. Under specific sets of conditions B-cell tumors can form.
- Shingles is the painful rash that develops when latent HHV-3 (the chickenpox virus) reactivates. The shingles vaccine is used to stimulate immune responses to the virus that tend to wane among the elderly. The same attenuated virus used for the shingles vaccine is found in the chickenpox vaccine recommended for children.
- Acyclovir is competitive inhibitor of dGTP. It competes with dGTP for incorporation into herpesviral DNA by the viral DNA polymerase. It is a DNA chain-terminator. Acyclovir is very specific to herpesvirus-infected cells because it must be phosphorylated to be active. Acyclovir is a substrate for herpesvirus TK but is not converted to the monophosphate form in uninfected cells.

References

Ensser, A., Thurau, M., Wittmann, S., Fickenscher, H., 2003. The genome of herpesvirus saimiri C488 which is capable of transforming human T cells. Virology. 314, 471–487. Available from: http://dx.doi.org/10.1016/S0042-6822(03)00449-5.

CHAPTER

35

Family *Poxviridae*

OUTLINE

After reading this chapter, you should be able to answer the following questions:

- What are the general characteristics of poxviruses?
- Where in the cell do poxviruses replicate?
- Are host or viral enzymes used for mRNA synthesis and genome replication?
- Compared to other enveloped viruses, what is unique about poxvirus morphogenesis?
- What are virokines and viroceptors and what is their function in poxvirus replication?
- What was "variolation" and why was Jenner's "vaccination" an improvement?
- Explain the outcome of the introduction of rabbit myxoma virus into Australia. What does this show about the interactions of viruses and their hosts?

Poxviruses are large, complex, DNA viruses whose various members infect animals and insects. The term "pox" derives from the English *pocks,* referring to blister-like skin lesions that are the hallmark of smallpox and other poxvirus infections. Smallpox is an ancient disease, as evidenced by scarred Egyptian mummies. By the middle ages, the disease was common in Europe and from there it traveled to North and South America with European explorers and colonists in the 16th and 17th centuries. Smallpox was notable for many reasons. Mortality varied, depending on the particular strain, but the average was ~25%. The disease was severe and painful, culminating in pus-filled blisters that sometimes covered the entire body. The resulting scars permanently marked smallpox survivors, sometimes causing blindness or limb deformities. Smallpox spread most readily from person to person, but could also be spread via contaminated clothes, bedding, and household items that might be infectious for weeks or months.

Smallpox is also notable for its place in the history of vaccines and vaccination. As disease survivors were clearly protected from further infection, by about the 10th century dried scab material was being collected from mild cases of disease and used to infect naïve individuals. In China the process involved blowing powdered material into the nostrils while in India infection was accomplished by inoculating scratched skin. Both methods usually resulted in mild disease with mortality significantly lower than natural infection (although these procedures were certainly not risk-free; the fatality rate was ~2%).

By the late 18th century, cutaneous variolation was a common practice in Europe and North America and this set the stage for English physician Edward Jenner to develop an even safer method to achieve protection. Jenner knew that milkmaids, who often experienced mild cowpox lesions on their hands, believed that they

Viruses. DOI: http://dx.doi.org/10.1016/B978-0-12-803109-4.00035-0

were resistant to both variolation and smallpox infection. In 1796 Jenner used material from a cowpox lesion to variolate a child. Jenner reported the resulting reaction as "barely perceptible." To test for protection, Jenner challenged the child by variolation with smallpox, and he was indeed resistant to infection. Jenner called his technique vaccination (*vacca*, Latin for cow) to distinguish it from the older procedure. Vaccination was an effective and much safer procedure than variolation. Safety was key to its widespread use, and the eventual elimination of smallpox.

While Jenner's original vaccine was cowpox, other poxviruses (i.e., horse pox and buffalo pox) were used as the source of vaccine as the procedure became widely accepted. By the late 19th century, the virus we know today as "vaccinia virus" (VACV) was widely used as the smallpox vaccine. We now know, through genome sequencing, that origin of VACV was not Jenner's cowpox.

In 1958 the World Health Organization began the (then) unprecedented task of eliminating smallpox by launching the Global Smallpox Eradication Program. While a daunting undertaking, circumstances favoring success included the following:

- The smallpox or variola virus was exclusively a human virus.
- A safe, effective, and stable vaccine was available (lyophilized preparations were developed by 1950).
- Smallpox did not cause silent or persistent infections.
- The disease was easy to recognize, diagnose, and track.

In the United States, routine use of smallpox vaccination ended in 1972, when the small risk of serious complications came to exceed the risk of infection. The last natural case of smallpox in the world was in 1977 and the disease was declared eradicated in 1979. Routine vaccination was discontinued worldwide in 1982. Variola virus now exists only in two high-containment laboratories. However development and production of safer smallpox vaccines continues to this day, and vaccine stockpiles serve to guard against accidental or intentional release of this deadly and disfiguring virus.

GENOME ORGANIZATION

Poxviruses are large DNA viruses that replicate in the cell cytoplasm. Genomes are a linear, base paired, molecule of DNA, with linked ends. Genome sizes range from 130 to 300 kbp. The genome has a unique central region flanked by inverted repeats of about 10 kbp. Very close to each end is a series of short (~70 bp) tandem repeats. Genes encoded in the central region are well conserved and essential to basic replication, while genes in the repeat regions are more virus-specific and tend to function in virulence, immune evasion, and host range. VACV is the best studied of the poxviruses. It encodes about 200 genes, densely packed on the genome. Genes have short promoters and no introns (as there is no splicing apparatus in the cell cytosol). Proteins are encoded from both strands of DNA and some genes are overlapping. Genome replication and virion assembly occur in discrete virus "factories" in the cell cytosol.

VIRION STRUCTURE

Poxvirions are large, enveloped, ovoid, or brick shaped (dimensions of VACV are about $360 \times 270 \times 250\ nm^3$). Virion morphology is often referred to as complex and they contain 70–80 structural proteins. Virions have multiple envelope proteins, with obvious transmembrane domains, but notably, none are glycosylated. This may be a consequence of the unusual mechanism of virion assembly described later. The VACV virion is covered with a random pattern of proteinaceous surface ridges (Fig. 35.1). Other poxviruses (e.g., parapoxviruses) have a more regular (criss-crossed) surface pattern.

Viewed in thin section, virions contain a central tubular region (often dumbbell-shaped) containing DNA surrounded by a thick core wall. Electron dense, proteinaceous lateral bodies are positioned in the curves of the core and are surrounded by the lipid envelope. The majority of poxvirions (so-called mature virions or MV) are retained within infected cells, only released after cell death. However some poxvirions have two envelope layers. These specialized "extracellular virions" (EVs) are released from (living) infected cells, and remain associated with cellular structures (microvilli) that actively transport them to nearby uninfected cells (Box 35.1).

REPLICATION

The family *Poxiviridae* contains viruses that infect vertebrates (subfamily *Chordopoxvirinae*) and insects (subfamily *Entomopoxvirinae*). VACV, genus *Orthopoxvirus*, is the best studied of the poxviruses and is the focus of this chapter (Box 35.2).

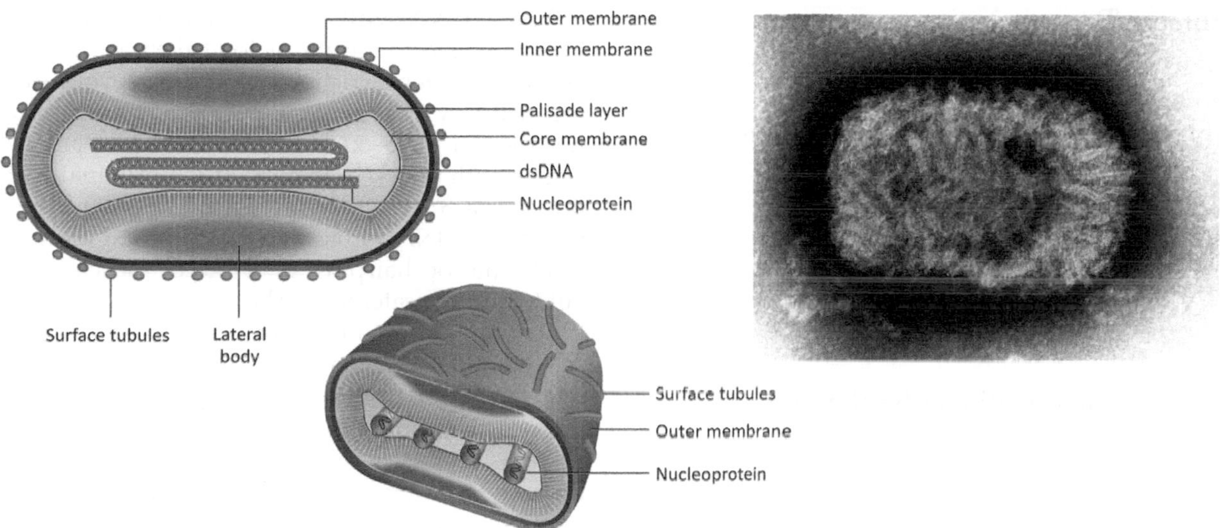

FIGURE 35.1 Virion structure (left). Transmission electron microscopic image depicting a smallpox (variola) virus particle (right). *From CDC/J. Nakano. Public Health Image Library Image ID#8392.*

BOX 35.1

GENERAL CHARACTERISTICS

Poxviruses are large DNA viruses that replicate in the cell cytoplasm. Genomes are linear, base paired, molecules of DNA, with linked ends. Genomes range in size from 130 to 375 kbp and encode hundreds of polypeptides. DNA synthesis occurs in a discrete location in the cytoplasm to form a region called a "virus factory," an electron dense region of cytosol devoid of cellular organelles. Poxviruses encode both RNA and DNA polymerases. They also encode many proteins that inhibit innate and adaptive immune responses. Virions are brick shaped or ovoid; complex-enveloped structures that range in size from 140 to 260 nm in height and 220 to 450 nm in length. Although poxviruses encode all of the enzymes they require for replication in the cytoplasm, naked DNA is poorly infectious due to the need for packaged viral proteins that can initiate transcription.

BOX 35.2

TAXONOMY

Family *Poxviridae*
Subfamily: *Chordopoxvirinae*
Genus: *Avipoxvirus*
Genus: *Capripoxvirus*
Genus: *Cervidpoxvirus*
Genus: *Crocodylipoxvirus*
Genus: *Leporipoxvirus* (myxamatosis virus, rabbits)
Genus: *Molluscipoxvirus* (moluscum contagiosum virus)
Genus: *Orthopoxvirus* (variola (smallpox virus), cowpox, VACV, monkeypox virus)
Genus: *Parapoxvirus* (contageous exthyma virus, sheep, and goats)
Genus: *Suipoxvirus*
Genus: *Yatapoxvirus*
Subfamily: *Entomopoxvirinae* (insect poxviruses)
Genus: *Alphaentomopoxvirus*
Genus: *Betaentomopoxvirus*
Genus: *Gammaentomopoxvirus*

Attachment/Penetration

VACV has a *very* broad cell tropism (seemingly able to infect any cultured cell) and at least four viral proteins have been implicated in entry. It is likely that multiple cell receptors are used for attachment. Endocytosis by various pathways has been described. Membrane fusion releases cores (and lateral bodies) into the cytoplasm, the site of poxvirus replication.

Transcription and Genome Replication

The first synthetic event in the replication cycle is transcription of a set of early genes. All proteins required for synthesis of early mRNAs are present in the lateral bodies and core. These include a nine-subunit RNA polymerase, a virus transcription factor that activates early promoters, a capping enzyme, and a poly(A) polymerase (Fig. 35.2).

Early gene transcription occurs prior to the complete uncoating of the genome and sets the stage for viral DNA replication. Functions of early gene products include the following:

- Regulate (promote) cell growth.
- Suppress various innate and adaptive immune pathways.
- Support DNA replication (including DNA polymerase and primase, enzymes needed for synthesis of dNTPs, DNA topoisomerase, helicase, and ligase among others).

VACV DNA synthesis occurs in a discrete location in the cytoplasm to form a region called a "virus factory," an electron dense region of cytosol devoid of cellular organelles. VACV replication is rapid, with virus factories formed by 2–3 hours postinfection. DNA synthesis begins with a nick in the one of the terminal hairpins to produce a 3′ end that serves as a primer. The overall process of DNA replication involves the formation and resolution of hairpins to produce head-to-head and tail-to-tail concatemers. The last step in the process is ligation to form a new covalently closed molecule.

Newly replicated genomes are templates for intermediate and late transcription. The products of intermediate and late transcription include structural proteins and assembly factors (scaffold and chaperone proteins) as well as enzymes and transcription factors that will be needed to initiate infection of a new cell.

Assembly/Release

Poxvirus assembly is a fascinating process that can be visualized by electron microscopy (Fig. 35.3). It begins with the appearance of rigid, crescent-shaped membrane structures that contain a lipid bilayer and protein. In the case of VACV, crescents can be visualized by about 5–6 h postinfection. The membrane fragments that form the crescents are likely derived from preexisting cellular membranes that rupture to generate these open fragments. Virion proteins are closely associated with the membrane fragments, forming an organized scaffold.

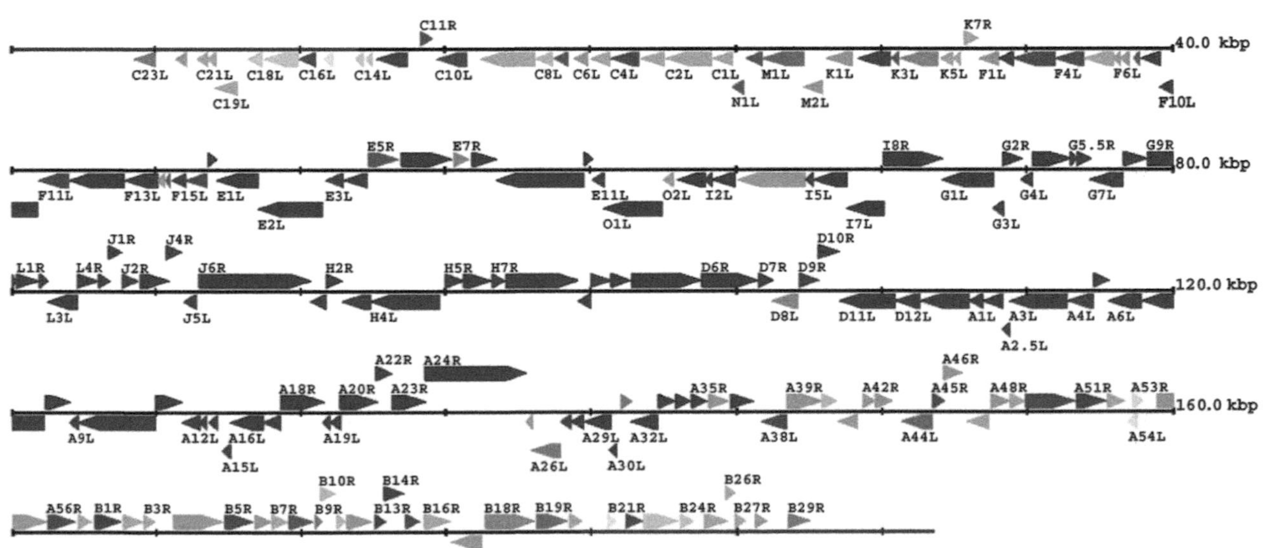

FIGURE 35.2 Diagram of the VACV strain Copenhagen genome. Genes are colored using a sliding scale, with dark blue representing those that are present in all poxviruses and yellow those that are in only one to three poxviruses. *From Lefkowitz, E.J., Wang, C., Upton, C., 2006. Poxviruses: past, present and future. Virus Res. 117, 105–118. http://dx.doi.org/10.1016/j.virusres.2006.01.016.*

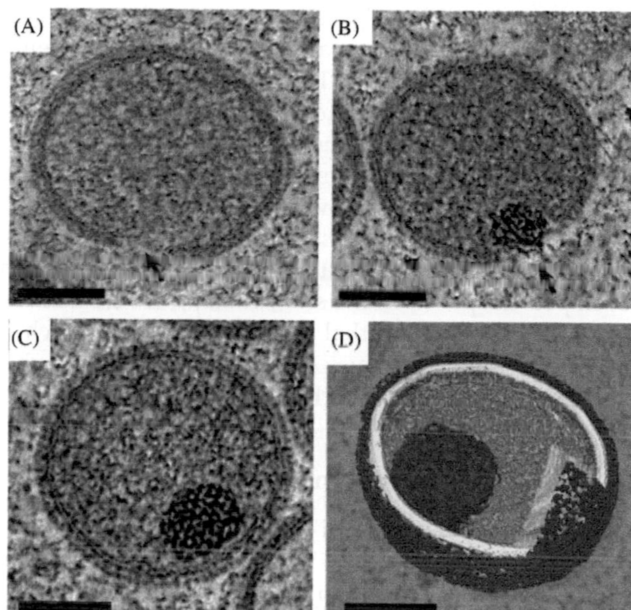

FIGURE 35.3 Poxvirus morphogenesis. After DNA uptake, the VACV crescent is sealed and forms a closed compartment. (A) Late stage crescent. (B) Immature virus in the process of DNA uptake. (C) The membrane is sealed and the DNA is condensed. (D) Yellow indicates membrane, red indicates the outer protein scaffold, blue indicates viral DNA. *From Chlanda, P., Carbajal, M.A., Cyrklaff, M., Griffiths, G., Krijnse-Locker, J., 2009. Membrane rupture generates single open membrane sheets during vaccinia virus assembly. Cell Host Microbe 6, 81–90.*

Virions form during a process in which crescents close around electron dense material that contains viral core proteins. DNA enters the crescents before they close completely. After the envelope closes, the captured material rearranges and coalesces to form the core and lateral bodies. Finally the ovoid or brick shape of the MV is formed. MVs have a single envelope layer and most remain within the cell.

A small number of MVs are further processed to become extracellular virions (EV). EVs are distinct from MVs and they play a special role in poxvirus spread. EVs are specifically designed for export from the infected cell and transport to nearby uninfected cells. Formation of EVs begins when MVs are wrapped in an additional layer of lipid (derived from the Golgi and containing a special set of viral proteins) to form the so-called wrapped virus (WV). WVs are particles with three lipid bilayers. WVs are transported to, and fuse with, the plasma membrane, releasing EVs, particles with two lipid bilayers. EVs remain associated with the cell membrane and serve as a focus for assembly of actin structures within the cell, to form EV-tipped microvilli that extend into the extracellular space. These virion-tipped microvilli contact and infect nearby cells (Fig. 3.11).

Immune Avoidance

Poxviruses encode a variety of proteins that interfere with the innate and adaptive immune responses. Many of these have been described for VACV (and homologs have been found among many sequenced orthopoxviruses). The number and type of immune antagonists vary and they contribute significantly to host range and virulence. Among host defense systems compromised by VACV are the interferon system, complement system, inflammatory responses, and apoptosis.

Examples of intracellular signaling pathways affected by VACV-encoded proteins include the interleukin 1 receptor signaling pathway and multiple toll-like receptor signaling pathways. Blocking these pathways ultimately inhibits activation of nuclear factor kappa B (NF-kB) and interferon regulatory factors such as IRF3 thus inhibiting synthesis of interferon-regulated genes (Fig. 6.3). VAVC also encodes proteins that interfere with mitochondrial components of the apoptotic cascade and inhibit activation of caspases. Another layer of protection is provided by a VACV protein that binds directly to double-stranded RNA to inhibit activation of dsRNA-dependent protein kinase and 2′-5′-oligoadenylate synthetase (see Chapter 6: Immunity and Resistance to Viruses).

In addition to intracellular proteins that disrupt cell-signaling pathways, poxviruses encode secreted proteins and peptides in their battle with the immune system. *Virokines* are a group of secreted peptides that resemble (and thus interfere with) host cell cytokines (Fig. 35.4). Another group of secreted proteins, the

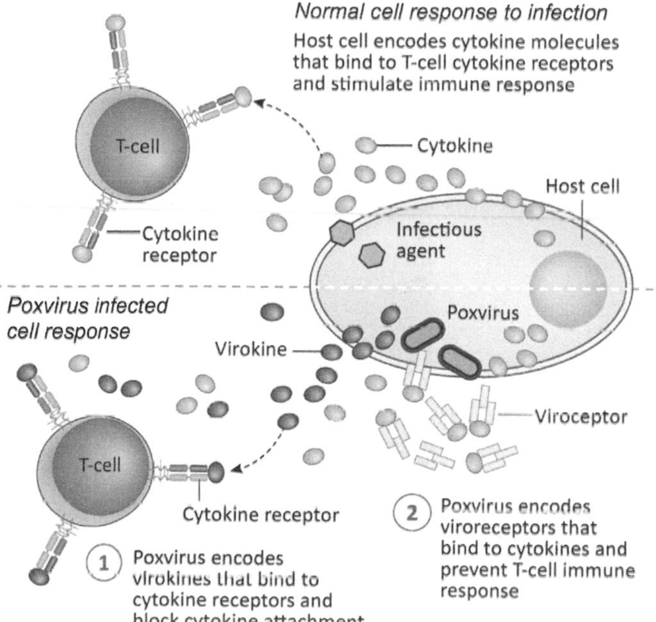

FIGURE 35.4 Poxvirus evasion of host responses to infection.

viroceptors, bind and sequester host cell ligands. VACV produces soluble products that bind to cytokines (including TNF, IL-1β, IFN-γ, and IL-18), chemokines, interferon receptors, and components of the complement cascade.

DISEASES

Smallpox

Smallpox was an exclusively human disease. The earliest recorded descriptions of smallpox date back ~3000 years and it is likely that smallpox emerged as a human disease ~5000–10,000 years ago. One scenario has the virus moving from rodents to domesticated cattle to humans. By AD 700 smallpox is recorded in China, Japan, Europe, and Northern Africa. The virus was a relative "late-comer" to the Caribbean (AD ~1500), North America (AD ~1600), South Africa (AD ~1700), and Australia (AD 1789). Disease severity varied somewhat, but 20%–30% case fatality was very common in Europe and Asia. When the virus moved to Americas and Australia, it was even more lethal among populations newly exposed to the virus. Smallpox (and other infections, such as measles) played a critical role in the ability of Europeans to settle the new world, killing large numbers of native inhabitants with mortality rates of 90%–95%. In the early 1900s a milder form of variola emerged (called variola minor) and eventually spread throughout the United States, South America, and Europe. The cases fatality rate of variola minor was often less than 1%.

During natural infection, transmission of smallpox was by the aerosol route, inhaled into the lungs. After primary replication, the virus became systemic and caused a febrile illness typified by high fever, malaise, and prostration. The digestive tract could be affected, causing nausea, vomiting, and backache. Virus eventually moved to the skin (~14 days postinfection) and a rash developed over a 1- to 4-day period. Parts of the body involved included mucosa of the mouth and pharynx, face, arms, legs, and soles of the hands and feet. The rash progressed slowly, through characteristic stages of papules (raised spots), firm fluid-filled vesicles, pus-filled lesions (pustules), and scabs. Lesions developed deep within the skin, causing severe scarring (Fig. 35.5). The most lethal cases of smallpox (such as those seen among native Americans) occurred when deep lesions caused hemorrhage and sloughing of skin.

Molluscum Contagiosum

Molluscum contagiosum is caused by another exclusively human poxvirus (molluscum contagiosum virus). In sharp contrast to smallpox, infection results in a *mild, localized skin disease* that presents as small, raised, pearly looking lesions (called mollusca). Lesions range in size from that of a pinhead to 5 mm in diameter and may occur anywhere on the body. Transmission is via direct contact or fomites. The virus is not systemic, but can be spread to other body parts as a result of scratching the itchy lesion or by activities such as shaving.

Monkeypox

Monkeypox is an orthopoxvirus endemic to parts of Africa. Despite the name, the natural hosts appear to be rodents. Genetic analysis suggests that there are two distinct lineages of monkeypox (with different virulence for humans). Monkeypox virus infection of humans causes a smallpox-like disease and the

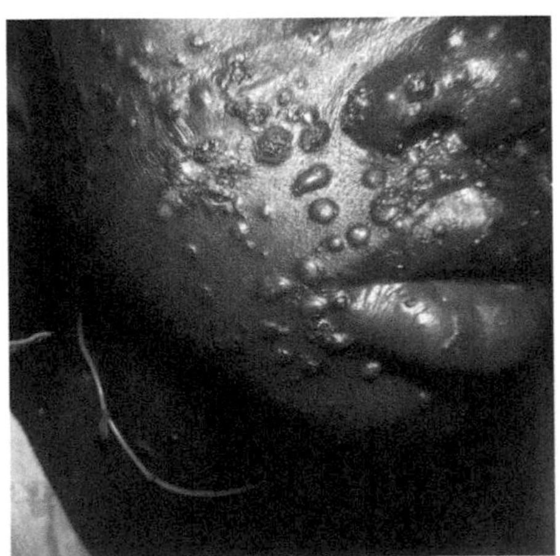

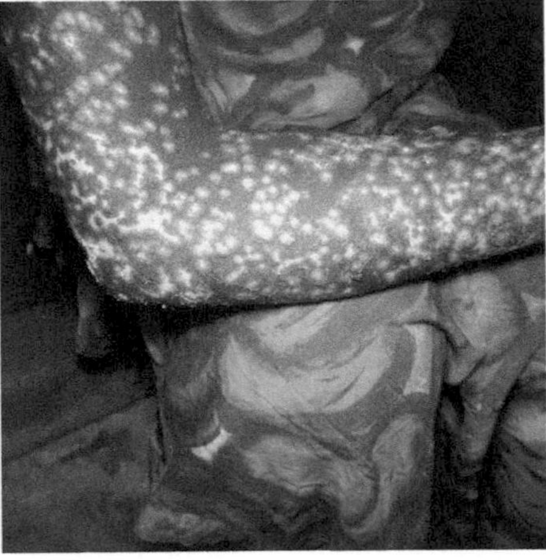

FIGURE 35.5 Smallpox lesions. Left: Close up of smallpox lesions on the face of a patient living in Accra, Ghana, 1967. Right: Arm of an adult woman, which displays the characteristic maculopapular rash due to the smallpox virus. *From CDC/Dr. J. Noble Jr. CDC Public Health Image Library Image ID #1994 (left) and #1990 (right).*

virus was first isolated from humans in 1970 as part of a smallpox surveillance campaign. An important distinction between smallpox and monkeypox viruses is that the latter is not efficiently transmitted among humans. Most human cases involve direct contact with infected animals (rodents or monkeys).

In the United States in 2003 an outbreak of monkeypox was associated with pet prairie dogs. By the end of the outbreak, the CDC reported 47 cases (37 confirmed and an additional 10 probable human cases). The cases were relatively mild (no deaths) but the outbreak prompted a significant response and investigation. How did monkeypox, an African virus, come to find its way into prairie dogs in the United States?

The introduction on monkeypox was traced to importation of a large shipment of wild caught small mammals (including rodents), from West Africa. Some of those rodents found their way to a prairie dog breeder and all infected persons had purchased pet prairie dogs. This incident serves as an example of the danger of importing exotic wild animals for the pet trade.

RABBIT MYXOMA VIRUS

Rabbit myxoma virus is a poxvirus that causes localized skin infections in American rabbits but can cause high mortality among European breeds. The virus is mechanically transmitted by mosquitoes. Rabbit myxoma virus is best known for its introduction to Australia in 1950 to control huge populations of imported European rabbits. While the virus initially produced high mortality (estimated to have reduced the rabbit population from ~600 million to ~100 million) within about 20 years, the virus became notably less virulent and Australian rabbits become much more resistant (allowing the rabbit population to rebound somewhat). This is an excellent example of host-virus coevolution. An elegant set of experiments demonstrated that viral and host genetics had both changed. (Viruses from Australia were demonstrated to be less virulent when inoculated into European rabbits and Australian rabbits proved to be more resistant when infected with highly virulent myxoma virus strains.)

In this chapter we have learned that:

- Poxviruses are large, enveloped DNA viruses that encode hundreds polypeptides. They have a complex structure.
- Poxviruses encode both RNA and DNA polymerases, which allows them to replicate in the cell cytoplasm.
- Poxviruses do not obtain their envelopes by a budding process. Instead, discrete membrane fragments appear to coalesce around viral proteins with a virus factory in the infected cells.
- In addition to encoding intracellular proteins that interfere with antiviral responses, poxviruses encode two different types of secreted protein. Virokines are peptides similar to host cytokines and they interfere with normal cytokine signaling. Viroceptors are secreted proteins that bind and sequester host ligands (cytokines, for example). This prevents ligands from initiating signals to activate antiviral or inflammatory pathways.
- Variolation was use of dried smallpox lesions to intentionally infect people with smallpox. The process used scabs from mild cases of smallpox that were inoculated into scratches in the skin to produce a localized infection. Jenner used cowpox lesions to the same effect. However, cowpox was naturally much more attenuated for humans thus vaccination produced less disease, but achieved the desired protection.
- The introduction of rabbit myxoma virus into Australia was intentional, to control the rabbit population. Initially mortality was very high, however over time, selection resulted in changes in the genetics of both virus and host. Viruses became less virulent and the Australian rabbit population became more resistant.

References

Chlanda, P., Carbajal, M.A., Cyrklaff, M., Griffiths, G., Krijnse-Locker, J., 2009. Membrane rupture generates single open membrane sheets during vaccinia virus assembly. Cell Host Microbe 6, 81–90.

Lefkowitz, E.J., Wang, C., Upton, C., 2006. Poxviruses: past, present and future. Virus Res. 117, 105–118. http://dx.doi.org/10.1016/j.virusres.2006.01.016.

CHAPTER

36

Family *Retroviridae*

OUTLINE

After reading this chapter, you should be able to answer the following questions:

- What are the major steps in a retrovirus replication cycle?
- Which step in the retrovirus replication cycle is unique to this virus family?
- What are the enzymatic activities of reverse transcriptase, RNaseH and integrase?
- What is the function of retroviral protease (PR) and when is it active?
- Explain the difference between exogenous and endogenous retroviruses.
- What is insertional activation and how does it lead to cell transformation?
- How are retroviral oncogenes and cellular proto-oncogenes related?

Retroviruses are enveloped viruses with RNA genomes (Fig. 36.1). The name retrovirus was coined when it was discovered that the RNA genome is a template for synthesis of double-stranded (ds) DNA. This process was counter to the prevailing dogma that DNA was a template for RNA synthesis (the process of transcription), never the reverse. An older name for the group of viruses we now call retroviruses was "RNA tumor virus" because they were associated with transmissible cancers in mammals and birds. Retroviruses can transform cells because they *must* insert or integrate their genomes into the host cell chromosome, an activity that is inherently mutagenic (Fig. 36.2). But as we see in this chapter, the retroviruses are much more than just cancer agents. About 8% of the human genome is comprised of retroviral sequences; retroviruses have shaped our genomes and driven our evolution. The human immunodeficiency virus (HIV), discovered only a few decades ago, infects millions of people worldwide. It causes disease primarily by damaging of a subset of T-lymphocytes that are central regulators of the immune response. Modified retroviruses are commonly used for gene delivery, genetic engineering, and gene therapy; they are often used as tools to alter animal genomes. Clearly there is a lot to know about these unique viruses!

All of the viruses discussed to this point in this textbook are transmitted via infectious particles (virions). However retroviruses can be transmitted *either* as infectious particles or as heritable genes. While some viruses occasionally insert their genomes into host DNA, this is usually a "mistake" that does not benefit the virus. In contrast, retroviruses *must* insert their genomes into host DNA, in order to complete a single replication cycle (Box 36.1).

There is important, but sometimes confusing, terminology used to describe retroviruses. Retroviruses transmitted as virions are called *exogenous* (think *extra*-cellular) while retroviruses transmitted exclusively via

Viruses. DOI: http://dx.doi.org/10.1016/B978-0-12-803109-4.00036-2

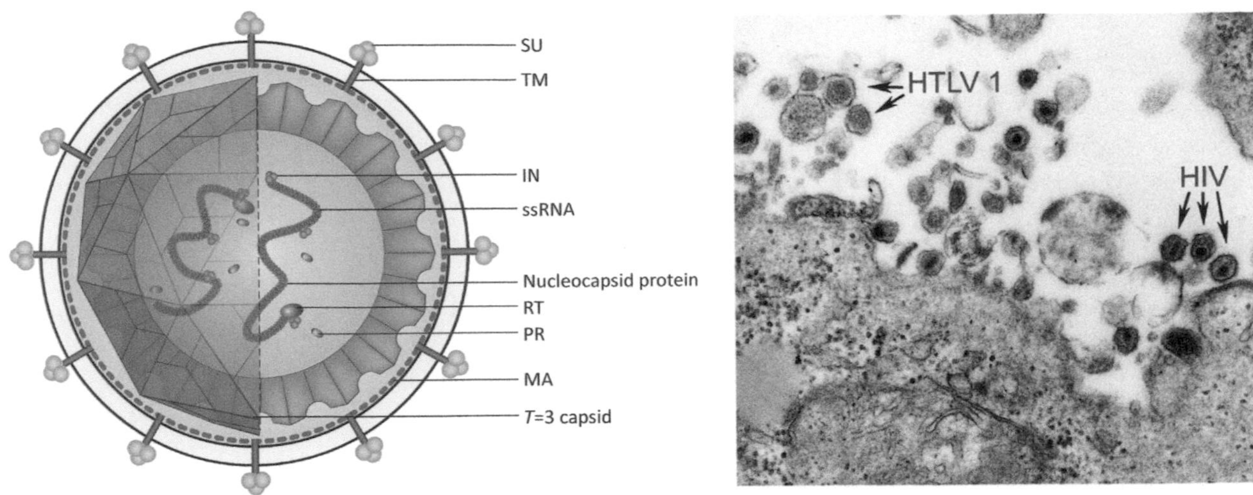

FIGURE 36.1 Left: Retrovirus structure. Right: Micrograph showing both the HTLV-1 and the HIV. *From CDC/Cynthia Goldsmith. CDC Public Health Image Library, Image ID# 8241.*

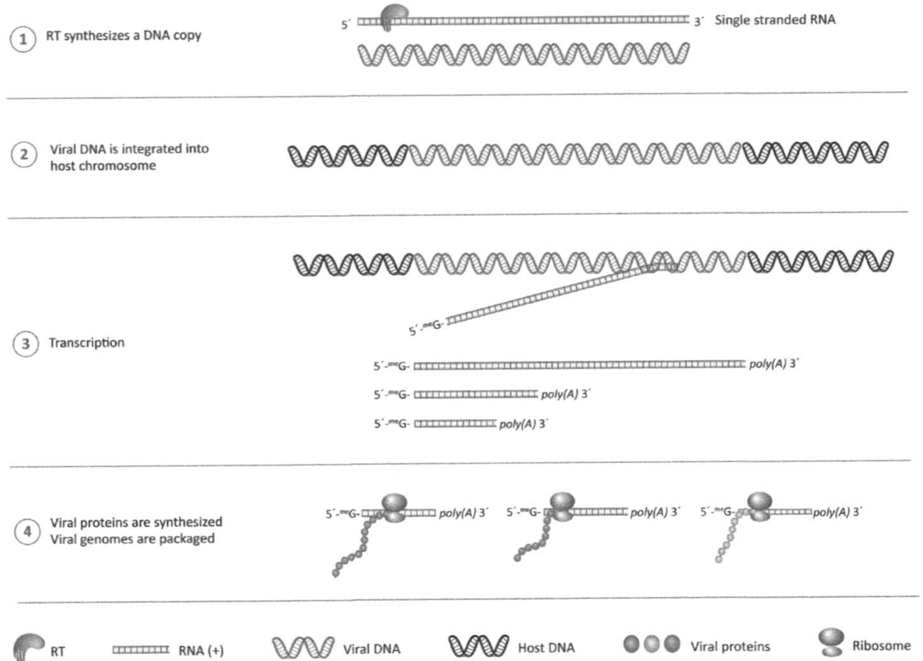

FIGURE 36.2 Retrovirus replication process showing steps of reverse transcription, integration, transcription, and translation of viral proteins.

the germ-line are called endogenous retroviruses. The many retroviral sequences that populate our genomes are endogenous retroviruses. Most are defective or fragmented but some do express proteins. Endogenous retroviruses are essentially cellular genes; they rarely, if ever, are transmitted as virus particles. The presence of large numbers of endogenous retroviruses in our genomes is a reminder that we have coexisted with these viruses for hundreds of millions of years. Retroviruses are one group of a larger family of replicating molecules called retrotransposons.

HIV is an exogenous retrovirus; it is transmitted from person to person via infectious virions in blood and body fluids. But confusion often stems from the following: Exogenous retroviruses *must* integrate their genomes into host DNA for replication. We call the *integrated* versions of retroviral genomes *proviruses*. RNA transcribed from the provirus is packaged into particles for transmission. In order for an exogenous retrovirus to become endogenous, it must integrate into *germ-line* cells and become *fixed* in the host genome. The remainder of this

BOX 36.1

GENERAL CHARACTERISTICS OF RETROVIRUSES

Packaged within retroviruses are two identical copies of capped, polyadenylated, positive-strand RNA. Upon penetration/uncoating, the RNA molecules are not translated, instead they are temples for synthesis of a molecule of ds DNA that is integrated into the host chromosome. Transcription of mRNAs from the integrated DNA genome (called the provirus) is catalyzed by host RNA Pol II. Transcripts are capped and polyadenylated and some are spliced.

Retroviral genomes are 7–12 kb. All retroviral genomes encode three precursor polyproteins (GAG, GAG-POL, and ENV). Some retroviruses encode additional proteins with a variety of regulatory and auxiliary activities.

Virions are enveloped with icosahedral-shaped capsids or cores. Their envelopes contain two glycoproteins produced by cleavage of the ENV precursor. One is a transmembrane (TM) glycoprotein with a cytoplasmic tail. The second glycoprotein is the surface (SU) glycoprotein. SU is the attachment protein and TM is the fusion protein. capsids are assembled from the capsid protein (CA). A basic, RNA-binding nucleocapsid (NC) protein binds the genomic RNA. A membrane-associated matrix (MA) protein associates with the cytoplasmic tails of TM.

Upon penetration into the cell, the RNA genome serves as a template for the synthesis of a molecule of ds DNA. This is accomplished by reverse transcriptase (RT) a viral enzyme that is an RNA-dependent DNA polymerase. ds DNA is then inserted (integrated) into the cell genome by the viral integrase (IN) protein.

chapter deals almost exclusively with exogenous retroviruses.

VIRION STRUCTURE

Retrovirus particles are enveloped with icosahedral (or icosahedral-like) cores (Fig. 36.1). Two glycoproteins are associated with the viral envelope. A transmembrane (TM) glycoprotein is anchored into the viral envelope and a surface (SU) glycoprotein is found on the exterior of the virion. SU associates with TM via noncovalent interactions. Under the viral envelope, the matrix protein (MA) associates with the cytoplasmic tails of TM, as well as with the lipid envelope. Moving further into the virion, the capsid (or core) is assembled from capsid proteins (CA). Within the capsid are two identical copies of RNA (capped and polyadenylated) that serve as the retroviral genome (retroviruses therefore have diploid genomes). The RNA is bound by the positively charged (basic) nucleocapsid protein (NC). Also within the capsid are a few molecules of the enzymes reverse transcriptase (RT), integrase (IN), and protease (PR). Two molecules of tRNA are also found within the virion. They are associated with RT and are base paired to the genomic RNA near its 5′ end, at sequence called the primer-binding site (PBS). A molecule of tRNA is a primer for DNA synthesis (Box 36.2).

Some complex retroviruses, such as HIV, package additional proteins within their virions. Proteins such as HIV-1 viral protein R (VPR), viral protein U (VPU), and virus infectivity factor (VIF) play important roles in the HIV replication cycle. They are sometimes called "accessory" proteins and they serve to counter host cell defenses to retroviral infection. They will be discussed in more detail in Chapter 37, Replication and Pathogenesis of Human Immunodeficiency Virus (HIV).

GENOME ORGANIZATION

Retroviral genomes are 5′ capped and 3′ polyadenylated mRNAs. However, upon entry into the cell, the mRNA is not translated (in contrast to the replication cycle of plus-strand RNA viruses). Instead the RNA genome serves as the template for the synthesis of a molecule of ds DNA and the RNA genome is destroyed during this process.

All retroviruses have three major genes (Fig. 36.3). *Gag* (for group-associated antigen) encodes the GAG polyprotein. The *pol* gene encodes the polymerase (POL) polyprotein and *env* encodes the envelope (ENV) polyprotein. Retroviruses encoding only *gag*, *pol*, and *env* genes are sometimes called "simple" retroviruses. The order of these genes is always *gag-pol-env*. The oncogenic retroviruses of rodents and chickens are simple retroviruses. The so-called "complex" retroviruses such as HIV and human T-cell leukemia virus (HTLV) have additional genes encoding a variety of regulatory and accessory proteins (Fig. 36.4).

In addition to protein-coding regions, the retroviral genome contains a number of important regulatory sequences. We will start with a discussion of the

BOX 36.2

THE FAMILY *RETROVIRIDAE* CONTAINS SEVEN GENERA CONTAINED IN TWO SUBFAMILIES

Subfamily: *Orthoretrovirinae*
Genus: *Alpharetrovirus*. Examples include:

Avian leukosis virus
Rous sarcoma virus

Genus: *Betaretrovirus*. Examples include:

Mouse mammary tumor virus
Mason-Pfizer monkey virus
Jaagsiekte sheep retrovirus

Genus: *Deltaretrovirus*. Examples include:

Bovine leukemia virus
Human T-lymphotropic virus types 1, 2
Simian T-lymphotropic virus types 1, 2, 3

Genus: *Epsilonretrovirus*. Examples include:

Walleye dermal sarcoma virus
Walleye epidermal hyperplasia virus 1

Genus: *Gammaretrovirus*. Examples include:

Feline leukemia virus
Murine leukemia virus
Gibbon ape leukemia virus
Reticuloendotheliosis virus

Genus: *Lentivirus*. Examples include:

Human immunodeficiency virus 1
Human immunodeficiency virus 2
Simian immunodeficiency virus
Equine infectious anemia virus
Caprine arthritis encephalitis virus
Feline immunodeficiency virus
Visna/maedi virus

Subfamily: *Spumavirinae*
Genus: *Spumavirus*. Examples include:

Simian foamy virus.

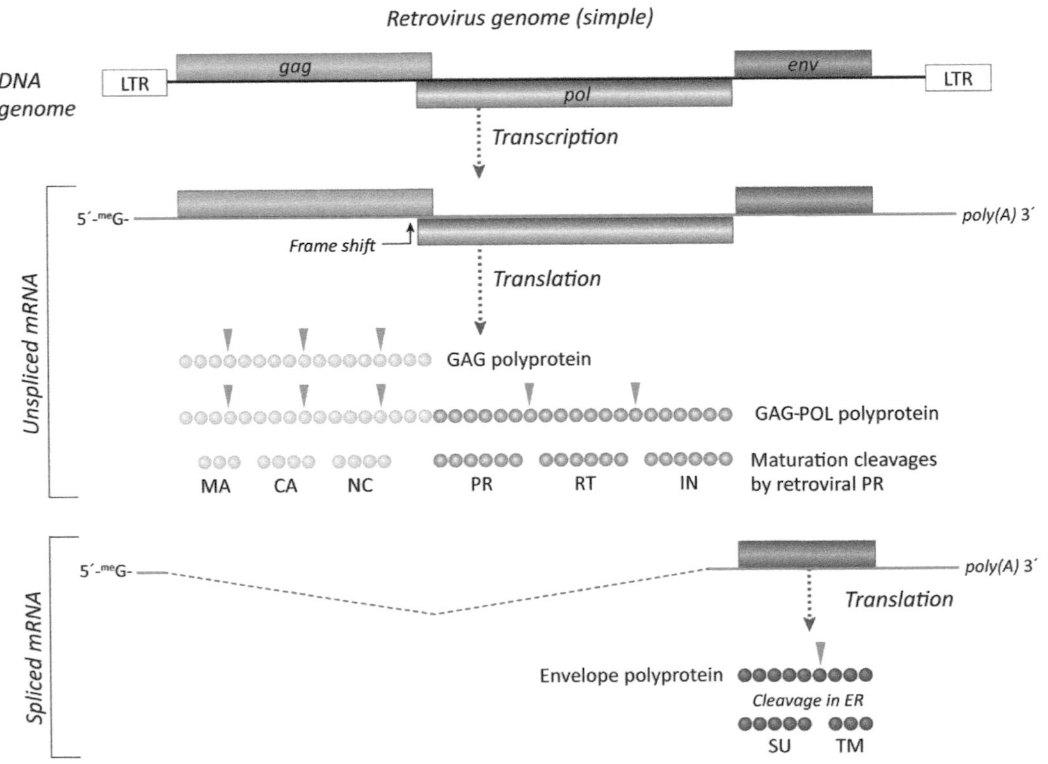

FIGURE 36.3 Genome of a simple retrovirus. Simple retroviruses have three open reading frames that encode three polyproteins.

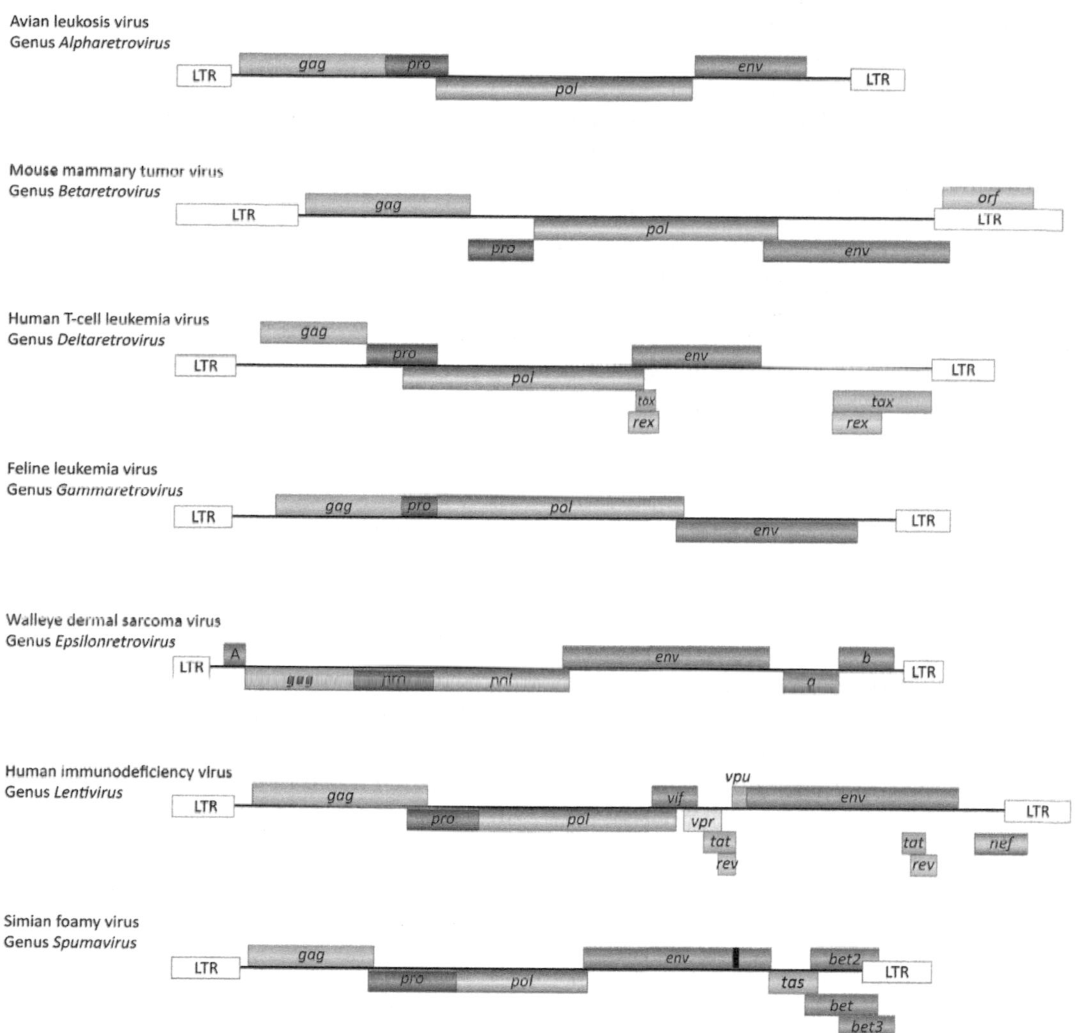

FIGURE 36.4 Comparison of retroviral genomes.

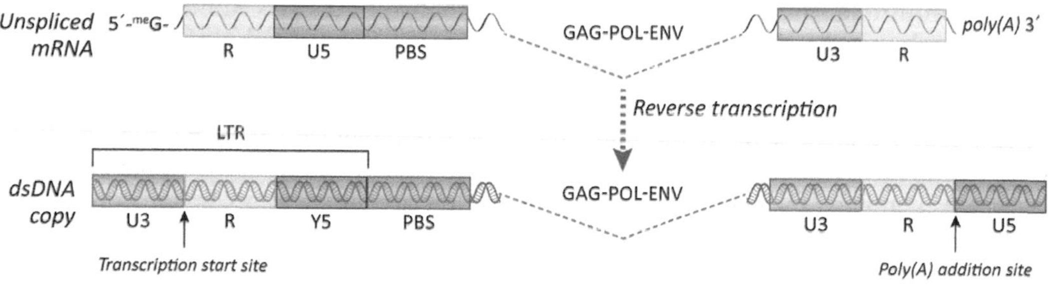

FIGURE 36.5 Comparison of RNA and DNA versions of a retroviral genome. During the process of reverse transcription the LTR is formed when U3 and U5 sequences are duplicated.

genomic *RNA* (Fig. 36.5). At each end of genome are 30–40 nucleotide long direct repeats (R) essential for the process of reverse transcription. One copy of the R sequence is found at the very 5′ end of the genome and the R sequence at the 3′ end precedes the poly(A) tail. Directly following R at the 5′ end of the genome is a short sequence called U5 (for unique at the 5′ end). Just upstream of R at the 3′ end of the genome is a region called U3 (unique at the 3′ end). Two additional cis-acting sites are present near the 5′ end of the genome. The short primer binding site (PBS) is base paired with a molecule of tRNA. Just following the PBS is the RNA packaging site or RNA encapsidation signal (also called ψ).

The product of reverse transcription, the DNA version of the genome, is longer than the RNA genome (Fig. 36.5). This occurs during the process of reverse transcription (described later) when the U5 sequence is copied to the 3′ end of the genome and the U3 sequence is copied to the 5′ end of the genome. This process forms the long terminal repeats (LTRs) that flank the protein-coding sequences. LTRs are organized as follows: U3-R-U5. LTRs are critical elements for transcription and polyadenylation of retroviral mRNA. The promoter and enhancer elements are often found in U3 and the upstream LTR the promoter for transcription of viral mRNA. A poly(A) addition site is present at the R-U5 junction of each LTR, but polyadenylation is suppressed at the upstream LTR.

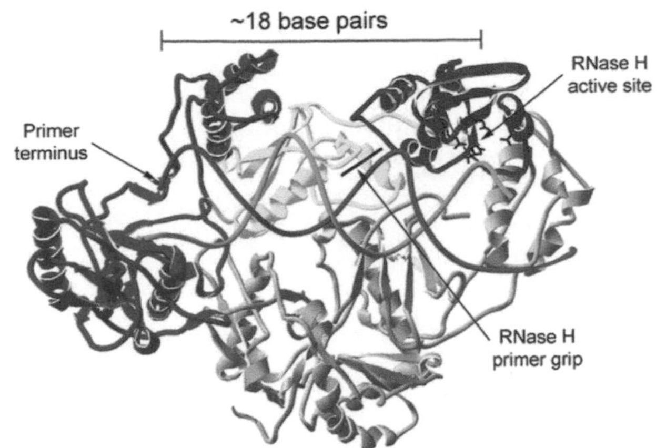

FIGURE 36.6 Molecular structure of RT. Cocrystal structure of HIV-1 RT and an RNA/DNA substrate. The polymerase and RNase H domain of p66 are shown in blue and red. The connection domain is yellow. p51 is shown in gray. The RNA template and DNA primer strands are shown in green and purple respectively. *From Schultz, S. J., Champoux, J.J., 2008. RNase H activity: structure, specificity, and function in reverse transcription. Virus Res. 134, 86–103.*

OVERVIEW OF THE RETROVIRAL REPLICATION CYCLE

The retroviral replication cycle begins with binding of SU to the receptor. Some retroviruses enter cells by fusion at the plasma membrane (for example, HIV-1) but others are endocytosed and the trigger for fusion is low pH. In either case, the hydrophobic amino terminus of the TM protein mediates membrane fusion. When the retroviral core is released into the cytosol, viral RT synthesizes a DNA copy of the RNA genome. The ds DNA molecule is then integrated into the host chromosome. The DNA genomes of some retroviruses (members of the genus *Lentivirus*) actively cross nuclear pores to gain access to the chromosome, but many retroviruses require a dividing cell, as their DNA genomes can only reach the host chromosome during mitosis (when the nuclear membrane breaks down). The next key player is the retrovirus replication cycle is IN. Recall that IN is present in the infecting virion. Dimers of IN interact with each end of the newly synthesized ds DNA genome. IN also interacts with host DNA and catalyzes cleavage and ligation steps that result in the covalent linkage of the viral DNA to the host DNA. The viral DNA is now, in essence, a cellular gene. Once integrated, the retroviral genome is referred to as the provirus. In order for the retroviral replication cycle to continue, the provirus must be transcribed by cellular RNA polymerase II to produce capped, polyadenylated mRNA. Spliced and unspliced viral mRNAs are transported out of the nucleus and are translated to produce a full complement of retroviral proteins. As the concentration of structural proteins increases, unspliced mRNAs are packaged to serve as the genomes of progeny virions.

The steps up to and including integration are sometimes called the early steps in the retroviral replication cycle, while those that occur after integration are called the late steps. In some cases integration may occur months to years before virions are ever produced, as expression of the provirus may require a specific set of circumstances. In the case of HIV, the infection of a T-cell may be silent until the cell is stimulated to divide upon recognizing a foreign antigen.

Some key steps in the retroviral replication cycle are now described in greater detail.

THE PROCESS OF REVERSE TRANSCRIPTION

The process of reverse transcription is carried out by a polymerase called RT. RT is a heterodimer. The longer polypeptide in the dimer contains the polymerization domain, a short linker domain, and an RNaseH domain. The shorter polypeptide in the dimer contains only the polymerization domain. The longer polypeptide provides the enzymatic functions of RT while the shorter polypeptide appears to play a strictly structural role. The structure of HIV-1 RT is shown in Fig. 36.6. HIV p66 folds to form the typical palm, fingers, and thumb domains of polymerases. The catalytic site is located in the palm and contains three critical aspartic acid residues and two coordinated Mg 2+ ions.

Reverse transcription is key to the replication cycle of a retrovirus. As described earlier, retroviral genomes are 5′ capped and 3′ polyadenylated mRNAs. The RNA genome serves as the template for the synthesis of a molecule of ds DNA. Although two copies of RNA enter the cell, only one copy of ds DNA is synthesized.

The steps in reverse transcription are outlined in Fig. 36.7. Note that in this model only one copy of RNA is shown. However there is good evidence that during the process of reverse transcription, RT may jump from one of the packaged RNA strands to another.

Reverse transcription begins with synthesis of a piece of DNA from the tRNA primer. The first strand of DNA synthesized is called "negative strand" because it is a copy of the mRNA (the positive strand). The first piece of DNA synthesized is rather short because RT reaches the 5′ end of the mRNA and can go no further. The RNA strand of this RNA:DNA hybrid molecule is digested away by the activity of the RNase H domain of RT. Note that the RNA:RNA duplex formed by the tRNA primer and the PBS is not digested as RNase H only degrades RNA that is part of an RNA:DNA hybrid. The short DNA product is called minus-strand strong-stop DNA.

The next step in the process is sometimes called a "jump" or a "strand transfer." In the infected cell this step probably requires the NC protein (although this is not shown in Fig. 36.7). Key to the jump is that the newly synthesized R DNA base pairs with the R RNA at the 3′ end of the genome. The negative strand of DNA is then synthesized through to the 5′ end of the RNA. Not shown in Fig. 36.7 is that most of the RNA template is degraded *during* the process of DNA synthesis. One piece of RNA remains undigested

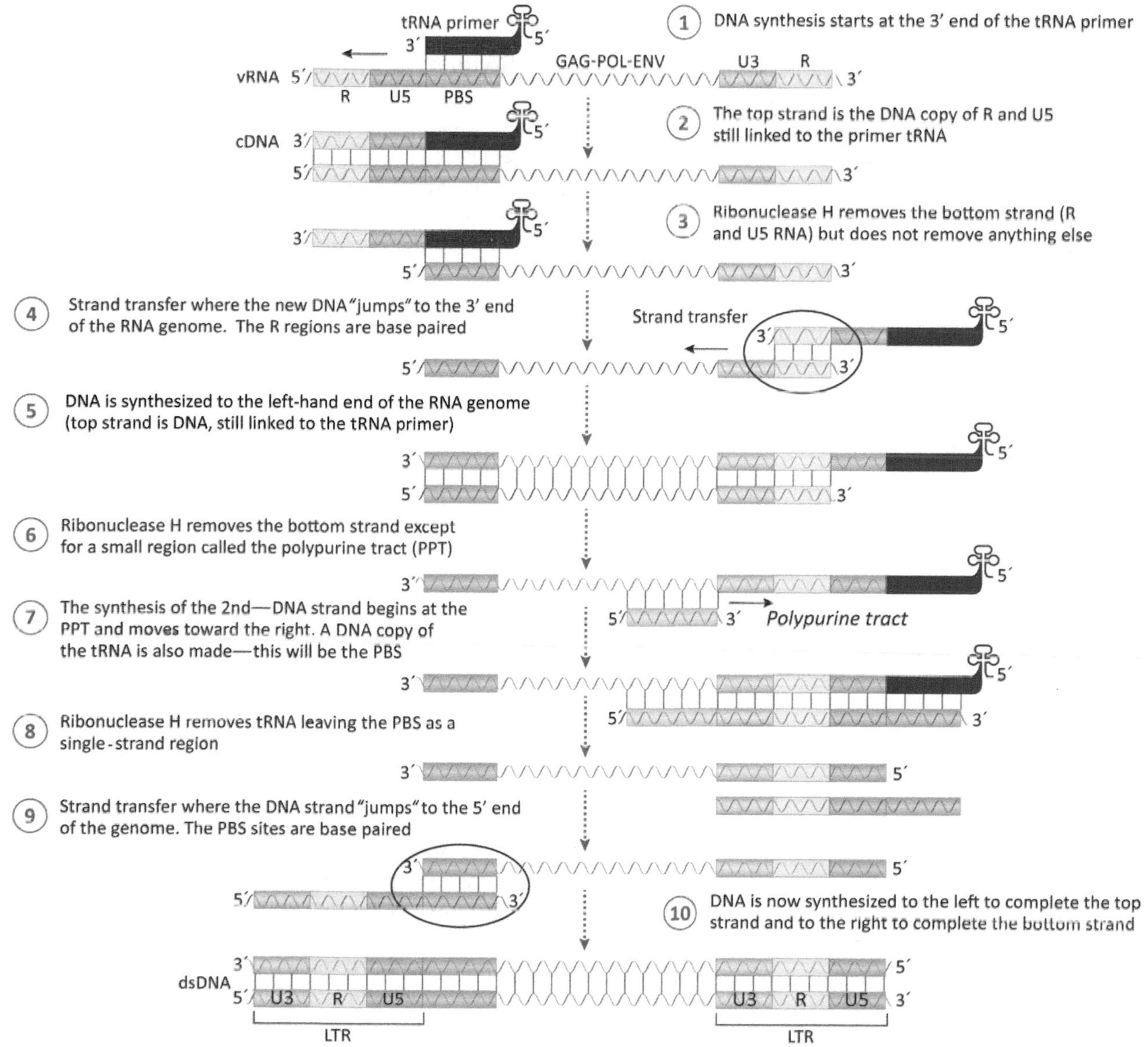

FIGURE 36.7 Steps in the reverse transcription process.

however: the so-called polypurine tract (PPT). The PPT is the RNA primer for synthesis of plus-strand DNA (Fig. 36.7, step 7). As we saw with minus-strand DNA synthesis, the first product of plus-strand DNA synthesis is a short fragment. As shown in step 8, part of the tRNA is copied, generating the PBS. Now the tRNA is digested away by RNase H and a second jump occurs: The PBS region on the plus strand is base paired to its complement on the minus strand. The ds DNA molecule is completed by extension of the two partial DNA strands to form the completed linear ds DNA. As we see later, the LTRs generated during reverse transcription are key regulatory regions of the provirus. It is important to note that RT is not the only viral player in the RT process. NC and other virion proteins (CA and IN) remain associated with the DNA as it is synthesized, forming the so-called preintegration complex (PIC). The PIC traffics to the chromosome. Many retroviruses replicate only in mitotically active cells as they require breakdown of the nuclear membrane to reach the host chromosome. However lentiviruses, such as HIV, can replicate in nondividing cells because the PIC actively crosses the nuclear pore.

INTEGRATION

IN is bound to the very ends of the linear ds DNA molecule (Fig. 36.8). IN polypeptides interact to form dimers and tetramers, bringing the two ends of the linear provirus together. IN carries out three enzymatic activities: First it is an exonuclease that cleaves 2–4 nt from the 3′ ends of the linear, unintegrated ds DNA molecule. Second, it is an endonuclease that makes a staggered cut in host cell DNA. Third it has ligase activity to join viral DNA to host cell DNA (ligation). Successful integration also requires host cell enzymes to repair gaps at the ends of the newly integrated molecule. The sites in the chromosome that serve as targets for integration are generally considered to be random. However some types of retroviruses preferentially integrate in region of active gene transcription, while others do not.

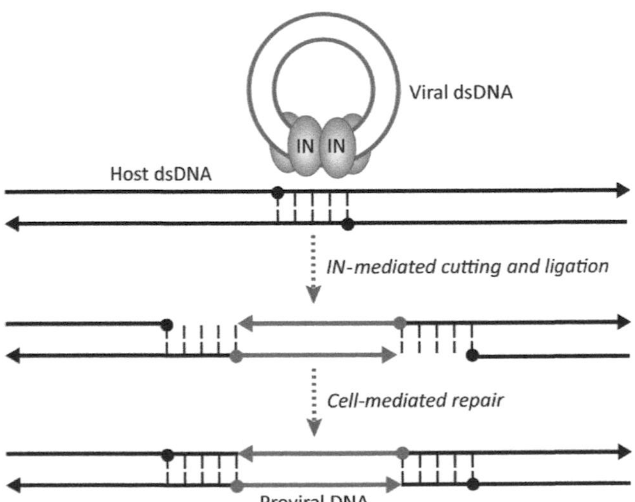

FIGURE 36.8 Integration. The ends of linear retroviral DNA are associated with IN protein. IN makes a staggered cut in target chromosomal DNA and ligates each strand. DNA repair (to fill gaps) is performed by host enzymes. This process generates short direct repeats at each end of the insertion.

Integration can be considered the last step in the *early phase* of the retrovirus replication cycle. The viral genome is now a host cell gene that can be passed to daughter cells. The provirus may remain silent for some time, but successful completion of the replication cycle requires that it eventually be transcribed.

TRANSCRIPTION OF RETROVIRAL MRNA

The retroviral LTR controls transcription of the provirus. U3 contains a core promoter as well as enhancer elements. Transcripts are synthesized by the host RNA Pol II complex. The efficiency of transcription is regulated by sequences in the LTR. Some retroviral promoters are active in many cell types while other promoters are only active in specific cells at specific times. An instructive example is the retrovirus mouse mammary tumor virus (MMTV). The MMTV promoter is active only in the mammary glands of a lactating mouse. Virus is produced during lactation, when it can be successfully transmitted to the suckling mouse pups. Some retroviral promoters, for example HIV, are relatively weak and require virally encoded transcription factors to increase transcription to sufficient levels for successful completion of the retrovirus replication cycle [Chapter 37: Replication and Pathogenesis of Human Immunodeficiency Virus (HIV)]. Oncogenic retroviruses often have strong promoters that drive transcription of not only the provirus, but of nearby cellular genes, a process called insertional activation.

All retroviral transcripts begin at the R region of the upstream LTR and are cleaved and polyadenylated at a site in the downstream LTR (Fig. 36.5). (Recall that the R sequences are found at the 5′ and 3′ ends of the RNA genome.) However before any RNA is packaged, structural proteins must be synthesized. The GAG polyprotein is translated from unspliced mRNA (Fig. 36.3). The *pol* gene products are also synthesized from the unspliced mRNA, but this requires either a ribosomal frame shift or suppression of a stop codon to produce the GAG-POL polyprotein.

The ENV precursor is synthesized from a spliced mRNA. (A mixture of spliced and unspliced mRNAs is needed to generate the full complement of retroviral proteins.) ENV is translated on rough ER and the precursor is processed in the ER and Golgi. Processing includes cleavage to produce SU and TM, as well as glycosylation.

Complex retroviruses also encode accessory and regulatory proteins from a variety of singly and multiply spliced mRNAs (see Fig. 37.3). In the case of HIV and other lentiviruses two key regulatory proteins are TAT and REV. TAT is a small protein that increases transcription from the HIV LTR. REV is a small protein that regulates transport of spliced and unspliced mRNAs from the nucleus. These two proteins and their activities will be described in more detail in Chapter 37, Replication and Pathogenesis of Human Immunodeficiency Virus (HIV).

STRUCTURAL PROTEINS

The full-length (unspliced) viral transcript is transported to the cytoplasm where it is translated to produce two precursor polyproteins, GAG and GAG-POL. GAG contains the domains for MA-CA-NC while the GAG-POL polyprotein also contains the RT and IN domains. As show in Fig. 36.4, the PR domain is sometimes part of the GAG precursor and sometimes part of GAG-POL. Thus retroviruses have evolved a variety of strategies for producing PR. The GAG and GAP-POL precursors remain uncleaved within the infected cell. As we see later, PR is not active until the budding process is complete.

The ENV polyprotein is translated from a singly spliced mRNA that lacks *gag* and *pol* sequences. The ENV polyprotein is synthesized on rough ER. Within the ER lumen, ENV is cleaved (by furin-like proteases) and the cleavage products are glycosylated as they travel through the ER and Golgi. For many retroviruses, the final destination of the ENV glycoproteins is the plasma membrane. Complex retroviruses like HIV use multiply spliced mRNAs to produce additional accessory proteins. Some of these are nonstructural proteins while others are packaged into new virions to exert their effects during the next cycle of replication (Chapter 37: Replication and Pathogenesis of Human Immunodeficiency Virus (HIV)).

ASSEMBLY, RELEASE, AND MATURATION

In the cytosol the RT domain of the GAG-POL precursor binds the tRNA that will be used to prime DNA synthesis in the next cell. Full-length retroviral mRNAs have a "packaging signal" near the 5′ end that allows specific incorporation of genomes into budding particles. GAG and GAG-POL polyproteins associate with a cell membrane (the PM in the case of HIV). That interaction is facilitated by a fatty acid molecule that is linked to the 5′ end of the MA domain of the GAG and GAG-POL polyproteins. These polyproteins also contain interaction domains that allow them to specifically interact with the cytoplasmic tail of TM (Fig. 36.9).

Mutational analysis of retroviral *gag* genes has revealed the presence of short amino acid motifs that promote the release of the budding particles from the PM. These motifs are called "late domains" because they are required in the very last steps of release from the cell. Three types of retroviral late domains have been identified: Pro-Thr/Ser-Ala-Pro (PTAP or PSAP), Pro-Pro-Pro-Tyr (PPPY), and Tyr-Pro-X_n-Leu (YPX_nL, where X is any amino acid and $n = 1-3$ residues). The motifs are found in various places in GAG: The HIV late domain is in p6 (near the carboxyl terminus of GAG) while the HTLV-1 late domain is in MA. Mutation of these motifs disrupts virion budding and release. Retroviral late domains function by interacting with components of the cellular endosomal sorting machinery. The so-called ESCRT complexes (endosomal sorting complexes required for transport) are multiprotein complexes that sort cellular proteins into vesicles; in addition to their "sorting" function they promote budding and membrane scission. There are four different ESCRT complexes that function together in cells; only a subset of ESCRT machinery is required for retrovirus budding.

HIV is an example of a retrovirus that buds from the plasma membrane. During budding, GAG and GAG-POL polyproteins become densely packed, triggering activation of the PR domain. PR cleaves the precursor proteins into their final products. MA remains associated with the inside of the viral envelope, CA rearranges and assembles to form a capsid that surrounds the genome, NC, RT, and IN. The virion is now ready to infect a new cell (Fig. 36.9). During an HIV infection, the virus infects several types of cell that are key players in the immune response. Disruption of the activities of these cells leads to immune suppression and disease. These processes will be discussed in more detail in Chapter 37, Replication and Pathogenesis of Human Immunodeficiency Virus (HIV). The remainder of this chapter will focus on retroviral oncogenesis.

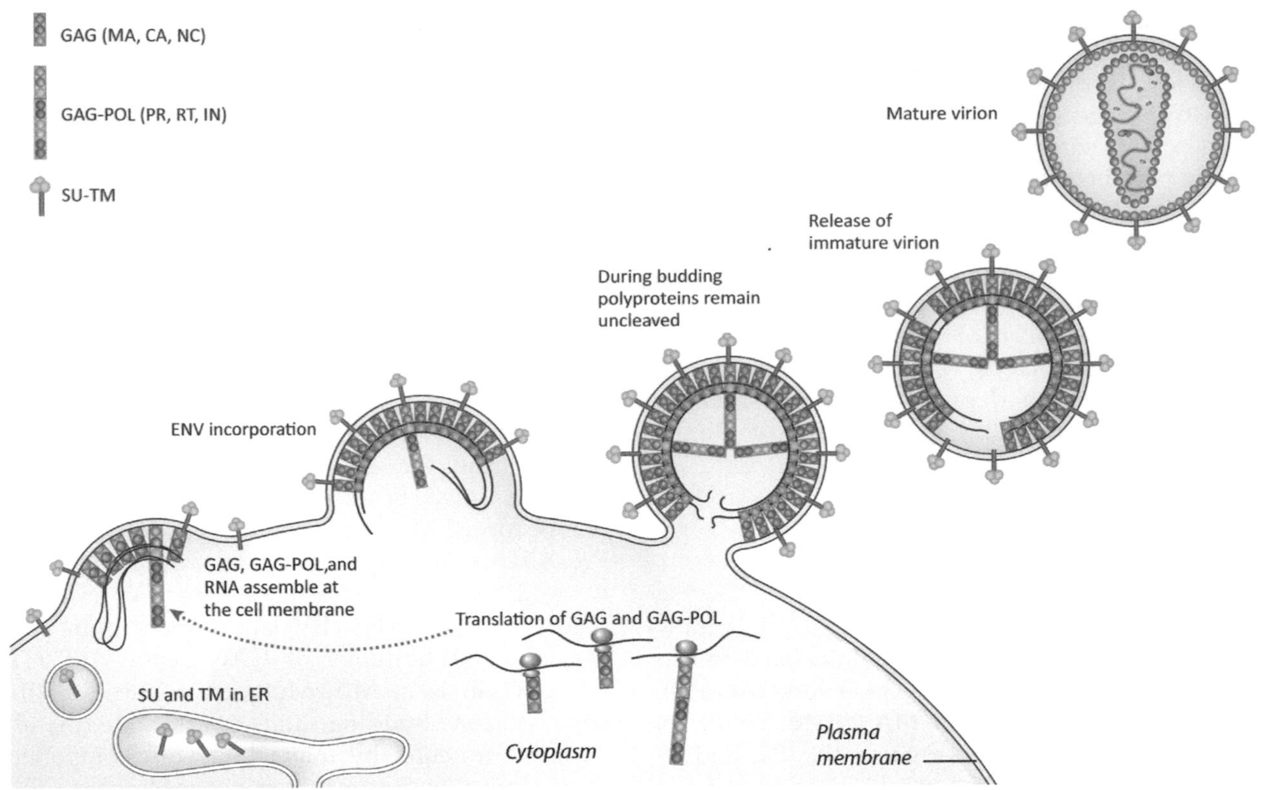

FIGURE 36.9 Assembly of a retrovirus. This figure illustrates assembly of HIV-1 at the plasma membrane of the infected cell. The immature virion assembles from polyprotein precursors that are cleaved to form infectious virions after release from the cell.

MECHANISMS OF RETROVIRAL ONCOGENESIS

Retroviruses *must* integrate (insert) their genomes into host DNA for the purposes of replication, and disruption of host DNA is an inherently mutagenic event (Box 36.3). Early studies of avian retroviruses revealed different mechanisms of transformation: Some retroviruses (for example, Rous sarcoma virus (RSV)) are *acutely* transforming and rapidly give rise to polyclonal tumors (each tumor is derived from a different infected cell). These so-called "acute" transforming retroviruses carry an oncogene into every infected cell (Fig. 36.10). In fact, these retroviral oncogenes were originally "captured" from cells.

Other oncogenic retroviruses are less "potent": Tumors are slower to develop and those that do are usually monoclonal. These tumors are caused by a process called *insertional* activation or mutagenesis (Fig. 36.10). Many cells are infected, but only rarely does chromosomal insertion provide a trigger for cell transformation. Studying tumors caused by insertional activation led to the discovery of dozens of "growth-promoting" cellular genes (so-called proto-oncogenes). These genes fall into two broad classes: Cell *receptors* that respond to extracellular signals to promote growth and *kinases* that activate signaling pathways to transmit messages to the cell nucleus.

Insertion of a provirus within a proto-oncogene may result in production of an aberrant protein: perhaps one with unregulated growth-promoting potential. But many insertions are near, not within, a proto-oncogene. These types of insertions may increase the *amount* of normal protein product. The expression of growth-promoting proteins is tightly regulated and insertion of a provirus, with its own promoter, enhancer, splicing, and polyadenylation signals can disrupt that regulation. The overexpression of a growth-promoting protein is often the first step in cell transformation. Retroviral insertions may also promote cell transformation by *inactivating* a growth-*suppressing* gene. However, this is much less common, as the inactivation of one allele leaves a second, intact allele to produce an active protein product. Finally, it is important to understand that cell transformation is a readily observable event: *most* insertional events are silent.

Jaagsiekte sheep retrovirus (JSRV) is an acutely transforming betaretrovirus that causes a contagious lung cancer in sheep called ovine pulmonary adenocarcinoma. JSRV brings an oncogene into every infected cell. However the JSRV "oncogene" is not a captured cellular gene, it is the viral TM protein. JSRV

BOX 36.3 A BRIEF HISTORY OF RETROVIRAL ONCOGENESIS

Studies of oncogenic retroviruses date back to the early 1900s with descriptions of "filterable agents" capable of causing avian leucosis and avian sarcoma. Over the next 20 years, many distinct avian cancers were attributed to transmissible agents (viruses). By the 1950s, a group of related, tumor-causing RNA viruses had been identified from birds, rodents, and other mammals. They shared biochemical features and were morphologically similar, but displayed different phenotypes as regarded host specificity, tumor types, and transforming ability. For many years, these viruses were collectively called "RNA tumor viruses." In general they caused noncytopathic infections, required dividing cells for establishment of productive infection, and were tumorigenic; once tumor cells were formed, the transformed phenotype was stably inherited by daughter cells.

In the 1960s, Howard Temin (working at the McArdle Laboratory for Cancer Research at the University of Wisconsin-Madison) began to suspect that after infection, the genetic information of RNA tumor viruses became *stably associated* with the cell genome. Temin thought that process might be similar to DNA bacteriophage such as lambda phage, known to integrate into bacterial genomes. But the RNA tumor viruses had *RNA genomes* and synthesis of DNA, from an RNA template was unheard of! Temin performed a variety of studies to support his idea, but the breakthrough came when he demonstrated that RNA tumor viruses package a DNA polymerase. Temin published this seminal work in 1970,[1] side-by-side with a paper from David Baltimore.[2] Thus in 1970, two laboratories independently and convincingly demonstrated that RNA tumor viruses package an enzymatically active DNA polymerase. Discovery of the polymerase we now call reverse transcriptase (*RT*) revolutionized the study of RNA tumor viruses. In 1975 the Karolinska Institute awarded the Nobel Prize in Physiology or Medicine jointly to David Baltimore, Renato Dulbecco,[3] and Howard Temin for their discoveries concerning interactions between cells and tumor viruses. In 1980, the first human oncogenic retrovirus (HTLV-1) was described.[4] HTLV-1 was discovered in the laboratory of Robert Gallo at the US National Cancer Institute. Also in the early 1980s a retroviral etiology was ascribed to lung tumors in sheep.[5] There are currently seven genera in the family *Retroviridae,* five of which contain oncogenic retroviruses.

1. Temin, H.M., Mizutani, S., 1970. RNA-dependent DNA polymerase in virions of Rous sarcoma virus. Nature 226, 1211–1213.
2. Baltimore, D. 1970. RNA-dependent DNA polymerase in virions of RNA tumour viruses. Nature 226, 1209–1211.
3. In 1975 the Karolinska Institute awarded the Nobel Prize in Physiology or Medicine jointly to David Baltimore, Renato Dulbecco, and Howard Temin for their discoveries concerning interactions between tumor viruses and cellular DNA. Dulbecco studied oncogenesis of the DNA virus SV40, a polyomavirus. Also in 1975, the family name *Retroviridae* was adopted at the first meeting of the International Committee on the Taxonomy of Viruses.
4. Poiesz, B.J., Ruscetti, F.W., Gazdar, A.F., Bunn, P.A., Minna, J.D., Gallo, R.C., 1980. Detection and isolation of type C retrovirus particles from fresh and cultured lymphocytes of a patient with cutaneous T-cell lymphoma. Proc. Natl. Acad. Sci. USA 1980, 77, 7415–7419.
5. Safran, N., Zimber, A., Irving, S.G., Perk, K., 1985. Transforming potential of a retrovirus isolated from lung carcinoma of sheep. Int. J. Cancer 35, 499–504.

TM directly activates cell-signaling cascades to promote cellular proliferation; additional mutations in proliferating cells lead to malignant transformation. Experimental infection of day-old lambs with JSRV induces tumor formation within 3–6 weeks.

Feline leukemia virus (FeLV) (genus *Gammaretrovirus*) causes a variety of diseases in infected cats, among them, tumors induced by insertional mutagenesis. The pathogenesis of FeLV is complex because the virus *mutates during infection* and can *recombine with endogenous FeLVs* to generate variants with different cell tropisms. FeLV infection of adult cats is well controlled by the immune system (Fig. 36.11) and very little, if any, replicating virus is detected after the acute infection; these cats do not transmit virus, nor do they develop FeLV-associated tumors. Infection of kittens often has a very different outcome. Kittens fail to control virus replication and become persistently infected (PI) cats. They shed virus and most will eventually become sick. Some will develop tumors, initiated by insertional mutagenesis. However the most interesting feature of PI cats (from the perspective of viral pathogenesis) is the emergence of mutated viruses with altered cell tropism. These variants arise by point mutations in SU and/or by recombination of the *env* gene of exogenous FeLV with *env* genes of endogenous copies of FeLV-like viruses. One variety of mutated FeLV is called the T subtype. FeLV-T is tropic for, and causes cytopathic infection of T cells,

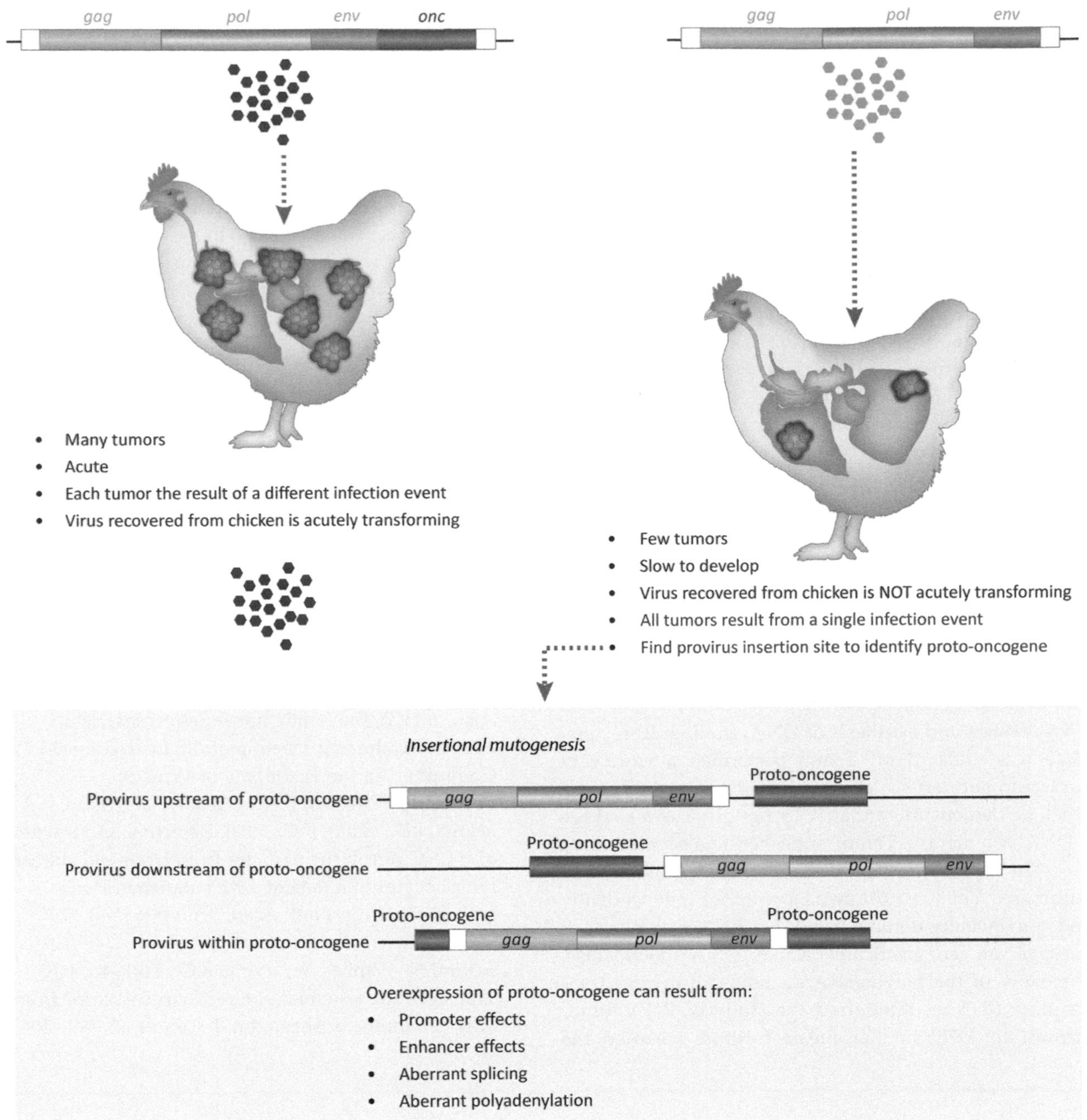

FIGURE 36.10 Two methods of retroviral oncogenesis. On the left, a chicken is infected with an acutely transforming retrovirus (for example, RSV). The viral genome encodes an oncogene that is delivered to every infected cell. Many cells become transformed to generate tumors. The tumors in the chicken on the left are polyclonal (each tumor originated from a different cell). The chicken on the right is infected with a retrovirus that has the potential to insertionally activate a chicken proto-oncogene. During infection the retrovirus inserts into chicken chromosomal DNA. In rare cases, insertion will activate a growth-promoting gene and the cell may be transformed. There are a variety of mechanisms by which a proto-oncogene can be activated. Often the gene is overexpressed by the strong retroviral promoter. The tumors in the chicken on the right at monoclonal. One cell was initially transformed.

leading to immunosuppression. Another variety, FeLV-C, causes fatal anemia. FeLV-T and C are rarely transmitted, but instead, arise from virus mutations within PI cats.

Some retroviruses that cause cancers do so by a completely different mechanism. They encode proteins that modify cell transcription and signaling pathways to promote cell growth. HTLV-1 is an oncogenic human retrovirus. HTLVs cause life-long infections that are largely asymptomatic but can increase risk of developing leukemia. Oncogenesis of HTLV-1 is thought to require two small viral proteins, TAX and HTLV-1 bZIP (HBZ)

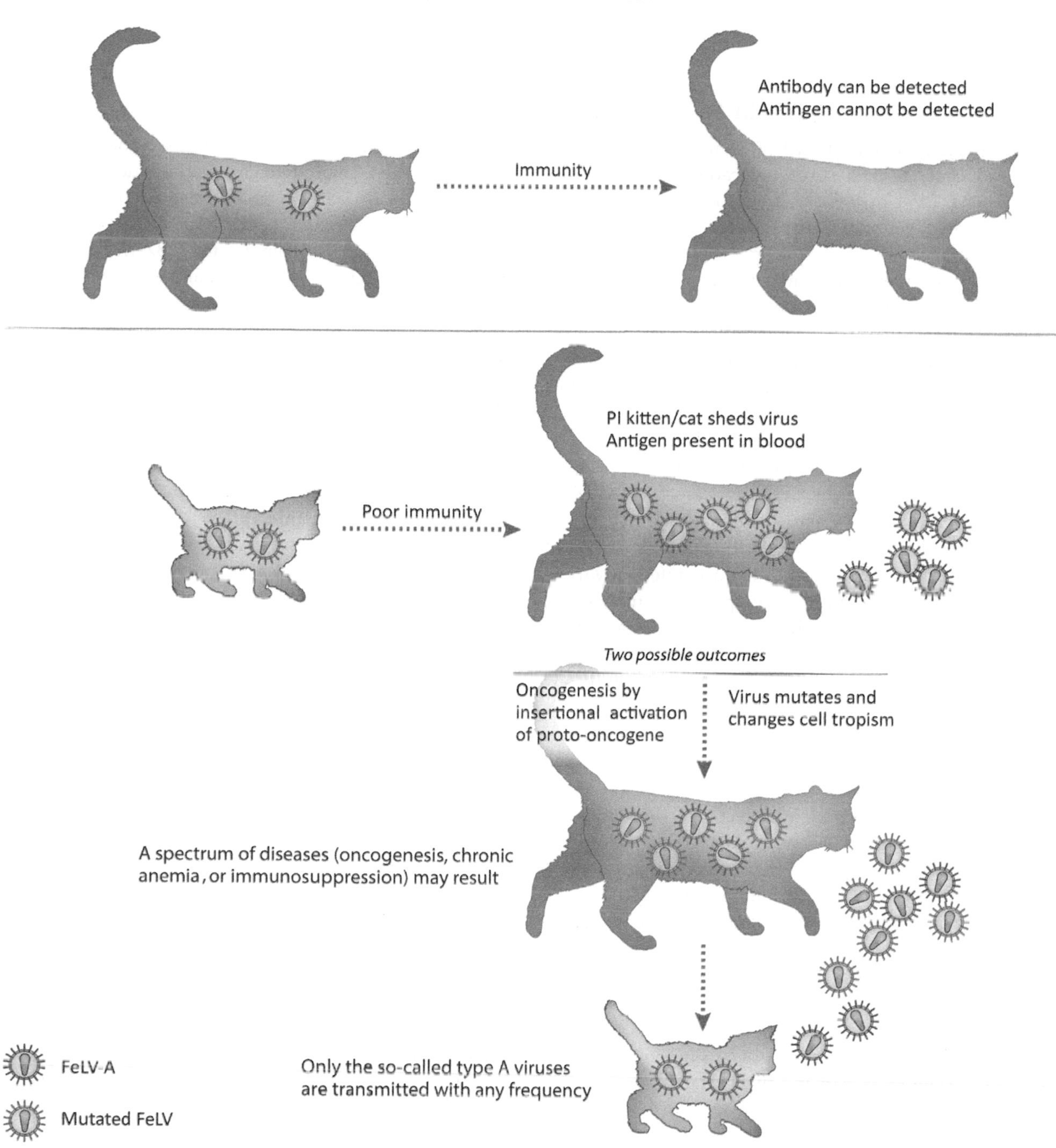

FIGURE 36.11 Pathogenesis of FeLV is complex. Infection of an adult cat often results in development of protective immunity. The cat is antibody positive but not viremic (viral antigen cannot be detected in the blood). If a kitten is infected with FeLV, it may not develop protective immunity. Instead it becomes persistently infected. The PI kitten or cat sheds virus and is identified by the presence of viral antigen in the blood. The antigen positive cat is a source of infectious virus and may transmit FeLV to other cats. The antigen positive cat may develop any number of diseases including cancers or acute (fatal) anemia. The viruses found in the sick cat often contain mutations (specific mutations are linked to different disease manifestations).

(Fig. 36.4). Both are multifunctional and interact with a variety of host cell proteins. TAX is thought to be a key player in tumor initiation but by the time cells are transformed, TAX is silenced, and HBZ is expressed (see Box 36.4).

Walleye dermal sarcoma virus (WDSV) is a member of the genus *Epsionretrovirus*. Infection is associated with proliferative lesions in walleye, a freshwater fish native to Canada and the Northern United States. In infected fish, proliferative lesions are seasonal. The

BOX 36.4

HUMAN T-CELL LEUKEMIA VIRUS-1

Oncogenic animal and bird retroviruses were discovered and characterized several decades before the discovery of HTLV-1 in 1980. HTLV-1 and its close relative HTLV-2 are complex retroviruses in the genus *Deltaretrovirus*.

HTLV-1 is endemic in parts of Japan, the Caribbean Islands, Africa, and South America. Mechanisms of virus transmission include sexual contact, breast-feeding, blood transfusion, and contaminated needles. It is estimated that 10–20 million people worldwide are infected. Most infections are asymptomatic; however, 3%–5% of infected persons develop HTLV-1-associated T-cell tumors. A smaller percentage of infected individuals develop a demyelinating disease called HTLV-1-associated myelopathy (HAM) or tropical spastic paraparesis (TSP).

In vivo HTLV-1 infects primarily T-lymphocytes but in vitro tropism is broader and the virus infects a variety of other cell types including B-lymphocytes, monocytes, endothelial cells, and fibroblasts. HTLV-1 is a unique retrovirus in several respects. First, HTLV-1 is a highly cell-associated virus. Whether in vitro or in vivo, few cell-free virions are released; for the most part, virus is transmitted directly from cell to cell via a virological synapse. Second, within an HTLV-1-infected individual, there is very little virus expression. Polymerase chain reaction (PCR) is required to detect very low levels of viral RNA; viral proteins are undetectable. (Note the contrast with HIV, another T-cell tropic human retrovirus.) When peripheral blood mononuclear cells from an infected patient are cultured, levels of viral RNA and protein increase significantly. This suggests that chronically infected patients maintain a pool of "latently" infected cells and HTLV-1 is passed to daughter cells via mitosis.

The HTLV-1 genome is shown in Fig. 36.4. *Gag*, *pro*, *pol*, and *env* genes overlap and their arrangement is reminiscent of simple retroviruses. However, an open reading frame called X, found at the 3′ end of the genome, encodes several additional proteins. The best studied of these are TAX and REX. TAX is a transactivating protein while REX increases nuclear export of unspliced and singly spliced viral mRNAs. TAX interacts with cellular proteins and forms a complex on the LTR thereby increasing transcription. The mechanism of transcription activation seems to involve changes to chromatin structure.

HTLV-1 bZIP factor (HBZ) is also encoded in the X region but is transcribed in the opposite direction to all other transcripts. The promoter for HBZ transcription is the 3′ LTR. HBZ is a Basic Leucine Zipper Domain (bZIP) protein that interacts with many cellular proteins including several important transcription factors. HBZ activities are complex: (1) HBZ is required for efficient viral infectivity and persistence in a rabbit model; (2) HBZ is dispensable for HTLV-1-mediated cellular transformation in cultured cells but (3) HBZ by itself promotes proliferation of T-cell lines; (4) HBZ suppresses TAX-mediated viral gene transcription.

Adult T-Cell Leukemia

Adult T-cell leukemia (ATL) is a very aggressive T-cell cancer that may arise decades after infection with HTLV-1. The leukemic cells of ATL are monoclonal and harbor HTLV-1 proviral DNA at random chromosomal integration sites. This suggests an important role for HTLV-1 in tumor development but also suggests that insertional activation of a cellular proto-oncogene does not drive cell proliferation. Models of ATL focus on the activities of TAX and HBZ. The process is clearly complex and multistep, but some common findings are: (1) ATL cells are genetically unstable and TAX promotes chromosomal damage. This suggests an early role for TAX in tumor development. (2) TAX expression is largely silenced in ATL cells. As TAX is a target of cytotoxic T cells, lack of expression would protect proliferating tumor cells. (3) HBZ mRNA and protein are found in most ATL tumors. (iv) Continued expression of HBZ is necessary to maintain tumor cell proliferation.

HTLV-1-Associated Myelopathy/Tropical Spastic Paraparesis

It is estimated that between 0.25% and 3% of HTLV-1-infected persons may eventually develop HAM/TSP, a slowly progressive, chronic disease of the spinal cord. Spinal cord damage results from inflammatory responses directed against infected cells. Unfortunately, nerve cells are also damaged in the process. The main symptoms are painful stiffness and weakness of the legs.

highest incidence of disease occurs in the Fall/Winter seasons and lesions regress as water temperatures warm up in the Spring/Summer seasons. WDSV is a complex retrovirus that encodes auxiliary proteins believed to be responsible for tumor development. One of these is a cyclin-like protein [retrovirus (rv)-cyclin] that may stimulate cell growth. The other (ORF B) protects cells from apoptotic stimuli. The specific factors involved in the seasonality of tumor formation are unclear and this interesting retrovirus has yet to reveal all of its secrets.

In this chapter we learned that:

- The major steps in the retrovirus replication cycle are: attachment, penetration, reverse transcription, integration, transcription, translation, assembly, release, and maturation.
- Integration of the retroviral genome (the ds DNA version) is a unique to the family *Retroviridae*. Reverse transcription is *almost* unique to this family but the hepadnaviruses also use reverse transcription for genome synthesis (see Chapter 38: Family *Hepadnaviridae*).
- Retroviral PR cleaves the GAG and GAG-POL precursors to generate MA, CA, NC, PR, RT, and IN. PR is active after budding. Budded virions are initially immature and become mature after PR cleavages are complete.
- Exogenous retroviruses are transmitted horizontally, by infectious particles, from one animal or person to another. Endogenous retroviruses are transmitted through the germ-line (sperm and eggs) from parent to offspring. They are seldom found as particles and many are defective. About 8% of the human genome consists of endogenous retroviruses (or their remains).
- Insertional activation describes the process by which integration of a retrovirus activates a cellular proto-oncogene. Proto-oncogene is a general term to describe any growth-promoting gene. Retroviral integration can directly mutate a proto-oncogene or it can increase the level of expression of a normal protein product. Expression of growth-promoting genes is tightly regulated. Insertion of a provirus may provide a strong promoter, may provide enhancer elements, may change splicing patters, change use of a polyadenylation signal or RNA stability element. Insertional activation of a proto-oncogene is often the first step in the transformation process. Often additional mutations are required to confer a fully transformed phenotype.
- The acutely transforming retroviruses of chickens encode oncogenes. These growth-promoting genes were picked up from the host during the process of reverse transcription. Thus a retroviral oncogene was originally a host proto-oncogene.

CHAPTER

37

Replication and Pathogenesis of Human Immunodeficiency Virus

OUTLINE

After reading this chapter, you should be able to answer the following questions:

- What are the most common routes of transmission of human immunodeficiency virus (HIV)?
- How can transmission of HIV be avoided?
- HIV infects what types of cells?
- How does the cell tropism of HIV relate to its pathogenesis?
- What is the cell receptor for HIV?
- What proteins serve as coreceptors for HIV?
- What is the function of HIV TAT?
- What are some common features of the auxiliary proteins encoded by HIV and simian immunodeficiency virus (SIV)?
- What is a viral "restriction factor"?
- What classes of antiretroviral drugs are available for treatment of HIV infection?

Viruses. DOI: http://dx.doi.org/10.1016/B978-0-12-803109-4.00037-4

- Explain why it is important to use drug "cocktails" to treat HIV infection.
- What are some of the obstacles to development of an effective HIV vaccine?

HISTORY OF HIV AND ACQUIRED IMMUNODEFICIENCY SYNDROME

The modern history of HIV-AIDS began in June 1981 with a paper in Morbidity and Mortality Reports Weekly published by the Centers for Disease Control. The paper described an unusual cluster of cases of severe immunodeficiency in previously healthy men. In the weeks and months that followed, additional cases were reported. All involved the development of severe immunodeficiency in previously healthy individuals. By 1982 over 800 cases of severe immunodeficiency had been reported in the United States (Box 37.1). Persons at highest risk were intravenous drug users, hemophiliacs, transfusion recipients, homosexual men, and children born to mothers at risk. The epidemiology was consistent with an *infectious agent* transmitted via *blood and body fluids* as the cause of this new disease. The name given to the syndrome was acquired immunodeficiency syndrome (AIDS). By 1984

BOX 37.1

THE ORIGINS OF HIV/AIDS

AIDS was first recognized in the United States and other developed countries in the early 1980s. The origins of the disease were unknown at that time. Once it was shown to be caused by HIV, a retrovirus, it was possible to trace back its origins. For example, it was soon shown that similar retroviruses were present in numerous species of African monkeys. Genomic sequence analysis showed that some were much more closely related to HIV than others. Thus the chimpanzee retroviruses were most closely related, suggesting that the human virus originated from chimpanzees. Within African chimpanzee populations, there are multiple subpopulations separated from each other by broad rivers. (Chimpanzees do not like to swim and little mixing occurs.) The most closely related virus to HIV could thus be tracked down to a forest in central Africa by the Congo river. By estimating the mutation rate of HIV, and comparing chimpanzee strains with human strains, it has been calculated that the chimpanzee virus entered humans between 1900 and 1920.

Chimpanzees are quite good to eat and are a significant source of bush meat for African villagers. They are, however, fierce and a bow-and-arrow is not the best way to kill them. Only when guns became available to villagers around 1900 did chimpanzees become a significant food source. Once killed, the meat had to be carried back to the village. It was therefore necessary to butcher the carcass. This was a bloody business and inevitably some chimpanzee blood could contaminate cuts or scratches on a hunter. Once infected, the hunter would return to the village and perhaps infect his wife. Their deaths would have attracted little attention and provided they were monogamous, the infection would not have spread.

However, both the French and Belgian colonists in the Congo employed forced labor and many villagers would have been compelled to work in rubber plantations and other activities. To ensure the health of their workers, the colonial authorities made widespread use of antimalarial drugs, usually administered by injection. It is highly likely that HIV was thus spread from isolated villages into a wider work force. Nevertheless, the number of HIV cases would have remained low and unremarkable against a background of many other tropical diseases.

Beginning in the 1950s, both the Belgian and French Congos gained their independence but were politically highly unstable. Huge numbers of villagers moved to the cities in search of better opportunities. With so many single men in the cities prostitution thrived, and as a result the prevalence of HIV in the urban population of cities like Kinshasa grew rapidly in the 1950s and 1960s. Because of instability and chronic warfare, few Western aid workers entered these countries and they relied on development aid and training from countries such as Haiti. Sometime in the 1970s, the virus probably infected Haitian workers and was introduced into the Haitian sex-trade. From Haiti it spread to the United States and Western Europe by way of the trade in blood products as well as sex-tourism. Eventually it spread across the globe. It should be pointed out, however, that 75% of AIDS cases are still occurring in Sub-Saharan Africa.

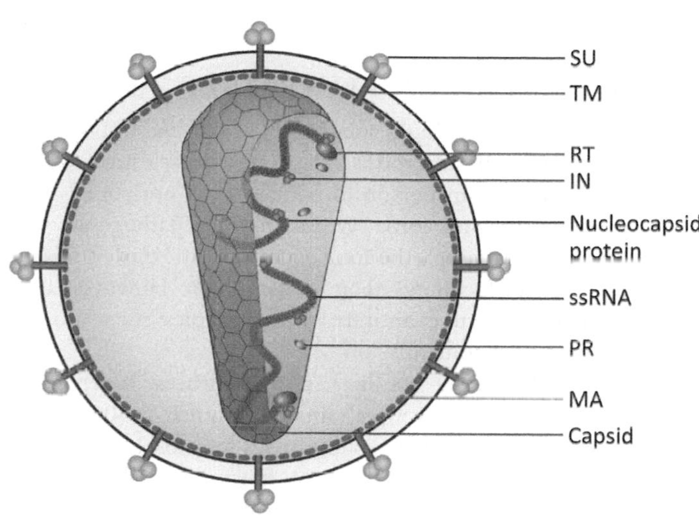

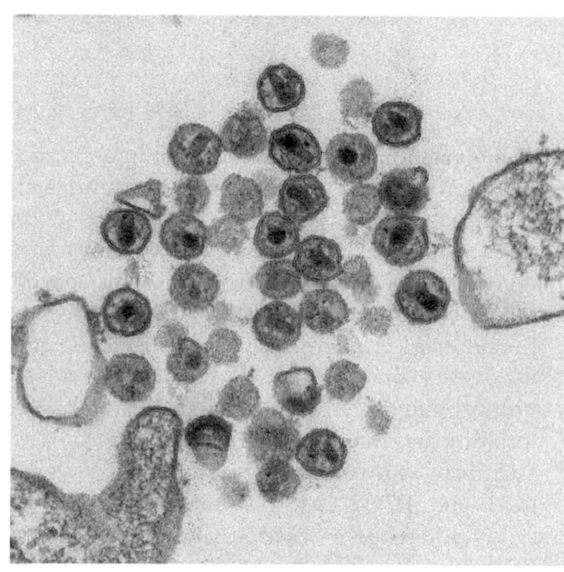

FIGURE 37.1 Retrovirus structure (left). Thin section transmission electron microscopic image depicts numerous HIV virions (right). *From CDC/Maureen Metcalfe, Tom Hodge. CDC Public Health Image Library Image ID# 13472.*

a lymphotropic retrovirus had been isolated by a number of laboratories (Fig. 37.1). Blood screening tests were quickly developed to help eliminate HIV from the blood supply. Blood screening tests were also used to identify healthy persons in high-risk groups, with antibodies to HIV. These patients were then closely followed in an effort to understand the progression of the disease.

In the 1980–90s a diagnosis of AIDS or HIV infection was considered a death sentence. However, improvements in drug therapy changed the prognosis drastically. While it would be a mistake to minimize the costs and potential side effects of drug treatment, the condition is now considered a chronic disease, more along the lines of a cancer diagnosis. HIV has taken a huge toll on human life and many billions of dollars have been spent developing vaccines (mostly unsuccessful to date) and treatments, as well as to probe the human immune system and unravel the disease process.

As treatment has improved, the number of persons living with HIV/AIDS has increased and the number of deaths has decreased. In 2012 the World Health Organization estimated over 35 million HIV infections worldwide, with 1.6 million deaths. Estimates in the United States are ~1.2 million infected persons. Unfortunately, one in seven infected people do *not* know that they carry the virus and they pose a significant risk for transmitting the disease. Although the earliest reports of HIV/AIDS in the United States focused on homosexual men, it is quite clear that HIV does not discriminate: It is an equal opportunity infection for those who ignore the hazards of unsafe sexual practices and illicit drug use.

Where did HIV come from and how did it emerge so rapidly? The closest relatives to HIV are the SIVs that infect a variety of Old World monkeys. The virus we call HIV-1 jumped from simian to human hosts at least four times. These cross species jumps likely occurred through hunting and consuming monkeys. The vast majority of human HIV-1 infections are caused by the so-called M group viruses. M group viruses emerged in Central Africa and are now found worldwide. The spread of HIV-1 was certainly facilitated by the long asymptomatic period during which the virus can be silently and unknowingly transmitted.

HIV IS A LENTIVIRUS

When HIV was identified in the early 1980s, it was clear that it was an unusual retrovirus. It was soon determined that the closest known relatives of HIV were an obscure group of animal retroviruses called lentiviruses (lenti = slow) that cause chronic life-long infections of horses, sheep, and goats. Today we know that HIV's closest relatives are other primate lentiviruses Box 37.2).

BOX 37.2

RELATIONSHIPS AMONG PRIMATE LENTIVIRUSES

Understanding the taxonomy of the primate lentiviruses provides an understanding of the origins of HIV. All primate lentiviruses have a common ancestor. It appears that these viruses evolved with their hosts, such that today, each species of monkey has its own SIV. SIVs are known to naturally infect at least 40 different species of nonhuman primates in Africa. Most SIVs cause minimal disease in their natural host, but a cross species infection (naturally or experimentally) may cause high morbidity and mortality in the new host.

Humans can be infected with two different viruses, HIV-1 and HIV-2. An examination of their genomes reveals differences in their auxiliary proteins (Fig. 37.4). HIV-1 has spread widely while HIV-2 has remained more restricted to countries in West Africa. The closest relative of HIV-1 is chimpanzee SIV (SIV_{CPZ}) while the closest relative of HIV-2 is the sooty mangabey virus, SIV_{SM}. In addition to genomic differences, the human epicenters of HIV-1 and HIV-2 are discrete. Thus HIV-1 and HIV-2 are two different viruses that jumped independently into the human population (Fig. 37.5).

In fact, both HIV-1 and HIV-2 have jumped into the human population *multiple* times. In the case of HIV-1, four distinct "groups" have been identified. Each group is the result of an *independent* introduction into the human population. The M (main) group of HIV-1 is without a doubt the *main* group. M group viruses emerged in Central Africa and have spread worldwide, infecting many millions of people. In addition, sublineages or clades have developed within HIV-1 M during worldwide spread. Sublineages are important because these are genetically distinct enough that vaccines developed for a single subgroup are not like to be effective across all subgroups. M group viruses are readily transmitted from person to person (both horizontal and vertical transmission). While highly pathogenic (in untreated patients), the long asymptomatic period (clinical latency) facilitates their transmission. However, use of highly effective antiretroviral therapies seems to be slowing their transmission.

The HIV-1 O (outlier) group viruses diverge in sequence from M group strains by as much as 50% in the envelope gene. The O group viruses are closely related to chimpanzee and gorilla SIVs. A third genetically distinct group of HIV-1 was discovered in 1998; only a handful of human infections has been identified to date. This group is called N, for non-M, non-O. In 2009 another HIV-1 "group" was defined. This group is tentatively called P (P for "pending the identification of further human cases") has only been recovered from a single patient.

HIV-2 entered the human population in West Africa. Both HIV-1 and HIV-2 have the same modes of transmission and are associated with similar opportunistic infections and development of AIDS. However, in persons infected with HIV-2, immunodeficiency seems to develop more slowly and to be milder. HIV-2 may be less transmissible than HIV-1. Currently two groups of HIV-2 have been identified, Groups A and B.

The identification of genetically distinct groups of HIV-1 and HIV-2 is medically important. Current tests to screen patients and blood may not identify HIV-2 or non-M group HIV-1 infections. In addition, drugs developed to treat HIV-1 may be ineffective against HIV-2.

As presented in Chapter 36, Family *Retroviridae*, the family *Retroviridae* contains seven genera within two subfamilies. Members of the genus *Lentivirus* include HIV, SIVs, equine infectious anemia virus (EIAV), visna-maedi viruses, feline immunodeficiency virus (FIV), and bovine immunodeficiency virus. Lentiviruses are among the so-called complex retroviruses because they encode a number of gene products in addition to the GAG, POL, and ENV polyproteins (Fig. 37.2). All known lentiviruses have a preference for infecting cells of the immune system. HIV and SIV use a receptor (the CD4 molecule) present on T-helper cells, monocytes, macrophages, and dendritic cells (DCs). FIV also infects feline B-cells while EIAV is more restricted, infecting only equine monocytes and macrophages. One characteristic shared by these cell types is that they are often found in resting state, although they can be activated by immune stimulation. Lentiviruses are unique among the retroviruses in their ability to infect nondividing cells. Lentiviruses encode proteins that actively move a preintegration complex (PIC) across a nuclear pore to gain access to host DNA.

Another common trait of lentiviruses is their high mutation rate. Even before the discovery of HIV, studies of horse, sheep, and goat lentiviruses suggested that these viruses persisted for years in their

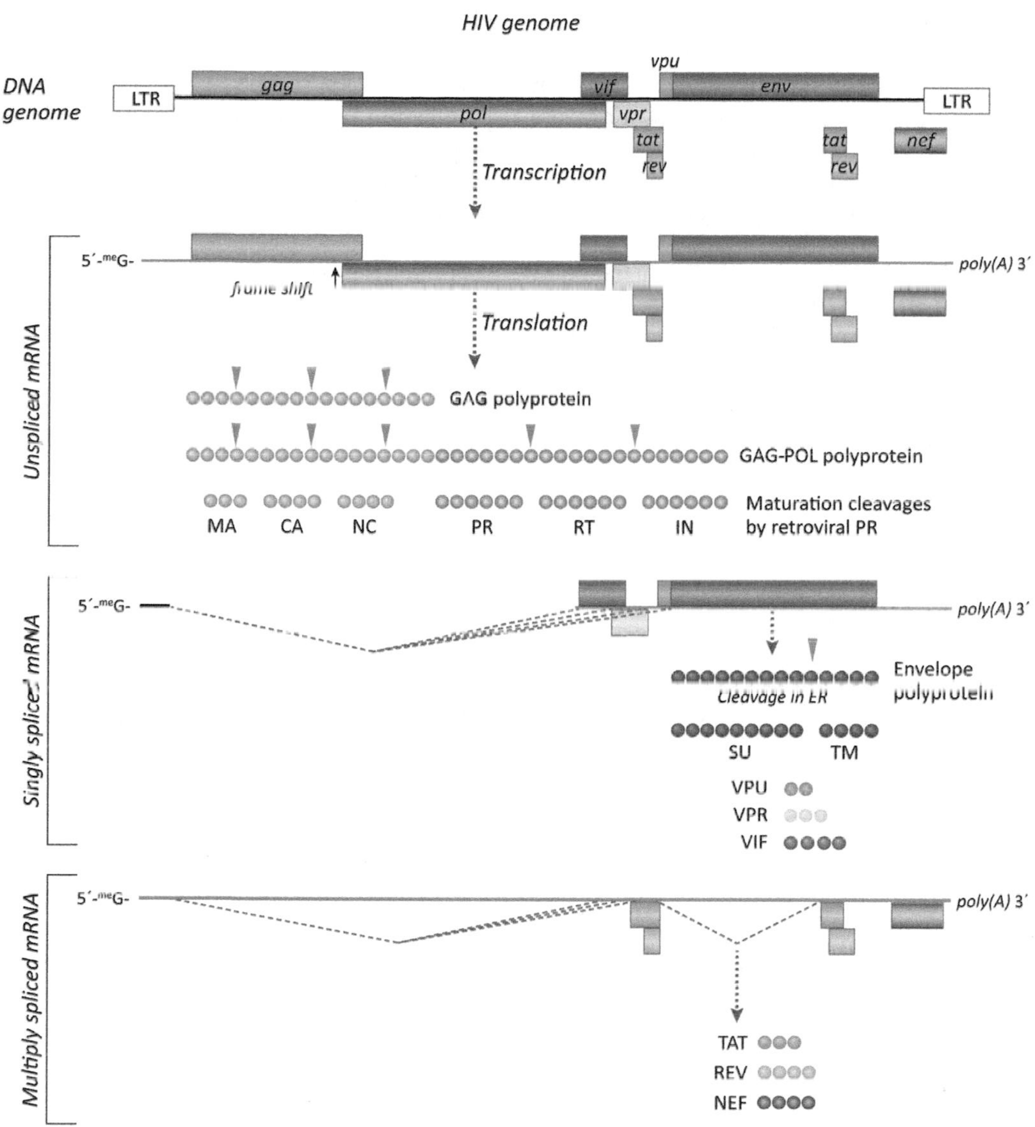

FIGURE 37.2 HIV genome and gene expression strategy.

infected hosts, in part, because they mutated to escape immune responses. Sequencing thousands of HIV genomes confirmed that genome diversity is indeed a common trait of lentiviruses. It has been estimated that during an infection, viral genetic diversity in HIV patients increases by ~1% per year from the initial founder virus strain (the infecting virus). Several factors contribute to the genetic heterogeneity of HIV-1:

- Reverse transcriptase (RT) does not have a proofreading/correction function. Estimates of mutation rate in cell culture are $1-2\times10^{-5}$ mutations/nucleotide synthesized/replication cycle. Estimates of mutation rates in patients are up to five times higher.
- Within an infected patient, there are millions of HIV replication cycles per year.
- Virus production in vivo can be an astounding 10^9 particles per day.
- Recombination by template switching (during reverse transcription) increases viral genetic diversity.

HIV PROTEINS

HIV and other lentiviruses are complex retroviruses, but their basic genome organization and replication cycle have much in common with other retroviruses. The basic steps in lentivirus replication are:

Attachment/entry, reverse transcription, integration, transcription, protein synthesis, assembly, budding, and maturation. These processes were described in Chapter 36, Family *Retroviridae*. In this section we review HIV proteins and their functions.

The three longest open reading frames (ORFs) of all retroviruses encode precursor polyproteins. HIV encodes the GAG precursor and the GAG-POL precursor from unspliced mRNA. The *env* ORF is expressed from a singly spliced mRNA; it encodes the HIV surface (SU) and transmembrane (TM) glycoproteins.

GAG Precursor ($Pr55^{GAG}$)

$Pr55^{Gag}$ is translated from unspliced mRNA. During virion maturation $Pr55^{Gag}$ is cleaved by the vial protease (PR) to produce four structural proteins (MA, CA, NC, and p6) and two spacer peptides (SP1 and SP2). The positions of the structural proteins in the precursor reflect their final locations in the mature virion. MA, at the amino-terminus of $Pr55^{Gag}$, is found in association with the lipid envelope. CA is the capsid protein that forms the elongated icosahedral shell around the genome (this is positioned in the center of the mature virion). The positively charged NC protein binds to RNA.

MA domain of $Pr55^{Gag}$ functions during the late phase of the HIV replication cycle and is a key player is assembly. The MA domain directs and localizes $Pr55^{Gag}$ to the plasma membrane. MA is linked to a molecule of myristic acid (a 14-carbon fatty acid) via a glycine near its amino-terminus. The fatty acid moiety inserts into the plasma membrane to form an anchor for $Pr55^{Gag}$.

CA domain has several discrete functions that can be identified by mutagenesis of different regions of the protein. One region of CA promotes oligomerization of $Pr55^{Gag}$, and mutations in the oligomerization domain block production of immature particles. Other CA mutations interfere with formation of mature capsids. Still other CA mutants produce viruses that bud and form capsids, but are noninfectious (these virions are probably deficient in a function needed for successful entry into a new cell).

Another important biological activity of HIV-1 CA is that it binds to the CypA protein during virus assembly. CypA is a member of the cyclophilin family of peptidyl-prolyl *cis-trans* isomerases. HIV CA binds CypA via a proline rich loop and the interaction leads to incorporation of CypA into virions. It appears that CypA may counteract a host restriction factor (TRIM5α) early in the infection process.

NC domain interacts with the retroviral genome (unspliced mRNA) and is critical for its encapsidation. HIV-1 NC has two zinc finger motifs that specifically interact with HIV-1 RNA at the packaging signal. NC also assists in placement of the tRNA primer on the primer-binding site and participates in oligomerization of $Pr55^{Gag}$. After particle budding and maturation, NC remains bound to the genome. NC also functions during the early stage of infection when its activity as a nucleic acid chaperone greatly facilitates successful reverse transcription. NC appears to remain associated with newly synthesized DNA and plays a role in integration as well.

p6 domain is located at the very carboxyl terminus of $Pr55^{Gag}$. The p6 domain is a key player in release of immature virions from the budding site. Some mutations in p6 yield particles that are tethered at the PM. The critical site for particle release is a highly conserved amino acid motif called PTAP (so named because its amino acid sequence is Pro-Thr-Ala-Pro). PTAP is a so-called *late* domain signifying its role in the very last stages of virion production. The p6 domain interacts with host proteins of the cellular endosomal sorting machinery (ESCRT machinery). P6 also interacts with HIV-1 viral protein R (VPR) (described later) facilitating its incorporation into virions.

GAG-POL Polyprotein Precursor ($Pr160^{GAG\text{-}pol}$)

$Pr160^{Gag\text{-}pol}$ is a 160 kDa polyprotein synthesized as the result of a ribosomal frame-shift in CA. It is estimated that frame-shifting occurs about 20% of the time. HIV $Pr160^{Gag\text{-}pol}$ contains domains for PR, RT, and integrase (IN).

Protease

PR is an aspartic protease that cleaves $pr55^{Gag}$ and $pr160^{Gag\text{-}pol}$. PR is active as a dimer and the active site contains two opposed aspartic acids that coordinate a molecule of water to catalyze hydrolysis of a peptide bond. The regions flanking the active site contribute to substrate recognition. PR catalytic activity is activated upon budding and is absolutely required for production of mature, infectious virions. HIV-1 PR inhibitors are effective antivirals that are incorporated into immature particles and inhibit processing of precursor polyproteins.

Reverse Transcriptase

HIV RT is a heterodimer. One subunit is 66 kDa (p66) and the other is 51 kDa (p51). P66 contains an RNase H domain missing from p51. In the mature RT, p66 appears to contribute catalytic activity while p51 plays a structural role (Fig. 36.6).

Integrase

IN is located at the carboxy terminus of $Pr160^{Gag\text{-}pol}$. HIV IN is 32 kDa and is the major player in integration. Upon infection, IN remains associated with the core during reverse transcription and is part of the PIC. Enzymatic activities of IN include exonuclease, endonuclease, and ligase. IN form tetramers that bind to the ends of the linear double-stranded DNA (Fig. 36.8).

ENV Precursor

The ENV precursor polyprotein is a 160 kDa glycoprotein called gp160. It is cleaved to produce the gp120 SU and gp41 TM proteins. Gp160 is translated from a singly spliced mRNA. The ENV precursor is translated on rough ER and within the ER gp160 is glycosylated and forms trimers. Gp160 is cleaved by furin or a furin-like PR in the Golgi. SU and TM remain associated during transit through the Golgi to the PM. TM is anchored in the PM by a single TM domain, but SU is maintained on the surface of cells or virions by relatively weak (noncovalent) interactions with TM.

Gp120 is highly glycosylated. In fact, ~50% of the total mass of SU is carbohydrate. The carbohydrate shields the polypeptide backbone, limiting immune recognition. The gp120 molecule has alternating regions of conserved (C) and variable (V) amino acids. The so-called variable domains are regions that form flexible loops on the more conserved backbone of the molecule. SU is the attachment protein and the CD4-binding site maps to highly conserved regions within SU. CD4 binding induces conformational changes in gp120 that allows binding to coreceptors (Fig. 37.3). The coreceptors for HIV are members of the G protein-coupled receptor superfamily of seven-TM domain proteins. The two major coreceptors for HIV are CCR5 and CXCR4. CCR5 is found on monocytes and macrophages; HIVs that use CCR5 are called R5 isolates. Strains that bind CXCR4 are referred to as X4 viruses. Other chemokine receptors can also serve as HIV coreceptors. CCR5 appears to play a major role in establishment of infection as individuals with a mutant version of CCR5 (called CCR5/Δ32) do not express the protein on the surface of cells. Individuals homozygous for this mutation are only rarely infected with HIV-1.

Gp41 or TM is the HIV fusion protein. Upon binding to CD4 and a chemokine coreceptor, conformational changes in SU activate the membrane fusion potential of TM. A stretch of hydrophobic amino acids at the N-terminus of TM initiate fusion by inserting into the PM (Fig. 37.3). Mutational analyses have also identified additional regions TM ectodomain that are critical for

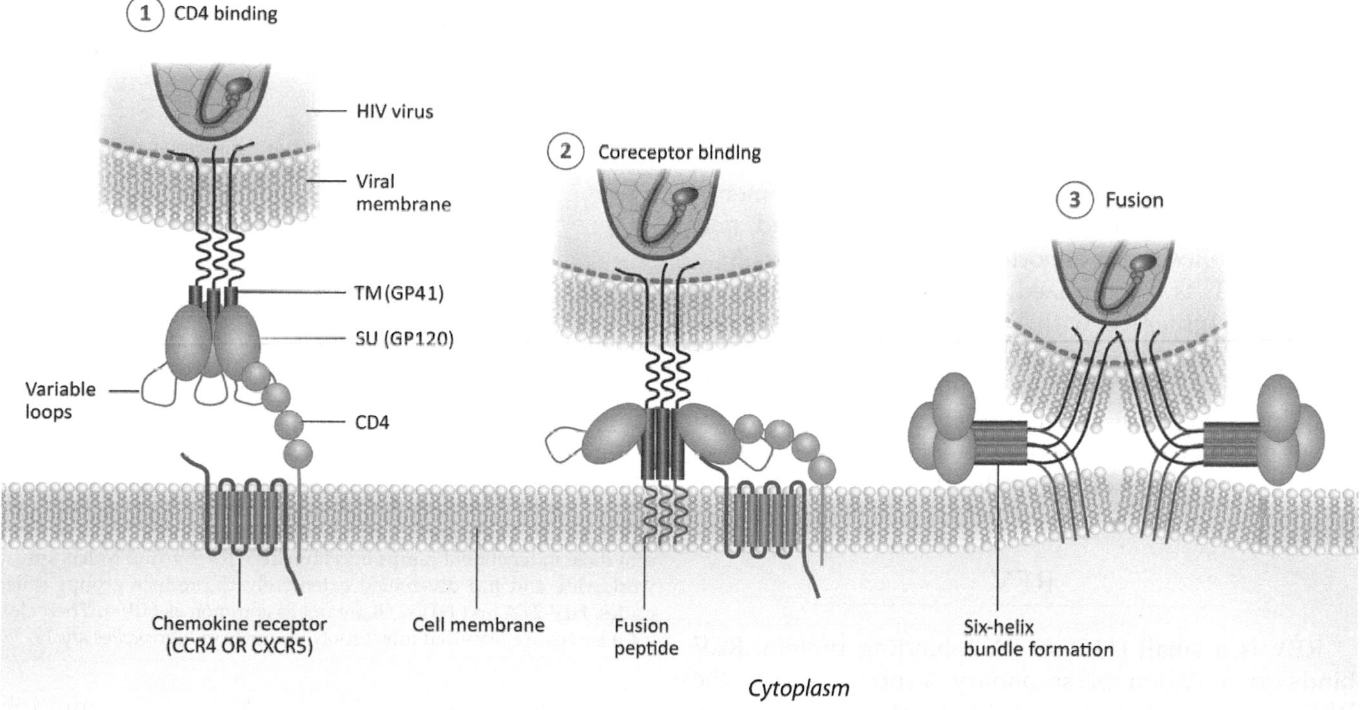

FIGURE 37.3 HIV attachment and fusion. Attachment starts with interactions between SU and the CD4 protein receptor. This is followed by additional interactions between SU and a chemokine receptor that allow the fusion domain of TM to interact with the plasma membrane.

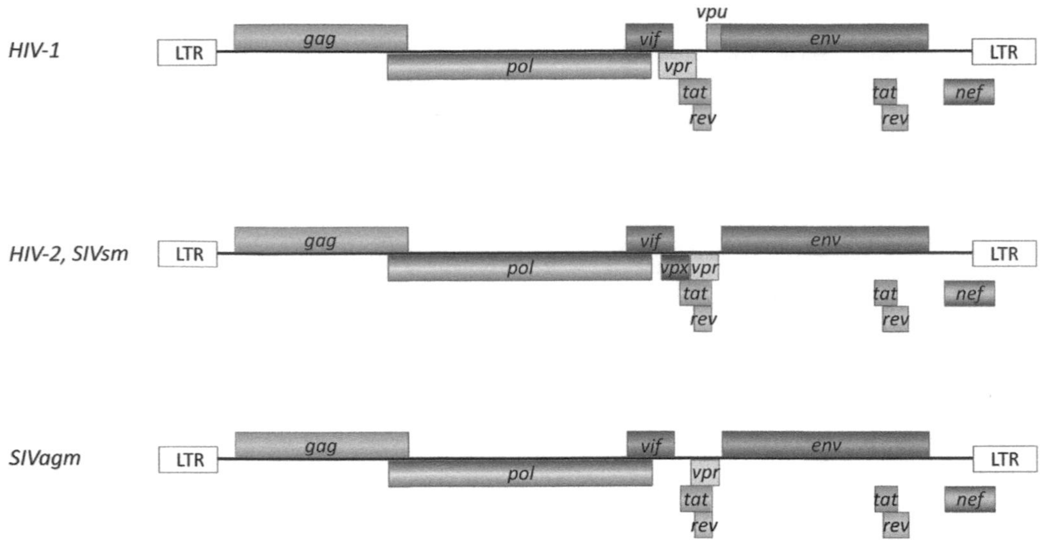

FIGURE 37.4 Comparison of primate lentivirus genomes showing differences in genes for auxiliary proteins.

membrane fusion. Identification of regions critical for fusion provides targets for development of novel drugs.

TAT

Complex retroviruses encode a variable of number of proteins in addition to those produced from GAG, GAG-POL, and ENV precursors. Two of these, TAT and REV, are encoded by all lentiviruses. TAT is a small protein (101 aa) with a critical role as an activator of transcription (Fig. 37.6). In the absence of TAT, RNA pol II is poorly processive; while transcripts are initiated, most terminate close to a stem-loop region called TAR (transactivating response element). However, a few transcripts will be completed and these are spliced and exported from the nucleus. As a result, TAT is synthesized, imported back into the nucleus and interacts with TAR. TAT also binds to cellular proteins, thus forming a protein complex that modifies host RNA pol II to significantly increase transcription elongation. In the presence of TAT the number of completed transcripts is increased by ~100-fold. TAT is essential for HIV-1 replication.

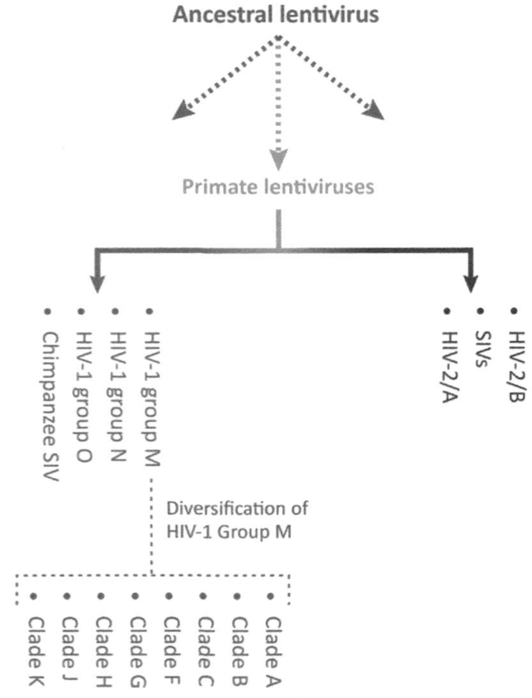

FIGURE 37.5 Schematic representation of major relationships among primate lentiviruses and diversification. Nonhuman primate lentiviruses have entered the human population at least five times. HIV-1 groups M, N, and O are related to chimpanzee SIV and represent three independent jumps into humans. HIV-1 group M has spread worldwide and has diversified extensively to produce groups called clades. HIV-2/A and HIV-2/B are not as common as HIV-1. Their closest relatives are SIVs that infect sooty mangabeys (*Cercocebus atys*).

REV

REV is a small (116 aa) RNA-binding protein. REV binds to a region of secondary structure called the REV response element or RRE. In HIV-1 the RRE is within the *env* gene (near the junction of the SU- and TM-coding regions). The position of the RRE means that it is present in unspliced mRNA as well as in singly spliced mRNA but is absent from multiply spliced mRNAs (Fig. 37.7). When REV is present in the nucleus, it binds to the RRE, oligomerizes and facilitates export of unspliced and singly spliced mRNA

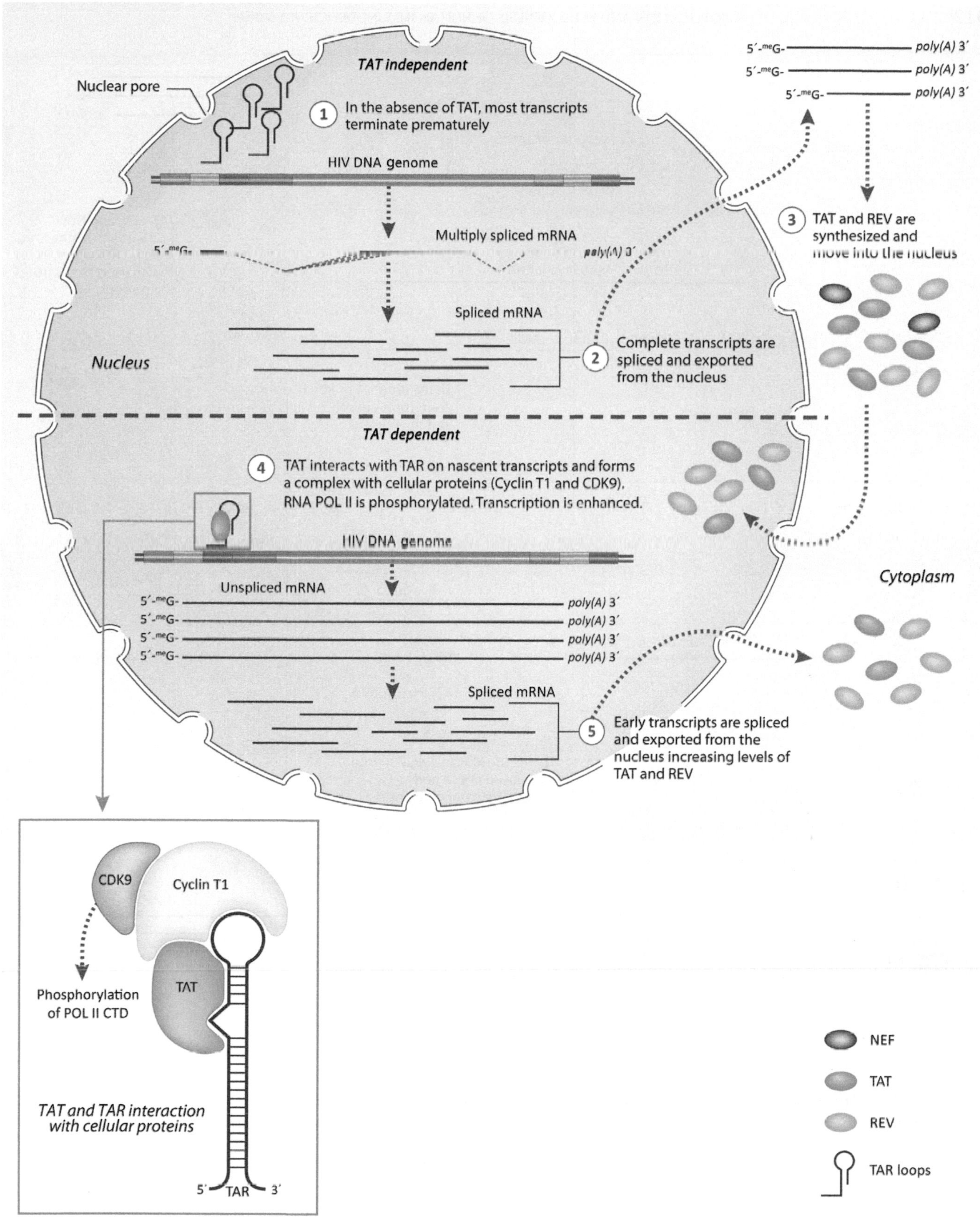

FIGURE 37.6 TAT transactivation. TAT is one of the first proteins expressed from the integrated HIV genome. In the absence of TAT, transcription initiates but is poorly processive and most transcripts terminate close to the promoter. The few full-length transcripts are multiply spliced to produce mRNAs that encode TAT. After translation, TAT moves into the nucleus and binds the TAR RNA stem-loop. Cellular proteins cyclin T1 and CDK9 join the complex and activate (by phosphorylation) RNA pol II. Transcription becomes highly processive.

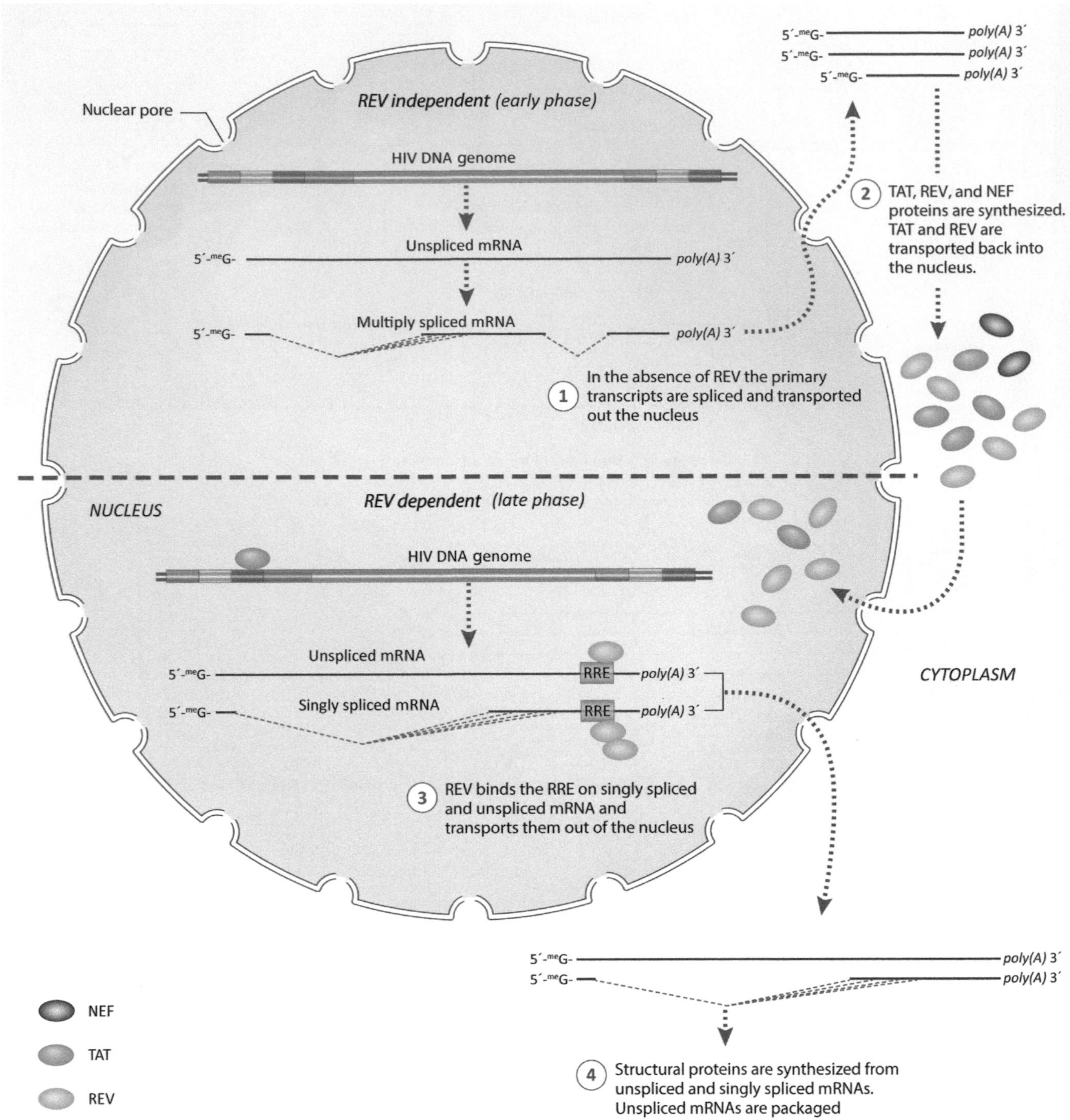

FIGURE 37.7 HIV REV mediates export of unspliced and singly spliced mRNA from the nucleus. REV protein is expressed from spliced mRNAs, early in the replication cycle. REV moves into the nucleus and binds to viral mRNAs containing an RNA element called the RRE. The RRE is found on unspliced and singly spliced mRNAs. Multiple copies of REV bind to the RRE. In the cytoplasm, REV is released from the RRE and cycles back to the nucleus. REV is required for the production of structural proteins.

from the nucleus. In the absence of REV all viral mRNAs are multiply spliced before leaving the nucleus. Therefore, without REV, no structural protein can be synthesized. REV disassembles from mRNA in the cytoplasm and moves back into the nucleus to bind additional mRNA targets. Like TAT, REV is essential for HIV replication. Neither TAT nor REV is a structural protein, neither is found in virus particles.

OTHER HIV-1 PROTEINS

HIV-1 and other primate lentiviruses encode additional proteins required for efficient replication and a comparison of primate lentiviruses genomes (Fig. 37.4) reveals some variation. A common trait of these auxiliary genes and proteins is their small size and lack on enzymatic activity. They tend to interact with multiple cellular proteins. All auxiliary proteins appear to be required for optimal replication in the natural host, but all are expendable in some types of cultured cells (although virus infectivity may be affected up to several 1000-fold, depending on the accessory gene and the type of infected cell). A primary (although not exclusive) feature of primate lentivirus auxiliary proteins is their role in antagonizing host antiviral proteins. Among the primate lentiviruses, HIV-1 encodes virus infectivity factor (VIF), viral protein R (VPR), viral protein U (VPU), and NEF proteins; HIV-2 lacks VPU but encodes viral protein X (VPX); SIVagm lacks VPU and VPX.

Virus Infectivity Factor

VIF is required for production of infectious virions from some cells types. The cell-type-dependent phenotype of VIF mutants made it challenging to determine VIF function. A model of VIF activity emerged whereby *VIF was hypothesized to inhibit inclusion of an antiviral protein into virions*. The packaged antiviral protein then functioned to prevent infection of a new cell (Figs. 37.8 and 37.9).

The antiviral factor antagonized by HIV-1 VIF is a single-stranded DNA cytidine deaminase called APOBEC3G (A3G). A3G blocks retrovirus replication by damaging newly synthesized DNA during reverse transcription. The damage is caused by deamination of cytosine to generate uracil (Fig. 37.9). Key to its antiviral activity, A3G must be present at the "right time and place": It must be present within the infecting virion. HIV-1 VIF counteracts A3G by preventing its incorporation into virions (by targeting it for proteosomal degradation). The reason that VIF is cell-type-dependent is that some cells do not express A3G (and some species do not encode A3G). If a cell does not express A3G, there is no need for VIF. Thus we have a host antiviral factor (A3G) and viral protein (VIF) to counteract it. Defenses and counter-defenses in action!

Viral Protein U

VPU is encoded by HIV-1 but is absent from HIV-2 and most SIVs. This small (81 amino acid) protein is a key player in the evolution of HIV-1, because of its ability to *antagonize the interferon-inducible antiviral protein, tetherin, in human cells (Chapter 6: Immunity and Resistance to Viruses)*. VPU is synthesized at high levels in the infected cell and forms membrane-associated multimers. VPU has not been detected in virus particles. HIV-1 VPU mutants are defective late in the replication cycle: Particles fail to complete budding and maturation.

Early studies of VPU mutants revealed a cell-type-dependent phenotype, suggesting that it interacts with a cellular factor expressed in some, but not all human cells. Treatment of so-called "permissive" cells with interferon changes them to a restrictive type (an indication that the inhibitory cell factor in an interferon-inducible protein). Interestingly, VPU can also increase virus production of highly divergent retroviruses, an indication that it antagonizes a cellular factor that inhibits replication of many retroviruses. The molecular mechanism of VPU remained a puzzle until 2008, when two research groups identified a protein, now called tetherin, that is antagonized by VPU. A great deal of evidence points to tetherin as the mysterious restriction factor: VPU is required for HIV replication when cells express high levels of tetherin, but is dispensable in cells expressing low levels of tetherin; use of silencing RNA to reduce tetherin expression changes a restrictive cell to a permissive one; the ability of VPU to counteract tetherin is species specific. HIV-1 VPU specifically counteracts human tetherin. Tetherin from some nonhuman species can restrict HIV-1 release but cannot be counteracted by HIV-1 VPU. Finally, tetherin restricts the replication of many retroviruses as well as other enveloped viruses including ebolaviruses, arenaviruses, and herpesviruses, and these viruses have evolved their own tetherin antagonists. There is also evidence, from genetic studies of primates, that tetherin itself is undergoing *rapid positive selection*, possibly in response to lentivirus infections.

Another activity of VPU is its ability to bind to the CD4 receptor in the ER, mediating its degradation. This activity is important to the HIV replication cycle, as the presence of CD4 in the ER sequesters the envelope precursor gp160 therein.

Viral Protein R

VPR is a 96-amino acid protein incorporated into virus particles by interactions with GAG. The need for VPR is cell-type specific. HIV-1 VPR mutants replicate in proliferating peripheral blood mononuclear cells or cultured T-cell lines. But VPR mutants replicate poorly in nondividing monocyte-derived macrophages (MDM). VPR is part of the PIC and this may explain some of its activity in nondividing cells.

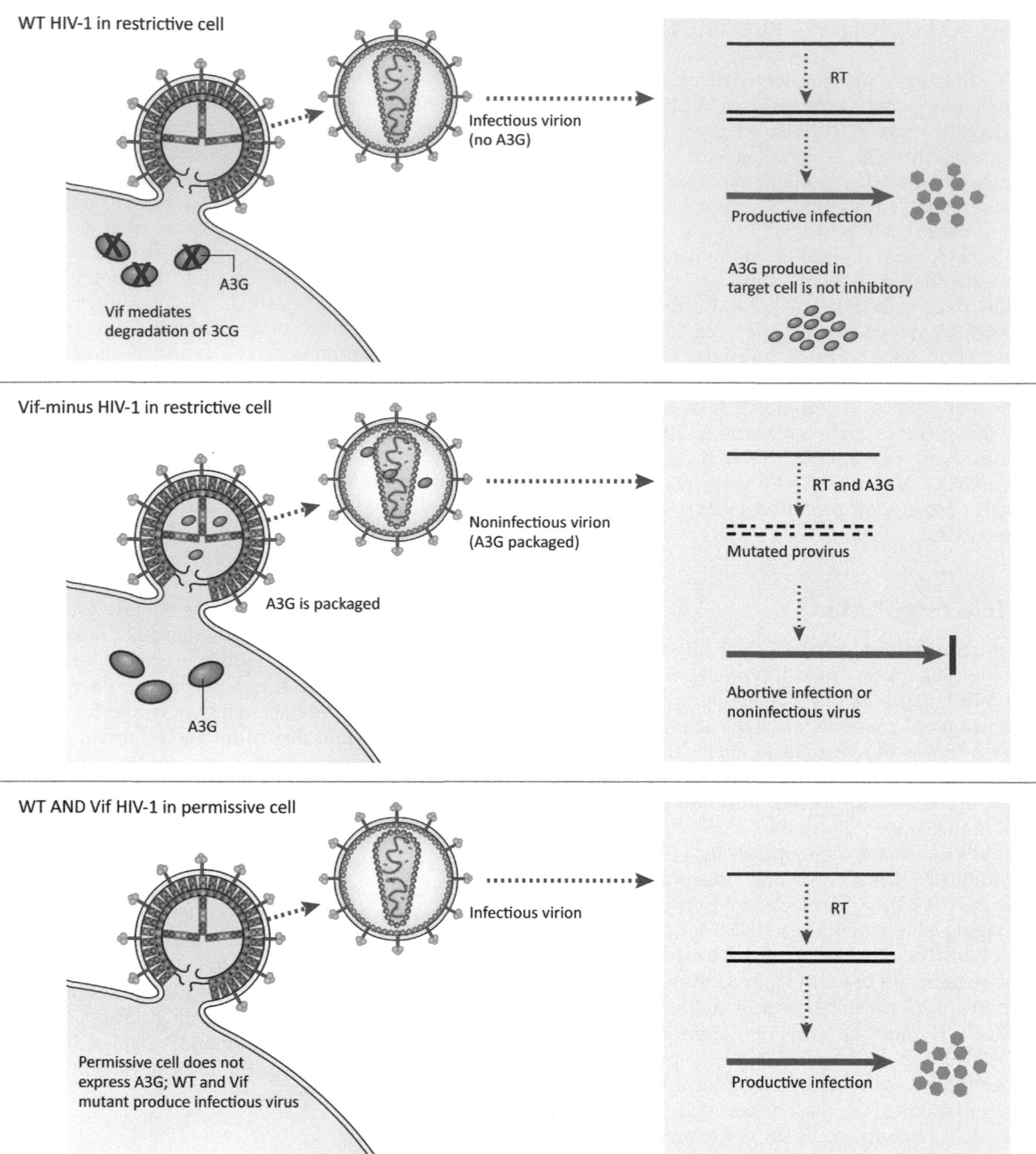

FIGURE 37.8 Interactions of HIV VIF and host protein APOBEC3G (A3G). (APOBEC stands for apolipoprotein B mRNA editing enzyme, catalytic polypeptide-like.) A3G is a single-stranded DNA cytidine deaminase that restricts retroviral replication. If A3G is present within an infecting virion, it deaminates newly synthesized retroviral DNA, a process that is mutagenic. In the top panel HIV VIF mediates degradation of A3G, inhibiting its incorporation into virions. In the middle panel A3G is incorporated into virions and deaminates cytidine during reverse transcription. In the bottom panel the cell producing virions does not express A3G, thus wild-type or mutant viruses lacking VIF produce infectious virions.

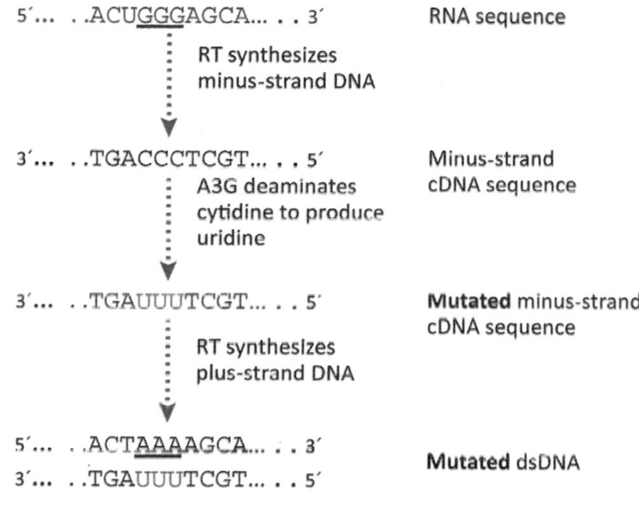

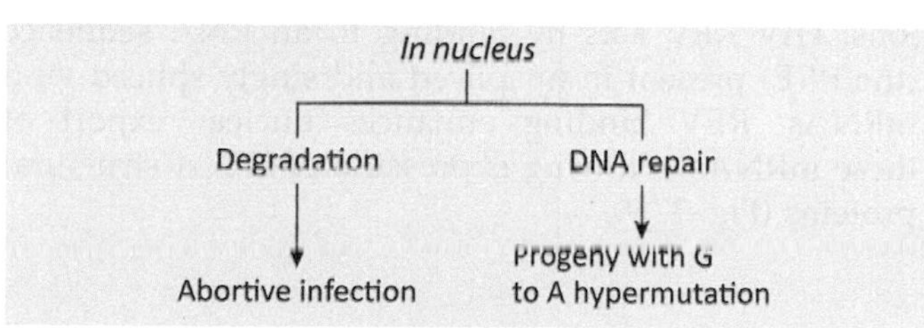

FIGURE 37.9 Antiretroviral activity of A3G. If A3G is present during reverse transcription, it deaminates cytosine (C) to generate uracil (U) in newly synthesized first-strand cDNA. During synthesis of the second DNA strand uracil base pairs with adenine (A). The ds DNA product containing uracil may be degraded by repair enzymes in the nucleus. If the retroviral DNA is integrated, U will be removed and replaced with thymine (T). Thus the original bases GGG become mutated to AAA.

However, VPR also acts by usurping the host ubiquitin-proteasome-mediated protein degradation pathway, modulating the activities multiple host proteins. Targets of VPR-mediated degradation appear to include proteins involved in DNA repair, although how this facilitates HIV-1 replication in MDM is unclear.

Viral Protein X is Unique to HIV-2

HIV-2 encodes auxiliary protein VPX. VPX is related to VPR (and may have arisen through gene duplication of VPR). Like VPR, VPX is incorporated into virions. HIV-2 VPX is required for efficient infection of nondividing cells (e.g., MDM and DCs). Like VPR, VPX activates/modulates a ubiquitin ligase complex. One target of VPX appears to be an interferon-inducible host cell restriction factor called Sterile Alpha Motif and HD Domain Protein 1 (SAMHD1), an enzyme with important roles in nucleic acid metabolism. SAMHD1 specifically blocks reverse transcription of retroviruses in MDM and DCs. In fact, expression of VPX in such cells renders them susceptible to infection by retroviruses other than HIV-2.

NEF

NEF is small protein encoded only by primate lentiviruses. HIV NEF was named after early studies suggested that it was a negative factor for HIV-1 replication. Although subsequent studies did not support the early data, the name stuck. The *nef* gene is located at the 3′ end of the HIV genome and partially overlaps the 3′ LTR. One domain important for NEF activity is its amino-terminus. NEF is myristylated and contains a cluster of basic amino acids that mediate membrane interactions. Another important NEF domain is a cluster of amino acids (Pro-X-X-Pro; a so-called Src homology 3 domain) that mediates binding to several important cellular kinases (e.g., Lyn, Hck, and Lck); NEF binding modulates the catalytic activities of these kinases leading to global changes in cell signaling and gene expression. NEF also disrupts endosomal trafficking in infected cells, resulting in down-regulation of important cell surface molecules, including CD4, MHC-I, CD8, and the CD3 T-cell receptor complex.

If NEF modulates signaling and protein expression patterns of immune cells, we might expect that NEF mutants would differ from wild-type virus as regards pathogenesis, and this is indeed the case. An intact *nef* gene is crucial for pathogenesis of HIV and SIV. In studies of SIV a *nef* deletion was used to generate an attenuated vaccine. The vaccine induced strong antibody responses and the attenuated virus was cleared. Direct evidence for a role of NEF in HIV-1 pathogenesis comes from studies of long-term nonprogressors, a small cohort of HIV-1-infected patients that develop disease very slowly. The HIVs infecting these patients have defective *nef* genes. The ability of NEF to modulate cell signaling and receptor proteins appears to activate key cells in the immune system. Continuous activation of immune cells is detrimental and is linked to the pathogenesis of HIV-1.

Another activity of the NEF proteins of many SIVs (*but not HIV*) is to antagonize tetherin. HIV NEF is unable to antagonize tetherin as it lacks a "tetherin-interacting" domain. However, recall that HIV VPU does antagonize tetherin. This points to the importance of tetherin as an antiviral factor, and the flexibility of viruses to counter its inhibition.

SUMMARY OF HIV REPLICATION

Before discussing HIV-1 pathogenesis, a brief review of the major steps in the HIV replication cycle is warranted. Details of these processes have been described elsewhere.

Binding/Entry

SU is the HIV attachment protein. It binds to its primary receptor, the immunoglobulin family member protein, CD4. CD4 expression restricts HIV replication to CD4+ T-lymphocytes, monocytes, and DCs. Following interaction with CD4, SU undergoes conformational changes that unmask binding sites for coreceptors CCR5 and CXCR4. Coreceptor interaction is required for entry and probably serves to place the virion in close proximity to the PM. The next step in the process is insertion of the fusion domain of TM into the PM. Fusion takes place at the PM, releasing the core into the cytoplasm.

Reverse Transcription

The first viral product to be synthesized is the double-stranded DNA copy of the genomic RNA. The process is similar to other retroviruses, as described in Chapter 36, Family *Retroviridae*. Key proteins in reverse transcription include heterodimeric RT and NC. However, the process takes place in association with other proteins in the infecting core.

Nuclear Entry and Integration

Lentiviruses are unique among retroviruses in their ability to replicate in nondividing cells. This is a result of the ability of the ~56 nm diameter PIC to actively move across nuclear pores to gain access to cellular chromatin. It is not entirely clear which viral components are involved the process. Suggested players include IN, VPR, and CA.

The IN protein is the key player in integration of all retroviruses. HIV IN enters the nucleus as part of the PIC, in association with ds DNA. IN catalyzes removal of two nucleotides from the 3′ end of each viral DNA strand and makes a staggered double-stranded cut in the target host cell DNA. Finally IN mediates DNA strand ligation.

Integration is generally considered to be the final step in the *early* phase of the retrovirus life cycle. In the case of HIV-1, entry into quiescent cells may result in a long "latent" period before the replication cycle continues. This is a problem in terms of treatment and "elimination" of HIV infection, as cells may harbor transcriptionally inactive proviruses for many years.

Transcription and Translation

Retroviral transcription is dependent on recognition of the retroviral promoter by host cell transcription machinery. The HIV promoter is relatively weak (compared to the strong promoters of some oncogenic retroviruses). Transcripts are initiated but not efficiently extended in the absence of the HIV-1 TAT protein (Fig. 37.3). The level of transcription in the absence of TAT is not sufficient to support virus replication. When present, TAT binds to the TAR region of the nascent viral transcripts and promotes assembly of host proteins that activate transcription.

Early HIV transcripts are multiply spliced to produce mRNAs that encode a variety of regulatory and auxiliary proteins. But unspliced and single-spliced HIV-1 transcripts are required for production of virions. HIV REV acts by binding to an RNA sequence (the RRE) present in unspliced and singly spliced viral mRNAs. REV binding enhances nuclear export of these mRNAs, allowing expression of HIV-1 structural proteins (Fig. 37.9).

Assembly/Release/Maturation

HIV-1 assembly takes place at the plasma membrane (Fig. 36.9). Budding particles contain uncleaved $pr55^{Gag}$ and $pr160^{Gag\text{-}pol}$ at membrane sites decorated by SU and TM. Other viral and cellular proteins, for example, CypA and VPR associate with $pr55^{Gag}$. Two copies of unspliced, capped, and polyadenylated HIV-1 mRNA are brought to the assembly site by interaction with the NC domain.

HIV-1 release requires cooperation of cellular and viral proteins. HIV-1 p6 contains a highly conserved amino acid motif PTAP called a "late domain" because it functions at the very end of the replication cycle. Without a late domain, immature HIV particles remain attached to the plasma membrane. Late domains function by interacting with a subset of components of the cellular ESCRT machinery.

Assembly and release of HIV-1 clearly takes place at the PM of T-cells. However, the process is not as clear in MDM. In these cells HIV-1 appears to assemble at internal sites as well as at the PM. However, it has recently been determined that the "intracellular" locations of HIV assembly in MDM are actually deep invaginations of the PM.

HIV virions become infectious upon maturation, a process that occurs after release. The PR domain of $pr160^{Gag\text{-}pol}$ is activated, generating mature structural proteins. MA remains associated with the envelope, CA forms the capsid. Within the capsid, RNA is bound by NC. RT and IN are also present within the capsid.

OVERVIEW OF HIV PATHOGENESIS

HIV pathogenesis is complex and is impacted by both host and viral genetics. Some untreated, infected individuals remain clinically healthy for decades while others develop AIDS within months. What is very clear is that inhibiting HIV replication, by use of highly effective drug cocktails, is profoundly successful in preventing both disease and transmission! In the following sections we outline some of the basic interactions between HIV and host cells that contribute to the disease process in untreated individuals.

HIV-1 Infection of CD4+ T-Lymphocytes

T-lymphocytes are major players in the immune response. They are called "T" cells because they develop in the thymus. There are different types of T-cell; those most important in HIV replication and pathogenesis are the so-called T-helper cells (Th cells). Th cells express the CD4 protein on their cell surface (thus another name for a Th cell is CD4+ T-cell).

Key to their role in immunity, Th cells display receptors that recognize foreign antigens. A naïve CD4+ T-cell is one that has not yet encountered its cognate foreign antigen. When a naïve Th cell binds to its specific antigen, it proliferates (antigen recognition is a complex process that is best reviewed in an immunology textbook). Proliferating Th cells are in an activated state; they secrete cytokines and productively interact with other cells. After many generations, activated T-cells differentiate into memory T-cells. Memory Th cells are long-lived cells that recognize the same antigen as their progenitors.

All Th cells express the CD4 protein on their surface, thus all can bind HIV. However, the results of infection vary, depending on whether or not the Th cell is a naïve or memory cell. Memory CD4+ T-cells express the chemokine receptors CXCR4 and CCR5. Memory CD4+ T-cells are present at high levels in the gastrointestinal tract and lung (where they are likely to contact microbial invaders). Most memory CD4+ T-cells are activated and highly susceptible to HIV-1 infection. The result is production of large numbers of virions and death of the infected cells. Within a few days of initial infection, there is wholesale depletion of gut-associated T-cells.

Some memory Th cells are resting cells. They can also be infected by HIV-1 but the result is integration, followed by a latent infection (no viral proteins are expressed). These cells serve as important long-term reservoirs for the virus. These latently infected cells are directly responsible for the inability to "cure" most HIV-1 infections even after prolonged treatment with antiretroviral drugs.

Naïve Th cells are the predominant type found in peripheral blood. Naïve Th cells express the chemokine receptor CXCR4. Thus at early times postinfection most naïve Th cells are resistant to HIV-1 infection. However, in some infected individuals, X4 variants of HIV-1 will emerge, a process called coreceptor switching. These viruses infect and deplete the naïve T-cell population; this is associated with accelerated progression to AIDS.

What about the innate immune responses that should control early HIV-1 replication? Unfortunately, early, innate responses (Chapter 6: Immunity and Resistance to Viruses) induce production of inflammatory molecules that increase the numbers and activation status of Th cells at the site of infection.

HIV-1 Infection of Monocytes and Macrophages

Monocytes and macrophages express CD4 and CCR5 on their surface. However, not all HIV isolates replicate well in these cells. Isolates from the blood usually replicate poorly in cultured macrophages; in contrast isolates from tissue or the central nervous system replicate much more productively in these cells. In cultured cells, vigorous replication of HIV requires fully differentiated macrophages; these are generally considered to be analogous to tissue macrophages in vivo. Brain-resident macrophages are called microglia cells and their infection may lead to HIV-associated encephalitis.

HIV-1 Infection of Dendritic Cells

DCs capture, transport, and present antigens to CD4+ and CD8+ T-lymphocytes. There are two main types of DCs, myeloid DCs (MDCs) and plasmacytoid DCs (PDCs). MDCs are found in tissues, including skin and intestinal and genital tract mucosa. (MDCs in the skin are called Langerhans cells.) MDCs at mucosal surfaces are the first line of defense against infectious agents encountered at these sites. Their job is to engage infectious agents, degrade them, and carry processed antigens to draining lymph nodes rich in T-lymphocytes. Unfortunately, MDCs do not degrade HIV-1 efficiently and thus bring infectious virus to CD+ T-cells in the lymph nodes.

Disruption of Immune Function

Regulation of the immune system is a very complex process. Not only are many different cell types involved, but gene expression patterns change in response to various stimuli. Th cells, monocytes, macrophages, and dendritic cells are all susceptible to HIV infection. But depending on the activation state of the cell, infection may be productive or latent. A confounding factor in truly understanding HIV pathogenesis is that even in the absence of direct cell killing, infection can change the ability of immune cells to do their jobs. For example, CD8+ T-cells may be less able to kill their targets, even though they are not directly infected by HIV-1.

Immune Activation

Another factor in HIV-1 pathogenesis involves its ability to persistently activate the immune system. As mentioned earlier, activated Th cells are targets for productive infection. But persistent immune activation is damaging, even in the absence of direct infection. There is evidence that uninfected, but persistently activated CD4+ T lymphocytes undergo enhanced apoptosis, contributing to overall depletion of T-cells.

There are several lines of evidence-linking HIV-1 NEF protein to immune activation and pathogenesis. It was shown early in the AIDS epidemic that a few patients were so-called long-term nonprogressors (infected individuals that did not develop immunodeficiency). Analysis of their HIVs revealed naturally occurring NEF mutations. Another line of evidence for the role of immune activation in disease development comes from studies of SIV infections of natural hosts. Where SIV and monkey host are well adapted to one another, there is little evidence of generalized immune activation and little disease.

HIV TRANSMISSION

HIV is transmitted by infected body fluids. Thus HIV is transmitted most often by sexual contact, sharing needles, or from mother to child (during birth or through breast milk). Sexual transmission can be avoided by abstinence or greatly reduced by practicing safe sex (i.e., using condoms). Risk of sexual transmission increases between persons with other sexually transmitted diseases. Treatment of HIV-infected women can greatly reduce risk of transmission to their babies. Prophylactic treatment with antiretroviral drugs is often recommended after a possible exposure. There is increasing evidence that treatment to reduce viral loads also reduces risk of transmission.

DISEASE COURSE

The disease course of HIV-infected individuals is variable, but for the most part, untreated HIV infection leads to AIDS and death. Within 2–4 weeks after infection with HIV, a person will usually develop a high-titer viremia and a generalized febrile or flu-like illness that may be accompanied by lymphadenopathy. However, some patients will be viremic, but experience no symptoms. During acute infection, much of the gut-associated lymphoid tissue is lost and circulating CD4+ T-cell numbers are reduced. Immune responses develop but do not clear the infection; in particular, levels of neutralizing antibody are slow to develop and titers remain quite low. T-cell numbers rebound, but usually remain below preinfection levels. Patients recover from the acute illness but virus continues to replicate and can be readily transmitted.

The best predictors of the course of infection are numbers of circulating CD4+ T-cells and virus load. Lower CD4+ T-cell counts and higher virus loads are predictors of a more rapid onset of AIDS-defining opportunistic infections or cancers. In terms of T-cell counts, the definition of AIDS is a CD4+ T-cell count of less than 200 cells/μL of blood. The time to development of AIDS may be as short as 6 months or as long as several decades (even in some untreated patients). Treatment with antiretroviral drugs drastically changes the progress of the infection. As highly effective drugs, with fewer side effects, have been developed, the trend is to treat patients earlier. At one time the trigger for treatment was a CD4+ T-cell count of <350 cells/ μL blood; however, some patients may choose to begin drug treatment when their T-cell counts are higher. It is clear that antiretroviral drugs drastically reduce viral load, preserving healthy levels of CD4+ T-cells. There is also accumulating evidence that reducing virus load significantly reduces transmission!

ANTIRETROVIRAL DRUGS

Development of drugs to treat HIV infection is an ongoing process. Currently, there are six classes of drugs actively used to treat HIV infection. Most are used in multidrug cocktails. Many of the newest drugs used to treat HIV were developed in a targeted manner, based on understanding the specific molecular interactions required for each step in the replication process.

Entry/Fusion Inhibitors

Maraviroc binds to the CCR5 coreceptor preventing an interaction with HIV gp120. Maraviroc does not inhibit strains of HIV that bind to other chemokine coreceptors.

Enfuvirtide (brand name Fuzeon) is a 36-amino acid peptide based on the sequence of HIV-1 TM. The peptide interacts with TM (after attachment) thereby inhibiting formation of the specific structure required for fusion.

Nucleoside/Nucleoside Reverse Transcriptase Inhibitors

There are many approved nucleoside reverse transcriptase inhibitors (NRTIs), including azidothymidine or AZT, the first drug approved to treat HIV. NRTIs are molecules recognized by RT and inserted into the growing DNA chain during reverse transcription. After insertion of an NRTI, the DNA strand cannot be elongated. Therefore NRTIs can prevent establishment of an infection.

Nonnucleoside Reverse Transcriptase Inhibitors

These are compounds that directly bind to RT and inhibit its activity. They are not incorporated into DNA. There are currently four approved drugs in this class.

Integrase Inhibitors

These drugs inhibit insertion of HIV DNA into the host chromosome. Recall that integration is an obligatory step in the HIV replication cycle.

Protease Inhibitors

PR inhibitors were the first successful "designer" antiviral drugs. PR inhibitors were developed based upon a clear understanding of the *molecular structure and catalytic activity of PR*. They are competitive inhibitors of HIV-1 PR and function by inserting into the substrate-binding cleft of PR. The first PR inhibitors were derivatives of small peptides. There are eight approved PR inhibitors and in the presence of these drugs, noninfectious virions are produced from cells containing integrated copies of the HIV genome. PR inhibitors are extremely effective when used in combination NRTIs, perhaps because they target a different step in the virus replication cycle.

Combination Drugs

Combination HIV medicines contain two or more HIV medicines from one or more drug classes. Some of the currently approved combinations include up to four active compounds.

VACCINES

It has been over 30 years since the discovery of HIV and a vaccine remains elusive, despite considerable effort and investment by vaccine companies, governments, and a variety of nonprofit agencies. Challenges to development of an efficacious vaccine include:

- *Correlates of protection remain undefined.* Because the healthy immune system does not control HIV during a natural infection, we have no road map to the types of immune responses that might be protective. There are small groups of individuals that resist HIV infection or remain healthy without treatment. These individuals are intensively studied for clues about natural mechanisms of resistance that could be "mimicked" by a vaccine.
- *Genome variation.* HIV is genetically variable and evolves within patients and on a worldwide population scale. It is likely that no single vaccine will provide protection against all groups or clades of HIV. RT has a high intrinsic mutation rate as it does not have a proofreading mechanism. RNA recombination is frequent (perhaps because the process of reverse transcription requires translocation of newly synthesized products from one position to another on the template). Additional mutation can be generated by the antiviral activities of cellular cytidine deaminases and cellular DNA repair machinery.
- *Immune escape.* Genome variation gives rise to mutants that escape established immune responses in infected individuals. Viruses that escape vaccine-induced immune responses might emerge if protection is not complete.
- *Primate models are expensive, ethically challenging* and have not yet provided a clear path to an effective vaccine. However, they have provided a wealth of information about pathogenesis of, and immune responses to lentiviral infection.
- *Regulatory hurdles for human vaccine trials are high and trials are expensive.* Over 200 vaccines have been tested for safety and immunogenicity but only five have been tested for efficacy and for the most part the results have been disappointing. To determine efficacy, half of the individuals in a trial will receive a placebo, not the candidate vaccine. Given the lifelong consequences of HIV infection, it is imperative that all individuals enrolled in such trials be provided with education and support aimed at avoiding exposure to HIV. Volunteers also need access to follow-up care and treatment should they become infected.

In this chapter we have learned that:

- The most common routes of HIV transmission are sexual contact and shared use of contaminated needles. HIV can also be transmitted from mother to child (during birth or by breast feeding). HIV can also be transmitted through nonsterile medical instruments or blood transfusions (screening of blood and blood products has virtually eliminated this mode of transmission).
- Virus transmission can be avoided by practicing "safe sex" (using condoms) and not sharing needles. Being aware of one's infection status (by getting tested) and seeking treatment can be very effective in reducing spread. Strict attention to infection control in medical settings is important to avoid spread of many infectious agents, including HIV. Blood and blood products should be tested.
- HIV infects cells that display the CD4 receptor on their surface (a coreceptor is also needed). Cells that express CD4 include a subset of T-lymphocytes, monocytes, macrophage, and dendritic cells.
- Coreceptors for HIV are chemokine and cytokine receptors including CCR5 and CXCR4.
- HIV is tropic for cells that are major players in the immune response. HIV kills some infected cells and alters the function of others. HIV causes disease by damaging the immune system; this leads to immunodeficiency (AIDS). Immunodeficient persons are at high risk for, and cannot control, many viral, bacterial, and fungal diseases.
- HIV TAT is a protein that is required for replication. It is a transcriptional activator that increases the production of HIV transcripts by ~100-fold.
- HIV and SIV encode a variety of auxiliary proteins. While they have different activities, they also share some features: They are small proteins that lack enzymatic activity. They work by interacting with various cellular proteins. Interaction with cellular proteins can result in antagonizing antiviral defenses, altering cell growth or signaling pathways, and inhibiting certain proteins from being expressed on the cell surface.
- Viral restriction factors are host-encoded proteins that have evolved to inhibit virus replication. Some examples include tetherin, A3 family proteins and SAMHD1. Retroviruses can be inhibited by restriction factors but some have evolved specific proteins (i.e., HIV VIF and VPU) to antagonize them. Retroviral restriction factors are important in limiting cross species transmission.
- Different classes of drugs target different events in the HIV replication cycle. Entry inhibitors interfere with attachment, fusion inhibitors interfere with fusion, inhibitors of reverse transcription include nucleoside/nucleotide RT inhibitors. PR inhibitors block virion maturation.
- It is important to use drug cocktails to treat HIV infection because of the combination of high mutation rate and high virus load. It is less likely that drug resistant mutants will arise when 2 or 3 different viral proteins are targeted at the same time.

CHAPTER

38

Family *Hepadnaviridae*

OUTLINE

After reading this chapter, you should be able to answer the following questions:

- What are the major characteristics of the members of the family *Hepadnaviridae*?
- By what process is the hepatitis B virus (HBV) genome synthesized?
- Describe the three particle types found in the blood of persons with chronic HBV infection.
- What are the major host cells for HBV replication? How does this impact the ability to study HBV in cultured cells?
- What animal models are used to study hepadnavirus pathogenesis?
- Why is HBV vaccination recommended for infants?
- What groups of adults are at highest risk for acquiring HBV?

The World Health Organization (WHO) estimates that 240 million people worldwide are chronically infected with hepatitis B virus (HBV) and that more than 686,000 die every year due to complications that include cirrhosis and liver cancer. HBV, a member of the family *Hepadnaviridae,* is one of several important human hepatitis viruses. HBV is the etiologic agent of hepatitis B, or blood-borne hepatitis. HBV is highly infectious and is transmitted by blood and body fluids. HBV pathogenesis is variable and the outcome of infection largely depends on host immune responses. A robust immune response clears the infection. This process is mediated by cytotoxic T cells, causing transient liver damage. Weaker immune responses may fail to clear the virus, resulting in a long-term, chronic infection. Chronic infection is often asymptomatic as HBV is noncytopathic for hepatocytes. Thus infected individuals may remain healthy carriers for years or decades, but may eventually develop liver damage or cancer.

Hepadnaviruses are enveloped viruses with $T=4$ icosahedral capsids (Fig. 38.1). Genomes are relaxed circles of partially double-stranded DNA (Box 38.1). Only a handful of hepadnaviruses have been identified, and known hosts include birds, mammals, fish, and frogs. The most unique aspect of hepadnavirus replication is the synthesis of genomes by reverse transcription. Hepadnaviruses have a definite preference for replication in differentiated hepatocytes thus it was a long and somewhat difficult process to determine even the most basic aspects of the hepadnavirus replication cycle. HBV does not infect animals (other than apes) so natural infections of woodchucks, ground squirrels and ducks, with their respective hepadnaviruses have provided important insights into replication and pathogenesis (Box 38.2).

GENOME ORGANIZATION

The HBV genome is quite small, only ~3.2 kbp. Within virions, hepadnavirus genomes have an

Viruses. DOI: http://dx.doi.org/10.1016/B978-0-12-803109-4.00038-6

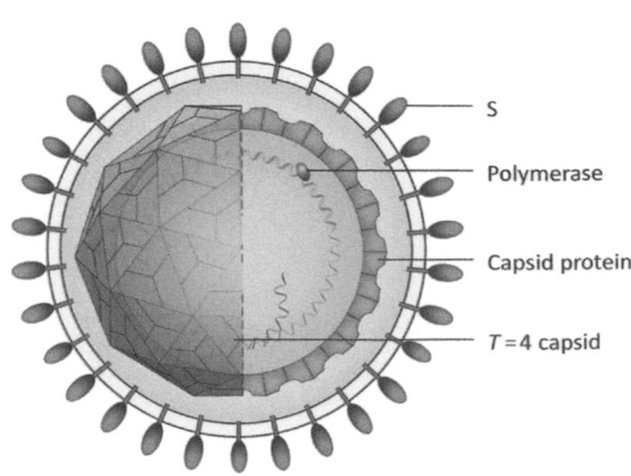

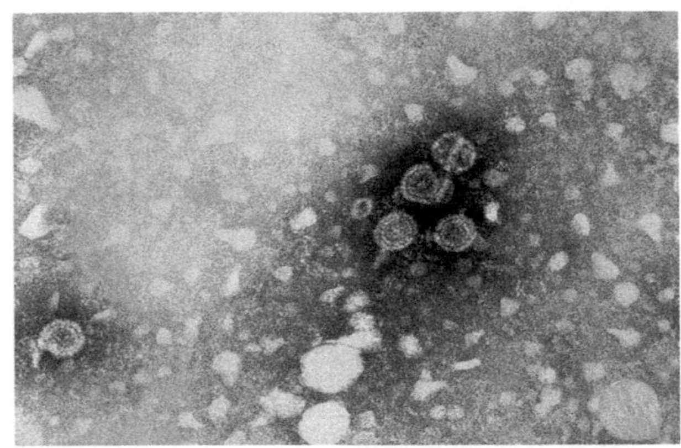

FIGURE 38.1 Virus structure (left). The transmission electron microscopic image (right) shows hepatitis B virions. The large round virions are known as Dane particles. *Content providers: CDC/Dr. Erskine Palmer. CDC Public Health Image Library Image ID# 270.*

BOX 38.1

GENERAL CHARACTERISTICS

Hepadnaviruses are enveloped viruses with icosahedral capsids/cores. The HBV virion is ~42 nm in diameter with a core of ~34 nm. A single capsid (C) protein assembles to make a structure with $T=4$ icosahedral symmetry. A set of overlapping envelope glycoproteins (Large, Medium, and Small S) are associated with the envelope.

Genomes are relaxed circles of partially double-stranded DNA (~3.2 kbp) that are synthesized by reverse transcription. Early after infection genomes are converted to cccDNA in the nucleus. There are four ORFs. The largest encodes the polymerase (P). P has three domains: The amino terminus encodes a protein primer (terminal protein) used for priming minus-strand DNA synthesis. P also has a reverse transcriptase domain and an RNase H domain. The compact nature of the genome is achieved by extensive use of overlapping reading frames.

Hepadnavirus mRNAs are capped and polyadenylated. The genome has multiple promoters but only a single poly(A) addition site.

BOX 38.2

TAXONOMY

Family: *Hepadnaviridae*

Genus: *Orthohepadnavirus* (HBV and isolates from woodchuck, ground squirrels, and bats)

Genus: *Avihepadnavirus* (Duck and heron hepatitis B viruses)

Other hepadnaviruses: Several isolates have been recovered from fish, and one from a frog.

unusual form: They are partially double-stranded, relaxed circles of DNA (Fig. 38.2) and the two stands are of unequal length. The longer DNA strand is the noncoding or negative (−) strand; it contains short direct repeats (*r*) at its ends. The negative strand of DNA is covalently linked to a protein at the 5′ end (the result of protein primed DNA synthesis). The coding or plus (+) strand of DNA is of variable length, but is always incomplete (shorter than the negative strand). The coding strand is linked to an RNA oligonucleotide at its 5′ end, as plus-strand DNA synthesis is primed with an RNA primer. Upon delivery to the nucleus, the relaxed circular DNA genome is converted to a covalently closed circle (ccc) of DNA that associates with histones and other cell proteins. This form of the genome is called cccDNA. It serves as a template for transcription, but is *not* replicated by DNA polymerases.

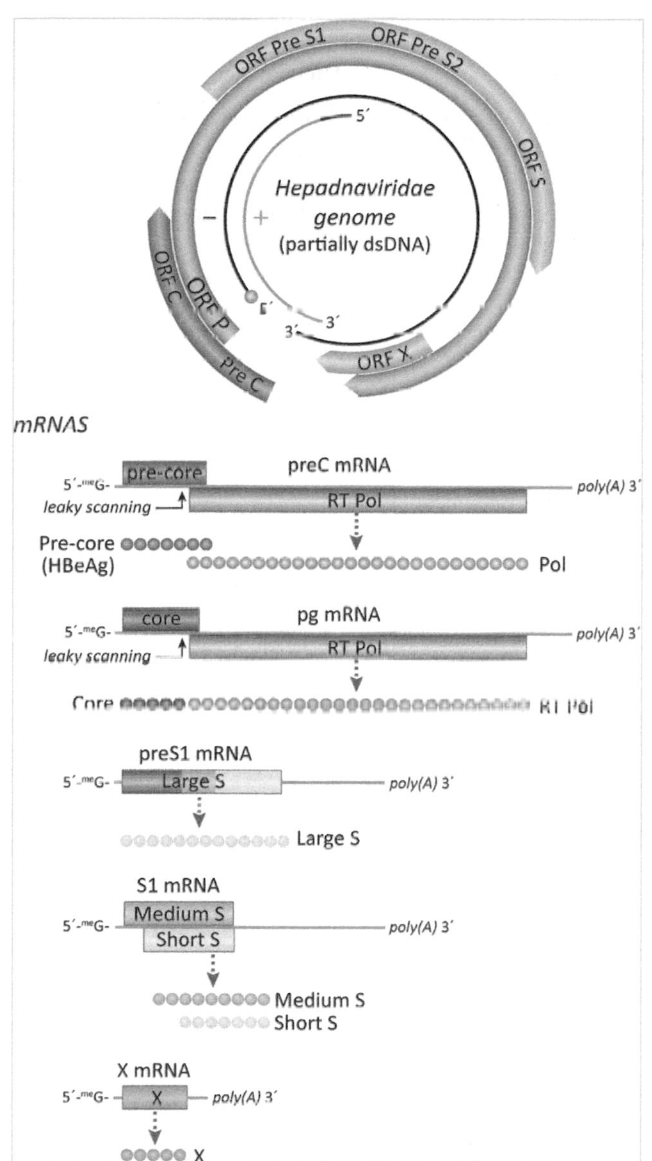

FIGURE 38.2 Genome organization. The genome is partially ds DNA. The longer, (−) strand is covalently linked to a protein primer. The shorter, (+) strand is primed by an RNA oligonucleotide. Note the extensive use of overlapping ORFs.

The HBV genome has four open reading frames (ORFs). The longest ORF encodes the polymerase (P), a reverse transcriptase (Fig. 38.2). A set of surface (S) glycoproteins is encoded by the S ORF. The C ORF encodes the core protein [and a longer "preC" ORF encodes a secreted protein (E) not present in the virion]. The small X ORF encodes a nonstructural protein with regulatory functions. An examination of the genome reveals the large extent of gene overlap characteristic of the hepadnaviruses. The S ORF is contained completely within the P ORF and all ORFs share some overlap with P. Table 38.1 lists HBV ORFs and their protein products and functions. As a number of naming schemes have been used over the years, alternative names are provided in Table 38.1.

Hepadnavirus mRNAs are capped and polyadenylated. The genome has multiple promoters but only a single poly(A) addition site. One mRNA, called pregenomic RNA (pgRNA), is longer than the cccDNA, thus contains repeated regions at the ends. pgRNA is the template for reverse transcription (by P) to form a new molecule of relaxed circular DNA. Reverse transcription occurs within a newly assembled capsid.

VIRION STRUCTURE

Hepadnaviruses are enveloped particles with a diameter of ~42 nm. The envelope contains three related glycoproteins, the large (L), medium (M), and small (S) surface proteins. (In this chapter they refer to as Large S, Medium S, and Small S.) The envelope surrounds an icosahedral, $T=4$ capsid. Capsids are ~34 nm in diameter. The partially double-stranded DNA genome, associated with a molecule of P, is found within the core.

In addition to infectious virions, two other types of particles are seen in the blood of persistently infected patients. These can be present in very large amounts. The most abundant (up to 10^{12} particles per mL serum) are 20 nm particles that are largely comprised of surface proteins. Longer filaments or rods are also ~20 nm in diameter and contain lipid and surface proteins. Neither of these particle types contains cores or DNA, thus they are noninfectious.

REPLICATION CYCLE

The general steps in the hepadnavirus replication are shown in Fig. 38.3.

Attachment/Penetration

HBV infects differentiated liver cells (hepatocytes) to establish a persistent infection. Initial interactions are based upon weak interactions with heparan sulfate proteoglycans. Stronger binding results from interaction of Large S with the liver-specific hepatic bile acid transporter: sodium-taurocholate cotransporting polypeptide. HBV uptake is not well characterized but probably occurs via an endocytic pathway followed by fusion to release the DNA-containing core into the cytosol. The core traffics to the nuclear pore and the genome enter the nucleus.

TABLE 38.1 Hepadnavirus ORF and Protein Products

ORF	Protein products	Functions	Alternative names
P	Polymerase (three domains from N to C are: terminal protein- RT-RNaseH)	Terminal protein domain primes DNA synthesis; RT synthesizes DNA; RNAse H cleaves RNA during reverse transcription process	Reverse transcriptase, pol
S	Small S	Envelope glycoprotein. Shares a common carboxyl terminus (with Medium S and Large S proteins).	Small HB surface protein (SHBs), S, HBs, S-HBsAg
	Medium S	Envelope glycoprotein.	Medium HB surface protein (MHBs), preS1, M, M-HBsAg
	Large S	Envelope glycoprotein. Large S contains the receptor-binding domain.	Large HB surface protein (LHBs), preS2, L, L-HBsAg
C/pg mRNA	Core	Core/capsid formation	HBcAg, C
preC	E	Secreted from cells, not found virion	HBe-protein, HBeAg
X	X	Regulatory protein, important in development of hepatocellular carcinoma	

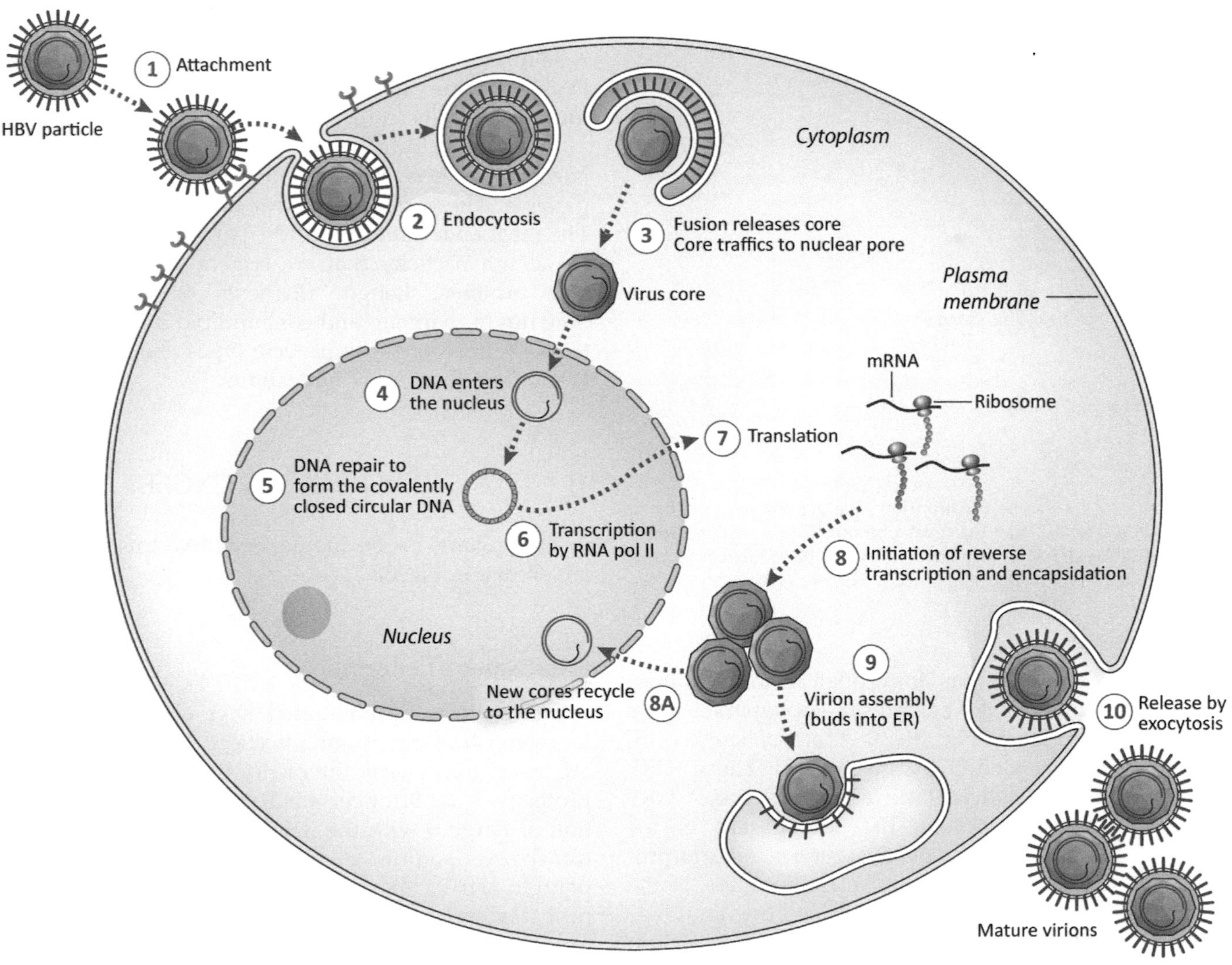

FIGURE 38.3 Hepadnavirus replication cycle.

Transcription

Within the nucleus the HBV genome is converted to a covalently closed circular molecule (cccDNA) that associates with histones and other cellular proteins. While the details are not known, it is thought that cellular DNA repair enzymes are responsible for generating cccDNA.

cccDNA is maintained as stable episome within the nucleus. It is transcribed by host-cell RNA polymerase II to generate a set of capped, polyadenylated mRNAs. Four promoters have been identified on the HBV genome but transcription start sites are variable, forming a more complex set of mRNAs. Transcription depends, in part, on liver-specific transcription factors.

preC mRNA is the longest mRNA and encodes the E protein, a secreted protein that is not found in virions, but is present in blood of infected patients (Table 38.1). This mRNA is also a source of P protein, by result of leaky ribosome scanning.

pg mRNA is slightly smaller than preC mRNA and used to produce C and P proteins. Pg mRNA is also the template for reverse transcription, the process by which relaxed circular DNA genomes are produced.

preS1 and S mRNAs overlap each other and completely overlap the P ORF. PreS1 mRNA is used to produce the Large S protein. S mRNA is used to produce Medium and Small S proteins. All S proteins share the same C-terminal amino acid sequences.

X mRNA. The promoter for X mRNA is within the P ORF. X mRNA is ~ 0.8 kbp and is used to produce the X protein. X is small nonstructural protein required for efficient virus replication. It can be expressed at high levels and appears have a variety of cell-signaling and regulatory functions. Its functions have been studied in a variety of cultured cells (as well as in transgenic animal models), but none of these models accurately reflects the natural infection of differentiated hepatocytes. X interacts with many different cellular proteins, explaining the wide variety of activities that have been ascribed to it.

Genome Replication

Hepadnavirus genomes are produced by reverse transcription (Fig. 38.4). Thus genome synthesis actually begins with transcription of capped, polyadenylated pg mRNA, the template for the hepadnavirus P. The process of reverse transcription occurs in the cytoplasm and is tightly linked to capsid formation. Genomes may be synthesized within newly formed capsids and it has been proposed that they are incomplete due to size limits of the capsid.

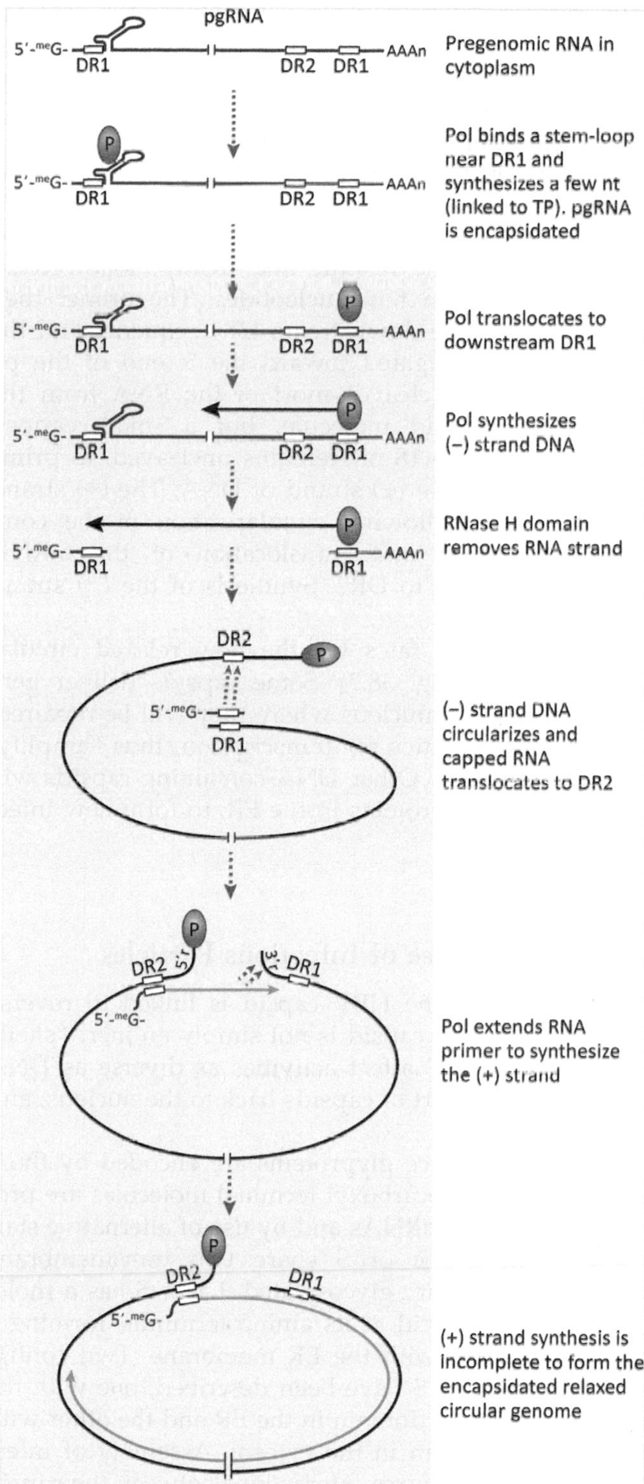

FIGURE 38.4 Process of reverse transcription to generate HBV genomes. The gray-shaded box indicates events that occur in association with capsid proteins.

HBV P, key to the process of genome synthesis, is a multidomain protein. The so-called terminal domain is nearest the amino terminus and serves as the primer for (−) strand DNA synthesis. The terminal domain is followed by a spacer region that links it to reverse transcriptase and ribonuclease H domains. Hepadnavirus RT and RNaseH function in a similar manner to their retroviral counterparts.

Cellular chaperones participate in the initial association between P and pg mRNA. The priming reaction occurs near direct repeat 1 (DR1). The initial priming reaction leads to the formation of a covalent bond between a tyrosine residue and dGMP, followed by synthesis of just a few nucleotides. The primer then translocates to the downstream DR1 sequence and the (−) strand is elongated toward the 5′ end of the pg mRNA. RNaseH cleaves most of the RNA from the DNA/RNA hybrid molecule, but a small, capped RNA fragment (~18 nt) remains uncleaved, to prime for synthesis of the (+) strand of DNA. The (+) strand is synthesized following circularization of the complete (−) strand and translocation of the mRNA primer from DR1 to DR2. Synthesis of the (+) strand is not complete.

There are two fates for the new relaxed circular DNA genome (Fig. 38.3). Some capsids deliver genomes back to the nucleus where they will be repaired, to serve as templates for transcription, thus "amplifying" the infection. Other DNA-containing capsids will associate with S proteins in the ER, to form new infectious virions.

Assembly/Release of Infectious Particles

Formation of the HBV capsid is linked to reverse transcription. The capsid is not simply an inert "shell" as mutations in C affect activities as diverse as DNA synthesis, transport of capsids back to the nucleus, and envelopment.

The HBV surface glyproteins are encoded by the S ORF. The three cocarboxyl terminal molecules are produced from two mRNAs and by use of alternative start codons. All forms of S share two transmembrane domains and all are glycosylated. Large S has a molecule of myristic acid at its amino terminus, forming a third interaction with the ER membrane. Two configurations of Large S have been described, one with the capsid interacting domain in the ER and the other with this critical domain in the cytosol. Assembly of infectious particles requires interactions between the capsid and Large S in the cytosol. Envelopment seems to occur by budding into the ER, with accumulation of virions in mutlivesicular bodies (MVB) of the late endosomal compartment. Virion release would then occur upon fusion of the MVB with the plasma membrane.

HEPATITIS B VIRUS AND DISEASE

The course of HBV infection is highly dependent on the age of infection. Among adults, ~90% will clear the infection within 6 months. Virus clearance is accomplished by the immune system. Disease is usually mild and acute infection may be asymptomatic. Approximately 10% of infected adults do not clear the infection and have chronic (usually lifelong) HBV infection. In contrast, infection of infants usually results in development of chronic HBV infection. Only ~10% of infected infants will clear the infection. The rate and extent of liver damage among persistently infected individuals varies widely; some will experience cirrhosis after a few decades while others will remain healthy. Chronic HBV infection is a significant risk factor for development of hepatocellular carcinoma. It is estimated that among chronically infected patients, 25% will develop liver cirrhosis and 5% will develop HCC.

In the United States there are an estimated 1 million HBV carriers. The WHO estimates 200 million carriers worldwide, and asymptomatic carriers serve as a large virus reservoir. However, vaccination is reversing the trend. Since 1991 the number of new infections among infants in the United States has declined by over 80% (Fig. 38.5), as vaccinating infants has become a common practice. If this trend continues, the overall burden of HBV infection in the United States will continue to decline, as chronic infection becomes a rare event. However, there is still a significant risk of contracting HBV infection among certain groups, and vaccination is highly recommended for high-risk adults (including unvaccinated health care workers or any

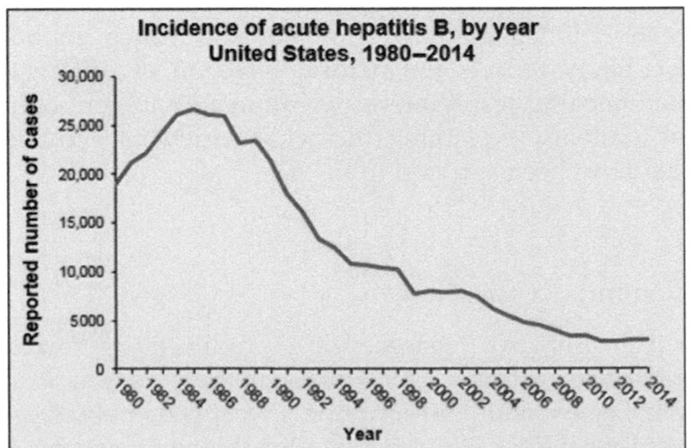

FIGURE 38.5 Reduction in cases of acute HBV infection in the United States as reported by the CDC. *From https://www.cdc.gov/hepatitis/hbv/hbvfaq.htm#ref2.*

individuals whose job involved potential contact with human blood). According to the CDC, adults at highest risk for catching HBV:

- have sex with an HBV-infected partner,
- inject drugs,
- live in the same house with someone who has lifelong HBV infection,
- have a job that involves possible exposure to human blood,
- travel to areas where hepatitis B is common.

The first generation HBV vaccine was prepared from pooled plasma from chronically infected HBV patients; taking advantage of the large quantities of noninfectious 20 nm particles and filaments found therein. Second generation HBV vaccine is a recombinant product. When expressed in yeast, the S protein assembles into particles that are antigenically similar to those found in blood. In the United States, the recombinant vaccine is recommended for all infants and anyone under 18 years of age who has not been vaccinated, persons of any age whose behavior puts them at high risk for HBV infection and persons whose jobs expose them to human blood.

In this chapter we have learned that:

- Members of the family *Hepadnaviridae* are small, enveloped viruses with ds DNA genomes. They infect differentiated hepatocytes, causing persistent, noncytopathic infection.
- Hepadnaviruses genomes are synthesized by reverse transcription. Thus every ds DNA genome is produced by reverse transcription of a molecule of capped, polyadenylated viral mRNA. DNA synthesis is tightly linked to encapsidation. The products of reverse transcription are relaxed circles of DNA. One strand (the plus strand) is usually incomplete.
- HBV P is a reverse transcriptase. The amino terminal domain is noncatalytic and serves as the primer for minus-strand DNA synthesis. The catalytic RT and RNaseH domains are similar to those found among retroviruses.
- Blood of persons with chronic HBV infection contains large numbers of noninfectious 20 nm particles, mostly consisting of S protein. There are also longer 20 nm diameter filaments. Least abundant is the Dane particle, the 42 nm infectious particle.
- HBV replicates in differentiated hepatocytes, which cannot be maintained in culture. Thus cell culture systems for HBV are of limited utility. HBV infection is also limited to apes (expensive and morally questionable animal models). Thus pathogenesis is often studied in woodchucks (woodchuck hepatitis virus), ground squirrels (ground squirrel hepatitis virus), and ducks (duck hepatitis virus).
- HBV vaccination is recommended for infants as they are at highest risk of developing chronic infection if exposed to HBV. At particular risk are infants of HBV positive mothers.
- Adults at highest risk for acquiring HBV are nonvaccinated persons with HBV-infected sexual partners, i.v. drug users, and those with jobs involving possible contact with human blood.

Index

Note: Page numbers followed by "f," "t," and "b" refer to figures, tables, and boxes, respectively.

H

Q

R

S

W

X

Y

Z